The Topology of Chaos

The Topology of Chaos

Alice in Stretch and Squeezeland

Robert Gilmore

Marc Lefranc

Wiley-VCH Verlag GmbH & Co. KGaA

Library of Congress Card No.:
Applied for

British Library Cataloging-in-Publication Data:
A catalogue record for this book is available from the British Library

Bibliographic information published by Die Deutsche Bibliothek
Die Deutsche Bibliothek lists this publication in the Deutsche Nationalbibliografie; detailed bibliographic data is available in the Internet at <http://dnb.ddb.de>.

Printed in the Federal Republic of Germany
Printed on acid-free paper

Printing and Bookbinding buch bücher dd ag, birkach

ISBN-13: 978-0-471-40816-1
ISBN-10: 0-471-40816-6

Preface

Before the 1970s opportunities sometimes arose for physicists to study nonlinear systems. This was especially true in fields like fluid dynamics and plasma physics, where the fundamental equations are nonlinear and these nonlinearities masked (and still mask) the full spectrum of spectacularly rich behavior. When possible we avoided being dragged into the study of abstract nonlinear systems. For we believed, to paraphrase a beautiful generalization of Tolstoy, that

> All linear systems are the same.
> Each nonlinear system is nonlinear in its own way.

At that time we believed that one could spend a whole lifetime studying the nonlinearities of the van der Pol oscillator [Cartwright and Littlewood (1945), Levinson (1949)] and wind up knowing next to nothing about the behavior of the Duffing oscillator.

Nevertheless, other intrepid researchers had been making an assault on the complexities of nonlinear systems. Smale (1967) described a mechanism responsible for generating a great deal of the chaotic behavior which has been studied up to the present time. Lorenz, studying a drastic truncation of the Navier-Stokes equation, discovered and described 'sensitive dependence on initial conditions' (1963). The rigid order in which periodic orbits are created in the bifurcation set of the logistic map, and in fact any unimodal map of the interval to itself, was described by May (1976) and by Metropolis, Stein, and Stein (1973).

Still, $\cdots$. There was a reluctance on the part of most scientists to indulge in the study of nonlinear systems.

This all changed with Feigenbaum's discoveries (1978). He showed that scaling invariance in period doubling cascades leads to quantitative (later, qualitative) predictions. These are the scaling ratios

$$\begin{array}{rcll} \delta & = & 4.66920\,16091\,029\cdots & \text{control parameter space} \\ \alpha & = & -2.50290\,78750\,959\cdots & \text{state variable space} \end{array}$$

which are eigenvalues of a renormalization transformation. The transformation in the attitude of scientists is summarized by the statement (Gleick, 1987)

> "It was a very happy and shocking discovery that there were structures in nonlinear systems that are always the same if you looked at them the right way."

This discovery launched an avalanche of work on nonlinear dynamical systems. Old experiments, buried and forgotten because of instabilities or unrepeatability due to incompetent graduate students (in their advisor's opinions) were resurrected and pushed as ground-breaking experiments exhibiting 'first observations' of chaotic behavior (by these same advisors). And many new experiments were carried out, at first to test Feigenbaum's scaling predictions, then to test other quantitative predictions, then just to see what would happen.

Some of the earliest experiments were done on fluids, since the fundamental equations are known and are nonlinear. However, these experiments often suffered from the long time scales (days, weeks, or months) required to record a decent data set. Oscillating chemical reactions (e.g., the Belousov-Zhabotinskii reaction) yielded a wide spectrum of periodic and chaotic behavior which was relatively easy to control and to tune. These data sets could be generated in hours or days. Nonlinear electric circuits were also extensively studied, although there was (and still is) a prejudice to regard them with a jaundiced eye as little more than analog computers. Such data sets could be generated very quickly (seconds to minutes) — almost as fast as numerical simulations. Finally, laser laboratories contributed in a substantial way to build up extensive and widely varying data banks very quickly (milleseconds to minutes) of chaotic data.

It was at this time (1988), about 10 years into the 'nonlinear science' revolution, that one of the authors (RG) was approached by his colleague (J. R. Tredicce, then at Drexel, now at the Institut Non Linéaire de Nice) with the proposition: "Bob, can you help me explain my data?" (See Chapter 1). So we swept the accumulated clutter off my desk and deposited his data. We looked, pushed, probed, discussed, studied, ··· for quite a while. Finally, I replied: "No." Tredicce left with his data. But he is very smart (he is an experimentalist!), and returned the following day with the same pile of stuff. The conversation was short and effective: "Bob," "Still my name," "I'll bet that you *can't* explain my data." (Bob sees red!) We sat down and discussed further. At the time two tools were available for studying chaotic data. These involved estimating Lyapunov exponents (dynamical stability) and estimating fractal dimension (geometry). Both required lots of very clean data and long calculations. They provided real number(s) with no convincing error bars, no underlying statistical theory, and no independent way to verify these guesses. And at the end of the day neither provided any information on "how to model the dynamics."

Even worse: before doing an analysis I would like to know what I am looking for, or at least know what the spectrum of possible results looks like. For example, when we analyze chemical elements or radionuclides, there is a Periodic Table of the Chemical Elements and another for the Atomic Nuclei which accommodate any such analyses. At that time, no classification theory existed for strange attractors.

In response to Tredicce's dare, I promised to (try to) analyze his data. But I pointed out that a serious analysis couldn't be done until we first had some handle on the classification of strange attractors. This could take a long time. Tredicce promised to be patient. And he was.

Our first step was to consider the wisdom of Poincaré. He suggested about a century earlier that one could learn a great deal about the behavior of nonlinear systems by studying their unstable periodic orbits which

> "... yield us the solutions so precious, that is to say, they are the only breach through which we can penetrate into a place which up to now has been reputed to be inaccessible."

This observation was compatible with what we learned from experimental data: the most important features that governed the behavior of a system, and especially that governed the perestroikas of such systems (i.e., changes as control parameters are changed) are the features that you can't see — the unstable periodic orbits.

Accordingly, my colleagues and I studied the invariants of periodic orbits, their (Gauss) linking numbers. We also introduced a refined topological invariant based on periodic orbits — the relative rotation rates (Chapter 4). Finally, we used these invariants to identify topological structures (branched manifolds or templates, Chapter 5) which we used to classify strange attractors "in the large." The result was that "low dimensional" strange attractors (i.e., those that could be embedded in three-dimensional spaces) could be classified. This classification depends on the periodic orbits "in" the strange attractor, in particular, on their organization as elicited by their invariants. The classification is topological. That is, it is given by a set of integers (also by very informative pictures). Not only that, these integers can be extracted from experimental data. The data sets do not have to be particularly long or particularly clean — especially by fractal dimension calculation standards. Further, there are built in internal self-consistency checks. That is, the topological analysis algorithm (Chapter 6) comes with reject/fail to reject test criteria. This is the first — and remains the only — chaotic data analysis procedure with rejection criteria.

Ultimately we discovered, through analysis of experimental data, that there is a secondary, more refined classification for strange attractors. This depends on a "basis set of orbits" which describes the spectrum of all the unstable periodic orbits "in" a strange attractor (Chapter 9).

The ultimate result is a doubly discrete classification of strange attractors. Both parts of this doubly discrete classification depend on unstable periodic orbits. The classification depends on identifying:

Branched Manifold - which describes the stretching and squeezing mechanisms that operate repetitively on a flow in phase space to build up a (hyperbolic) strange attractor and to organize all the unstable periodic orbits in the strange attractor in a unique way. The branched manifold is identified by the spectrum of the invariants of the periodic orbits that it supports.

Basis Set of Orbits - which describes the spectrum of unstable period orbits in a (nonhyperbolic) strange attractor.

The perestroikas of branched manifolds and of basis sets of orbits in this doubly discrete classification obey well-defined topological constraints. These constraints provide both a rigidity and a flexibility for the evolution of strange attractors as control parameters are varied.

Along the way we discovered that dynamical systems with symmetry can be related to dynamical systems without symmetry in very specific ways (Chapter 10). As usual, these relations involve both a rigidity and a flexibility which is as surprising as it is delightful.

Many of these insights are described in the paper *Topological analysis of chaotic dynamical systems*, Reviews of Modern Physics, **70**(4), 1455-1530 (1998), which forms the basis for part of this book. We thank the editors of this journal for their policy of encouraging transformation of research articles to longer book format.

The encounter (fall in love?) of the other author (M.L.) with topological analysis dates back to 1991, when he was a PhD student at the University of Lille, struggling to extract information from the very same type of chaotic laser that Tredicce was using. At that time, Marc was computing estimates of fractal dimensions for his laser. But they depended very much on the coordinate system used and gave no insight into the mechanisms responsible for chaotic behavior, even less into the succession of the different behaviors observed. This was very frustrating. There had been this very intringuing paper in Physical Review Letters about a "Characterization of strange attractors by integers", with appealing ideas and nice pictures. But as with many short papers, it had been difficult to understand how you should proceed when faced with a real experimental system. Topological analysis struck back when Pierre Glorieux, then Marc's advisor, came back to Lille from a stay in Philadephia, and handed him a preprint from the Drexel team, saying: "You should have a look at this stuff". The preprint was about topological analysis of the Belousov-Zhabotinskii reaction, a real-life system. It was the Rosetta Stone that helped put pieces together. Soon after, pictures of braids constructed from laser signals were piling up on the desk. They were absolutely identical to those extracted from the Belousov-Zhabotinskii data and described in the preprint. There was universality in chaos if you looked at it with the right tools. Eventually, the system that had motivated topological analysis in Philadephia, the CO_2 laser with modulated losses, was characterized in Lille and shown to be described by a horseshoe template. Indeed, Tredicce's laser could not be characterized by topological analysis because of long periods of zero output intensity that prevented invariants from being reliably estimated. The high signal to noise ratio of the laser in Lille allowed us to use a logarithmic amplifier and to resolve the structure of trajectories in the zero intensity region.

But a classification is only useful if there exist different classes. Thus, one of the early goals was to find experimental evidence of a topological organization that would differ from the standard Smale's horseshoe. At that time, some regimes of the modulated CO_2 laser could not be analyzed for lack of a suitable symbolic encoding. The corresponding Poincaré sections had peculiar structures that, depending on the observer's mood, suggested a doubly iterated horsehoe or an underlying three-branch manifold. Since the complete analysis could not be carried out, much time was spent on trying to find at least one orbit that could not fit the horseshoe template. The result was extremely disappointing: For every orbit detected, there was at least one horseshoe orbit with identical invariants. One of the most important lessons of Judo is that if you experience resistance when pushing, you should pull (and vice versa). Similarly, this failed attempt to find a nonhorseshoe template turned into techniques to determine underlying templates when no symbolic coding is available and to construct such codings using the information extracted from topological invariants.

But the search for different templates was not over. Two of Marc's colleagues, Dominique Derozier and Serge Bielawski, proposed for him to study a fiber laser they had in their laboratory (that was the perfect system to study knots). This system exhibits chaotic tongues when the modulation frequency is near a subharmonic of its relaxation frequency: It was tempting to check whether the topological structures in each tongue differed. That was indeed the case: The corresponding templates were basically horseshoe templates but with a global torsion increasing systematically from one tongue to the other. A Nd:YAG laser was also investigated. It showed similar behavior, until the day where Guillaume Boulant, the PhD student working on the laser, came to Marc's office and said: "I have a weird data set". Chaotic attractors were absolutely normal, return maps resembled the logistic map very much, but the invariants were simply not what we were used to. This was the first evidence of a reverse horseshoe attractor. How topological organizations are modified as a control parameter is varied was the subject of many discussions in Lille in the following months, a rather accurate picture finally emerged and papers began to be written. In the last stages, Marc did a bibliographic search just to clear his mind and... a recent 22-page Physical Review paper, by McCallum and Gilmore, turned up. Even though it was devoted to the Duffing attractor, it described with great detail what was happening in our lasers as control parameters are modified. Every occurence of "We conjecture that" in the papers was hastily replaced by "Our experiments confirm the theoretical prediction...", and papers were sent to Physical Review. They were accepted 15 days later, with a very positive review. Soon after, the Referee contacted us and proposed a joint effort on extensions of topological analysis. The Referee was Bob, and this was the start of a collaboration that we hope will last long.

It would indeed be very nice if these techniques could be extended to the analysis of strange attractors in higher dimensions (than three). Such an extension, if it is possible, cannot rely on the most powerful tools available in three dimensions. These are the topological invariants used to tease out information on how periodic orbits are organized in a strange attractor. We cannot use these tools (linking numbers, relative rotation rates) because knots "fall apart" in higher dimensions. We explore (Chapter 11) an inviting possibility for studying an important class of strange attractors in four dimensions. If a classification procedure based on these methods is successful, the door is opened to classifying strange attractors in R^n, $n > 3$. A number of ideas that may be useful in this effort have already proved useful in two closely-related fields (Chapter 12): Lie group theory and singularity theory.

Some of the highly technical details involved in extracting templates from data have been archived in the appendix. Other technical matter is archived at our web sites.[1]

Much of the early work in this field was done in response to the challenge by J. R. Tredicce and carried out with my colleagues and close friends: H. G. Solari, G. B. Mindlin, N. B. Tufillaro, F. Papoff, and R. Lopez-Ruiz. Work on symmetries was done with C. Letellier. Part of the work carried out in this program has been supported by the National Science Foundation under grants NSF 8843235 and NSF 9987468. Similarly, Marc would like to thank colleagues and students with whom he enjoyed working and exchanging ideas about topological analysis: Pierre Glorieux, Ennio Arimondo, Francesco Papoff, Serge Bielawski, Dominique Derozier, Guillaume Boulant, and Jérôme Plumecoq. Stays of Bob in Lille were partially funded by the University of Lille, the Centre National de la Recherche Scientifique, Drexel University under sabbatical leave, and by the NSF.

Last and most important, we thank our wives Claire and Catherine for their warm encouragement while physics danced in our heads, and our children, Marc and Keith, Clara and Martin, who competed with our research and demanded our attention, and by doing so, kept us human.

R.G. AND M.L.

Lille, France, Jan. 2002

[1]`http://einstein.drexel.edu/directory/faculty/Gilmore.html` and `http://www.phlam.univ-lille1.fr/perso/lefranc.html`

Contents

1

Introduction

The subject of this book is the analysis of data generated by a dynamical system operating in a chaotic regime. More specifically, we describe how to extract, from chaotic data, topological signatures that determine the stretching and squeezing mechanisms which act on flows in phase space and which are responsible for generating chaotic data.

In the first section of this introductory chapter we describe, for purposes of motivation, a laser that has been operated under conditions in which it behaved chaotically. The topological methods of analysis that we describe in this book were developed in response to the challenge of analyzing chaotic data sets generated by this laser.

In the second section we list a number of questions which we would like to be able to answer when analyzing a chaotic signal. None of these questions can be addressed by the older tools for analyzing chaotic data: estimates of the spectrum of Lyapunov exponents and estimates of the spectrum of fractal dimensions. The question that we would particularly like to be able to answer is this: How does one model the dynamics? To answer this question we must determine the stretching and squeezing mechanisms that operate together—repeatedly—to generate chaotic data. The stretching mechanism is responsible for *sensitivity to initial conditions* while the squeezing mechanism is responsible for *recurrent nonperiodic behavior*. These two mechanisms operate repeatedly to generate a strange attractor with a self-similar structure.

A new analysis method, topological analysis, has been developed to respond to the fundamental question just stated [1,2]. At the present time this method is suitable only for strange attractors that can be embedded in three-dimensional spaces. However, for such strange attractors it offers a complete and satisfying resolution to this question. The results are previewed in the third section of this chapter. In the final section we provide a brief overview of the organization of this book. In particular, we summarize the organization and content of the following chapters.

It is astonishing that the topological analysis tools that we describe have provided answers to more questions than we asked originally. This analysis procedure has also raised more questions than we have answered. We hope that the interaction between experiment and theory and between old questions answered and new questions raised will hasten evolution of the field of nonlinear dynamics.

1.1 LASER WITH MODULATED LOSSES

The possibility of observing chaos in lasers was originally demonstrated by Arecchi et al. [3] and by Gioggia and Abraham [4]. The use of lasers as a testbed for generating deterministic chaotic signals has two major advantages over fluid and chemical systems, which until that time had been the principal sources for chaotic data:

1. The time scales intrinsic to a laser (10^{-7} to 10^{-3} s) are much shorter than the time scales in fluid experiments and oscillating chemical reactions. This is important for experimentalists, since it is possible to explore a very large parameter range during a relatively short time.

2. Reliable laser models exist in terms of a small number of ordinary differential equations whose solutions show close qualitative similarity to the behavior of the lasers that are modeled [5,6].

The topological methods described in the remainder of this work were originally developed to understand the data generated by a laser with modulated losses [6]. A schematic of this laser is shown in Fig. 1.1. A CO_2 gas tube is placed between two infrared mirrors (M). The ends of the tube are terminated by Brewster angle windows, which polarize the field amplitude in the vertical direction. Under normal operating conditions, the laser is very stable. A Kerr cell (K) is placed inside the laser cavity. The Kerr cell modifies the polarization state of the electromagnetic field. This modification, coupled with the polarization introduced by the Brewster windows, allows one to change the intracavity losses. The Kerr cell is modulated at a frequency determined by the operating conditions of the laser. When the modulation is small, the losses within the cavity are small, and the laser output tracks the input from the signal generator. The input signal (from the signal generator) and the output signal (the measured laser intensity) are both recorded in a computer (C). When the modulation crosses a threshold, the laser output can no longer track the signal input.

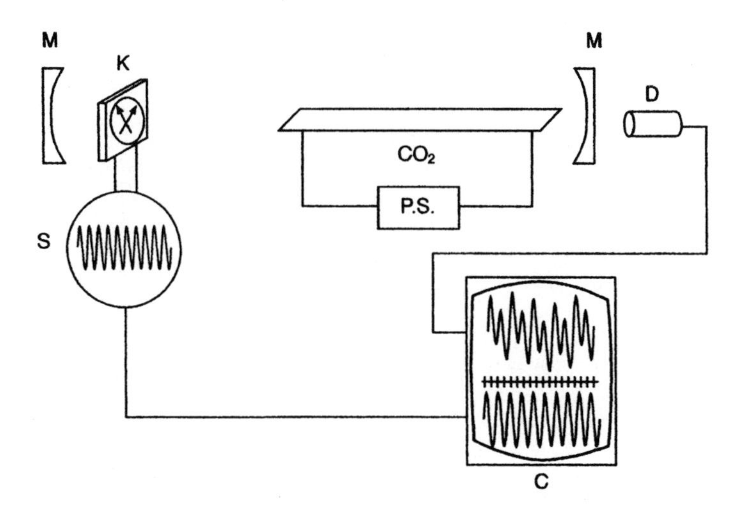

Fig. 1.1 This schematic representation of a laser with modulated losses shows the carbon dioxide tube (CO_2); power source (P.S.); mirrors (M); Kerr cell (K); signal generator (S); detector (D); and computer, oscilloscope, and recorder (C). A variable electric field across the Kerr cell rotates its polarization direction and modulates the electric field amplitude within the cavity.

At first every other output peak has the same height, then every fourth peak, then every eighth peak, and so on.

In Fig. 1.2 we present some of the recorded and processed signals from this part of the period-doubling cascade and beyond [6]. The signals were recorded under different operating conditions and are displayed in five lines, as follows: (a) period 1; (b) period 2; (c) period 4; (d) period 8; (e) chaos. Each of the four columns presents a different representation of the data. In the first column the intensity output is displayed as a function of time. In this presentation the period-1 and period-2 behaviors are clear but the higher-period behavior is not.

The second column displays a projection of the dynamics into a two-dimensional plane, the dI/dt vs. $I(t)$ plane. In this projection, periodic orbits appear as closed loops (deformed circles) which go around once, twice, four times, ... before closing. In this presentation the behavior of periods 1, 2, and 4 is clear. Period 8 and chaotic behavior is less clear. The third column displays the power spectrum. Not only is the periodic behavior clear from this display, but the relative intensity of the various harmonics is also evident. Chaotic behavior is manifest in the broadband power spectrum. Finally, the last column displays a stroboscopic sampling of the output. In this sampling technique, the output intensity is recorded each time the input signal reaches a maximum (or some fixed phase with respect to the maximum). There is one sample per cycle. In period-1 behavior, all samples have the same value. In period-2 behavior, every other sample has the same value. The stroboscopic display clearly distinguishes between periods 1, 2, 4, and 8. It also distinguishes periodic behavior from chaotic behavior. The stroboscopic sampling technique is equivalent to the construction of a Poincaré section for this periodically driven dynamical system. All four of these display modalities are available in real time, during the experiment.

The laser with modulated losses has been studied extensively both experimentally [3–9] and theoretically [10–12]. The rate equations governing the laser intensity I and the population inversion N are

$$\begin{aligned} \frac{dI}{dt} &= -k_0 I[(1-N) + m\cos(\omega t)] \\ \frac{dN}{dt} &= -\gamma[(N-N_0) + (N_0-1)IN] \end{aligned} \tag{1.1}$$

Here m and ω are the modulation amplitude and angular frequency, respectively, of the signal to the Kerr cell; N_0 is the pump parameter, normalized to $N_0 = 1$ at the threshold for laser activity; and k_0 and γ are loss rates. In dimensionless, scaled form this equation is

$$\begin{aligned} \frac{du}{d\tau} &= [z - A\cos(\Omega\tau)]u \\ \frac{dz}{d\tau} &= (1-\epsilon_1 z) - (1+\epsilon_2 z)u \end{aligned} \tag{1.2}$$

The scaled variables are $u = I, z = k_0\kappa(N-1), t = \kappa\tau, A = k_0 m, \epsilon_1 = 1/\kappa k_0$, and $\kappa^2 = 1/\gamma k_0(N_0-1)$. The bifurcation behavior exhibited by the simple models (1.1) and (1.2) is qualitatively, if not quantitatively, in agreement with the experimentally observed behavior of this laser.

A bifurcation diagram for the laser model (1.2) is shown in Fig. 1.3. The bifurcation diagram is constructed by varying the modulation amplitude A and keeping all other parameters fixed. The overall structures of the bifurcation diagrams are similar to experimentally observed bifurcation diagrams.

This figure shows that a period-1 solution exists above the laser threshold ($N_0 > 1$) for $A = 0$ and remains stable as A is increased until $A \sim 0.8$. It becomes unstable above $A \sim 0.8$, with a stable period-2 orbit emerging from it in a period-doubling bifurcation. Contrary to what might be expected, this is not the early stage of a period-doubling cascade, for the period-2 orbit is annihilated at $A \sim 0.85$ in an inverse saddle-node bifurcation with a period-2 regular saddle. This saddle-node bifurcation destroys the basin of attraction of the period-2 orbit. Any point in that basin is dumped into the basin of a period $4 = 2 \times 2^1$ orbit, even though there are two other coexisting basins of attraction for stable orbits of periods $6 = 3 \times 2^1$ and 4 at this value of A.

Subharmonics of period n ($P_n, n \geq 2$) are created in saddle-node bifurcations at increasing values of A and I (P_2 at $A \sim 0.1$, P_3 at $A \sim 0.3$, P_4 at $A \sim 0.7$, P_5 and higher shown in the inset). All subharmonics in this series up to period $n = 11$ have been seen both experimentally and in simulations of (1.2). The evolution (*perestroika* [13]) of each of these subharmonics follows a standard scenario as T increases [14]:

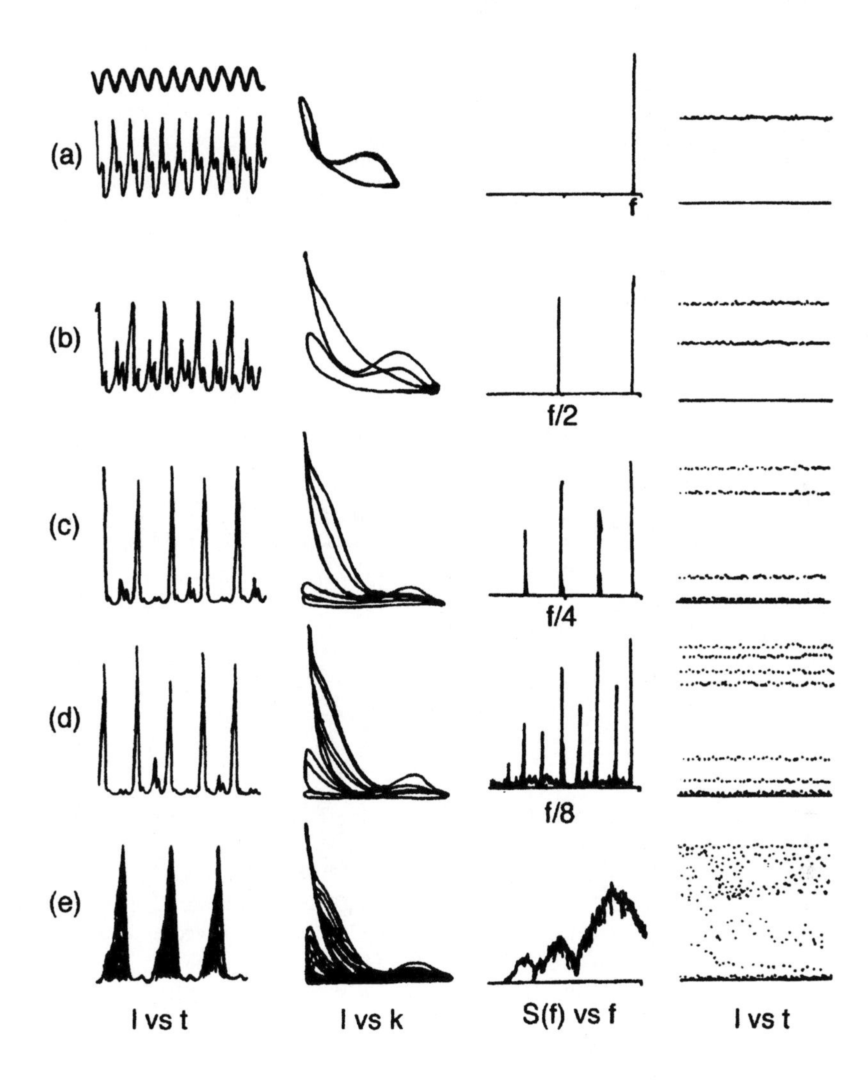

Fig. 1.2 Each column provides a different representation of the experimental data. Each row describes different experimental conditions. The first column shows the recorded intensity time signal, $I(t)$. The second column presents the phase-space projection, $dI(t)/dt$ vs. $I(t)$. The third column shows the power spectrum of the recorded intensity signal. The frequencies of the Fourier components in the signal, and their relative amplitudes, jump out of this plot. The last column presents a stroboscopic plot (Poincaré section). This is a record of the intensity output at each successive peak (or more generally, at some constant phase) of the input signal. The data sets were recorded under the following experimental conditions: (a) period 1; (b) period 2; (c) period 4; (d) period 8; (e) chaotic. Reprinted with permission from Tredicce et al. [6].

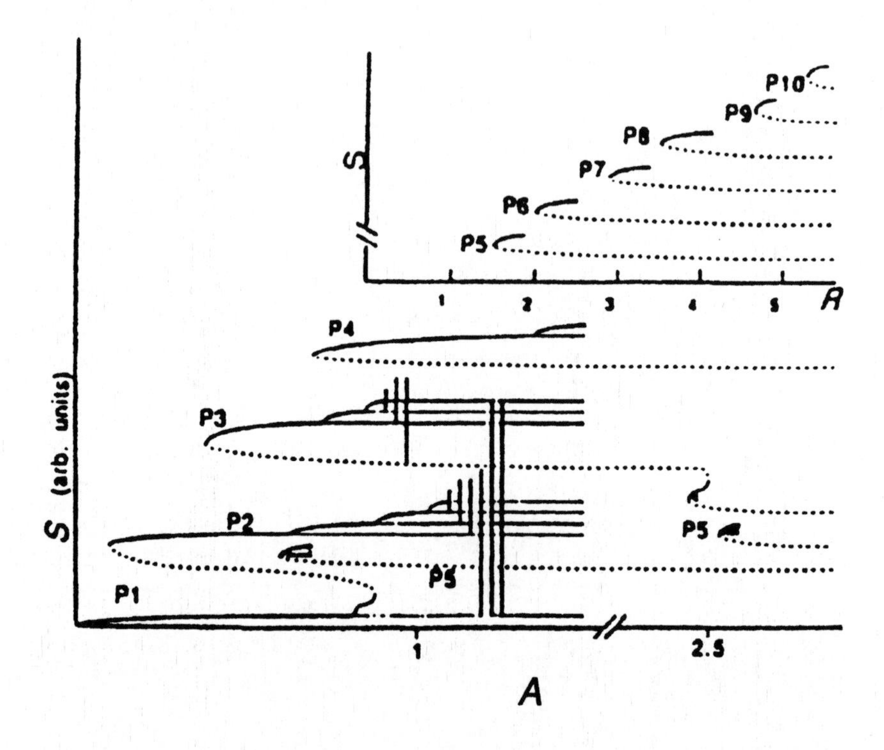

Fig. 1.3 The bifurcation diagram for the laser model (1.2) is computed by varying the modulation amplitude A. Stable periodic orbits (solid lines), regular saddles (dotted lines), and strange attractors are shown. Period n branches ($P_n \geq 2$) are created in saddle-node bifurcations and evolve through the Feigenbaum period-doubling cascade as the modulation amplitude increases. There are two apparently distinct stable period-2 orbits. However, these are connected by an unstable period-2 orbit (dotted, extending from $A \simeq 0.1$ to $A \simeq 0.8$) and thus constitute a single period-2 orbit which is a *snake*. A period-3 snake is also present. Two distinct stable period-4 orbits are present and coexist over a short range of parameter values ($0.7 < A < 0.8$). The inset shows a sequence of period-n orbits (*Newhouse orbits*) for $n \geq 5$. The Smale horseshoe mechanism predicts that as many as three inequivalent pairs of period-5 orbits could exist. The locations of the two additional pairs have been shown in this diagram at $A \simeq 0.65$ and $A \simeq 2.5$. Parameter values: $\epsilon_1 = 0.03$, $\epsilon_2 = 0.009$, $\Omega = 1.5$.

1. A saddle-node bifurcation creates an unstable saddle and a node that is initially stable.

2. Each node becomes unstable and initiates a period-doubling cascade as A increases. The cascade follows the standard Feigenbaum scenario [15–18]. The ratio of A intervals between successive bifurcations, and of geometric sizes of the stable nodes of periods $n \times 2^k$, have been estimated up to $k \leq 6$ for some

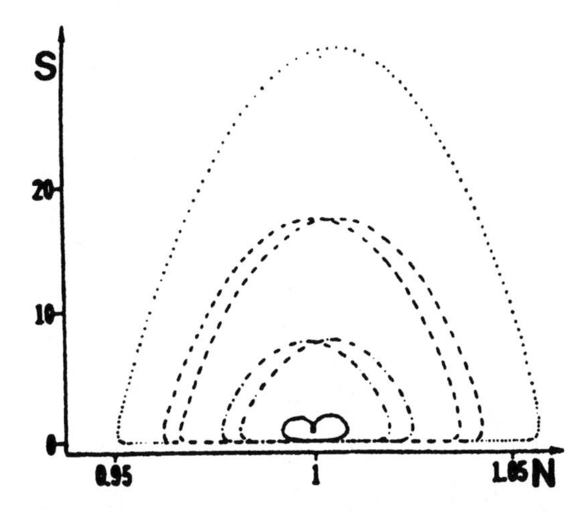

Fig. 1.4 Multiple basins of attraction can coexist over a broad range of parameter values. The stable periodic orbits and the strange attractors within these basins have a characteristic organization. The coexisting orbits shown above are, from the inside to the outside: period 2 bifurcated from a period 1 branch; period 2; period 3; period 4. The two inner orbits are separated by an unstable period-2 orbit (not shown); all three are part of a snake.

of these subharmonics, both from experimental data and from the simulations. These ratios are compatible with the universal scaling ratios.

3. Beyond accumulation, there is a series of noisy orbits of period $n \times 2^k$ that undergo inverse period-halving bifurcations. This scenario has been predicted by Lorenz [19].

Additional systematic behavior has been observed. Higher subharmonics are generally created at larger values of A. They are created with smaller basins of attraction. The range of A values over which the Feigenbaum scenario is played out becomes smaller as the period n increases. In addition, the subharmonics show an ordered pattern in phase space. In Fig. 1.4 we show four stable periodic orbits that coexist under certain operating conditions. Roughly speaking, the larger period orbits exist "outside" the smaller period orbits. These orbits share many other systematics, which have been described by Eschenazi, Solari, and Gilmore [14]. In Fig. 1.5 we show an example of a chaotic time series taken for $A \sim 1.3$. The chaotic attractor based on the period-2 orbit (the period-1 orbit) has just collided with the period-3 regular saddle.

The period-doubling, accumulation, inverse noisy period-halving scenario described above is often interrupted by a crisis (Grebogi and Ott [20]) of one type or another:

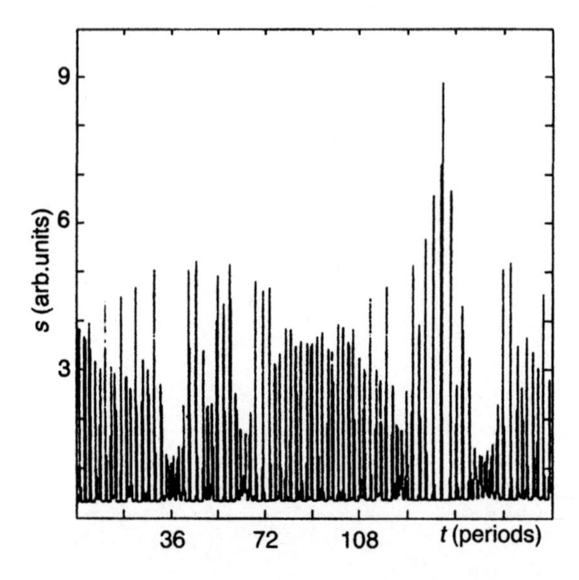

Fig. 1.5 This time series from a laser with modulated losses was taken at a value of $A \sim 1.3$, which is just beyond the collision (crisis) of the strange attractors based on the period-2 and period-3 orbits. There is an alternation in this time series between noisy period-2 and noisy period-3 behavior.

Boundary Crisis: A regular saddle on a period-n branch in the boundary of the basin of attraction surrounding either the period-n node or one of its periodic or noisy periodic progeny collides with the attractor. The basin is annihilated or enlarged.

Internal Crisis: A flip saddle of period $n \times 2^k$ in the boundary of a basin surrounding a noisy period $n \times 2^{k+1}$ orbit collides with the attractor to produce a noisy period-halving bifurcation.

External Crisis: A regular saddle of period n' in the boundary of a period n ($P_n \neq P_{n'}$) strange attractor collides with the attractor, thereby annihilating or enlarging the basin of attraction.

Figure 1.6(a) provides a schematic representation of the bifurcation diagram shown in Fig. 1.3. The different kinds of bifurcations encountered in both experiments and simulations are indicated here. These include both direct and inverse saddle-node bifurcations, period-doubling bifurcations, and boundary and external crises. As the laser operating parameters (k_0, γ, ω) change, the bifurcation diagram changes. In Fig. 1.6(b) and (c) we show the schematics of bifurcation diagrams obtained for slightly different values of these operating—or control—parameters.

In addition to the subharmonic orbits of period n created at increasing values of A (Fig. 1.3), there are orbits of period n that do not appear to belong to that series

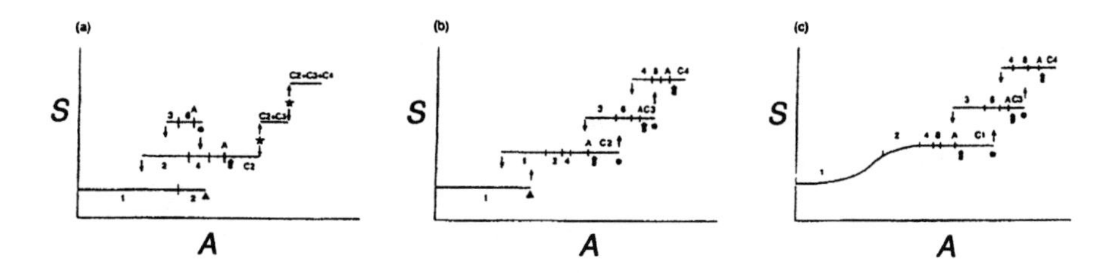

Fig. 1.6 Schematics of three bifurcation diagrams for three different operating conditions of Eqs. (1.2). As control parameters change, the bifurcation diagram is modified. Slow change in control parameter values deforms the bifurcation diagram from (a) to (b) to (c). The sequence (a) to (c) shows the unfolding of the snake in the period-2 orbit. The unstable period-2 orbit connecting the two lowest branches is invisible in (a) and (b) since only stable attractors are shown. In each diagram the bifurcations are: $\downarrow$, saddle-node; Δ, inverse saddle-node; $\blacktriangle$, inverse saddle-node; doesn't work $\bullet$, boundary crisis; $\star$, external crisis. Period-doubling bifurcations are indicated by a small vertical line separating stable orbits of periods differing by a factor of 2. Accumulation points are indicated by A. Strange attractors based on period-n orbits are indicated by the Cn.

(Newhouse series [21]) of subharmonics. The clearest example is the period-2 orbit, which bifurcates from period 1 at $A \sim 0.8$. Another is the period-3 orbit pair created in a saddle-node bifurcation that occurs at $A \sim 2.45$. These bifurcations were seen in both experiments and simulations. It was possible to trace the unstable orbits of period 2 ($0.1 \leq A \leq 0.85$) and period 3 ($0.4 \leq A \leq 2.5$) in simulations and find that these orbits are components of an orbit snake (Alligood [22]; Alligood, Sauer, and Yorke [23]). This is a single orbit that folds back and forth on itself in direct and reverse saddle-node bifurcations as A increases (this is not unlike a Feynman diagram for hard scattering of an electron by a photon, which scatters the electron backward in time, creating a positron ...). The unstable period-2 orbit ($0.1 \leq T \leq 0.85$) is part of a snake. By changing operating conditions, both snakes can be eliminated [see Fig. 1.6(c)]. As a result, the "subharmonic P_2" is really nothing other than the period-2 orbit, which bifurcates from the period-1 branch P_1. Furthermore, instead of having saddle-node bifurcations creating four inequivalent period-3 orbits (at $A \sim 0.4$ and $A \sim 2.45$) there is really only one pair of period-3 orbits, the other pair being components of a snake.

Topological tools (relative rotation rates; Solari and Gilmore [12]) were first developed to determine which orbits might be equivalent, or components of a snake, and which are not. Components of a snake have the same topological invariants (cf. Chapter 4). These tools suggested that the Smale horseshoe mechanism [24] was responsible for generating the nonlinear phenomena observed in both the experiments and the simulations. This mechanism predicts that additional inequivalent subharmonics of period n can exist for $n \geq 5$. Since the size of a basin of attraction decreases rapidly with n, a search was made for additional inequivalent basins of attraction of period 5. Two additional stable period-5 orbits (besides P_5) were located in simulations. Their locations are shown in Fig. 1.3 at $A \sim 0.6$ and $A \sim 2.45$. One was also located experimentally. The other may also have been seen, but its basin was too small to be certain of its existence.

Bifurcation diagrams had been observed for a variety of physical systems at that time: other lasers [25,26]; electric circuits [27–30]; a biological model [31]; and a bouncing ball [32]. Their bifurcation diagrams are similar but not identical to those shown above. This raised the question of whether similar processes were governing the description of this large variety of physical systems.

During these analyses, it became clear that the standard tools for analyzing chaotic data—estimates of the spectrum of Lyapunov exponents and estimates of the various fractal dimensions—were not sufficient for a satisfying understanding of the stretching and squeezing processes that occur in phase space and which are responsible for generating chaotic behavior. In the laser we found many coexisting basins of attraction, some containing a periodic attractor, others containing a strange attractor. The rapid alternation between periodic and chaotic behavior as control parameters (e.g., A and Ω) were changed meant that Lyapunov exponents and fractal dimensions depended on the basins and varied at least as rapidly.

For this reason we sought to develop additional tools for the analysis of data generated by dynamical systems that exhibit chaotic behavior. The objective was to develop measures that were invariant under control parameter changes.

1.2 OBJECTIVES OF A NEW ANALYSIS PROCEDURE

In view of the experiments just described and the data that they generated, we hoped to develop a procedure for analyzing data that achieved a number of objectives. These included an ability to answer the following questions:

1. Is it possible to develop a procedure for understanding dynamical systems *and their evolution* (perestroikas) as the control parameters (e.g., k_0, m, γ, or A, Ω, ϵ_1, ϵ_2) change?

2. Is it possible to identify a dynamical system by means of topological invariants, following suggestions proposed by Poincaré?

3. Can selection rules be constructed under which it is possible to determine the order in which periodic orbits can be created and/or annihilated by standard bifurcations? Or when different orbits might belong to a single snake?

4. Is it possible to determine when two strange attractors are (a) equivalent in the sense that one can be transformed into the other without creating or annihilating orbits; or (b) adiabatically equivalent (one can be deformed into the other by changing parameters to create or annihilate only a small number of orbit pairs below any period); or (c) inequivalent (there is no way to transform one into the other)?

1.3 PREVIEW OF RESULTS

A new topological analysis procedure was developed in response to the questions asked of the data initially. These questions are summarized in Section 1.2. The remarkable result is that there is now a positive and constructive answer to the question: How can I look at experimental data, such as shown in Fig. 1.2 or 1.5, and extract useful information, let alone information about stretching and squeezing, let alone a small set of integers?

This new analysis procedure answered more questions than were asked originally. It also raised a great many additional questions. This is one of the ways we know that we are on the right track.

The results of this new topological analysis procedure are presented throughout this book. Below we provide a succinct preview of the major accomplishments of this topological analysis tool.

- It is possible to classify low-dimensional strange attractors. These are strange attractors that exist in three-dimensional spaces.
- This classification is topological in nature.
- This classification exists at two levels: a macroscopic level and a microscopic level.
- It is discrete at both levels. Thus, there exists a doubly discrete classification for low-dimensional strange attractors.
- This doubly discrete classification depends in an essential way on the (unstable) periodic orbits, which are embedded in strange attractors.
- At the macro level the classification is by means of a geometric structure that describes the topological organization of *all* the unstable periodic orbits that exist in a *hyperbolic* strange attractor. This geometric structure is called variously a (two-dimensional) *branched manifold*, *knot-holder*, or *template.*
- Branched manifolds can be identified by a set of integers. Thus, at the macro level the classification is discrete.
- At the micro level the classfication is by means of a set of orbits in a *nonhyperbolic* strange attractor whose existence implies the presence of all the other orbits that can be found in the nonhyperbolic strange attractor. This subset of orbits is called a *basis set of orbits.*
- To any given period, a basis set of orbits is also discrete.
- As control parameters change, the basis set of orbits changes. The changes that are allowed are limited by topological arguments.
- Each different sequence of basis sets describing the transition from the laminar to the hyperbolic limit describes a different route to chaos. Each different route to chaos is a different path in a forcing diagram, shown in Fig. 9.8.

- During this transition the underlying branched manifold is robust: It generally does not change.
- Large changes in control parameter values can cause changes in the underlying branched manifold.
- These changes occur by adding branches to or removing branches from the branched manifold. The branch changes that are allowed are also limited by topological and continuity arguments.
- The information required for this doubly discrete classification of strange attractors can be extracted from experimental data.
- The data requirements are not heavy. Data sets of limited length are required.
- The data need not be exceptionally clean. Only a modest signal-to-noise level is required. The analysis method degrades gracefully with noise. Specifically, as the noise level degrades the data, it becomes more difficult to identify the higher-period orbits, which are the least important for this analysis. The most important orbits, those of lowest period, persist longest with increasing noise. As a result, "Murphy is on vacation" (author of the famous law).
- The data analysis method comes endowed with a rejection criterion.
- The branched manifold identifies the stretching and squeezing mechanisms that generate chaotic behavior.
- Thus, this doubly discrete classification describes "how to model the dynamics."

1.4 ORGANIZATION OF THIS WORK

The best way to summarize the organization of this book is to provide a brief summary of each chapter.

Chapter 1—Introduction. Data generated by a laser with modulated losses are used to illustrate the complexity of the problem. The problem—in a nutshell—is this: How do you let data speak to you and tell you what mechanisms are generating chaotic behavior? The basic objectives of a new analysis method are outlined. Also summarized are the accomplishments of the new topological analysis method.

Chapter 2—Discrete Dynamical Systems: Maps. Mappings have been excellent surrogates for continuous dynamical systems. They were introduced by Poincaré as a way to reduce the dimension of the space needed to study the properties of chaotic dynamical systems. We introduce the idea of a map (of a Poincaré section to itself) as a dynamical system in this chapter. The chapter contains many familiar results useful for the topological analyses that follow. These include review of the logistic and circle maps; the interplay between periodic orbits and chaotic behavior; the implications of topology on the order in which periodic orbits are created as control parameters vary; symbolic dynamics; kneading theory; and several ways to compute topological entropy. We emphasize that noninvertible maps are characterized by their singularity structure. Invertible maps are not, but if they are dissipative, their large n iterates are also characterized by their singularities.

Chapter 3—Continuous Dynamical Systems: Flows. The principal subject of this book is the analysis of dynamical systems. These are systems of coupled (nonlinear) first-order ordinary differential equations. We review the properties of such systems in this chapter, such as the laser equations (1.1) and (1.2). In particular, we introduce the existence and uniqueness theorem, on which all else rests. All the other important concepts related to flows are also introduced here. These include especially the spectrum of Lyapunov exponents and the concepts of stretching and squeezing in phase space. Four of the basic theoretical testbeds used in the study of dynamical systems theory are introduced here. These are the Duffing and van der Pol equations for periodically driven two-dimensional nonlinear oscillators and the autonomous Lorenz and Rössler equations.

Chapter 4—Topological Invariants. If we want to understand chaos in dynamical systems, we must understand strange attractors. In particular, we must understand the (unstable) periodic orbits "in" strange attractors. How do we know this? Poincaré [33] told us so over a century ago in an oft quoted and more often neglected statement:

> [periodic orbits] yield us the solutions so precious, that is to say, they are the only breach through which we can penetrate into a place which up to now has been reputed to be inaccessible.

Our contribution to this understanding is the introduction of Gaussian linking numbers between pairs of orbits into the study of chaotic dynamical systems. These, and an even more refined invariant for orbits and pairs of orbits, the relative rotation rates, are introduced in this chapter. These topological invariants exist only in R^3 or more generally, three-dimensional manifolds. This presently limits the results of the topological analysis method to strange attractors that live in three-dimensional inertial manifolds.

Chapter 5—Branched Manifolds. Different stretching and squeezing mechanisms generate different (inequivalent) strange attractors. The topological organization of the unstable periodic orbits in a strange attractor provides a clear fingerprint for the strange attractor. In fact, the organization identifies the stretching and squeezing mechanisms that generate the strange attractor while organizing all the unstable periodic orbits in the strange attractor in a unique way. There is a geometric structure that supports all the unstable periodic orbits in a strange attractor with the same unique organization that they possess in the strange attractor. This structure is variously called a knot-holder or branched manifold or template. These structures are introduced in this chapter. They can be identified by a set of integers. Therefore, they are discretely classifiable. As a result, strange attractors are discretely classifiable.

Chapter 6—Topological Analysis Program. There is a straightforward procedure for extracting the signature of a strange attractor from experimental data. This consists of a number of simple and easily implementable steps. The input to this analysis procedure consists of experimental time series. The output consists of a branched manifold, or more abstractly, a set of integers. The results of this analysis are subject to a follow-on rejection or "confirmation" test. In Chapter 6 we present a step-by-step account of the topological analysis method.

Chapter 7—Folding Mechanisms: $\mathbf{A}_2$. The topological analysis procedure has been applied to a large number of data sets. Most of them revealed stretching and squeezing mechanisms which were variations on a single theme. The basic theme in its most elementary form is the Smale horseshoe stretch and fold mechanism (without global torsion). Other variations include reverse horseshoes, horseshoes with global torsion, and stretch and role mechanisms, variously called gâteau roulé or jelly role mechanisms. These form a large subset of the possible mechanisms that can occur when the phase space containing the attractor (the inertial manifold) is $R^2 \times S^1$. The topological structure of the phase space is evident by inspection of the attractor: If it has a hole in the middle, it has this property and the dynamics can be formed only by stretch and folding mechanisms, although the folding can sometimes take on imaginative forms.

Chapter 8—Tearing Mechanisms: $\mathbf{A}_3$. Dynamical systems that possess twofold symmetry sometimes exhibit a different stretching and squeezing mechanism. This mechanism involves tearing the flow apart in phase space. Two physical systems whose strange attractors are generated by this mechanism are analyzed in this chapter. The Lorenz system is a standard representation for this process. Folding and tearing are intimately related to the two simplest catastrophes, the fold A_2 and the cusp A_3. Other mechanisms are related to other catastrophes and singularities.

Chapter 9—Unfoldings. One of our objectives is to be able to predict what will—or can—happen as control parameters are changed. This question reduces to the problem of determining the unfoldings of the "germs" of dynamical systems. If control parameters are changed a little, there is a perturbation in the spectrum of orbits in a strange attractor, but the underlying branched manifold remains unchanged (is robust). The change in the specturm of orbits can occur in a number of ways limited by topological considerations. If the change in control parameters is very large, the branched manifold itself can change. In the latter case, new branches can be added or old branches removed in a number of ways limited by topological and continuity considerations. As a result, the perestroikas in strange attractors can take place in a number of ways that is large but constrained. Put another way: Given any point in this doubly discrete classification representing a strange attractor with a specific spectrum of periodic orbits, under perturbation of the control parameters it can only move to its neighboring points in this classification. Its neighbors are large in number but constrained by both continuity and topological arguments.

Chapter 10—Symmetry. The measurements described in the first section are of the laser output intensity. We understand that the electric field amplitude $E(t)$ is more fundamental than the intensity, since $I(t) = E(t)^2$. We also feel (correctly) that if the intensity behaves chaotically, so must the amplitude. But is the chaos the same? In other words, is the strange attractor generated by the amplitude $E(t)$ (were we able to measure it) equivalent to the strange attractor generated by the intensity $I(t)$? This chapter is devoted to an analysis of dynamical systems that are locally equivalent but not globally so (E is a two-branch cover of I: $E =$ "$\pm$"$\sqrt{I}$). We present unexpected, elegant, and powerful results relating cover and image dynamical systems, even when the symmetry group relating them has only two group elements. These results allow us, for example, to take the square root of the Duffing and van der Pol dynamical systems, simplifying the analyses of these two systems dramatically.

Chapter 11—Flows in Higher Dimensions. What happens when a strange attractor cannot be shoehorned into a three-dimensional inertial manifold? This question is asked for four-dimensional strange attractors. Although periodic orbits can no longer be the working tool in this dimension, it appears that the singularity structure of a return map of a certain manifold into itself provides information needed to classify higher-dimensional attractors. The classification can be discrete when the number of unstable directions in the flow is large and the number of stable directions is small. We survey the rich spectrum of possibilities that exist for four-dimensional chaotic flows with two unstable directions and one stable direction. The results in this chapter are few—rather, the chapter lays the groundwork for future developments in this field.

Chapter 12—Program for Dynamical Systems Theory. When this work was initiated, one of the authors was surprised that a classification theory for strange attractors did not exist. He was very frustrated by the fact that nobody was even asking the question: How do you classify strange attractors? The first thing you do when confronted with new physics is to try to understand and to classify. The situation has not improved with time. Many are the works devoted to observation or the analysis of small problems. The response of many in the field of dynamical systems to the lack of rapid understanding and development of the field in low dimensions was to look for and to describe more complicated problems in higher dimensions! If we don't have a theory in low dimensions, how are we to find one in higher dimensions? Chapter 12 addresses this problem in an indirect way. There are many similarities between the older fields of Lie group theory and singularity theory (catastrophe theory) and the newer field of dynamical systems theory. It is a hope that simply asking the same questions of the newer field that were asked of the older fields will at least provide a structure in which advances can more rapidly be made. The entire chapter is devoted to comparing similar questions in these three fields, and suggesting how some answers might hasten the development of a theory for dynamical systems which is as mature as Lie group theory and singularity theory are today.

Appendix A—Determining Templates from Topological Invariants. A characterization method is most useful when it can be routinely and effortlessly applied to many different systems. In Appendix A, we describe computational techniques useful for determining the simplest template compatible with a given set of topological invariants. These techniques have been implemented in computer programs. We also discuss how the information about the symbolic dynamics of the periodic orbits extracted from invariants can be used to construct symbolic encodings for chaotic attractors.

And now, on with the details.

2

Discrete Dynamical Systems: Maps

2.1 INTRODUCTION

Many physical systems displaying chaotic behavior are accurately described by mathematical models derived from well-understood physical principles. For example, the fundamental equations of fluid dynamics, namely the Navier–Stokes equations, are obtained from elementary mechanical and thermodynamical considerations. The simplest laser models are built from Maxwell's laws of electromagnetism and from the quantum mechanics of a two-level atom.

Except for stationary regimes, it is in general not possible to find closed-form solutions to systems of nonlinear partial or ordinary differential equations (PDEs and ODEs). However, numerical integration of these equations often reproduces surprisingly well the irregular behaviors observed experimentally. Thus, these models must have some mathematical properties that are linked to the occurrence of chaotic behavior. To understand what these properties are, it is clearly desirable to study chaotic dynamical systems whose mathematical structure is as simple as possible.

Because of the difficulties associated with the analytical study of differential systems, a large amount of work has been devoted to dynamical systems whose state is known only at a discrete set of times. These are usually defined by a relation

$$X_{n+1} = f(X_n) \tag{2.1}$$

where $f : M \to M$ is a map of a state space into itself and X_n denotes the state at the discrete time n. The sequence $\{X_n\}$ obtained by iterating (2.1) starting from an initial condition X_0 is called the *orbit* of X_0, with $n \in \mathbb{N}$ or $n \in \mathbb{Z}$, depending on whether or not the map f is invertible. If X_0 is chosen at random, one observes generally that its orbit displays two distinct phases: A *transient* that is limited in time and specific to the orbit, and an *asymptotic regime* that persists for arbitrarily long times and is qualitatively the same for different typical orbits. Rather than study individual orbits, we are interested in classifying all the asymptotic behaviors that can be observed in a dynamical system.

To understand generic properties of chaotic behavior, there is no loss of generality in restricting ourselves to discrete-time systems: It is often easy to extract such a system from a continuous-time system. A common example is provided by the technique of Poincaré sections, where the study is narrowed to the dynamics of the map relating the successive intersections of a trajectory with a given surface of phase space, which is usually called the *first return map*. Another example is the *time-one map* $X(t+1) = \phi^1(X(t))$ of a differential system, which relates two states located one time unit apart along the same trajectory.

It might seem natural to restrict our study to discrete-time dynamical systems sharing some key properties with differential systems. For example, the solution of a system of ODEs depends continuously on initial conditions, so that continuous maps are naturally singled out. Invertibility is also a crucial property. Given an initial condition, the state of an ODE system can in principle be determined at any time in the future but also in the past; thus, we must be able to go backward in time. Maps satisfying these two requirements (i.e., continuous maps with a continuous inverse) are called *homeomorphisms*. Another important class of maps is made of *diffeomorphisms*: These are homeomorphisms that are differentiable as well as their inverse.

The most important aspects of chaotic behavior should appear in systems of lowest dimension. Thus, we would like in a first step to reduce as much as possible the dimension of state space. However, this quickly conflicts with the requirement of invertibility. On the one hand, it can be shown that maps based on a one-dimensional homeomorphism can only display stationary or periodic regimes, and hence cannot be chaotic. On the other hand, if we sacrifice invertibility temporarily, thereby introducing *singularities*, one-dimensional chaotic systems can easily be found, as illustrated by the celebrated logistic map. Indeed, this simple system will be seen to display many of the essential features of deterministic chaos.

It is, in fact, no coincidence that chaotic behavior appears in its simplest form in a noninvertible system. As emphasized in this book, singularities and noninvertibility are intimately linked to the mixing processes (stretching and squeezing) associated with chaos.

Because of the latter, a dissipative invertible chaotic map becomes formally noninvertible when infinitely iterated (i.e., when phase space has been infinitely squeezed). Thus, the dynamics is, in fact, organized by an underlying singular map of lower dimension, as can be shown easily in model systems such as the horseshoe map. A classical example of this is the Hénon map, a diffeomorphism of the plane into itself that is known to have the logistic map as a backbone.

2.2 LOGISTIC MAP

A noninvertible one-dimensional map has at least one point where its derivative vanishes. The simplest such maps are quadratic polynomials, which can always be brought to the form $f(x) = a - x^2$ under a suitable change of variables. The logistic map[1]

$$x_{n+1} = a - x_n^2 \tag{2.2}$$

which depends on a single parameter a, is thus the simplest one-dimensional map displaying a singularity. As can be seen from its graph [Fig. 2.1(a)], the most important consequence of the singularity located at the critical point $x = 0$ is that each value in the range of the map f has exactly two preimages, which will prove to be a key ingredient to generate chaos. Maps with a single critical point are called *unimodal*. It will be seen later that all unimodal maps display very similar dynamical behavior.

As is often the case in dynamical systems theory, the action of the logistic map can not only be represented algebraically, as in Eq. (2.2), but also geometrically. Given a point x_n, the graph of the logistic map provides $y = f(x_n)$. To use y as the starting point of the next iteration, we must find the corresponding location in the x space, which is done simply by drawing the line from the point $[x_n, f(x_n)]$ to the diagonal $y = x$. This simple construction is then repeated ad libitum, as illustrated in Fig. 2.1(b).

The various behaviors displayed by the logistic map are easily explored, as this map depends on a single parameter a. As illustrated in Fig. 2.2, one finds quickly that two main types of dynamical regimes can be observed: stationary or periodic regimes on the one hand, and "chaotic" regimes on the other hand. In the first case, iterations eventually visit only a finite set of different values that are forever repeated in a fixed order. In the latter case, the state of the system never repeats itself exactly

[1] A popular variant is $x_{n+1} = \lambda x_n(1 - x_n)$, with parameter λ.

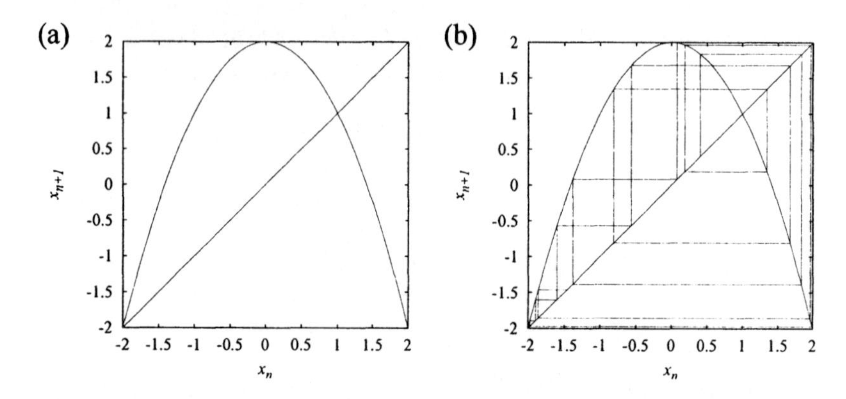

Fig. 2.1 (a) Graph of the logistic map for $a = 2$. (b) Graphical representation of the iteration of (2.2).

and seemingly evolves in a disordered way, as in Fig. 2.1(b). Both types of behaviors have been observed in the experiment discussed in Chapter 1.

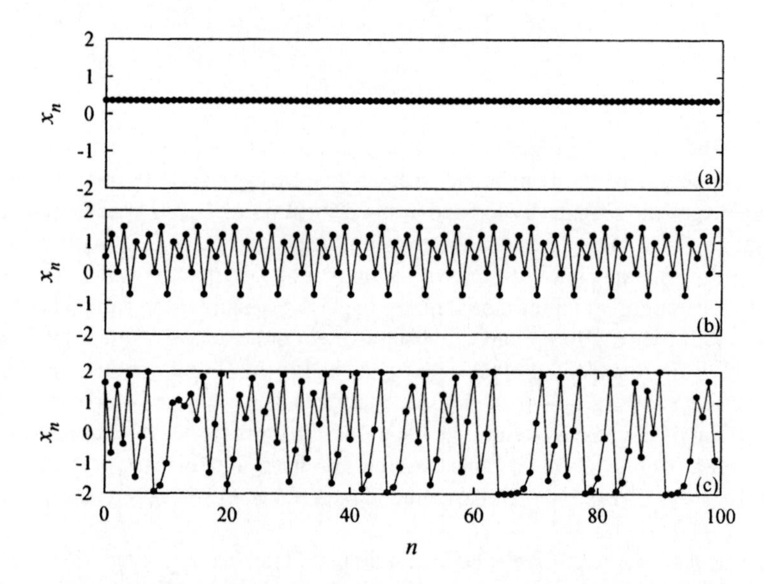

Fig. 2.2 Different dynamical behaviors observed in the logistic map system are represented by plotting successive iterates: (a) stationary regime, $a = 0.5$; (b) periodic regime of period 5, $a = 1.476$; (c) chaotic regime, $a = 2.0$.

What makes the study of the logistic map so important is not only that the organization in parameter space of these periodic and chaotic regimes can be completely understood with simple tools, but that despite of its simplicity it displays the most important features of low-dimensional chaotic behavior. By studying how periodic and chaotic behavior are interlaced, we will learn much about the mechanisms responsible for the appearance of chaotic behavior. Moreover, the logistic map is not only a paradigmatic system: One-dimensional maps will later prove also to be a fundamental tool for understanding the topological structure of flows.

2.3 BIFURCATION DIAGRAMS

A first step in classifying the dynamical regimes of the logistic map is to obtain a global representation of the various regimes that are encountered as the control parameter a is varied. This can be done with the help of *bifurcation diagrams*, which are tools commonly used in nonlinear dynamics. Bifurcation diagrams display some characteristic property of the asymptotic solution of a dynamical system as a function of a control parameter, allowing one to see at a glance where qualitative changes in the asymptotic solution occur. Such changes are termed *bifurcations*.

In the case of the logistic map that has a single dynamical variable, the bifurcation diagram is readily obtained by plotting a sample set of values of the sequence (x_n) as a function of the parameter a, as shown in Fig. 2.3.

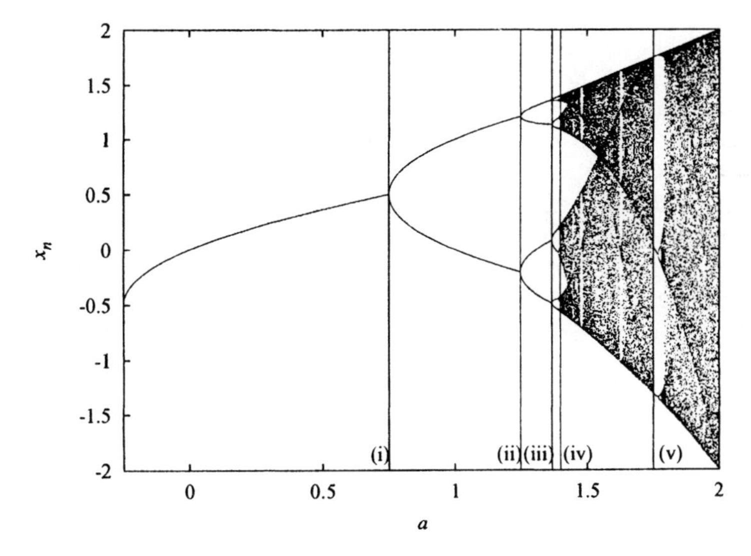

Fig. 2.3 Bifurcation diagram of the logistic map. For a number of parameter values between $a = -0.25$ and $a = 2.0$, 50 successive iterates of the logistic map are plotted after transients have died out. From left to right, the vertical lines mark the creations of (i) a period-2 orbit; (ii) a period-4 orbit; (iii) a period-8 orbit, and (iv) the accumulation point of the period-doubling cascade; (v) the starting point of a period-3 window.

For $a < a_0 = -\frac{1}{4}$, iterations of the logistic map escape to infinity from all initial conditions. For $a > a_R = 2$ almost all initial conditions escape to infinity.

The bifurcation diagram is thus limited to the range $a_0 < a < a_R$, where bounded solutions can be observed.

Between $a_0 = -\frac{1}{4}$ and $a_1 = \frac{3}{4}$, the limit set consists of a single value. This corresponds to a stationary regime, but one that should be considered in this context as a period-1 *periodic orbit*. At $a = a_1$, a bifurcation occurs, giving birth to a period-2 *periodic orbit*: Iterations oscillate between two values. As detailed in Section 2.4.2, this is an example of a *period-doubling bifurcation*. At $a = a_2 = \frac{5}{4}$, there is another period-doubling bifurcation where the period-2 orbit gives place to a period-4 orbit.

The period-doubling bifurcations occurring at $a = a_1$ and $a = a_2$ are the first two members of an infinite series, known as the *period-doubling cascade*, in which an orbit of period 2^n is created for every integer n. The bifurcation at $a = a_3$ leading to a period-8 orbit is easily seen in the bifurcation diagram of Fig. 2.3, the one at $a = a_4$ is hardly visible, and the following ones are completely indiscernible to the naked eye.

This is because the parameter values a_n at which the period-2^n orbit is created converge geometrically to the accumulation point $a_\infty = 1.401155189\ldots$ with a convergence ratio substantially larger than 1:

$$\lim_{n \to \infty} \frac{a_n - a_{n-1}}{a_{n+1} - a_n} = \delta \sim 4.6692016091\ldots \tag{2.3}$$

The constant δ appearing in (2.3) was discovered by Feigenbaum [15, 16] and is named after him. This distinction is justified by a remarkable property: Period-doubling cascades observed in an extremely large class of systems (experimental of theoretical, defined by maps of by differential equations...) have a convergence rate given by δ.

At the accumulation point a_∞, the period of the solution has become infinite. Right of this point, the system can be found in chaotic regimes, as can be guessed from the abundance of dark regions in this part of the bifurcation diagram, which indicate that the system visits many different states. The period-doubling cascade is one of the best-known *routes to chaos* and can be observed in many low-dimensional systems [34]. It has many universal properties that are in no way restricted to the case of the logistic map.

However, the structure of the bifurcation diagram is more complex than a simple division between periodic and chaotic regions on both sides of the accumulation point of the period-doubling cascade. For example, a relatively large *periodic window*, which corresponds to the domain of stability of a period-3 orbit, is clearly seen to begin at $a = \frac{7}{4}$, well inside the chaotic zone. In fact, periodic windows and chaotic regions are arbitrarily finely interlaced as illustrated by Fig. 2.4. As will be shown later, there are infinitely many periodic windows between any two periodic windows. To interpret Fig. 2.4, it should be noted that periodic windows are visible to the naked eye only for very low periods. For higher periods, (1) the periodic window is too narrow compared to the scale of the plot, and (2) the number of samples is sufficiently large that the window cannot be distinguished from the chaotic regimes.

Ideally, we would like to determine for each periodic solution the range of parameter values over which it is stable. In Section 2.4 we will perform this analysis for

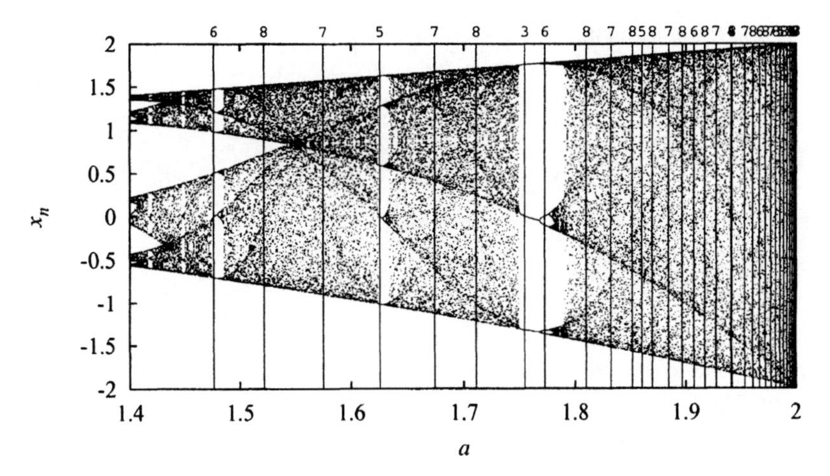

Fig. 2.4 Enlarged view of the chaotic zone of the bifurcation diagram of Fig. 2.3. Inside periodic windows of period up to 8, vertical lines indicate the parameter values where the corresponding orbits are most stable, with the period indicated above the line.

the simple cases of the period-1 and period-2 orbits, so that we get a better understanding of the two types of bifurcation that are encountered in the logistic map. This is motivated by the fact that these are the two bifurcations that are generically observed in low-dimensional dynamical systems (omitting the Hopf bifurcation, which we discuss later).

However, we will not attempt to go much further in this direction. First, the complexity of Figs. 2.3 and 2.4 shows that this task is out of reach. Moreover, we are only interested in properties of the logistic map that are shared by many other dynamical systems. In this respect, computing exact stability ranges for a large number of regimes would be pointless.

This does not imply that a deep understanding of the structure of the bifurcation diagram of Fig. 2.3 cannot be achieved. Quite to the contrary, we will see later that simple topological methods allow us to answer precisely the following questions: How can we classify the different periodic regimes? Does the succession of different dynamical regimes encountered as the parameter a is increased follow a logical scheme? In particular, a powerful approach to chaotic behavior, *symbolic dynamics*, which we present in Section 2.7, will prove to be perfectly suited for unfolding the complexity of chaos.

2.4 ELEMENTARY BIFURCATIONS IN THE LOGISTIC MAP

2.4.1 Saddle-Node Bifurcation

The simplest regime that can be observed in the logistic map is the period-1 orbit. It is stably observed on the left of the bifurcation diagram of Fig. 2.3 for $a_0 < a < a_1$.

It corresponds to a *fixed point* of the logistic map (i.e., it is mapped onto itself) and is thus a solution of the equation $x = f(x)$. For the logistic map, finding the fixed points merely amounts to solving the quadratic equation

$$x = a - x^2 \tag{2.4}$$

which has two solutions:

$$x_-(a) = \frac{-1 - \sqrt{1+4a}}{2} \qquad x_+(a) = \frac{-1 + \sqrt{1+4a}}{2} \tag{2.5}$$

The fixed points of a one-dimensional map can also be located geometrically, since they correspond to the intersections of its graph with the diagonal (Fig. 2.1).

Although a single period-1 regime is observed in the bifurcation diagram, there are actually two period-1 orbits. Later we will see why. Expressions (2.5) are real-valued only for $a > a_0 = -\frac{1}{4}$. Below this value, all orbits escape to infinity. Thus, the point at infinity, which we denote x_∞ in the following, can formally be considered as another fixed point of the system, albeit unphysical.

The important qualitative change that occurs at $a = a_0$ is our first example of a ubiquitous phenomenon of low-dimensional nonlinear dynamics, a *tangent*, or *saddle-node*, bifurcation: The two fixed points (2.5) become simultaneously real and are degenerate: $x_-(a_0) = x_+(a_0) = -\frac{1}{2}$. The two designations point to two different (but related) properties of this bifurcation.

The saddle-node qualifier is related to the fact that the two bifurcating fixed points have different stability properties. For a slightly above a_0, it is found that orbits located near x_+ converge to it, whereas those starting in the neighborhood of x_- leave it to either converge to x_+ or escape to infinity, depending on whether they are located right or left of x_-. Thus, the fixed point x_+ (and obviously also x_∞) is said to be *stable* while x_- is *unstable*. They are called the *node* and the *saddle*, respectively.

Since trajectories in their respective neighborhoods converge to them, x_+ and x_∞ are *attracting sets*, or *attractors*. The sets of points whose orbits converge to an attractor of a system is called the *basin of attraction* of this point. From Fig. 2.5 we see that the unstable fixed point x_- is on the boundary between the *basins of attraction* of the two stable fixed points x_+ and x_∞. The other boundary point is the preimage $f^{-1}(x_-)$ of x_- (Fig. 2.5).

It is easily seen that the stability of a fixed point depends on the derivative of the map at the fixed point. Indeed, if we perturb a fixed point $x_* = f(x_*)$ by a small quantity δx_n, the perturbation δx_{n+1} at the next iteration is given by

$$\delta x_{n+1} = f(x_* + \delta x_n) - x_* = \left.\frac{\mathrm{d}f(x)}{\mathrm{d}x}\right|_{x_*} \delta x_n + O(\delta x_n^2) \tag{2.6}$$

If we start with an infinitesimally small δx_0, the perturbation after n iterations is thus $\delta x_n \approx (\mu_*)^n \delta x_0$, where μ_*, the *multiplier* of the fixed point, is given by the map derivative at $x = x_*$.

A fixed point is thus stable (resp., unstable) when the absolute value of its multiplier is smaller (resp., greater) than unity. Here the multipliers $\mu_\pm$ of the two fixed points

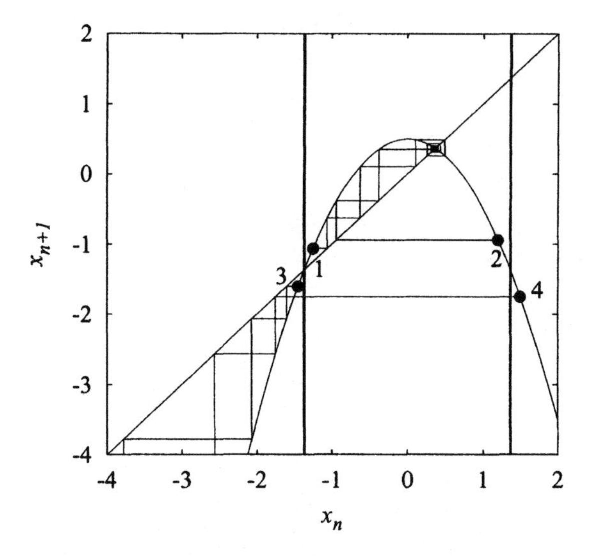

Fig. 2.5 The basin of attraction of the x_+ fixed point is located between the left fixed point x_- and its preimage, indicated by two vertical lines. The orbits labeled 1 and 2 are inside the basin and converge towards x_+. The orbits labeled 3 and 4 are outside the basin and escape to infinity (i.e., converge to the point at infinity x_∞.)

of the logistic map are given by

$$\mu_- = \left.\frac{\mathrm{d}f(x)}{\mathrm{d}x}\right|_{x_-} = -2x_- = 1 + \sqrt{1+4a} \tag{2.7a}$$

$$\mu_+ = \left.\frac{\mathrm{d}f(x)}{\mathrm{d}x}\right|_{x_+} = -2x_+ = 1 - \sqrt{1+4a} \tag{2.7b}$$

Equation (2.7a) shows that x_- is unconditionally unstable on its entire domain of existence, and hence is generically not observed as a stationary regime, whereas x_+ is stable for parameters a just above $a_0 = -\frac{1}{4}$, as mentioned above. This is why only x_+ can be observed on the bifurcation diagram shown in Fig. 2.3.

More precisely, x_+ is stable for $a \in [a_0, a_1]$, where $a_1 = \frac{3}{4}$ is such that $\mu_+ = -1$. This is consistent with the bifurcation diagram of Fig. 2.3. Note that at $a = 0 \in [a_0, a_1]$, the multiplier $\mu_+ = 0$ and thus perturbations are damped out faster than exponentially: The fixed point is then said to be *superstable*.

At the saddle-node bifurcation, both fixed points are degenerate and their multiplier is $+1$. This fundamental property is linked to the fact that at the bifurcation point, the graph of the logistic map is tangent to the diagonal (see Fig. 2.6), which is why this bifurcation is also known as the tangent bifurcation. Tangency of two smooth curves (here, the graph of f and the diagonal) is generic at a multiple intersection point. This is an example of a *structurally unstable situation*: An arbitrarily small perturbation of f leads to two distinct intersections or no intersection at all (alternatively, to two real or to two complex roots).

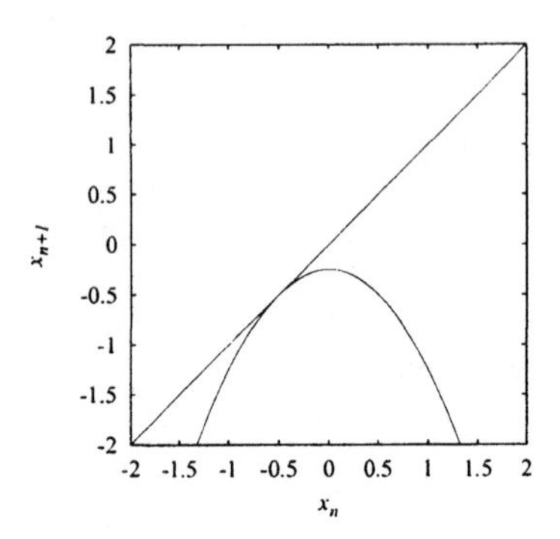

Fig. 2.6 Graph of the logistic map at the initial saddle-node bifurcation.

It is instructive to formulate the intersection problem in algebraic terms. The fixed points of the logistic equations are zeros of the equation $G(x, a) = f(x; a) - x = 0$. This equation defines implicit functions $x_+(a)$ and $x_-(a)$ of the parameter a. In structurally stable situations, these functions can be extended to neighboring parameter values by use of the implicit function theorem.

Assume that $x_*(a)$ satisfies $G(x_*(a), a) = 0$ and that we shift the parameter a by an infinitesimal quantity δa. Provided that $\partial G(x_*(a), a)/\partial x \neq 0$, the corresponding variation δx_* in x_* is given by

$$G(x, a) = G(x_* + \delta x_*, a + \delta a) = G(x, a) + \frac{\partial G}{\partial x}\delta x_* + \frac{\partial G}{\partial a}\delta a = 0 \tag{2.8}$$

which yields:

$$\delta x_* = -\frac{\partial G/\partial a}{\partial G/\partial x}\delta a \tag{2.9}$$

showing that $x_*(a)$ is well defined on both sides of a if and only if $\partial G/\partial x \neq 0$. The condition

$$\frac{\partial G(x_*(a), a)}{\partial x} = 0 \tag{2.10}$$

is thus the signature of a bifurcation point. In this case, the Taylor series (2.8) has to be extended to higher orders of δx_*. If $\partial^2 G(x_*(a), a)/\partial x^2 \neq 0$, the variation δx_* in the neighborhood of the bifurcation is given by

$$(\delta x_*)^2 = -2\,\frac{\partial G/\partial a}{\partial^2 G/\partial x^2}\delta a \tag{2.11}$$

From (2.11), we recover the fact that there is a twofold degeneracy at the bifurcation point, two solutions on one side of the bifurcation and none on the other side. The stability of the two bifurcating fixed points can also be analyzed: Since $G(x, a) =$

$f(x; a) - x$, their multipliers are given by $\mu_* = 1 + \partial G(x_*, a)/\partial x$ and are thus equal to 1 at the bifurcation.

Just above the bifurcation point, it is easy to show that the multipliers of the two fixed points x_+ and x_- are given to leading order by $\mu_\pm = 1 \mp \alpha\sqrt{|\delta a|}$, where the factor α depends on the derivatives of G at the bifurcation point. It is thus generic that one bifurcating fixed point is stable while the other one is unstable. In fact, this is a trivial consequence of the fact that the two nondegenerate zeros of $G(x, a)$ must have derivatives $\partial G/\partial x$ with opposite signs.

This is linked to a fundamental theorem, which we state below in the one-dimensional case but which can be generalized to arbitrary dimensions by replacing derivatives with Jacobian determinants. Define the degree of a map f as

$$\deg f = \sum_{f(x_i)=y} \operatorname{sgn} \frac{\mathrm{d}f}{\mathrm{d}x}(x_i) \tag{2.12}$$

where the sum extends over all the preimages of the arbitrary point y, and $\operatorname{sgn} z = +1$ (resp., -1) if $z > 0$ (resp., $z < 0$). It can be shown that $\deg f$ does not depend on the choice of y provided that it is a regular value (the derivatives at its preimages x_i are not zero) and that it is invariant by homotopy. Let us apply this to $G(x, a)$ for $y = 0$. Obviously, $\deg G = 0$ when there are no fixed points, but also for any a since the effect of varying a is a homotopy. We thus see that fixed points must appear in pairs having opposite contributions to $\deg G$. As discussed above, these opposite contributions correspond to different stability properties at the bifurcation.

The discussion above shows that although we have introduced the tangent bifurcation in the context of the logistic map, much of the analysis can be carried to higher dimensions. In an n-dimensional state space, the fixed points are determined by an n-dimensional vector function $\mathbf{G}$. In a structurally stable situation, the Jacobian $\partial\mathbf{G}/\partial\mathbf{X}$ has rank n. As one control parameter is varied, bifurcations will be encountered at parameter values where $\partial\mathbf{G}/\partial\mathbf{X}$ is of lower rank. If the Jacobian has rank $n-1$, it has a single null eigenvector, which defines the direction along which the bifurcation takes place. This explains why the essential features of tangent bifurcations can be understood from a one-dimensional analysis.

The theory of bifurcations is in fact a subset of a larger field of mathematics, the *theory of singularities* [35, 36], which includes *catastrophe theory* [13, 37] as a special important case. The tangent bifurcation is an example of the simplest type of singularity: The *fold singularity*, which typically corresponds to twofold degeneracies.

In the next section we see an example of a threefold degeneracy, the *cusp singularity*, in the form of the period-doubling bifurcation.

2.4.2 Period-Doubling Bifurcation

As shown in Section 2.4.1, the fixed point x_+ is stable only for $a \in [a_0, a_1]$, with $\mu_+ = 1$ at $a = a_1 = -\frac{1}{4}$ and $\mu_+ = -1$ at $a = a_1 = \frac{3}{4}$. For $a > a_1$, both fixed points (2.5) are unstable, which precludes a period-1 regime. Just above the bifurcation, what is observed instead is that successive iterates oscillate between two distinct values (see Fig. 2.3), which comprise a period-2 orbit. This could have been expected from the fact that at $a = a_1$, $\mu_+ = -1$ indicates that perturbations are reproduced every other period. The qualitative change that occurs at $a = a_1$ (a fixed point becomes unstable and gives birth to an orbit of twice the period) is another important example of bifurcation: The *period-doubling bifurcation*, which is represented schematically in Fig. 2.7. Saddle-node and period-doubling bifurcations are the only two types of local bifurcation that are observed for the logistic map. With the Hopf bifurcation, they are also the only bifurcations that occur generically in one-parameter paths in parameter space, and consequently, in low-dimensional systems.

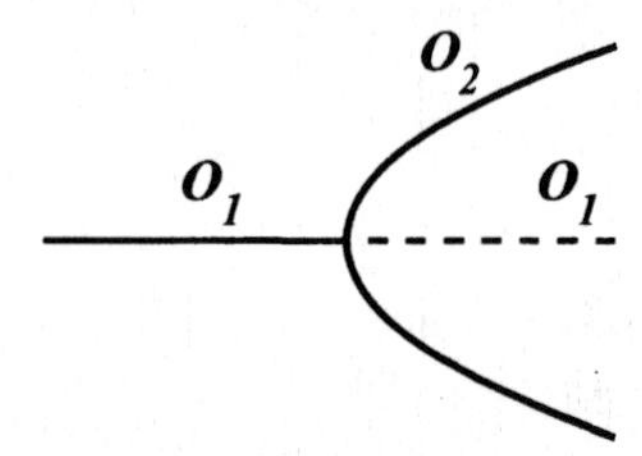

Fig. 2.7 Period-doubling bifurcation. The orbit O_1 becomes unstable in giving birth to an orbit O_2, whose period is twice that of O_1.

Before we carry out the stability analysis for the period-2 orbit created at $a = a_1$, an important remark has to be made. Expression (2.5) shows that the period-1 orbit x_+ exists for every $a > a_0$: Hence it does not disappear at the period-doubling bifurcation but merely becomes unstable. It is thus present in all the dynamical regimes observed after its loss of stability, including in the chaotic regimes of the right part of the bifurcation diagram of Fig. 2.3. In fact, this holds for all the periodic solutions of the logistic map. As an example, the logistic map at the transition to chaos ($a = a_\infty$) has an infinity of (unstable) periodic orbits of periods 2^n for any n, as Fig. 2.8 shows.

We thus expect periodic orbits to play an important role in the dynamics even after they have become unstable. We will see later that this is indeed the case and that much can be learned about a chaotic system from its set of periodic orbits, both stable and unstable.

Since the period-2 orbit can be viewed as a fixed point of the second iterate of the logistic map, we can proceed as above to determine its range of stability. The two periodic points $\{x_1, x_2\}$ are solutions of the quartic equation

$$x = f(f(x)) = a - \left(a - x^2\right)^2 \tag{2.13}$$

To solve for x_1 and x_2, we take advantage of the fact that the fixed points x_+ and x_- are obviously solutions of Eq. (2.13). Hence, we just have to solve the quadratic equation

$$p(x) = \frac{f(f(x)) - x}{f(x) - x} = 1 - a - x + x^2 = 0 \tag{2.14}$$

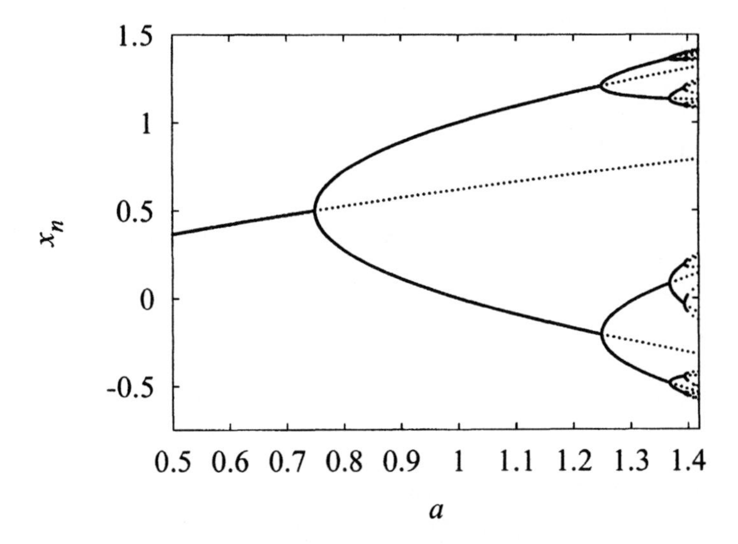

Fig. 2.8 Orbits of period up to 16 of the period-doubling cascade. Stable (resp., unstable) periodic orbits are drawn with solid (resp., dashed) lines.

whose solutions are

$$x_1 = \frac{1 - \sqrt{-3 + 4a}}{2} \qquad x_2 = \frac{1 + \sqrt{-3 + 4a}}{2} \tag{2.15}$$

We recover the fact that the period-2 orbit (x_1, x_2) appears at $a = a_1 = \frac{3}{4}$, and exists for every $a > a_1$. By using the chain rule for derivatives, we obtain the multiplier of the fixed point x_1 of f^2 as

$$\mu_{1,2} = \left.\frac{\mathrm{d}f^2(x)}{\mathrm{d}x}\right|_{x_1} = \left.\frac{\mathrm{d}f(x)}{\mathrm{d}x}\right|_{x_2} \times \left.\frac{\mathrm{d}f(x)}{\mathrm{d}x}\right|_{x_1} = 4x_1x_2 = 4(1-a) \tag{2.16}$$

Note that x_1 and x_2 viewed as fixed points of f^2 have the same multiplier, which is defined to be the multiplier of the orbit (x_1, x_2). At the bifurcation point $a = a_1$, we have $\mu_{1,2} = 1$, a signature of the two periodic points x_1 and x_2 being degenerate at the period-doubling bifurcation.

However, the structure of the bifurcation is not completely similar to that of the tangent bifurcation discussed earlier. Indeed, the two periodic points x_1 and x_2 are also degenerate with the fixed point x_+. The period-doubling bifurcation of the fixed point x_+ is thus a situation where the second iterate f^2 has *three* degenerate fixed points. If we define $G_2(x, a) = f^2(x; a) - x$, the signature of this threefold degeneracy is $G_2 = \partial G_2/\partial x = \partial^2 G_2/\partial x^2 = 0$, which corresponds to a higher-order singularity than the fold singularity encountered in our discussion of the tangent bifurcation. This is, in fact, our first example of the *cusp singularity*. Note that x_+ has a multiplier of -1 as a fixed point of f at the bifurcation and hence exists on both sides of the bifurcation: It merely becomes unstable at $a = a_1$. On the contrary, x_1 and x_2 have multiplier 1 for the lowest iterate of f of which they are fixed points, and thus exist only on one side of the bifurcation.

We also may want to verify that $\deg f^2 = 0$ on both sides of the bifurcation. Let us denote $d(x_*)$ as the contribution of the fixed point x_* to $\deg f^2$. We do not consider x_-, which is not invoved in the bifurcation. Before the bifurcation, we have $d(x_+) = -1$ $[df^2/dx(x_+) < 1]$. After the bifurcation, $d(x_+) = 1$ but $d(x_1) = d(x_2) = -1$, so that the sum is conserved.

The period-2 orbit is stable only on a finite parameter range. The other end of the stability domain is at $a = a_2 = \frac{5}{4}$ where $\mu_{1,2} = -1$. At this parameter value, a new period-doubling bifurcation takes place, where the period-2 orbit loses its stability and gives birth to a period-4 orbit. As shown in Figs. 2.3 and 2.8, period-doubling occurs repeatedly until an orbit of infinite period is created.

Although one might in principle repeat the analysis above for the successive bifurcations of the period-doubling cascade, the algebra involved quickly becomes intractable. Anyhow, the sequence of parameters a_n at which a solution of period 2^n emerges converges so quickly to the accumulation point a_∞ that this would be of little use, except perhaps to determine the exact value of a_∞, after which the first chaotic regimes are encountered.

A fascinating property of the period-doubling cascade is that we do not need to analyze directly the orbit of period 2^∞ to determine very accurately a_∞. Indeed, it can be remarked that the orbit of period 2^∞ is formally its own period-doubled orbit. This indicates some kind of scale invariance. Accordingly, it was recognized by Feigenbaum that the transition to chaos in the period-doubling cascade can be analyzed by means of renormalization group techniques [15, 16].

In this section we have analyzed how the periodic solutions of the logistic map are created. After discussing changes of coordinate systems in the next section, we shall take a closer look at the chaotic regimes appearing in the bifurcation diagram of Fig. 2.3. We will then be in position to introduce more sophisticated techniques to analyze the logistic map, namely *symbolic dynamics*, and to gain a complete understanding of the bifurcation diagram of a large class of maps of the interval.

2.5 MAP CONJUGACY

2.5.1 Changes of Coordinates

The behavior of a physical system does not depend on how we describe it. Equations defining an abstract dynamical system are meaningful only with respect to a given parameterization of its states (i.e., in a given coordinate system). If we change the parameterization, the dynamical equations should be modified accordingly so that the same physical states are connected by the evolution laws.

Assume that we have a system whose physical states are parameterized by coordinates $x \in X$, with an evolution law given by $f : X \to X$ [i.e., $x_{n+1} = f(x_n)$]. If we switch to a new coordinate system specified by $y = h(x)$, with $y \in Y$, the dynamical equations become $y_{n+1} = g(y_n)$, where the map $g : Y \to Y$ satisfies

$$h(f(x)) = g(h(x)) \tag{2.17}$$

Relation (2.17) simply expresses that on the one hand, $y_{n+1} = h(x_{n+1}) = h(f(x_n))$, and on the other hand, $y_{n+1} = g(y_n) = g(h(x_n))$. This is summarized by the following *commutative diagram*:

$$\begin{array}{ccc} x_n & \xrightarrow{f} & x_{n+1} \\ \downarrow{\scriptstyle h} & & \downarrow{\scriptstyle h} \\ y_n & \xrightarrow{g} & y_{n+1} \end{array} \tag{2.18}$$

where relation (2.17) is recovered by comparing the two paths from x_n to y_{n+1}.

Different types of conjugacy may be defined depending on the class of functions the transformation h belongs to (e.g., see [38]). *Conjugacy*, or *smooth conjugacy*, corresponds to the case where h is a diffeomorphism. If h is a homeomorphism, one has *topological conjugacy*. Note that in some cases, the transformation h can be $2 \to 1$. This is referred to as *semiconjugacy*.

2.5.2 Invariants of Conjugacy

Often, the problem is not to compute the evolution equations in a new coördinate system, but to determine whether two maps f and g correspond to the same physical system (i.e., whether or not they are conjugate). A common strategy to address this type of problem is to search for quantities that are invariant under the class of transformations considered. If two objects have different invariants, they cannot be transformed into each other. The knot invariants discussed later provide an important example of this. The ideal case is when there exists a complete set of invariants: Equality of the invariants then implies identity of the objects. In this section we present briefly two important invariants of conjugacy.

2.5.2.1 Spectrum of Periodic Orbits An important observation is that there is a one-to-one correspondence between periodic orbits of two conjugate maps. Assume that x_* is a period-p orbit of f: $f^p(x_*) = x_*$. If $f = h^{-1} \circ g \circ h$, we have

$$f^p = (h^{-1} \circ g \circ h)^p = h^{-1} \circ g^p \circ h \tag{2.19}$$

Thus, $y_* = h(x_*)$ satisfies

$$y_* = h(f^p(x_*)) = g^p(h(x_*)) = g^p(y_*) \tag{2.20}$$

This shows that y_* is itself a period-p orbit of g. If h is a one-to-one transformation, it follows immediately that f and g have the same number of period-p orbits.

Of course, this should have been expected: The existence of a periodic solution does not depend on the coordinate system. Yet this provides a useful criterion to test whether two maps are conjugate.

2.5.2.2 *Multipliers of Periodic Orbits* Similarly, the stability and the asymptotic evolution of a system are coordinate-independent. In algebraic terms, this translates into the invariance of the multipliers of a periodic orbit when the transformation h is a diffeomorphism.

To show this, let us compute the tangent map Dg^p of $g^p = h \circ f^p \circ h^{-1}$ at a point y_0, using the chain rule for derivatives:

$$Dg^p(y_0) = Dh\left((f^p \circ h^{-1})(y_0)\right) \times Df^p\left(h^{-1}(y_0)\right) \times Dh^{-1}(y_0) \tag{2.21}$$

To simplify notations, we set $x_0 = h^{-1}(y_0)$ and rewrite Eq. (2.21) as

$$Dg^p(y_0) = Dh\left(f^p(x_0)\right) \times Df^p\left(x_0\right) \times Dh^{-1}(y_0) \tag{2.22}$$

Relation (2.22) yields no special relation between $Df^p(x_0)$ and $Dg^p(y_0)$ unless x_0 is a period-p orbit and satisfies $f^p(x_0) = x_0$. In this case, indeed, we note that since $Dh(f^p(x_0)) = Dh(x_0)$, and because

$$Dh\left(h^{-1}(y_0)\right) \times Dh^{-1}(y_0) = \mathbf{1} \tag{2.23}$$

we have

$$Dg^p(y_0) = P \times Df^p(x_0) \times P^{-1} \tag{2.24}$$

where $P = Dh(x_0)$. Equation (2.24) indicates that the matrices $Df^p(x_0)$ and $Dg^p(y_0)$ are *similar*: They can be viewed as two representations of the same linear operator in two bases, with the matrix P (obviously nonsingular since h is a diffeomorphism) specifying the change of basis. Two matrices that are similar have, accordingly, the same eigenvalue spectrum.

This shows that the multipliers of a periodic orbit do not depend on the coordinate system chosen to parameterize the states of a system, hence that they are invariants of (smooth) conjugacy. Note, however, that they need not be preserved under a topological conjugacy.

2.6 FULLY DEVELOPED CHAOS IN THE LOGISTIC MAP

The first chaotic regime that we study in the logistic map is the one observed at the right end of the bifurcation diagram, namely at $a = 2$. At this point, the logistic map is surjective on the interval $I = [-2, 2]$: Every point $y \in I$ is the image of two different points, $x_1, x_2 \in I$. Moreover, I is then an invariant set since $f(I) = I$.

It turns out that the dynamical behavior of the surjective logistic map can be analyzed in a particularly simple way by using a suitable change of coordinates, namely $x = 2\sin(\pi x')/4$. This is a one-to-one transformation between I and itself, which is a diffeomorphism everywhere except at the endpoints $x = \pm 2$, where the inverse function $x'(x)$ is not differentiable. With the help of a few trigonometric identities, the action of the logistic map in the x' space can be written as

$$x'_{n+1} = g(x'_n) = 2 - 2\,|x'_n| \tag{2.25}$$

a piecewise linear map known as the *tent map*.

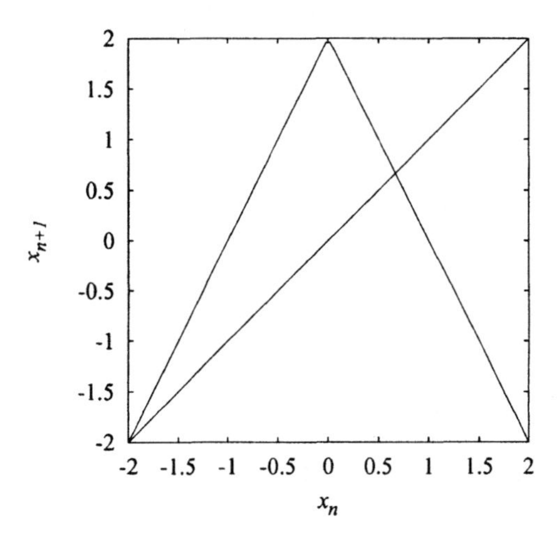

Fig. 2.9 Graph of the tent map (2.25).

Figure 2.9 shows that the graph of the tent map is extremely similar to that of the logistic map (Fig. 2.1). In both cases, the interval I is decomposed into two subintervals: $I = I_0 \cup I_1$, such that each restriction $f_k : I_k \to f(I_k)$ of f is a homeomorphism, with $f(I_0) = f(I_1) = I$. Furthermore, f_0 (resp., f_1) is orientation-preserving (resp., orientation-reversing).

In fact, these topological properties suffice to determine the dynamics completely and are characteristic features of what is often called a *topological horseshoe*. In the remainder of Section 2.6, we review a few fundamental properties of chaotic behavior that can be shown to be direct consequences of these properties.

2.6.1 Iterates of the Tent Map

The advantage of the tent map over the logistic map is that calculations are simplified dramatically. In particular, higher-order iterates of the tent map, which are involved in the study of the asymptotic dynamics, are themselves piecewise-linear maps and are easy to compute. For illustration, the graphs of the second iterate g^2 and of the fourth iterate g^4 are shown in Fig. 2.10. Their structure is seen to be directly related to that of the tent map.

Much of the structure of the iterates g^n can be understood from the fact that g maps linearly each of the two subintervals I_0 and I_1 to the whole interval I. Thus,

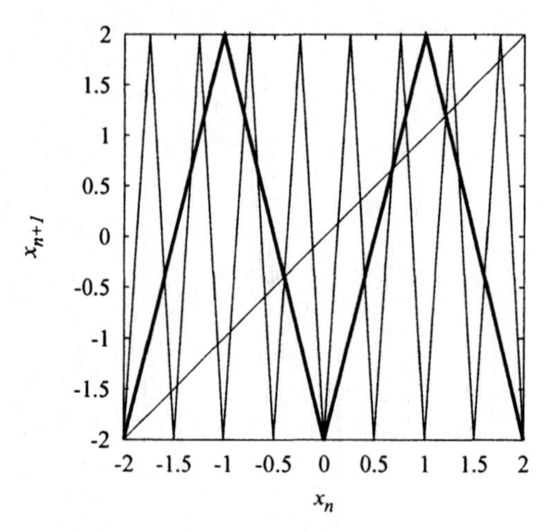

Fig. 2.10 Graphs of the second (heavy line) and fourth (light line) iterates of the tent map (2.25).

the graph of the restriction of g^2 to each of the two components I_k reproduces the graph of g on I. This explains the two-"hump" structure of g^2. Similarly, the trivial relation

$$\forall x \in I_k, \quad g^n(x) = g(g^{n-1}(x)) = g^{n-1}(g(x)) \tag{2.26}$$

shows that the graph of g^n consists of two copies of that of g^{n-1}. Indeed, Eq. (2.26) can be viewed as a semiconjugacy between g^n and g^{n-1} via the 2-to-1 transformation $x' = g(x)$.

By recursion, the graph of g^n shows 2^{n-1} scaled copies of the graph of g, each contained in a subinterval $I_k^n = [X_k - \epsilon_n, X_k + \epsilon_n]$ $(0 \leq k < 2^{n-1})$, where $\epsilon_n = 1/2^{n-2}$ and $X_k = -2 + (2k+1)\epsilon_n$. The expression of g^n can thus be obtained from that of g by

$$\forall x \in I_k^n = [X_k - \epsilon_n, X_k + \epsilon_n] \quad g^n(x) = g\left(\epsilon_n(x - X_k)\right) \tag{2.27}$$

An important consequence of (2.27) is that each subinterval I_k^n is mapped to the whole interval I in no more than n iterations of g:

$$\forall k = 0 \ldots 2^{n-1} \quad g^n\left(I_k^n\right) = g(I) = I \tag{2.28}$$

More precisely, one has $g(I_k^n) = I_{k'}^{n-1}$, where $k' = k$ (resp., $k' = 2^{n-1} - k$) if $k < 2^{n-2}$ (resp., $k \geq 2^{n-2}$). Note also that each I_k^n can itself be split into two intervals $I_{k,i}^n$ on which g^n is monotonic and such that $g^n(I_{k,i}^n) = I$.

Because the diameter of I_k^n is $|I_k^n| = 2^{3-n}$ and can be made arbitrarily small if n is chosen sufficiently large, this implies that an arbitrary subinterval $J \subset I$, however small, contains at least one interval I_k^n:

$$\forall J \subset I \quad \exists N_0 \quad n > N_0 \Rightarrow \exists k \quad I_k^n \subset J \tag{2.29}$$

Thus, how the iterates g^n act on the intervals I_k^n can help us to understand how they act on an arbitrary interval, as we will see later. In general, chaotic dynamics is better characterized by studying how sets of points are globally mapped rather than by focusing on individual orbits.

2.6.2 Lyapunov Exponents

An important feature of the tent map (2.25) is that the slope $|dg(x)/dx| = 2$ is constant on the whole interval $I = [-2, 2]$. This simplifies significantly the study of the stability of solutions of (2.25). From (2.6), an infinitesimal perturbation δx_0 from a reference state will grow after n iterations to $|\delta x_n| = 2^n|\delta x_0|$. Thus, any two distinct states, however close they may be, will eventually be separated by a macroscopic distance. This shows clearly that no periodic orbit can be stable (see Section 2.4.1) and thus that the asymptotic motion of (2.25) is aperiodic.

This exponential divergence of neighboring trajectories, or *sensitivity to initial conditions*, can be characterized quantitatively by *Lyapunov exponents*, which correspond to the average separation rate. For a one-dimensional map, there is only one Lyapunov exponent, defined by

$$\lambda = \lim_{n\to\infty} \frac{1}{n} \sum_{i=0}^{n-1} \log \frac{|\delta x_{n+1}|}{|\delta x_n|} = \lim_{n\to\infty} \frac{1}{n} \sum_{i=0}^{n-1} \log \left| \frac{df}{dx}(f^i(x_0)) \right| \tag{2.30}$$

which is a geometric average of the stretching rates experienced at each iteration. It can be shown that Lyapunov exponents are independent of the initial condition x_0, except perhaps for a set of measure zero [39].

Since the distance between infinitesimally close states grows exponentially as $\delta x_n \sim e^{n\lambda}\delta x_0$, sensitivity to initial conditions is associated with a strictly positive Lyapunov exponent. It is easy to see that the Lyapunov exponent of surjective tent map (2.25) is $\lambda = \ln 2$.

2.6.3 Sensitivity to Initial Conditions and Mixing

Sensitivity to initial conditions can also be expressed in a way that is more topological, without using distances. The key property we use here is that any subinterval $J \subset I$ is eventually mapped to the whole I:

$$\forall J \subset I, \quad \exists N_0, \quad n > N_0 \Rightarrow g^n(J) = I \tag{2.31}$$

This follows directly from the fact that J contains one of the basis intervals I_k^n, and that these expand to I under the action of g [see (2.28) and (2.29)].

We say that a map is *expansive* if it satisfies property (2.31). In plain words, the iterates of points in any subinterval can take every possible value in I after a sufficient number of iterations. Assume that J represents the uncertainty in the location of an initial condition x_0: We merely know that $x_0 \in J$, but not its precise position. Then (2.31) shows that chaotic dynamics is, although deterministic, asymptotically unpredictable: After a certain amount of time, the system can be anywhere in the state space. Note that the time after which all the information about the initial condition has been lost depends only logarithmically on the diameter $|J|$ of J. Roughly, (2.28) indicates that $N_0 \simeq -\ln|J|/\ln 2 \sim -\ln|J|/\lambda$.

In the following, we use property (2.31) as a topological definition of chaos in one-dimensional noninvertible maps. To illustrate it, we recall the definitions of various properties that have been associated with chaotic behavior [40] and which can be shown to follow from (2.31).

A map $f : I \to I$:

- Has *sensitivity to initial conditions* if $\exists\delta > 0$ such that for all $x \in I$ and any interval $J \ni x$, there is a $y \in J$ and an $n > 0$ such that $|f^n(x) - f^n(y)| > \delta$.
- Is *topologically transitive* if for each pair of open sets $A, B \subset I$, there exists n such that $f^n(A) \cap B \neq \emptyset$.
- Is *mixing* if for each pair of open sets $A, B \subset I$, there exists $N_0 > 0$ such that $n > N_0 \Rightarrow f^n(A) \cap B \neq \emptyset$. A mixing map is obviously topologically transitive.

Sensitivity to initial conditions trivially follows from (2.31), since any neighborhood of $x \in I$ is eventually mapped to I. The mixing property, and hence transitivity, is also a consequence of expansiveness because the N_0 in the definition can be chosen so that $f^{N_0}(A) = I$ intersects any $B \subset I$. It can be shown that a topologically transitive map has at least a dense orbit (i.e., an orbit that passes arbitrarily close to any point of the invariant set).

Note that (2.31) precludes the existence of an invariant subinterval $J \subset I$ other than I itself: We would have simultaneously $f(J) = J$ and $f^{N_0}(J) = I$ for some N_0. Thus, invariant sets contained in I necessarily consist of isolated points; these are the periodic orbits discussed in the next section.

2.6.4 Chaos and Density of (Unstable) Periodic Orbits

It has been proposed by Devaney [40] to say that a map f is chaotic if it:

- Displays sensitivity to initial conditions
- Is topologically transitive
- Has a set of periodic orbits that is dense in the invariant set

The first two properties were established in Section 2.6.3. It remains to be proved that (2.31) implies the third. When studying the bifurcation diagram of the logistic map (Section 2.4.2), we have noted that chaotic regimes contain many (unstable) periodic orbits. We are now in a position to make this observation more precise. We begin by showing that the tent map $x' = g(x)$ has infinitely many periodic orbits.

2.6.4.1 Number of Periodic Orbits of the Tent Map A periodic orbit of g of period p is a fixed point of the pth iterate g^p. Thus, it satisfies $g^p(x) = x$ and is associated with an intersection of the graph of g^p with the diagonal. Since g itself has exactly two such intersections (corresponding to period-1 orbits), (2.27) shows that g^p has

$$N_f(p) = 2^p \tag{2.32}$$

fixed points (see Fig. 2.10 for an illustration).

Some of these intersections might actually be orbits of lower period: For example, the four fixed points of g^2 consist of two period-1 orbits and of two points constituting a period-2 orbit. As another example, note on Fig. 2.10 that fixed points of g^2 are also fixed points of g^4. The number of periodic orbits of lowest period p is thus

$$N(p) = \frac{N_f(p) - \sum_q qN(q)}{p} \tag{2.33}$$

where the q are the divisors of p. Note that this a recursive definition of $N(p)$. As an example, $N(6) = [N_f(6) - 3N(3) - 2N(2) - N(1)]/6 = (2^6 - 3\times 2 - 2\times 1 - 2)/6 = 9$ with the computation of $N(3)$, $N(2)$, and $N(1)$ being left to the reader. As detailed in Section 2.7.5.3, one of these nine orbits appears in a period doubling and the eight others are created by pairs in saddle-node bifurcations. Because $N_f(p)$ increases exponentially with p, $N(p)$ is well approximated for large p by $N(p) \simeq N_f(p)/p$.

We thus have the important property that there are an infinite number of periodic points and that the number $N(p)$ of periodic orbits of period p increases exponentially with the period. The corresponding growth rate,

$$h_P = \lim_{p\to\infty} \frac{1}{p} \ln N(p) = \lim_{p\to\infty} \frac{1}{p} \ln \frac{N_f(p)}{p} = \ln 2 \tag{2.34}$$

provides an accurate estimate of a central measure of chaos, the *topological entropy* h_T. In many cases it can be proven rigorously that $h_P = h_T$. Topological entropy itself can be defined in several different but equivalent ways.

2.6.4.2 Expansiveness Implies Infinitely Many Periodic Orbits We now prove that if a continuous map $f : I \to I$ is expansive, *its unstable periodic orbits are dense in* I: Any point $x \in I$ has periodic points arbitrarily close to it. Equivalently, any subinterval $J \subset I$ contains periodic points.

We first note that if $J \subset f(J)$ (this is a particular case of a topological covering), then J contains a fixed point of f as a direct consequence of the intermediate value theorem.[2] Similarly, J contains at least one periodic point of period p if $J \subset f^p(J)$.

Now, if (2.31) is satisfied, every interval $J \subset I$ is eventually mapped to I: $f^n(J) = I$ [and thus $f^n(J) \subset J$], for $n > N_0(J)$. Using the remark above, we deduce that J

[2] Denote $x_a, x_b \in J$ the points such that $f(J) = [f(x_a), f(x_b)]$. If $J \subset f(J)$, one has $f(x_a) \leq x_a$ and $x_b \leq f(x_b)$. Thus, the function $F(x) = f(x) - x$ has opposite signs in x_a and x_b. If f is continuous, F must take all the values between $F(x_a)$ and $F(x_b)$. Thus, there exists $x_* \in J$ such that $F(x_*) = f(x_*) - x_* = 0$: x_* is a fixed point of f.

contains periodic points of period p for any $p > N_0(J)$, but also possibly for smaller p. Therefore, *any interval contains an infinity of periodic points with arbitrarily high periods.* A graphical illustration is provided by Fig. 2.10: Each intersection of a graph with the diagonal corresponds to a periodic point.

Thus, the expansiveness property (2.31) implies that unstable periodic points are dense. We showed earlier that it also implies topological transitivity and sensitivity to initial conditions. Therefore, any map satisfying (2.31) is chaotic according to the definition given at the beginning of this section.

It is quite fascinating that sensitivity to initial conditions, which makes the dynamics unpredictable, and unstable periodic orbits, which correspond to perfectly ordered motion are so deeply linked: In a chaotic regime, order and disorder are intimately entangled.

Unstable periodic orbits will prove to be a powerful tool to analyze chaos. They form a skeleton around which the dynamics is organized. Although they can be characterized in a finite time, they provide invaluable information on the asymptotic dynamics because of the density property: The dynamics in the neighborhood of an unstable periodic orbit is governed largely by that orbit.

2.6.5 Symbolic Coding of Trajectories: First Approach

We showed above that because of sensitivity to initial conditions, the dynamics of the surjective tent map is asymptotically unpredictable (Section 2.6.3). However, we would like to have a better understanding of how irregular, or random, typical orbits can be. We also learned that there is a dense set of unstable periodic orbits embedded in the invariant set I, and that this set has a well-defined structure. What about the other orbits, which are aperiodic?

In this section we introduce a powerful approach to chaotic dynamics that answers these questions: *symbolic dynamics*. To do so as simply as possible, let us consider a dynamical system extremely similar to the surjective tent map, defined by the map

$$x_{n+1} = 2x_n \pmod 1 \tag{2.35}$$

It only differs from the tent map in that the two branches of its graph are both orientation-preserving (Fig. 2.11). As with the tent map, the interval $[0, 1]$ is decomposed in two subintervals I_k such that the restrictions $f_k : I_k \to f_k(I_k)$ are homeomorphisms.

The key step is to recognize that because the slope of the graph is 2 everywhere, the action of (2.35) is trivial if the coordinates $x \in [0, 1]$ are represented in base 2. Let x_n have the binary expansion $x_n = 0.d_0d_1 \ldots d_k \ldots$, with $d_k \in \{0, 1\}$. It is easy to see that the next iterate will be

$$x_{n+1} = (d_0.d_1d_2 \ldots d_k \ldots) \pmod 1 = 0.d_1d_2 \ldots d_k \ldots \tag{2.36}$$

Thus, the base-2 expansion of x_{n+1} is obtained by dropping the leading digit in the expansion of x_n. This leading digit indicates whether x is greater than or equal to $\frac{1}{2} = 0.1\overline{0}$ ($\overline{s}$ represents an infinite repetition of the string s), thus which interval

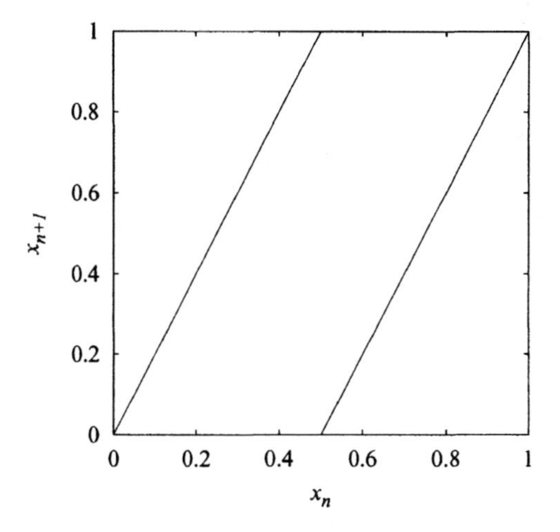

Fig. 2.11 Graph of map (2.35).

$I_0 = [0, 0.5)$ or $I_1 = [0.5, 1]$ the point belongs to. Note that in the present case $0.1\overline{0}$ and $0.0\overline{1}$, which usually represent the same number $\frac{1}{2}$, correspond here to different trajectories because of the discontinuity. The former is located at $(\frac{1}{2})^+$ and remains on the fixed point $x = 1$ forever, while the latter is associated with $(\frac{1}{2})^-$ and converges to the fixed point $x = 0$.

Thus, there is a 1:1 correspondence between orbits of the dynamical system (2.35) (parameterized by their initial condition x) and infinite digit sequences $(d_k) \in \{0, 1\}^{\mathrm{N}}$. Furthermore, the action of the map in the latter space has a particularly simple form. This correspondence allows one to establish extremely easily all the properties derived for the tent map in previous sections.

- **Sensitivity to initial conditions.** Whether the nth iterate of x falls in I_0 or I_1 is determined by the nth digit of the binary expansion of x. A small error in the initial condition (e.g., the nth digit is false) becomes macroscopic after a sufficient amount of time (i.e., after n iterations).

- **Existence of a dense orbit.** Construct an infinite binary sequence such that it contains all possible finite sequences. For example, concatenate all sequences of length $1, 2, \ldots, n$ for arbitrarily large n. The iterates of the associated point $x = 0.0|1|00|01|10|11|000|001\ldots$ will pass arbitrarily close to any point of the interval. The existence of a dense orbit implies topological transitivity.

- **Density of periodic orbits.** Each periodic point of (2.35) obviously corresponds to a periodic binary sequence. It is known that a periodic or eventually periodic digit expansion is a characteristic property of rational numbers. Since it is a classical result that rational numbers are dense in $[0, 1]$, we deduce immediately that periodic or eventually periodic points are dense in the interval

$[0, 1]$. Alternatively, each point x can be approximated arbitrarily well by a sequence of periodic points $x_*(n)$ whose sequences consist of the infinitely repeated n first digits of x, with $n \to \infty$.

This analysis can easily be transposed to the case of the surjective tent map. Since its right branch is orientation-reversing, the action of this map on the binary expansion of a point x located in this branch differs slightly from that of (2.36). Assuming that the tent map is defined on $[0, 1]$, its expression at the right (resp., left) of the critical point is $x' = 2(1 - x)$ (resp., $x' = 2x$). Consequently, we have the additional rule that if the leading digit is $d_0 = 1$, all the digits d_i, $i \in \mathbb{N}$, should be replaced by $\tilde{d}_i = 1 - d_i$ before dropping the leading digit d_0 as with the left branch (in fact, the two operations can be carried out in any order). The operation $d_i \to \tilde{d}_i$ is known as *complement to one*.

Example: Under the tent map, $0.01001011 \to 0.1001011 \to 0.110100$. For the first transition, since $0.0100100 < \frac{1}{2}$, we simply remove the decimal one digit to the right. In the second transition, since $x = 0.1001011 > \frac{1}{2}$, we first complement x and obtain $x' = (1 - x) = 0.0110100$, then multiply by 2: $2x' = 0.1101100$.

Except for this minor difference in the coding of trajectories, the arguments used above to show the existence of chaos in the map $x' = 2x \pmod 1$ can be followed without modification. The binary coding we have used is thus a powerful method to prove that the tent map displays chaotic behavior.

The results of this section naturally highlight two important properties of chaotic dynamics:

- A series of coarse-grained measurements of the state of a system can suffice to estimate it with arbitrary accuracy if carried out over a sufficiently long time. By merely noting which branch is visited (one-bit digitizer) at each iteration of the map (2.35), all the digits of an initial condition can be extracted.

- Although a system such as (2.35) is perfectly deterministic, its asymptotic dynamics is as random as coin flipping (all sequences of 0 and 1 can be observed).

However, the coding used in these two examples (n-ary expansion) is too naive to be extended to maps that do not have a constant slope equal to an integer. In the next section we discuss the general theory of symbolic dynamics for one-dimensional maps. This topological approach will prove to be an extremely powerful tool to characterize the dynamics of the logistic map, not only in the surjective case but for any value of the parameter a.

2.7 ONE-DIMENSIONAL SYMBOLIC DYNAMICS

2.7.1 Partitions

Consider a continuous map $f : I \to I$ that is singular. We would like to extend the symbolic dynamical approach introduced in Section 2.6.5 in order to analyze its dynamics. To this end, we have to construct a coding associating each orbit of the map with a symbol sequence.

We note that in the previous examples, each digit of the binary expansion of a point x indicates whether x belongs to the left or right branch of the map. Accordingly, we decompose the interval I in N disjoint intervals I_α, $\alpha = 0 \ldots N-1$ (numbered from left to right), such that

- $I = I_0 \cup I_1 \cup \cdots I_{N-1}$
- In each interval I_α, the restriction $f|_{I_\alpha} : I_\alpha \to f(I_\alpha)$, which we denote f_α, is a homeomorphism.

For one-dimensional maps, such a *partition* can easily be constructed by choosing the critical points of the map as endpoints of the intervals I_α, as Fig. 2.12 illustrates. At each iteration, we record the symbol $\alpha \in \mathcal{A} = \{0, \ldots, N-1\}$ that identifies the interval to which the current point belongs. The alphabet $\mathcal{A}$ consists of the N values that the symbol can assume.

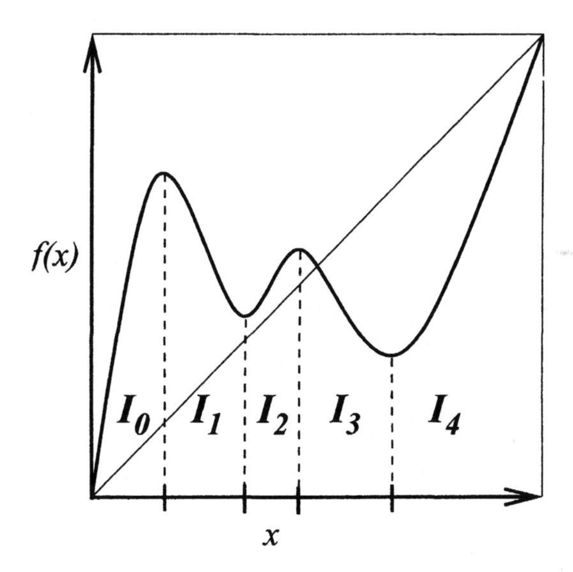

Fig. 2.12 Decomposition of the domain of a map f into intervals I_α such that the restrictions $f : I_\alpha \to f(I_\alpha)$ to the intervals I_α are homeomorphisms.

We denote by $s(x)$ the corresponding coding function:

$$s(x) = \alpha \Longleftrightarrow x \in I_\alpha \tag{2.37}$$

Any orbit $\{x, f(x), f^2(x), \ldots, f^i(x), \ldots\}$ of initial condition x can then be associated with the infinite sequence of symbols indicating the intervals visited successively by the orbit:

$$\Sigma(x) = \{s(x), s(f(x)), s(f^2(x)), \ldots, s(f^i(x)), \ldots\} \tag{2.38}$$

The sequence $\Sigma(x)$ is called the *itinerary* of x. We will also use the compact notation $\Sigma = s_0 s_1 s_2 \ldots s_i \ldots$ with the s_i being the successive symbols of the sequence (e.g., $\Sigma = 01101001\ldots$). The set of all possible sequences in the alphabet $\mathcal{A}$ is denoted $\mathcal{A}^{\mathrm{N}}$ and $\Sigma(I) \subset \mathcal{A}^{\mathrm{N}}$ represents the set of sequences actually associated with a point of I:

$$\Sigma(I) = \{\Sigma(x);\ x \in I\} \tag{2.39}$$

The finite sequence made of the n leading digits of $\Sigma(x)$ will later be useful. We denote it $\Sigma_n(x)$. For example, if $\Sigma(x) = 10110\ldots$, then $\Sigma_3(x) = 101$. Accordingly, the set of finite sequences of length n involved in the dynamics is $\Sigma_n(I)$.

An important property of the symbolic representation (2.38) is that the expression of the time-one map becomes particularly simple. Indeed, if we compare the symbolic sequence of $f(x)$:

$$\Sigma(f(x)) = \{s(f(x)), s(f^2(x)), s(f^3(x)), \ldots, s(f^{i+1}(x)), \ldots\} \tag{2.40}$$

with that of x given in (2.38), we observe that the former can be obtained from the latter by dropping the leading symbol and shifting the remaining symbols to the left. Accordingly, we define the *shift operator* σ by

$$\Sigma = \{s_0, s_1, s_2, \ldots, s_i, \ldots\} \overset{\sigma}{\rightarrow} \{s_1, s_2, \ldots, s_i, \ldots\} = \sigma\Sigma \tag{2.41}$$

Applying f to a point $x \in I$ is equivalent to applying the shift operator σ on its symbolic sequence $\Sigma(x) \in \Sigma(I)$:

$$\Sigma(f(x)) = \sigma\Sigma(x) \tag{2.42}$$

which corresponds to the commutative diagram

$$\begin{array}{ccc} x & \xrightarrow{f} & f(x) \\ \downarrow{\scriptstyle \Sigma} & & \downarrow{\scriptstyle \Sigma} \\ \{s_i\}_{i\in\mathrm{N}} & \xrightarrow{\sigma} & \{s_{i+1}\}_{i\in\mathrm{N}} \end{array} \tag{2.43}$$

Note that because only forward orbits $\{f^n(x)\}_{n\geq 0}$ can be computed with a noninvertible map, the associated symbolic sequences are one-sided and extend to infinity only in the direction of forward time. This makes the operator σ noninvertible, as f is itself. Formally, we can define several "inverse" operators σ_α^{-1} acting on a sequence by inserting the symbol α at its head:

$$\Sigma = \{s_0, s_1, s_2, \ldots, s_i, \ldots\} \overset{\sigma_\alpha^{-1}}{\rightarrow} \{\alpha, s_0, s_1, \ldots, s_{i-1}, \ldots\} = \sigma_\alpha^{-1}\Sigma \tag{2.44}$$

However, note that $\sigma \circ \sigma_\alpha^{-1} = \mathrm{Id} \neq \sigma_\alpha^{-1} \circ \sigma$.

Periodic sequences $\Sigma = \{s_i\}$ with $s_i = s_{i+p}$ for all $i \in \mathbb{N}$ will be of particular importance in the following. Indeed, they satisfy $\sigma^p\Sigma = \Sigma$, which translates into $f^p(x) = x$ for the associated point that is thus periodic. Infinite periodic sequences will be represented by overlining the base pattern (e.g., $\overline{01011} = 010110101101011\ldots$). When there is no ambiguity, the base pattern will be used as the name of the corresponding periodic orbit (e.g., the orbit 01011 has sequence $\overline{01011}$).

2.7.2 Symbolic Dynamics of Expansive Maps

To justify the relevance of symbolic coding, we now show that it is a faithful representation. Namely, the correspondence $x \in I \leftrightarrow \Sigma(x) \in \Sigma(I)$ defined by (2.37) and (2.38) can under appropriate conditions be made a bijection, that is,

$$x_1 \neq x_2 \Longleftrightarrow \Sigma(x_1) \neq \Sigma(x_2) \tag{2.45}$$

We might additionally require some form of continuity so that sequences that are close according to some metric are associated with points that are close in space.

In plain words, the symbolic sequence associated to a given point is sufficient to distinguish it from any other point in the interval I. The two dynamical systems (I, f) and $(\Sigma(I), \sigma)$ can then be considered as equivalent, with $\Sigma(x)$ playing the role of a change of coordinate. Partitions of state space that satisfy (2.45) are said to be *generating*.

In Section 2.6.5, we have seen two particular examples of one-to-one correspondence between orbits and symbolic sequences. Here we show that such a bijection holds if the following two conditions are true: (1) the restriction of the map to each member of the partition is a homeomorphism (Section 2.7.1); (2) the map satisfies the expansiveness property (2.31). This will illustrate the intimate connection between symbolic dynamics and chaotic behavior.

In the tent map example, it is obvious how the successive digits of the binary expansion of a point x specify the position of x with increasing accuracy. As we show below, this is also true for general symbolic sequences under appropriate conditions.

As a simple example, assume that a point x has a symbol sequence $\Sigma(x) = 101\ldots$ From the leading symbol we extract the top-level information about the position of x, namely that $x \in I_1$. Since the second symbol is 0, we deduce that $f(x) \in I_0$ [i.e., $x \in f^{-1}(I_0)$]. This second-level information combined with the first-level information indicates that $x \in I_1 \cap f^{-1}(I_0) \equiv I_{10}$. Using the first three symbols, we obtain $x \in I_{101} = I_1 \cap f^{-1}(I_0) \cap f^{-2}(I_1) = I_1 \cap f^{-1}(I_0 \cap f^{-1}(I_1))$. We note that longer symbol sequences localize the point with higher accuracy: $I_1 \supset I_{10} \supset I_{101}$.

More generally, define the interval $I_\Lambda = I_{s_0 s_1 \ldots s_{n-1}}$ as the set of points whose symbolic sequence begins by the finite sequence $\Lambda = s_0 s_1 \ldots s_{n-1}$, the remaining part of the sequence being arbitrary:

$$\begin{aligned} I_\Lambda = I_{s_0 s_1 \dots s_{n-1}} &= \{x;\ \Sigma_n(x) = s_0 s_1 \dots s_{n-1}\} \\ &= \{x; s(f^i(x)) = s_i,\ i < n\} \\ &= \{x; f^i(x) \in I_{s_i},\ i < n\} \end{aligned} \tag{2.46}$$

Such sets are usually termed *cylinders*, with an n-cylinder being defined by a sequence of length n. We now show that cylinders can be expressed simply using inverse branches of the function f. We first define

$$\forall J \subset I \quad f_\alpha^{-1}(J) = I_\alpha \cap f^{-1}(J) \tag{2.47}$$

This is a slight abuse of notation, since we have only that $f_\alpha(f_\alpha^{-1}(J)) \subset J$ without the equality being always satisfied, but it makes the notation more compact. With this convention, the base intervals I_α can be written as

$$I_\alpha = \{x; s(x) = \alpha\} = f_\alpha^{-1}(I) \quad \forall \alpha \in \mathcal{A} \tag{2.48}$$

To generate the whole set of cylinders, this expression can be generalized to longer sequences by noting that

$$I_{\alpha\Lambda} = f_\alpha^{-1}(I_\Lambda) \tag{2.49}$$

which follows directly from definitions (2.46) and (2.47). Alternatively, (2.49) can be seen merely to express that $\alpha\Lambda = \sigma_\alpha^{-1}\Lambda$. By applying (2.49) recursively, one obtains

$$I_\Lambda = f_\Lambda^{-n}(I) \tag{2.50}$$

where f_Λ^{-n} is defined by

$$f_\Lambda^{-n} = f_{s_0 s_1 \dots s_{n-1}}^{-n} = f_{s_0}^{-1} \circ f_{s_1}^{-1} \circ \dots \circ f_{s_{n-1}}^{-1} \tag{2.51}$$

Just as the restriction of f to any interval I_α is a homeomorphism $f_\alpha : I_\alpha \to f(I_\alpha)$, the restriction of f^n to any set I_Λ with Λ of length n is a homeomorphism[3] $f_\Lambda^n : I_\Lambda \to f^n(I_\Lambda)$. The function f_Λ^{-n} defined by (2.51) is the inverse of this homeomorphism, which explains the notation. For a graphical illustration, see Fig. 2.10: Each interval of monotonic behavior of the graph of g^2 (resp., g^4) corresponds to a different interval I_Λ, with Λ of length 2 (resp., 4).

Note that because I is connected and the f_α^{-1} are homeomorphisms, all the I_Λ are connected sets, hence are intervals in the one-dimensional context. This follows directly from (2.49) and the fact that the image of a connected set by a homeomorphism is a connected set. This property will be important in the following.

[3]Note that f^n is a homeomorphism only on set I_Λ defined by symbolic strings Λ of length $p \geq n$ (e.g., f^2 has a singularity in the middle of I_0).

To illustrate relation (2.50), we apply it to the case $\Lambda = 101$ considered in the example above:

$$\begin{aligned} I_{101} = (f_1^{-1} \circ f_0^{-1} \circ f_1^{-1})(I) &= (f_1^{-1} \circ f_0^{-1})(I_1) \\ &= f_1^{-1}(I_0 \cap f^{-1}(I_1)) \\ &= I_1 \cap f^{-1}(I_0 \cap f^{-1}(I_1)) \end{aligned}$$

and verify that it reproduces the expression obtained previously.

The discussion above shows that the set of n-cylinders $\mathcal{C}_n = \{I_\Lambda; \Lambda \in \Sigma_n(I)\}$ is a partition of I:

$$I = \bigcup_{\Lambda \in \Sigma_n(I)} I_\Lambda \qquad I_\Lambda \cap I_{\Lambda'} = \emptyset \tag{2.52}$$

with $\mathcal{C}_n$ being a refinement of $\mathcal{C}_{n-1}$ (i.e., each member of $\mathcal{C}_n$ is a subset of a member of $\mathcal{C}_{n-1}$). As $n \to \infty$, the partition $\mathcal{C}_n$ becomes finer and finer (again, see Fig. 2.10). What we want to show is that the partition is arbitrarily fine in this limit, with the size of *each* interval of the partition converging to zero.

Consider an arbitrary symbolic sequence $\Sigma(x)$, with $\Sigma_n(x)$ listing its n leading symbols. Since by definition $I_{\Sigma_{n+1}(x)} \subset I_{\Sigma_n(x)}$, the sequence $(I_{\Sigma_n(x)})_{n \in \mathbb{N}}$ is decreasing, hence it converges to a limit $I_{\Sigma(x)}$. All the points in $I_{\Sigma(x)}$ share the same infinite symbolic sequence.

As the limit of a sequence of connected sets, $I_{\Sigma(x)}$ is itself a connected set, hence is an interval or an isolated point. Assume that $I_{\Sigma(x)}$ is an interval. Then, because of the expansiveness property (2.31), there is N_0 such that $f^{N_0}(I_{\Sigma(x)}) = I$. This implies that for points $x \in I_{\Sigma(x)}$, the symbol $s_{N_0} = s(f^{N_0})(x)$ can take any value $\alpha \in \mathcal{A}$, in direct contradiction with $I_{\Sigma(x)}$ corresponding to a unique sequence. Thus, the only possible solution is that the limit $I_{\Sigma(x)}$ is an isolated point, showing that the correspondence between points and symbolic sequences is one-to-one. Thus, the symbolic dynamical representation of the dynamics is faithful.

This demonstration assumes a partition of I constructed so that each interval of monotonicity corresponds to a different symbol (see Fig. 2.12). This guarantees that all the preimages of a given point are associated with different symbols since they belong to different intervals.

It is easy to see that partitions not respecting this rule cannot be generating. Assume that two points x_0 and x_1 have the same image $f(x_0) = f(x_1) = y$ and that they are coded with the same symbol $s(x_0) = s(x_1) = \alpha_k$ (Fig. 2.13). They are then necessarily associated with the same symbolic sequence, consisting of the common symbol α_k concatenated with the symbolic sequence of their common image: $\Sigma(x_0) = \Sigma(x_1) = \alpha_k \Sigma(y)$. In other words, associating x_0 and x_1 with different symbols is the only chance to distinguish them because they have exactly the same future.

In one-dimensional maps, two preimages of a given point are always separated by a critical point. Hence, the simplest generating partition is obtained by merely dividing the base interval I into intervals connecting two adjacent critical points, and associating each with a different symbol.

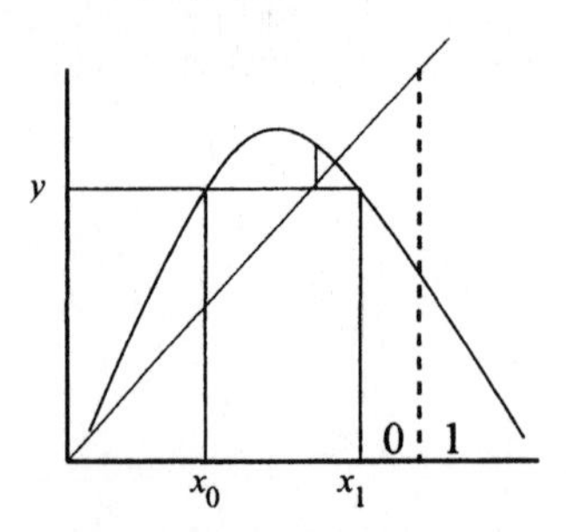

Fig. 2.13 The dashed line indicates the border of a partition such that the preimages x_0 and x_1 of the same point y are coded with the same symbol (0). As a consequence, the symbolic sequences associated to x_0 and x_1 are identical.

Remark 1: This is no longer true for higher-dimensional noninvertible maps, which introduces some ambiguity in the symbolic coding of trajectories.

Remark 2: Invertible chaotic maps do not have singularities, hence the construction of generating partitions is more involved. The forthcoming examples of the horseshoe and of the Hénon map will help us to understand how the rules established in the present section can be generalized.

2.7.3 Grammar of Chaos: First Approach

Symbolic dynamics provides a simple but faithful representation of a chaotic dynamical system (Section 2.7.2). It has allowed us to understand the structure of the chaotic and periodic orbits of the surjective tent map, and hence of the surjective logistic map (Section 2.6.5). But there is more.

As a control parameter of a one-dimensional map varies, the structure of its invariant set and of its orbits changes (perestroika). Symbolic dynamics is a powerful tool to analyze these modifications: As orbits are created or destroyed, symbolic sequences appear or disappear from the associated symbolic dynamics. Thus, a regime can be characterized by a description of its set of forbidden sequences. We refer to such a description as the *grammar of chaos*. Changes in the structure of a map are characterized by changes in this grammar.

As we illustrate below with simple examples, which sequences are allowed and which are not can be determined entirely geometrically. In particular, the orbit of the critical point plays a crucial role. The complete theory, namely *kneading theory*, is detailed in Section 2.7.4.

2.7.3.1 Interval Arithmetics and Invariant Interval We begin by determining the smallest invariant interval I [i.e., such that $f(I) = I$]. This is where the asymptotic dynamics will take place. Let us first show how to compute the image of an arbitrary interval $J = [x_l, x_h]$. If J is located entirely to the left or right of the critical point x_c, one merely needs to take into account that the logistic map is orientation-preserving (resp., orientation-reversing) at the left (resp., right) of x_c.

Conversely, if $x_c \in J$, J can be decomposed as $[x_l, x_h] = [x_l, x_c) \cup [x_c, x_h]$. This gives

$$f([x_l, x_h]) = \begin{cases} [f(x_l), f(x_h)] & \text{if} \quad x_l, x_h \leq x_c \\ [f(x_h), f(x_l)] & \text{if} \quad x_c \leq x_l, x_h \\ [\min\{f(x_l), f_h)\}, f(x_c)] & \text{if} \quad x_l \leq x_c \leq x_h \end{cases} \tag{2.53}$$

As was noted by Poincaré, the apparent complexity of chaotic dynamics is such that it makes little sense to follow individual orbits: What is relevant is how regions of the state space are mapped between each other. One-dimensional maps are no exception, and in fact many properties of the logistic map can be extracted from the interval arithmetics defined by (2.53). Here, we use them to show in a simple way that some symbolic sequences are forbidden.

Let us now determine $I = [x_{\min}, x_{\max}]$ such that $f(I) = I$. We are interested only in situations where this interval contains the critical point, so that the dynamics is nontrivial. Note that this implies that $x_c \leq f(x_c)$ because we must have $x_c = f(y) \leq f(x_c)$ (hence the top of the parabola must be above the diagonal). We use the third case of Eq. (2.53) to obtain the equation

$$[\min\{f(x_{\min}), f_{\max})\}, f(x_c)] = [x_{\min}, x_{\max}] \tag{2.54}$$

The upper bound is thus the image of the critical point: $x_{\max} = f(x_c)$. The lower bound $x_{\min}$ satisfies the equation

$$x_{\min} = \min\{f(x_{\min}), f(x_{\max})\} = \min\{f(x_{\min}), f^2(x_c)\} \tag{2.55}$$

An obvious solution is $x_{\min} = f^2(x_c)$, which is valid provided that $f(f^2(x_c)) > f^2(x_c)$. This is always the case between the parameter value where the period-1 orbit is superstable and the one where bounded solutions cease to exist. The other possible solution is the fixed point $x_- = f(x_-)$. In the parameter region of interest, however, one has $x_- < f^2(x_c)$, and thus the smallest invariant interval is given by

$$I = [f^2(x_c), f(x_c)] \tag{2.56}$$

That it depends only on the orbit of the critical point x_c is remarkable. However, this merely prefigures Section 2.7.4 where we shall see that this orbit determines the dynamics completely. Note that $I \neq \emptyset$ as soon as $f(x_c) > x_c$, which is the only interesting parameter region from a dynamical point of view.

2.7.3.2 *Existence of Forbidden Sequences* As shown previously, the set I_Λ of points whose symbolic sequence begins by the finite string Λ is given by $I_\Lambda = f_\Lambda^{-n}(I)$, where f_Λ^{-n} is defined by (2.51) and (2.47). It is easy to see that if $I_\Lambda = \emptyset$, the finite symbol sequence Λ is forbidden.

From the discussion above, the base intervals are

$$I_0 = [f^2(x_c), x_c] \quad I_1 = [x_c, f(x_c)] \tag{2.57}$$

which are nonempty for $f(x_c) > x_c$. The existence of symbolic sequences of length two is determined by the intervals

$$I_{00} = I_0 \cap f_0^{-1}(I_0) = [f^2(x_c), f_0^{-1}(x_c)] \quad (2.58a)$$
$$I_{01} = I_0 \cap f_0^{-1}(I_1) = [f_0^{-1}(x_c), x_c] \quad (2.58b)$$
$$I_{10} = I_1 \cap f_1^{-1}(I_0) = [f_1^{-1}(x_c), f(x_c)] \quad (2.58c)$$
$$I_{11} = I_1 \cap f_1^{-1}(I_1) = [x_c, f_1^{-1}(x_c)] \quad (2.58d)$$

which are computed by means of the interval arithmetics (2.53) but can also be obtained graphically (Fig. 2.14). The last three do not provide useful information: They are nonempty whenever $f(x_c) > x_c$, [i.e., as soon as I given by (2.56) is well-defined].

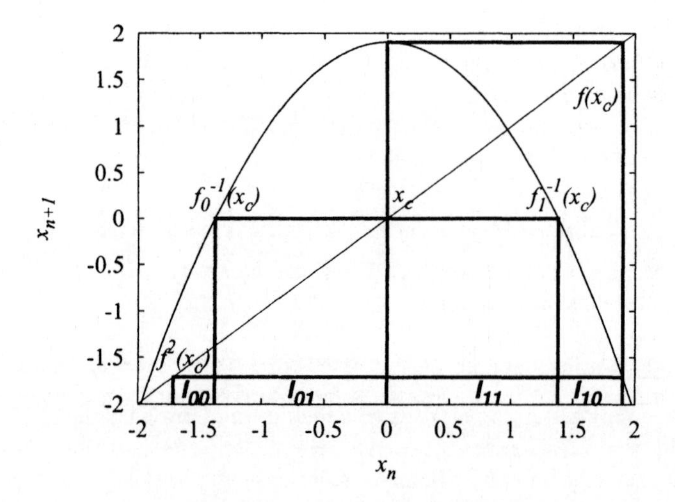

Fig. 2.14 Intervals I_{00}, I_{01}, I_{11}, and I_{10} defined in (2.58). In each interval, the itineraries have the same leading two symbols.

On the contrary, (2.58a) yields a nontrivial condition for I_{00} to be nonempty, namely that $f^2(x_c) < f_0^{-1}(x_c)$. This interval has zero width when its two bounds are equal, thus the string "00" becomes allowed when the critical point belongs to a period-3 orbit:

$$f^2(x_c) = f_0^{-1}(x_c) \Rightarrow f^3(x_c) = x_c \quad (2.59)$$

which is then superstable since the derivative of f is zero at the critical point. For the logistic map (2.2), this occurs precisely at $a = a_{00} = 1.75487766\ldots$, inside the unique period-3 window that can be seen in the bifurcation diagram of Fig. 2.3.

Since $I_{00} = \emptyset$ for $a < a_{00}$, we conclude that the symbolic string "00" never appears in the symbolic dynamics of regimes located at the left of the period-3 window. Thus, the presence or absence of this string suffices to distinguish regimes located before and after this window.

In particular, this has consequences on the order of the appearance of periodic orbits. The first periodic orbit carrying the "00" string is the period-3 orbit $\overline{001}$. Therefore, all other periodic orbits whose names contain "00," for example $\overline{001011}$, must appear after $\overline{001}$. This shows that the geometrical structure of the map has a deep influence on the order of appearance of periodic orbits, as detailed later in a more systematic way.

Reproducing the calculation above for longer symbolic strings, we would find that new symbolic sequences always appear when the critical point is part of a periodic orbit (i.e., at the parameter inside the periodic window where the orbit is superstable). This is not surprising if we note that the bounds of all the intervals I_Λ can be expressed in terms of the images and preimages of the critical point x_c.

As a result, the condition of zero width of these intervals can always be rewritten as an equation of the type $f_\Lambda^{-n}(x_c) = x_c$, expressing that x belongs to a periodic orbit of period n and of symbolic sequence Λ. For example, (2.59) corresponds to $f_{100}^{-3}(x_c) = x_c$. However, we will not proceed in this direction. The observation that the grammar of the symbolic dynamics is governed completely by the orbit of the critical point will lead us to a much more efficient framework for classifying symbolic sequences of orbits.

We conclude this section with the important remark that the symbolic dynamics of a chaotic dynamical system is in general *intimately related to its geometrical structure*. In the case of unimodal maps, the structure of the forbidden sequences depends only on the position of the image of the critical point organizing the dynamics. Thus, given an arbitrary symbolic sequence, it is in principle possible to determine whether it has been generated by a one-dimensional map. More generally, extracting the structure of a map from the grammar of the symbolic dynamics it generates is a fascinating problem. It has been much less explored for two-dimensional invertible maps than for maps of the interval, and even less for noninvertible maps of dimension 2 and higher.

2.7.4 Kneading Theory

Rather than solve algebraic equations such as (2.59) to determine forbidden sequences, it would be preferable to work completely in the space of symbolic sequences. Since the orbit of the critical point plays a crucial role in understanding which symbolic sequences are forbidden, it is natural to study more closely the distinguished symbolic sequence associated with the critical point.

Since the first symbol of this sequence does not carry any information (the critical point x_c is the border between intervals I_0 and I_1), we accordingly define the *kneading sequence* $K(f)$ as the itinerary of the image of x_c:

$$K(f) = \Sigma(f(x_c)) = \{s(f(x_c)), s(f^2(x_c)), \ldots\} \tag{2.60}$$

Note that the first two symbols are constant inside the parameter region where $f^2(x_c) < x_c < f(x_c)$: $f^2(x_c)$ and $f(x_c)$ are the left and right ends of the invariant interval I defined in (2.56) and are thus associated with the symbols 0 and 1, respectively. Since the value of the third symbol depends on whether $f^3(x_c)$ is located at the left or at the right of the critical point, it changes when $f^3(x_c) = x_c$ (i.e., when the string "00" becomes allowed), and thus

$$\begin{aligned} a < a_{00} \quad &\Rightarrow \quad K(f) = \{1,0,1,\ldots\} \\ a > a_{00} \quad &\Rightarrow \quad K(f) = \{1,0,0,\ldots\} \end{aligned} \tag{2.61}$$

This confirms the importance of the kneading sequence (2.60): The appearance of the symbolic string "00" in the symbolic dynamics of the logistic map coincides with its appearance in the kneading sequence.

To go beyond this observation, we need to be able to determine from $K(f)$ alone which sequences are allowed and which are not. The distinctive property of $f(x_c)$ is that it is the rightmost point of the invariant interval (2.56). To see that there is indeed a similar property for the kneading sequence, we first show that an order on itineraries can be defined.

2.7.4.1 Ordering of Itineraries In the example of the $x_{n+1} = 2x_n \pmod 1$ map (Section 2.6.5), the itinerary of a point (i.e., its binary expansion) not only identifies it uniquely, it also contains information about its position relative to the other points. In that case, the lexicographic order on symbolic sequences reflects exactly the order of the associated points on the interval. More generally, we would like to define for an arbitrary map an order relation $\prec$ on itineraries so that

$$\Sigma(x) \prec \Sigma(x') \Longleftrightarrow x < x' \tag{2.62}$$

Ordering two itineraries is easy when their leading symbols differ. If the base intervals I_α are numbered sequentially from left to right as in Fig. 2.12, the itinerary with the smallest leading symbol is associated with the leftmost point and should be considered "smaller" than the other.

If the two itineraries have a common leading substring, one has to take into account the fact that the map f can be orientation-reversing on some intervals I_α. For example, the two-symbol cylinders $I_{\alpha\alpha'}$ given by Eq. (2.58) and shown in Fig. 2.14 appear left to right in the order I_{00}, I_{01}, I_{11}, and I_{10}.

Thus, $11 \prec 10$ for the logistic map, which markedly differs from the lexicographic order. This is because both strings have a leading 1, which is associated with the orientation-reversing branch f_1. Indeed, assume that $x_{11} \in I_{11}, x_{10} \in I_{10}$. From the second symbol, we know that $f(x_{10}) < f(x_{11})$ because $0 < 1$. However, since f is orientation-reversing in I_1, this implies that $x_{11} < x_{10}$, hence $11 \prec 10$. With this point in mind, two arbitrary itineraries Σ, Σ' can be ordered as follows.

Assume that the two sequences $\Sigma = \Lambda s_m \ldots$ and $\Sigma' = \Lambda s'_m \ldots$ have a common leading symbolic string Λ of length m, and first differ in symbols s_m and s'_m. Thus, the corresponding points x and x' are such that $f^m(x)$ and $f^m(x')$ belong to different intervals I_α, hence can be ordered. As in the example above, it then suffices to determine whether the restriction f^m_Λ of f^m to the interval I_Λ is orientation-preserving or orientation-reversing (has a positive or a negative slope, respectively) to obtain the ordering of x and x', and thus that of Σ and Σ'. Define the branch parity

$$\epsilon(\alpha) = \begin{cases} +1 & \text{if} \quad f_\alpha : I_\alpha \to I \text{ is orientation-preserving} \\ -1 & \text{if} \quad f_\alpha : I_\alpha \to I \text{ is orientation-reversing} \end{cases} \tag{2.63}$$

The parity of the finite sequence $\Lambda = s_0 s_1 \ldots s_{m-1}$ is then given by

$$\epsilon(\Lambda) = \epsilon(s_0) \times \epsilon(s_1) \times \cdots \times \epsilon(s_{m-1}). \tag{2.64}$$

If the map $f_\Lambda^m = f_{s_{m-1}} \circ \ldots f_{s_1} \circ f_{s_0}$ (i.e., the restriction of f^m to the interval I_Λ) is orientation-preserving (resp., orientation-reversing), then $\epsilon(\Lambda) = +1$ (resp., -1). In the case of unimodal maps, $\epsilon(\Lambda) = +1$ if there is an even number of "1" (or of the symbol associated with the orientation-reversing branch), and -1 otherwise.

We can now define the order

$$\Sigma = \Lambda s \ldots \prec \Sigma' = \Lambda s' \ldots \Longleftrightarrow \begin{cases} s < s' \quad \text{and} \quad \epsilon(\Lambda) = +1 \\ \qquad \text{or} \\ s > s' \quad \text{and} \quad \epsilon(\Lambda) = -1 \end{cases} \tag{2.65}$$

This order satisfies condition (2.62). Let us illustrate these rules with the example of period-4 orbit $\overline{0111}$ of the logistic map. The relative order of the four periodic orbits is

$$\overline{0111} \prec \overline{1101} \prec \overline{1110} \prec \overline{1011} \tag{2.66}$$

as detailed in Fig. 2.15.

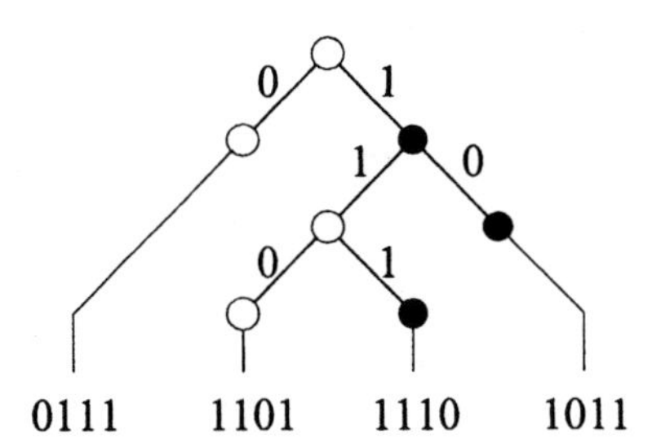

Fig. 2.15 Determination of the relative order of symbolic sequences. White (resp., black) nodes correspond to positive (resp., negative) parity. The topmost node corresponds to the empty sequence, and sequences are formed by following edges carrying the symbols 0 or 1. When an edge "1" is followed, the parity of the node changes. A white node has an edge "0" on its left and an edge "1" on its right (this is the lexicographic order). At a black node, these two edges are in the opposite order because of the negative parity. To order a set of symbolic sequences, one follows the edges corresponding to the successive symbols of the sequence until no other sequence remains in the branch. The ordered sequences can then be read from left to right.

Another common technique for ordering symbolic sequences is to use *invariant coordinates*. Given a sequence $\Sigma = s_0 s_1 s_2 \ldots s_k \ldots \in \{0, \ldots, N-1\}^{\mathbb{N}}$, we define its invariant coordinate $\theta(\Sigma)$ by

$$\theta(\Sigma) = \sum_{i=0}^{\infty} \frac{t_i}{N^{i+1}} \quad t_i = \begin{cases} s_i & \text{if} \quad \epsilon(s_0 \ldots s_{i-1}) = +1 \\ (N-1) - s_i & \text{if} \quad \epsilon(s_0 \ldots s_{i-1}) = -1 \end{cases} \tag{2.67}$$

so that $0 \leq \theta(\Sigma) \leq 1$. By inspecting (2.65) and (2.67), one easily verifies that two symbolic sequences can be ordered by comparing their invariant coordinates:

$$\Sigma \prec \Sigma' \Longleftrightarrow \theta(\Sigma) < \theta(\Sigma') \tag{2.68}$$

As an example, the invariant coordinate of the periodic point $\overline{1011}$ of the logistic map is

$$\begin{aligned} \theta(\overline{1011}) &= \left(\frac{1}{2^1} + \frac{\mathbf{1}}{2^2} + \frac{\mathbf{0}}{2^3} + \frac{1}{2^4} + \frac{\mathbf{0}}{2^5} + \frac{0}{2^6} + \frac{1}{2^7} + \frac{\mathbf{0}}{2^8} \right) \\ &\times \left(1 + \frac{1}{2^8} + \frac{1}{2^{16}} + \cdots \right) = \frac{105}{128} \times \frac{256}{255} = \frac{14}{17} \end{aligned} \tag{2.69}$$

where the digits in bold are those that have been inverted with respect to the original sequence. Because $\overline{1011}$ has negative parity, the binary digit sequence of $\theta(\Sigma)$ has period 8 instead of 4. The first factor in (2.69) corresponds to the basic pattern "11010010," and the second term comes from the infinite repetition of this pattern. Note that the fraction obtained is the position of the corresponding periodic point of the tent map defined on $[0, 1]$: The reader may verify as an exercise that $2 \times |1 - \theta(\overline{1011})| = \theta(\overline{0111})$ and that $2 \times \theta(\overline{0111}) = \theta(\overline{1110})$.

2.7.4.2 *Admissible Sequences* We showed earlier that each point x inside the invariant interval (2.56) satisfies $f^2(x_c) < x < f(x_c)$. Using (2.62), we can now translate this ordering relation between points into a ordering relation between symbolic sequences:

$$\forall x \in I \quad \sigma K(f) \prec \Sigma(x) \prec K(f) \tag{2.70}$$

since $K(f) = \Sigma(f(x_c))$, by definition. Moreover, the orbit of a point $x \in I$ is forever contained in I, by definition. A necessary condition for a sequence Σ to be the itinerary $\Sigma(x)$ of a point $x \in I$ is thus that (2.70) holds for any $\Sigma(f^n(x))$ and thus that

$$\forall n \geq 0 \quad \sigma^n \Sigma \preceq K(f) \tag{2.71}$$

One of the fundamental results of one-dimensional symbolic dynamics is that this is also a sufficient condition: Condition (2.71) completely determines whether a sequence occurs as the itinerary of a point [38,40]. A sequence satisfying it is said to be *admissible* [equivalently, one can test whether $\sigma K(f) < \sigma^n \Sigma$ for all n].

Therefore, all the information about the symbolic dynamics of a map is contained in its kneading sequence $K(f)$. As a matter of fact, it can be shown that if two unimodal maps have the same kneading sequence, and that if this sequence is dense (i.e., the orbit of x_c is aperiodic), the two maps are topologically conjugate.

Condition (2.71) is particularly simple to test when the symbolic sequence Σ is periodic, since the shifts $\sigma^n\Sigma$ are finite in number. For example, let us assume that $K(f) = 1001001\ldots$, and that we want to know whether the periodic sequences $\overline{01101101}$ and $\overline{00101}$ are admissible. We first determine the rightmost periodic points (for which $\sigma^n\Sigma$ is maximal) of the two orbits: These are $\overline{10110110}$ and $\overline{10010}$. We then compare them to the kneading sequence $K(f)$ and find that

$$\overline{10110110} \prec K(f) = 1001001\ldots \prec \overline{10010}$$

Thus, the period-8 sequence $\overline{10110110}$ is admissible, whereas the period-5 sequence $\overline{10010}$ is not. This indicates that the periodic orbit associated with the latter sequence does not exist in maps with the given $K(f)$. We also see that every map that has the second periodic orbit also has the first. Therefore, the order of appearance of periodic orbits is fixed, and the structure of the bifurcation diagram of Fig. 2.3 is universal for unimodal maps. We investigate this universality in the next section.

2.7.5 Bifurcation Diagram of the Logistic Map Revisited

We are now in a position to understand the structure of the bifurcation diagram shown in Fig. 2.3 using the tools of symbolic dynamics introduced in the previous sections. This bifurcation diagram displays two types of bifurcations: saddle-node and period-doubling bifurcations. Each saddle-node bifurcation creates a pair of periodic orbits of period p, one unstable (the saddle) and the other stable (the node). The latter is the germ of a period-doubling cascade with orbits of periods $p \times 2^n$.

As discussed in Section 2.7.4, the kneading sequence governs which symbolic sequences are admissible and which are forbidden, hence the order in which new sequences appear. Therefore, there must be a simple relation between the symbolic names of the orbits involved in a saddle-node or in a period-doubling bifurcation. Furthermore, the different saddle-node bifurcations and their associated period-doubling cascades must be organized rigidly.

2.7.5.1 Saddle-Node Bifurcations At a saddle-node bifurcation, the two newly born periodic orbits of period p are indistinguishable. Thus, they have formally the same symbolic name. This is not in contradiction with the one-to-one correspondence between orbits and itineraries that was shown to hold for chaotic regimes: At the bifurcation, the node is stable and there is no sensitivity to initial conditions.

When the two orbits are born, they have a multiplier of $+1$ (Section 2.4.1), which implies that f^p is orientation-preserving in the neighborhood of the orbit. Consequently, the common symbolic itinerary of the two orbits must contain an even number of symbols "1" (in the general case, an even number of symbols with negative parity).

A symbolic itinerary can change only if one of the periodic points crosses the critical point, which is the border of the partition. This happens to the stable node when it becomes superstable, changing its parity on its way to the period-doubling bifurcation where its multiplier crosses -1. Thus, its parity must be negative, and its final symbolic name (i.e., the one in the unstable regime) must differ from that of the saddle by a single symbol.

One can proceed as follows to see which symbol differs. Each periodic point is associated with a cyclic permutation of the symbolic name. For example, the orbit 01111 has periodic points with sequences $\overline{01111}$, $\overline{11110}$, $\overline{11101}$, $\overline{11011}$, and $\overline{10111}$. These periodic points can be ordered using the kneading order (2.65), here

$$\overline{01111} \prec \overline{11110} \prec \overline{11011} \prec \overline{11101} \prec \overline{10111}$$

The symbol that is flipped at the superstable parameter value is obviously associated with the point that is then degenerate with the critical point. The image of this point is thus the rightmost periodic point, corresponding to the highest sequence in the kneading order (Fig. 2.16). Consequently, a simple rule to obtain the symbolic name of the saddle-node partner of an orbit of given name is to flip the last symbol of the highest itinerary (i.e., of the itinerary of the rightmost point). Alternatively, one can flip the second to last symbol of the leftmost itinerary.

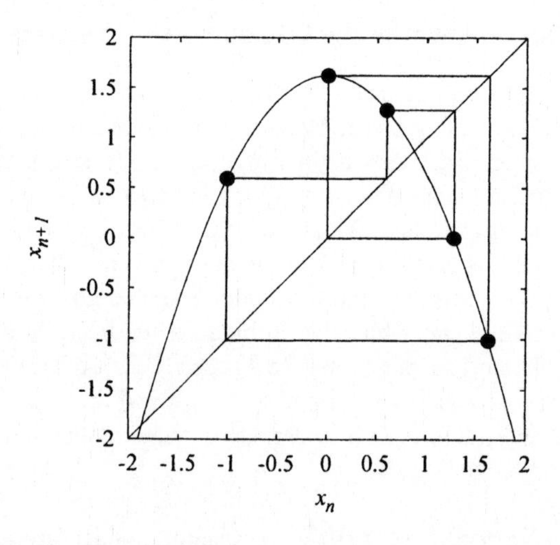

Fig. 2.16 The 01111 orbit becomes 01011 at a superstable point.

In the example above, the saddle-node partner of the 01111 orbit is thus 01101. Other examples of saddle-node pairs include $0\substack{0\\1}1$, $00\substack{0\\1}1$, and $01101\substack{0\\1}1$.

2.7.5.2 Period-Doubling Bifurcations When applying the algorithm above on an arbitrary symbolic name, it can occur that the result is not valid because it is the repetition of a shorter name. For example, 0111 leads to $0101 = (01)^2$. This indicates that the long name corresponds to the period-doubled orbit (the daughter) of the orbit identified by the short name (the mother). The 0111 orbit is the daughter of the 01 orbit. The latter is itself the daughter of the 1 orbit.

This can be understood with the same arguments as for the saddle-node bifurcations. When the period-doubled orbit is born, its itinerary is a double copy of that of the mother. Its parity is thus positive, which is consistent with the fact that the orbit is born with a multiplier of $+1$. The symbolic name of the mother at the bifurcation is its final name. As the daughter orbit proceeds to its own period doubling (and thus to its unstable domain), the preimage of its rightmost point crosses the critical point, changing the associated symbol.

Hence, we have a simple way to determine whether an orbit belongs to a period-doubling cascade and what is the name of its mother and all its ancestors. Other examples of mother–daughter pairs are $(001, 001011)$ and $(00101, 0010100111)$. We conclude with the period-doubling cascade originating from the period-1 orbit. The symbolic names of the successive period-doubled orbit can be constructed as

$$\mathbf{1} \xrightarrow{D} 11 \xrightarrow{F} \mathbf{01} \xrightarrow{D} 0101 \xrightarrow{F} \mathbf{0111} \xrightarrow{D} 01110111 \xrightarrow{F} \mathbf{01110101}, \cdots \tag{2.72}$$

where D and F represent the action of doubling the word and flipping the second-to-last symbol, respectively.

2.7.5.3 Universal Sequence Consider two periodic itineraries $\Sigma \prec \Sigma'$. For some parameter a, the kneading sequence $K(f_a)$ is such that $\Sigma \prec K(f_a) \prec \Sigma'$, so that Σ satisfies the admissibility condition (2.71) but not Σ'. Thus, the periodic orbit associated to Σ must be created before the one associated to Σ'.

This observation suffices to build a complete list of the successive bifurcations occurring in the bifurcation diagram of Fig. 2.3. Using the rules derived in previous sections, we can classify all the symbolic names according to which series of bifurcations they belong.

To this end, all periodic itineraries up to a given period p are sorted according to the kneading order, with saddle-node pairs and orbits of the same period-doubling cascade grouped together. We denote the ith bifurcation creating period-P orbits as P_i, with the node being called $P_i f$ (for flip) and the saddle $P_i r$ (for regular). This is illustrated in Table 2.1, which lists the symbolic names of all periodic orbits of period up to 8 of the logistic map. These names are sorted by order of appearance, and the bifurcation in which they appear is indicated.

This sequence of symbolic names, often referred to as the *universal sequence*, was discovered by Metropolis, Stein, and Stein [41]. It is universal in that it depends only on the kneading order (2.65): The bifurcation diagram of any unimodal map will display exactly the same bifurcations in exactly the same order.

Note, however, that this holds only for one-dimensional maps. If a two-dimensional map is sufficiently dissipative so that its return map can be well approximated by a one-dimensional map, most of the bifurcation sequences will occur in the order predicted by the universal sequence. However, there will be a few discrepancies, and the order of many bifurcations will be reversed as one decreases dissipation [42].

2.7.5.4 Self-Similar Structure of the Bifurcation Diagram In this section we mention briefly another surprising property of the bifurcation diagram of the logistic map that is unveiled by symbolic dynamics. Look at the period-3 window beginning at $a = 1.75$ in Fig. 2.4. There is a whole parameter range where the

Table 2.1 Sequence of bifurcations in the logistic map up to period 8 (from top and to bottom and left to right)[a]

Name	Bifurcation	Name	Bifurcation	Name	Bifurcation
${}^{0}_{1}$	$1_1[s_1]$	$00101\,{}^{0}_{1}1$	$7_3[s_7^3]$	$0001\,{}^{0}_{1}1$	$6_4[s_6^3]$
01	$2_1[s_1 \times 2^1]$	$001010\,{}^{0}_{1}1$	$8_5[s_8^4]$	$000111\,{}^{0}_{1}1$	$8_{11}[s_8^9]$
0111	$4_1[s_1 \times 2^2]$	$001\,{}^{0}_{1}1$	$5_2[s_5^2]$	$00011\,{}^{0}_{1}1$	$7_7[s_7^7]$
01010111	$8_1[s_1 \times 2^3]$	$001110\,{}^{0}_{1}1$	$8_6[s_8^5]$	$000110\,{}^{0}_{1}1$	$8_{12}[s_8^{10}]$
$0111\,{}^{0}_{1}1$	$6_1[s_6^1]$	$00111\,{}^{0}_{1}1$	$7_4[s_7^4]$	$000\,{}^{0}_{1}1$	$5_3[s_5^3]$
$011111\,{}^{0}_{1}1$	$8_2[s_8^1]$	$001111\,{}^{0}_{1}1$	$8_7[s_8^6]$	$000010\,{}^{0}_{1}1$	$8_{13}[s_8^{11}]$
$01111\,{}^{0}_{1}1$	$7_1[s_7^1]$	$0011\,{}^{0}_{1}1$	$6_3[s_6^2]$	$00001\,{}^{0}_{1}1$	$7_8[s_7^8]$
$011\,{}^{0}_{1}1$	$5_1[s_5^1]$	$001101\,{}^{0}_{1}1$	$8_8[s_8^7]$	$000011\,{}^{0}_{1}1$	$8_{14}[s_8^{12}]$
$01101\,{}^{0}_{1}1$	$7_2[s_7^2]$	$00110\,{}^{0}_{1}1$	$7_5[s_7^5]$	$0000\,{}^{0}_{1}1$	$6_5[s_6^4]$
$011011\,{}^{0}_{1}1$	$8_3[s_8^2]$	$00\,{}^{0}_{1}1$	$4_2[s_4^1]$	$000001\,{}^{0}_{1}1$	$8_{15}[s_8^{13}]$
$0\,{}^{0}_{1}1$	$3_1[s_3]$	00010011	$8_9[s_4^1 \times 2^1]$	$00000\,{}^{0}_{1}1$	$7_9[s_7^9]$
001011	$6_2[s_3 \times 2^1]$	$00010\,{}^{0}_{1}1$	$7_6[s_7^6]$	$000000\,{}^{0}_{1}1$	$8_{16}[s_8^{14}]$
$001011\,{}^{0}_{1}1$	$8_4[s_8^3]$	$000101\,{}^{0}_{1}1$	$8_{10}[s_8^8]$		

[a] The notation P_i refers to the ith bifurcation of period P. We also give inside brackets an alternative classification that distinguishes between saddle-node and period-doubling bifurcations. In this scheme, the ith saddle-node bifurcation of period P is denoted s_P^i, and $s_P^i \times 2^k$ is the orbit of period $P \times 2^k$ belonging to the period-doubling cascade originating from s_P^i.

attractor is contained in three disconnected pieces, before it expands suddenly. These pieces are visited successively in a fixed order. We call this parameter region the generalized period-3 window. Look more closely at, say, the middle branch: This is a complete copy of the whole bifurcation diagram! In particular, there is a period-9 window which is to the period-3 window what the period-3 is itself to the whole diagram.

To understand this, we note that the base symbols 0 and 1 can be viewed as the names of the period-1 orbits organizing the global dynamics. Similarly, let us denote by $X = 101$ and $Y = 100$ the symbolic names of the two period-3 orbits born in the saddle-node bifurcation initiating the period-3 window. All periodic orbits appearing in the generalized period-3 window can be written as words in the letters X and Y.

Indeed, since the attractor is split into three pieces visited successively, the dynamics can be simplified by considering the third iterate f^3. Each of the three pieces is a different attractor of f^3. The return map for each attractor is a unimodal map, with two "period-1" orbits that are in fact periodic points of the two period-3 orbits $\overline{100}$ and $\overline{101}$. Any pair of symbols X' and Y' such that the sequences $\overline{X'}$ and $\overline{Y'}$ correspond to periodic points that are degenerate at the period-3 saddle-node bifurcation can thus be used to code orbits of this map. Because we chose X and Y above to be higher in the kneading order than all their cyclic permutations, they satisfy this condition as well as any pair $\sigma^k X, \sigma^k Y$.

Since the two words $X = 101$ and $Y = 100$ are such that (1) $\overline{X} \prec \overline{Y}$ and (2) they have parities $\epsilon(X) = +1$ and $\epsilon(Y) = -1$, it is easy to see that the ordering of two sequences $W_1(X, Y)$ and $W_2(X, Y)$ will be exactly the same as for the corresponding sequences $W_1(0, 1)$ and $W_2(0, 1)$. For example,

$$\overline{YXYYXYYX} \prec \overline{YXXYX} \Longleftrightarrow \overline{10110110} \prec \overline{10010}$$

This explains why the bifurcation diagram in the generalized period-3 window has exactly the same structure as the whole diagram. Using the names of the standard period-doubling cascade given in (2.72), we find that the orbits involved in the period-doubling cascade of this window are Y, XY, $XYYY$, $XYYYXYXY$. The first orbits to appear in the window are the X and Y orbits (naming them after their sequences in the unstable regime), the last is the YX^{∞} orbit.

In fact, the results of this section could have been foreseen: They are a consequence of the qualitative universality of bifurcations in unimodal maps. Inside the period-3 window, the third return map is a unimodal map and therefore displays the same series of bifurcations as the first return map.

2.8 SHIFT DYNAMICAL SYSTEMS, MARKOV PARTITIONS, AND ENTROPY

In Section 2.7, we have seen how a chaotic system can be analyzed with the tools of symbolic dynamics. In particular, each regime of the logistic map is characterized by a different grammar (i.e., a set of forbidden symbolic sequences). Moreover, symbolic dynamics can be shown in some cases to provide a complete description of a dynamical system; for example, it is known that chaotic unimodal maps are conjugate if they have the same kneading sequence.

It is thus natural to study systems whose evolution laws are defined directly in a symbolic space by rules specifying which sequences are admissible. Such systems are usually referred to as *symbolic dynamical systems*, or as *shift dynamical systems* when they are based on the shift map [43]. Tools developed to characterize these systems can then be applied to any physical system for which a symbolic dynamical description has been obtained. This is illustrated by computations of entropy, an important measure of chaotic dynamics.

Here, we limit ourselves to *shifts of finite type*, which are characterized by a finite set of forbidden sequences. The interest of finite shifts is twofold. First, there are dynamical systems, those for which a *Markov partition* exists, that can be shown to be equivalent to a finite shift. Second, systems whose grammar cannot be specified by a finite set of rules can always be approximated with increasing accuracy by a sequence of finite shifts of increasing order.

2.8.1 Shifts of Finite Type and Topological Markov Chains

The natural phase space of a symbolic dynamical system is the set of infinite or bi-infinite sequences of symbols from an alphabet $\mathcal{A}$. Here we assume that the alphabet is finite and choose $\mathcal{A} = \{0, \ldots, N-1\}$, where N is the number of symbols. The systems we consider here share the same time-one map: the shift operator σ, which shifts symbols one place to the left (Section 2.7.1).

In the case of the logistic map, the symbolic space consisted of one-sided symbolic sequences. We noted in Section 2.7.1 that this makes σ noninvertible, since memory of the leading symbol is lost after each time step. If the shift operator has to be invertible, its action must not discard information. Thus, sequences must be bi-infinite (two-sided), for example,

$$\Sigma = \ldots s_{-3}s_{-2}s_{-1}.s_0s_1s_2 \tag{2.73}$$

with the dot separating the *forward sequence* $\Sigma_+ = s_0s_1\ldots$ from the *backward sequence* $\Sigma_- = s_{-1}s_{-2}\cdots$. These two sequences describe the future and the past of the point, respectively. The action of the shift operator on a sequence is then given by

$$\sigma(\ldots s_{-1}.s_0s_1\ldots) = \ldots s_{-1}s_0.s_1\ldots$$

The dot is merely moved to the right, which obviously preserves the information contained in the sequence. This is illustrated with the horseshoe map in Section 2.10.

The distinction between invertible and noninvertible dynamics is not made by most methods developed for characterizing symbolic dynamical systems. As we see below, they usually involve determining which finite blocks of symbols can appear in a typical sequence and which cannot. Thus, whether sequences are one- or two-sided is not relevant.

Full shifts are the simplest shift symbolic dynamical systems: Any sequence made of letters of the alphabet is allowed. Thus the symbolic space is $\mathcal{A}^{\mathbb{N}}$ or $\mathcal{A}^{\mathbb{Z}}$, depending on the invertibility of the dynamics. Two full shifts can differ only by the number of symbols of their alphabets. A full shift on r symbols is termed an r-shift.

In a general shift dynamical system, not all sequences are allowed: It is then called a *subshift*. We are interested only in subshifts whose set of allowed sequences $\mathcal{S}$ is shift-invariant (i.e., $\sigma\mathcal{S} = \mathcal{S}$). This implies that whether a finite symbol string can be found in a sequence $\Sigma \in \mathcal{S}$ does not depend on its position in the sequence but only in the content of the string. Finite symbol strings are also often referred to as *blocks*, with an n-block containing n symbols, or as *words*.

One could therefore provide a complete description of a dynamical system $(\mathcal{S}, \sigma)$ by specifying its *language* (i.e., the list of all finite strings of symbols that can be extracted from infinite sequences). It is usually much more convenient to specify its set $\mathcal{F}$ of *irreducible forbidden words* (IFWs). An IFW never appears in a sequence of $\mathcal{S}$ and does not contain any other forbidden word. For example, assume that $\mathcal{F} = \{00\}$. The word 001 is a forbidden word but is not irreducible because it contains 00. More generally, any word of the form $u00v$, where u and v are arbitrary words is not irreducible. By construction, any forbidden word of the language must have one of the elements of $\mathcal{F}$ as a substring, or is itself an IFW. IFWs are of length $l \geq 2$, since length-1 forbidden words can be removed by reducing the alphabet.

Of particular importance are the *shifts of finite type* (SFTs), which are described by a finite number of IFWs. Indeed, they can be specified with a finite amount of information, and invariant quantities such as the entropies described later can then be computed exactly. If the longest IFW of a shift of finite type is of length $L+1$, the *order* of the shift is L.

Shifts of finite type of order 1 are also called *topological Markov chains*. Since their set of IFWs contains only 2-blocks, the structure of this set can be described by a transition matrix M such that

$$M_{s_1,s_0} = \begin{cases} 0 & \text{if } s_0 s_1 \text{ is forbidden} \\ 1 & \text{if } s_0 s_1 \text{ is allowed} \end{cases} \tag{2.74}$$

That is, M_{s_1,s_0} is nonzero if s_1 is allowed to follow s_0 in a sequence (i.e., there is a transition $s_0 \to s_1$). A simple example is the transition matrix of the SFT with $\mathcal{F} = \{00\}$:

$$M = \begin{bmatrix} 0 & 1 \\ 1 & 1 \end{bmatrix} \tag{2.75}$$

which characterizes the symbolic dymamics of the logistic map immediately before the period-3 window (see Section 2.7.3.2).

Markov chains are all the more important as any shift of finite type of order L can be reformulated as a Markov chain by recoding sequences appropriately. Assume that there are N' allowed L-blocks, and denote $\mathcal{A}_L$ the alphabet made of these N' symbols. Any sequence $s_0 s_1 \ldots s_k \ldots$ can then be recoded as $S_0 S_1 \ldots S_k \ldots$, where the new symbol $S_k \in \mathcal{A}_L$ is the L-block starting at position k: $S_k = s_k s_{k+1} \ldots s_{L-1+k}$. For example, $S_0 = s_0 s_1 \ldots s_{L-1}$ and $S_1 = s_1 s_2 \ldots s_L$.

Now, assume that $S = s_0 s_1 \ldots s_{L-1}$ and $S' = s'_0 s'_1 \ldots s'_{L-1}$ are two symbols of $\mathcal{A}_L$. The element $M_{S',S}$ of the new transition matrix is 1 if:

- The head of S coincides with the tail of S': $s'_0 s'_1 \ldots s'_{L-2} = s_1 \ldots s_{L-1}$
- $s_0 s_1 \ldots s_{L-2} s'_{L-1}$ is an allowed L-block of the original shift

and is 0 otherwise. For example, assume that $\mathcal{F} = \{00, 0110\}$ and sequences are recoded using blocks of four symbols. Then one has $M_{101,011} = 0$ because 11 (tail of S) differs from 10 (head of S'), $M_{110,011} = 0$ because 0110 is an IFW, but $M_{111,011} = 1$ because 0111 does not contain any IFW.

Therefore, we see that any shift of finite type can be described completely by a transition matrix M. In the following section we show how to extract from this matrix information about the spectrum of periodic orbits of the dynamical system and an important measure of chaos, topological entropy.

2.8.2 Periodic Orbits and Topological Entropy of a Markov Chain

As discussed previously, periodic points of a symbolic dynamical system correspond to periodic sequences Σ satisfying $\sigma^p \Sigma = \Sigma$. Since periodic orbits play a crucial role, we want to be able to compute the number of periodic sequences of period p of an arbitrary Markov chain, given its transition matrix M. This problem is part of a more general one, which is to determine the number of allowed symbol strings of length n. An important measure of chaos, *topological entropy*, characterizes how this number increases when $n \to \infty$.

A transition matrix is conveniently represented by a directed graph. To each symbol corresponds a node, which can be viewed as a state. When the transition from "state" i to "state" j is allowed (i.e., $M_{ji} \neq 0$), there is a directed edge going from node i to node j (Fig. 2.17). The problems stated above can be reformulated as follows: How many distinct paths of length n does the graph have? How many of these paths are closed?

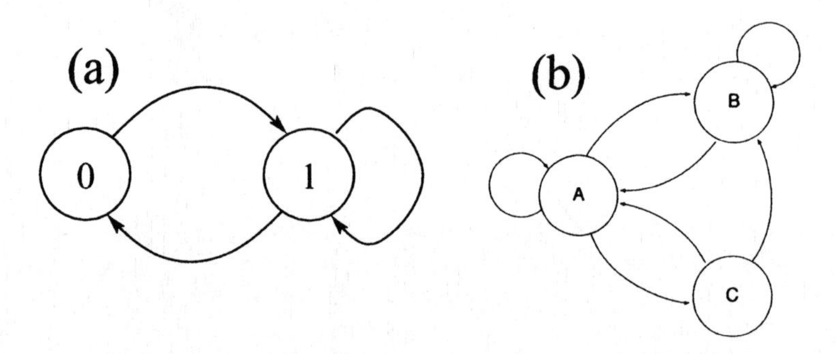

Fig. 2.17 A transition matrix can be represented by a directed graph. Nodes correspond to possible states and edges indicate whether transition from one state to another is possible. (a) This graph corresponds to the transition matrix (2.75). (b) This graph describes a grammar in which C must be preceded by A: The sequences BC and CC cannot occur.

We first compute the number P^n_{ji} of paths connecting node i to node j in exactly n steps. This can be done inductively. Obviously, one can go from i to j in one step only if there is an edge between the two sites, thus $P^1_{ji} = M_{ji}$. Then we note that each itinerary linking i to j in n steps goes from i to some site k in $n-1$ steps, then follows the edge from k to j in one step. By summing over all possible intermediary sites k, one obtains

$$P^n_{ji} = \sum_{k=0}^{N} M_{jk} P^{n-1}_{ki}$$

which is immediately seen to be the rule for matrix multiplication. Since $P^1_{ji} = M_{ji}$, it follows that

$$P^n_{ji} = (M^n)_{ji} \tag{2.76}$$

Hence all the relevant information is contained in the successive powers of the transition matrix.

Periodic sequences of period p correspond to paths of length p that begin and end at the same node. Thus, the number $N_f(p)$ of periodic points of period p is

$$N_f(p) = \sum_{k=0}^{N-1} M^p_{kk} = \text{tr}\, M^p \tag{2.77}$$

Similarly, the number $N_s(p)$ of p-symbol strings equals the total number of paths of length $p-1$ and thus is given by the sum of the elements of M^{p-1}. This can be formalized as follows. Let V^p be the vector whose ith component V_i^p is the number of length-p symbol strings beginning by symbol i. It is easy to see that the components of V^1 are all 1, and that $V^p = M^{p-1}V^1$. Thus,

$$N_s(p) = (V^1)^T V^p = (V^1)^T M^{p-1} V^1 \tag{2.78}$$

Expressions (2.77) and (2.78) show that $N_f(p)$ and $N_s(p)$ have the same asymptotic behavior. Indeed, the action of M^p for large p is determined by its largest eigenvalue $\lambda_{\max}$ and the associated eigenvector. It is easily shown that if $\lambda_{\max} > 1$, then

$$\lim_{p\to\infty} \frac{\log N_f(p)}{p} = \lim_{p\to\infty} \frac{\log N_s(p)}{p} = \log \lambda_{\max} \tag{2.79}$$

The growth rate of the number $N_s(p)$ of p-blocks

$$h_T = \lim_{p\to\infty} \frac{\log N_s(p)}{p} = \log \lambda_{\max} \tag{2.80}$$

is called the *topological entropy*. It measures the average amount of information that is extracted by reading one symbol of a typical sequence. Equation (2.79) shows that the topological entropy of a Markov chain depends in a very simple way on the transition matrix. It also illustrates the fact that in general the growth rate of the number of periodic points is equal to the topological entropy, as noted in Section 2.6.4.1. More sophisticated techniques to compute topological entropy are presented in Section 2.8.6. Let us consider two examples:

- The eigenvalues of the transition matrix (2.75) are $(1 \pm \sqrt{5})/2$. The largest one is $\lambda_{\max} \sim 1.6180339$ and is known as the *golden mean*. This Markov chain, accordingly called the *golden mean shift*, has topological entropy $h_T \sim 0.4812118$.

- The transition matrix of the full N-shift is filled with 1's. Its largest eigenvalue is N, and the topogical entropy is $h_T = \ln N$. In particular, the 2-shift has $h_T = 0.6931471$, which is greater than for the golden mean shift: Topological entropy increases as chaos becomes more developed.

2.8.3 Markov Partitions

Markov chains are not only interesting as model dynamical systems but also because there are some classical dynamical systems whose symbolic dynamics can be represented exactly by a topological Markov chain. We have already encountered a few examples of such systems. The simplest ones are the tent and logistic maps associated with a full 2-shift: Every sequence of "0" and "1" is associated with a physical orbit. The logistic map at the beginning of the period-3 window is another example of a finite shift; the only forbidden sequence is the string "00." A natural question then is: Under which conditions is a dynamical system described faithfully by a Markov chain?

The symbolic coding of a dynamical system relies on the existence of a partition $\mathcal{P} = \{\mathcal{P}_0, \mathcal{P}_1, \ldots, \mathcal{P}_{N-1}\}$ of phase space into N disjoint regions $\mathcal{P}_i$. At each time step, the current system state is coded with the symbol i associated with the region $\mathcal{P}_i$ to which it belongs. Because the partition and the time-one map determine completely the symbolic dynamics, it is not surprising that the condition for being describable by a Markov chain involves the partition and the images of the members of the partition.

We now state this condition without a proof. Assume that there exists a partition $\mathcal{P} = \{\mathcal{P}_i\}$ such that the intersection of any member with the image of another is either itself or is empty:

$$\forall i, j \quad \mathcal{P}_i \cap f(\mathcal{P}_j) = \begin{cases} \mathcal{P}_i \\ \emptyset \end{cases} \tag{2.81}$$

The structure of such a partition can be described concisely but faithfully by a transition matrix $M^{\mathcal{P}}$ defined by

$$M_{ij}^{\mathcal{P}} = \begin{cases} 1 & \text{if } \mathcal{P}_i \cap f(\mathcal{P}_j) \neq \emptyset \\ 0 & \text{otherwise} \end{cases} \tag{2.82}$$

It can be shown that the dynamical system coded by the partition $\mathcal{P}$ is then completely equivalent to the Markov chain of transition matrix $M^{\mathcal{P}}$. In particular, the topological entropy of the original dynamical system is equal to that of the Markov chain, and both systems have the same spectrum of periodic orbits. Accordingly, a partition $\mathcal{P}$ satisfying (2.81) is called a *Markov partition*. Examples of Markov partitions in the logistic map at special values of the parameter a are given in Section 2.9.2.

Note that even when there is a generating partition (such as the one based on critical points of a one-dimensional map) and it is not of Markov type, the existence of a Markov partition is not precluded. If the system is equivalent to a shift of finite type, an analysis of the symbolic dynamics obtained with the generating partition should reveal that there are a finite number of irreducible forbidden words. As described in Section 2.8.1, a Markov chain can then be obtained with a suitable recoding and the topogical entropy computed using the associated transition matrix.

2.8.4 Approximation by Markov Chains

In fact, only a small fraction of the regimes of the logistic map can be represented by a Markov chain exactly. Indeed, there is only a countable number of finite matrices of 0 and 1, whereas these regimes are indexed by the parameter a, which is a real number [38]. Moreover, chaotic regimes are associated with kneading sequences that are not eventually periodic, which makes it generally impossible to describe the symbolic dynamics by a finite number of IFWs.

However, this does not make shifts of finite type irrelevant. Indeed, it is not possible to analyze arbitrarily long symbol sequences. In practice, there is an upper bound on the length of the longest symbolic sequence that can be obtained in a reasonable time. This limits the search for forbidden symbol blocks to a maximal length. Otherwise, longer symbol blocks may be classified incorrectly as forbidden only because their probability of occurrence is too small. For example, assume that an orbit of 1 million ($\sim 2^{20}$) points has been recorded and coded on two symbols. It is certainly pointless to determine forbidden blocks longer than 20 symbols, since the least probable one will occur at most once in the best case, where all blocks are equiprobable.

Therefore, Markov chains are still relevant to characterize dynamical systems that are not conjugate to a shift of finite type, provided that a generating partition is known and a long symbolic sequence has been recorded. If the list of forbidden words has been determined for word lengths up to L, this gives a natural approximation of the system under study by a shift of finite type of order L, and hence by a Markov chain after a suitable recoding. Note that this systematically overestimates topological entropy estimates because higher-order forbidden sequences are neglected. The expression $h_T = \ln \lambda_{\max}$ for a Markov chain assumes that the number $N_s(p)$ of p-blocks can be determined for arbitrary p from the transition matrix. If there is more "pruning" than described by this transition matrix, the actual number of sequences will be lower, as well as the topological entropy. Carrying out this computation for increasing block lengths and comparing the results may help to estimate its accuracy.

2.8.5 Zeta Function

As shown in Section 2.8.2, the number of periodic points of period n, P_n, can be computed as $P_n = \text{tr } M^n$, where M is the Markov transition matrix. The information contained in the transition matrix can be transformed to a generating function for P_n by defining

$$\zeta(t) = \exp\left(\sum_{n=1}^{\infty} \frac{P_n}{n} t^n\right) \tag{2.83}$$

With a little bit of algebraic calisthenics, it is possible to show that

$$\zeta(M,t) = \frac{1}{\det(Id - tM)} \tag{2.84}$$

We illustrate one use of the zeta function in the example below.

Example: The spectrum of orbits forced by the period-three orbit 3_1 of the logistic map is computed using the Markov transition matrix M (cf. (2.75)). If the matrices $Id = \begin{bmatrix} 1 & 0 \\ 0 & 1 \end{bmatrix}$ and $M = \begin{bmatrix} 0 & 1 \\ 1 & 1 \end{bmatrix}$ have been defined previously, as well as the positive integer N ($= 12$ below), the generating function $\sum_n P_n/n\, t^n$ is given up to degree N by the simple Maple call

$$> \mathit{taylor}(-\mathit{log}(\mathit{det}(Id - t * M)), t = 0, N + 1);$$

$$t + \tfrac{3}{2}t^2 + \tfrac{4}{3}t^3 + \tfrac{7}{4}t^4 + \tfrac{11}{5}t^5 + \tfrac{18}{6}t^6 + \tfrac{29}{7}t^7 + \tfrac{47}{8}t^8 + \tfrac{76}{9}t^9 + \tfrac{123}{10}t^{10} + \tfrac{199}{11}t^{11} + \tfrac{322}{12}t^{12} + O(t^{13}) \tag{2.85}$$

Table 2.2 Number of orbits up to period $p = 12$ forced by 3_1 computed using the zeta function based on the golden mean matrix (2.75)

p	$N(p)$	Lower-Period Orbits	Period-Doubled	Number of Saddle-Node Pairs
1	1	1_1		
2	3	$(1_1+$	$2_1)$	
3	4	1_1+3_1		
4	7	$(1_1+2_1+$	$4_1)$	0
5	11	1_1		1
6	18	$1_1+2_1+3_1$		1
7	29	1_1		2
8	47	$(1_1+2_1+4_1$	$8_1)$	2
9	76	1_1+3_1		4
10	123	$1_1+2_1+(5_1+$	$10_4)$	5
11	199	1_1		9
12	322	$(1_1+2_1+4_1)+3_1+(6_1+$	$12_2)$	12

We read these results as follows:

1. There is one period-1 point 1_1.

2. There are three period-2 points. One is the period-1 point 1_1. The other two belong to the single period-2 *orbit* 2_1.

3. There are four period-3 points. One is 1_1. The other three belong to the *degenerate* saddle-node pair 3_1 (001 and 011).

4. There are seven period-4 points, which belong to the orbits 1_1, 2_1, and 4_1 of the initial period-doubling cascade.

5. There are 11 period-5 points. One belongs to 1_1. The remaining 10 belong to two period-5 orbits, which comprise the saddle-node pair 01111 and 01101.

Continuing in this way, we construct the remainder of the results. These are summarized in Table 2.2. It is a simple matter to verify that the results of this table are consistent with the results of Table 2.1 up to period 8.

2.8.6 Dealing with Grammars

At the saddle-node bifurcation of the period-three orbit, the adjacency matrix (Markov transition matrix) is given by Eq. (2.75). This matrix tells us that the symbol 0 must be followed by the symbol 1, and the symbol 1 can be followed by either of the symbols 0 or 1.

2.8.6.1 *Simple Grammars* It is useful to introduce an alternative representation for the dynamics. This involves introducing two symbols (words) $A = 01$ and $B = 1$. These two words have length 2 and 1, respectively. In this representation, A can be followed by either A or B, as is true also for B. The transition matrix is full:

$$M = \begin{bmatrix} 1 & 1 \\ 1 & 1 \end{bmatrix} \tag{2.86}$$

and the grammar is simple (no transitions are forbidden).

The periodic orbits in the dynamics can be constructed as follows. Replace the nonzero elements in the first row of M by the symbol A, those in the second row by the symbol B (A and B do not commute).

$$M = \begin{bmatrix} 1 & 1 \\ 1 & 1 \end{bmatrix} \longrightarrow \begin{bmatrix} A & A \\ B & B \end{bmatrix} = \mathcal{M} \tag{2.87}$$

Then the complete set of periodic orbits is constructed by computing $\text{tr}\,\mathcal{M}^n$, $n = 1, 2, 3, \ldots$.

Example: For $n = 1, 2$ and 3 we find

n	$\text{tr}\ \mathcal{M}^n$	Orbits
1	$A + B$	$01 + 1$
2	$2A^2 + AB + B^2$	$2(01)^2 + 011 + (1)^2$
3	$2A^3 + 2ABA + 2A^2B + AB^2 + B^3$	$2(01)^3 + 4(01101) + 0111 + (1)^3$

Reduction to simple grammars is often useful in analyzing experimental data. For example, chaotic data generated by the Belousov–Zhabotinskii reaction (cf. Chapter 7) have been reduced to a symbolic code sequence involving the two symbols 0 and 1. The rules of grammar observed in the experimental data are:

1. The symbol 0 must be followed by the symbol 1.
2. The symbol 1 can be followed by 0 or 1.
3. Symbol sequences $(11 \cdots 1)$ of length p can occur for $p = 1, 2, 3, 4$ but not for $p > 4$.

It appears that the vocabulary of this dynamics consists of the four words 01, 011, 0111, and 01111. It also appears that any of these words can be followed by any other word. The grammar is simple and represented by a 4×4 Markov transition matrix whose 16 elements are 1. The periodic orbits are obtained as described above for the golden mean case.

The topological entropy for dynamics consisting of a finite number of words of varying length obeying a simple grammar (full shift) is easily determined as follows. Assume that there are $w(p)$ words of length p, $p = 1, 2, 3, \ldots$. Then the number of ways, $N(T)$, of constructing a word of length T is determined by the difference equation

$$N(T) = w(1)N(T-1) + w(2)N(T-2) + \cdots = \sum_{p=1} w(p)N(T-p) \quad (2.88)$$

The number $N(T)$ behaves asymptotically like $N(T) \sim A(X_M)^T$, where A is some constant and X_M is the largest real root of the characteristic equation

$$X^T = \sum_{p=1} w(p)X^{T-p} \qquad \text{or} \qquad 1 = \sum_{p=1} \frac{w(p)}{X^p} \quad (2.89)$$

Example 1: For the full shift on two symbols 0 and 1,

$$1 = \frac{2}{X} \Longrightarrow h_T = \log 2$$

Example 2: For golden mean dynamics, $w(1) = 1$, $w(2) = 1$, and

$$1 = \frac{1}{X} + \frac{1}{X^2} \Longrightarrow X_M = \frac{1+\sqrt{5}}{2} \quad \text{and} \quad h_T = 0.481212$$

Example 3: For the Belousov data described above, $w(p) = 1$ for $p = 2, 3, 4, 5$ and $w(p) = 0$ for $p = 1$ and $p > 5$, so that

$$1 = \frac{1}{X^2} + \frac{1}{X^3} + \frac{1}{X^4} + \frac{1}{X^5} \qquad X_M = 1.534158 \text{ and } h_T = 0.427982$$

2.8.6.2 Complicated Grammars There are many cases in which the dynamics either consists of, or is well approximated by, a finite vocabulary with a nontrivial grammar. The two questions addressed above (description of periodic orbits, computation of entropy) are still of interest.

The spectrum of periodic orbits can be computed by following the algorithm used above.

1. Write out the Markov transition matrix M for the vocabulary.

2. Replace each nonzero matrix element 1 in row i by the noncommuting symbol w_i representing the ith word, effecting the transition $M \to \mathcal{M}$.

3. Compute tr $\mathcal{M}^n$ for $n = 1, 2, \ldots$.

4. Replace each word sequence by the appropriate sequence of symbols from the original alphabet (i.e., 0 and 1).

The problem of computing the topological entropy for this dynamics is more subtle. It is isomorphic to the problem of computing the capacity of a transmission channel. The capacity of a transmission channel is (Shannon, [44,45])

$$C = \lim_{T\to\infty} \frac{1}{T} \log N(T)$$

Here $N(T)$ is the number of allowed signals of duration T, and log is to base e.

In many grammars, not all symbol sequences are allowed (qu is OK, qv is not). In such cases, assume that there are m states $b_1, b_2, \ldots, b_m$. For each state only certain symbols from the set $S_1, S_2, \ldots, S_n$ can be transmitted (different subsets for different states). The transmission of symbol S_k from state b_i to state b_j (b_i may be the same as b_j) takes time $t_{ij}^{(k)}$, where k indexes all possible paths from b_i to b_j. This process is illustrated by a graph such as that shown in Fig. 2.17.

Theorem: The channel capacity C is $\log W_0$, where W_0 is the largest real root of the $m \times m$ determinantal equation

$$\left| \sum_k W^{-t_{ij}^{(k)}} - \delta_{ij} \right| = 0 \tag{2.90}$$

For our purposes, we can regard each state as a word and t_{ij} is the length of word i.

Example: Assume that the vocabulary has three words A, B, and C or w_1, w_2, and w_3 of lengths p, q, and r and a grammar defined by Fig. 2.17(b). The Markov matrix is

$$M = \begin{bmatrix} 1 & 1 & 1 \\ 1 & 1 & 0 \\ 1 & 1 & 0 \end{bmatrix}$$

The determinantal equation constructed from the Markov matrix and word lengths is

$$\begin{bmatrix} \frac{1}{W^p} - 1 & \frac{1}{W^p} & \frac{1}{W^p} \\ \frac{1}{W^q} & \frac{1}{W^q} - 1 & 0 \\ \frac{1}{W^r} & \frac{1}{W^r} & -1 \end{bmatrix} = 0$$

The characteristic equation for this dynamical system is

$$\frac{1}{W^p} + \frac{1}{W^q} + \frac{1}{W^{p+r}} = 1$$

The topological entropy is $h_T = \log W_0$, where W_0 is the largest real eigenvalue of this characteristic equation.

2.9 FINGERPRINTS OF PERIODIC ORBITS AND ORBIT FORCING

2.9.1 Permutation of Periodic Points as a Topological Invariant

Using kneading theory, periodic points with given symbolic itineraries can be ordered along the interval. This is not only useful to determine the order in which periodic orbits appear, but also to identify periodic orbits.

Indeed, consider the period-5 orbit with symbolic name 01011 of the logistic map (Fig. 2.16). Its five periodic points are associated with the five cyclic permutations of the symbolic name, ordered by kneading theory as follows:

$$\overline{01101} \prec \overline{01011} \prec \overline{11010} \prec \overline{10101} \prec \overline{10110} \tag{2.91}$$

Label the sequences in (2.91) from left to right by $\Sigma_i, i = 1 \ldots 5$, and express them in terms of the leftmost sequence $\Sigma_1 = 01101$ and of powers of the shift operator. We have

$$\Sigma_1 \prec \Sigma_2 = \sigma^3\Sigma_1 \prec \Sigma_3 = \sigma\Sigma_1 \prec \Sigma_4 = \sigma^2\Sigma_1 \prec \Sigma_5 = \sigma^4\Sigma_1 \tag{2.92}$$

We observe that under the action of the shift operator σ, these sequences are permuted: The lowest sequence becomes third, the second one the last, and so on. The corresponding permutation

$$\pi(01011) = (\pi_i) = (3, 5, 4, 2, 1) \tag{2.93}$$

such that $\sigma\Sigma_i = \Sigma_{\pi_i}$ provides crucial information about the orbit. Its dynamical relevance owes much to two fundamental properties.

First, *the permutation can be extracted directly from the periodic orbit without using any symbolic encoding and without having the graph of the map.* Indeed, consider the period-5 orbit in Fig. 2.16. If we label the periodic points $x_1, x_2, \ldots, x_5$ from left to right, we can determine as above a permutation π such that $f(x_i) = x_{\pi_i}$. Obviously, this permutation is identical to (2.93) (the image of the first point is the third, etc.), which can easily be checked in Fig. 2.16.

Second, the permutation (2.93) *remains identical on the entire domain of existence of the orbit in parameter space.* As a parameter is varied, the points x_i of a periodic orbit move along the interval, but they do so without ever becoming degenerate (otherwise, we would have one point with two images, in contradiction with the deterministic nature of the map). Thus the relative order of points is preserved, hence the corresponding permutation.

An important property of this invariance property is that the permutations associated with orbits interacting in a bifurcation will be strongly related. For example, saddle-node partners will have identical permutations since they are indistinguishable at the bifurcation. Similarly, the permutation associated with a period-doubled orbit can easily be obtained from that of its mother.

Since there is a definite relation between symbolic names and permutations on the one hand, and periodic orbits and permutations on the other hand, we see that the symbolic names of the orbits are more than a convenient labeling and that they carry topological information. To illustrate this, we now show that in one-dimensional maps, much about the symbolic name of an orbit can be recovered merely from the permutation. Consider the graphical representation in Fig. 2.18 of the permutation (2.93) extracted from Fig. 2.16.

The global shape of Fig. 2.18 is characteristic of unimodal permutations: The relative order of the leftmost points is preserved, while that of the rightmost points is reversed (in particular, the rightmost point is mapped into the leftmost). This is a signature of the existence of branches with different parities in the underlying map (assumed unknown). In fact, Fig. 2.18 can be viewed as a topological representation of this map in a coordinate system where the points x_i are equidistant.

Let us note the orientation-preserving (resp., orientation-reversing) branch 0 (resp., 1). It can be seen from Fig. 2.18 that x_1 is necessarily on branch 0, while x_3, x_4, and

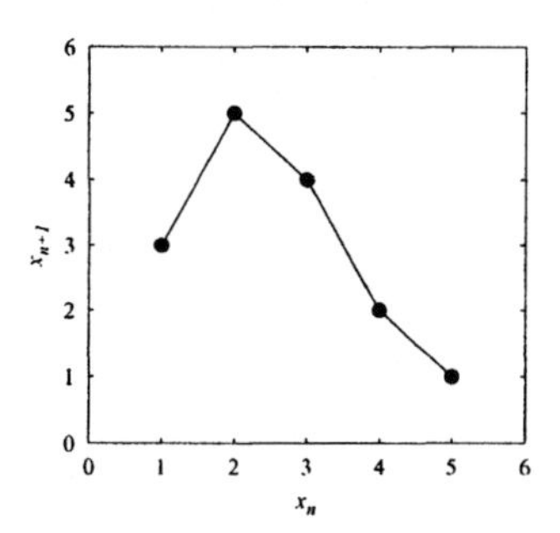

Fig. 2.18 Graphical representation (i, π_i) of the permutation (2.93). Orientation-preserving and orientation-reversing parts are easily distinguished.

x_5 must be on branch 1. The coding of x_2 is ambiguous: It can be on one branch or the other without modifying the permutation. Taking into account that the orbit of x_1 is $x_1 \to x_3 \to x_4 \to x_2 \to x_5$, the symbolic name of this orbit is thus $011\,{}^{0}_{1}1$ (i.e., any of the two saddle-node partners born in the 5_1 saddle-node bifurcation listed in Table 2.1). If additional information is available, such as parity (is the multiplier of the orbit positive or negative?), it is possible to distinguish between these two orbits.

By generalizing the example above, it is easy to see that every pair of saddle-node partners and every period-doubled orbit is associated with a different permutation. The permutation realized by a periodic orbit can thus be viewed *as a genuine fingerprint* of this orbit.

It is comforting to know that much of the discussion of this section is still relevant for orbits in three-dimensional flows. As discussed throughout this book, orbits in these systems will be associated with braids (a generalization of permutations deeply linked to knot theory). These braids will be characterized by topological invariants that do not depend on parameters and contain much information about the symbolic dynamics and the genealogy of periodic orbits. Just as the structure of unimodal maps governs that of the permutations (see Fig. 2.18), there is a systematic way to study the global organization of braids in three-dimensional systems.

2.9.2 Topological Entropy of a Periodic Orbit

The permutation associated with a periodic orbit not only provides qualitative information, it can also provide estimates of fundamental quantitative measures of chaotic dynamics, as we show next. The key idea is that if the underlying map is continuous, the way in which the points of the orbit are mapped onto each other provides information on orbits in an extended neighborhood of the orbit.

For simplicity, let us consider a superstable periodic orbit such as the period-5 orbit of Fig. 2.16. As previously, the points are numbered from left to right. Since x_2 is the critical point x_c, the leftmost and rightmost points x_1 and x_5 correspond to the lower and upper bounds of the invariant interval $I = [f(x_c), f^2(x_c)]$ where the relevant dynamics is confined. Using the fact that the x_i are mapped exactly onto each other, this will allow us to build a topological model of the dynamics.

To this end, consider the following partition of the invariant interval I:

$$I = [x_1, x_5] = I_1 \cup I_2 \cup I_3 \cup I_4 = [x_1, x_2] \cup [x_2, x_3] \cup [x_3, x_4] \cup [x_4, x_5] \quad (2.94)$$

Using the interval arithmetics (2.53) and the permutation (2.93), we find easily that

$$f(I_1) = f([x_1, x_2]) = [x_3, x_5] = I_3 \cup I_4 \quad (2.95a)$$
$$f(I_2) = f([x_2, x_3]) = [x_4, x_5] = I_4 \quad (2.95b)$$
$$f(I_3) = f([x_3, x_4]) = [x_2, x_4] = I_2 \cup I_3 \quad (2.95c)$$
$$f(I_4) = f([x_4, x_5]) = [x_1, x_2] = I_1 \quad (2.95d)$$

The set of relations (2.95) is the analog of the relations $f(I_1) = f(I_2) = I_1 \cup I_2$ which characterize the surjective logistic and tent maps (note that in a complete description of the map, the branch parities should also be specified). The key observation here is that both sets of relations define Markov partitions. In the example above, the Markov transition matrix as defined in (2.82) reads

$$A = \begin{bmatrix} 0 & 0 & 0 & 1 \\ 0 & 0 & 1 & 0 \\ 1 & 0 & 1 & 0 \\ 1 & 1 & 0 & 0 \end{bmatrix} \quad (2.96)$$

where the nonzero entries correspond to pairs (I_i, I_j) such that $I_i \cap f(I_j) = I_i$. Even though the regime under study corresponds to a superstable orbit, the matrix (2.96) is a signature of a chaotic dynamics. Its largest eigenvalue is $\lambda_{\text{max}} \sim 1.512876398$, which yields a topological entropy of $h_T = \ln \lambda_{\text{max}} \sim 0.4140127381$. Moreover, the matrix is transitive, which indicates that the associated Markov chain is topologically mixing.

In fact, the topological entropy as computed above characterizes the periodic orbit rather than the dynamical system to which it belongs: It is obtained from the permutation associated to the orbit,[4] not from the global structure of the system. However, the entire system cannot be less chaotic than implied by the periodic orbit. In particular, its topological entropy is necessarily greater than the entropy of the orbit. Thus, the observation of an orbit with a positive topological entropy (as obtained from its permutation), even in a window of stability, indicates the presence of chaos in the system under study, and in particular the existence of an infinity of periodic orbits. This is illustrated in the next section with the "period-3 implies chaos" theorem. A

[4]For the sake of simplicity, we have indicated how to compute the topological entropy of an orbit only at a special parameter value, where the orbit is superstable. However, the topological entropy of an orbit depends only on the permutation associated with it and should be considered as a topological invariant of the orbit, defined on its entire domain of existence.

similar statement will be made later for flows: Some periodic orbits have knot types that can exist only in a chaotic system.

Note that the tools introduced here show why one-dimensional diffeomorphisms are not chaotic: Since they globally preserve or reverse the order of points, the associated transition matrices cannot have eigenvalues larger than 1. In fact, the reader may want to check that a one-dimensional diffeomorphism can only have a period-1 orbit, possibly a period-2 orbit if it globally reverses orientation.

2.9.3 Period-3 Implies Chaos and Sarkovskii's Theorem

To illustrate the discussion above, we present now the famous statement "period-3 implies chaos," according to which the presence in a map of the interval of a periodic orbit of period 3 forces the presence of orbits of any other period [46].

If we carry out the same calculation for the superstable $0\,{}^0_1 1$ orbit as for the $011\,{}^0_1 1$ orbit in Section 2.9.2, we find a partition $I = [f^2(x_c), x_c] \cup [x_c, f(x_c)]$, whose transition matrix is the golden mean matrix (2.75) that we have already encountered:

$$A_{\text{GM}} = \begin{bmatrix} 0 & 1 \\ 1 & 1 \end{bmatrix} \tag{2.97}$$

It is easy to prove that $\forall n \geq 2$, $A^n_{ii} \neq 0$, hence there are fixed points for any period p. This is our first example of the general fact that the existence of some periodic orbits can force the existence of many other (here, an infinity) periodic orbits. This phenomenon is usually referred to as *orbit forcing*.

Note that the Markov partitions constructed from superstable orbits are refinements of the partition used for symbolic codings, since the critical point is one of the border points. As a result, the transition matrices contain all the information needed to determine whether a given symbolic name is admissible. The example of the period-3 orbit is particularly simple since the Markov partition coincides with the coding partition: The transition matrix (2.97) indicates that all itineraries are allowed except those containing the string "00" (showing again that orbits of all periods exist). One may check in Table 2.1 that this is indeed what distinguishes orbits born before the 3_1 saddle-node bifurcation from orbits created after.

That the existence of an orbit of period 3 implies the existence of orbits of any other period can in fact be viewed as a particular case of a more general theorem due to Sarkovskii [47] (see also [38]). Consider the following ordering of the natural integers, written as the product $2^k \times (2n+1)$ of a power of 2 by a prime number:

$$\begin{aligned} 2^k \times 1 &\lhd \quad 2^l \times 1 && (k < l) && \text{(2.98a)} \\ 2^k \times 1 &\lhd \quad 2^l \times (2n+1) && (n > 0) && \text{(2.98b)} \\ 2^l \times (2n+1) &\lhd \quad 2^k \times (2n+1) && (k < l, n > 0) && \text{(2.98c)} \\ 2^k \times (2n+1) &\lhd \quad 2^k \times (2m+1) && (m < n) && \text{(2.98d)} \end{aligned}$$

For example,

$$1 \lhd 2 \lhd 2^2 \lhd 2^3 \lhd 2^4 \times 7 \lhd 2^4 \times 3 \lhd 2^2 \times 5 \lhd 2^2 \times 3 \lhd 2 \times 3 \lhd 7 \lhd 5 \lhd 3$$

Sarkovskii's theorem states that if a continuous map of an interval into itself has an orbit of lowest period p and $q \lhd p$, it also has an orbit of lowest period q. It is easily seen that (1) if there are infinitely many different periods in a map, all the periods corresponding to the period-doubling cascade of the period-1 orbit must be present; (2) since 3 comes last in (2.98), the presence of a period-3 orbit forces orbits of all other periods as shown above. It can be checked that the succession of bifurcations given in Table 2.1 satisfies the Sarkovskii theorem. In particular, the first period-7 orbit is created before the first period-5 orbit, which itself is created before the period-3 orbit. Moreover, all even periods are present when the first odd period appears.

2.9.4 Period-3 Does Not Always Imply Chaos: Role of Phase-Space Topology

Since so many properties of unimodal maps hold at least approximately for higher-dimensional systems, it might be troubling that numerous apparent counterexamples to the "period-3 implies chaos" theorem can be found in physical systems. In the modulated CO_2 lasers described in Chapter 1 and in Section 7.5.2, for example, it is quite common to observe multistability, with a period-3 orbit coexisting with the initial period-1 orbit, no other periodic orbit being present. Thus, a period-3 orbit does not necessarily imply chaos.

The clue to this paradox is that the modulated CO_2 laser can be described by a three-dimensional flow, hence by a two-dimensional Poincaré map, while the Sarkovskii theorem holds for a map of a one-dimensional interval into itself. It turns out that phase-space topology has a dramatic influence on orbit forcing.

The key topological difference between the two geometries is that in the two-dimensional case, it is possible to connect each of the three periodic points of the period-3 orbit to any other without encountering the third one. In the one-dimensional case, the two extreme points are isolated by the middle one.

Therefore, the three periodic points must be considered as arranged along a topological circle (obviously, this also applies to maps of a circle into itself; see Section 2.12). The three points divide this circle into three intervals versus two in the one-dimensional case. When the map is applied, the three points are cyclically permuted and the three intervals accordingly exchanged. As a result, the associated transition matrix is a simple permutation matrix, with zero topological entropy. The action of the map on the periodic orbit is equivalent to a pure rotation, which does not itself imply the existence of a chaotic dynamics. However, although it appears that orbit forcing can be modified dramatically when the topology of phase space is changed, it remains true in all cases that some orbits can force the existence of an infinity of other orbits.

2.9.5 Permutations and Orbit Forcing

The statements of the Li-Yorke theorem and of the Sarkovski theorem seem to imply that period is the fundamental property in implication chains among orbits. However, forcing relations are more fine-grained. Admittedly, the existence of a period-5 orbit forces the existence of at least a period-7 orbit. However, an orbit of the period-5 pair 5_1 of the logistic map does not force every period-7 orbit of this map. Actually, it forces the two 7_1 orbits *but is forced*by the two 7_2 orbits. As we see below, it is in fact the permutation associated with a given orbit that determines which orbits it forces.

In Section 2.9.1, we noted that the permutation associated with a periodic orbit $\mathcal{O}$ can be represented graphically by a piecewise-linear map such as shown in Fig. 2.18. This map is conjugate to the Markov chain that describes the action of the map on the periodic points of the orbit (see Section 2.9.2). If the Markov chain has positive topological entropy, it is chaotic and so is the piecewise-linear map which has thus an infinite number of unstable periodic orbits. These orbits comprise the minimal set of orbits that must exist in a map having the original orbit $\mathcal{O}$ as one of its periodic orbits. The existence of these orbits is forced by that of $\mathcal{O}$.

As an example, we show that the period-5 orbit 01011 (born in bifurcation 7_1) forces exactly one pair of period-7 orbits, and determine the symbolic code of this orbit. To this end, we apply the techniques outlined in Section 2.8.6 to the transition matrix (2.96). Using the non-commutative symbols I_1, I_2, I_3 and I_4, we first rewrite it as

$$\mathcal{A} = \begin{bmatrix} 0 & 0 & 0 & I_1 \\ 0 & 0 & I_2 & 0 \\ I_3 & 0 & I_3 & 0 \\ I_4 & I_4 & 0 & 0 \end{bmatrix} \tag{2.99}$$

Then we compute

$$\operatorname{tr} \mathcal{A}^7 = [I_1 I_3 I_3 I_2 I_4 I_1 I_4] + [I_1 I_3 I_3 I_3 I_3 I_2 I_4] + (I_3)^7 \tag{2.100}$$

where $[W]$ represents the n cyclic permutations of a length-n word W. The minimal model compatible with the period-5 orbit 01011 has thus 2 period-7 orbits, whose itineraries in the Markov partition can be read from (2.100). Since the interval I_1 is located left of the critical point (Sec. 2.9.2) and the intervals $I_2, \ldots, I_4$ right of it, the canonical symbolic names of these orbits can be obtained trough the recoding $I_1 \to 0$, $\{I_2, I_3, I_4\} \to 1$. The two period-7 orbits forced by the period-5 orbit under study are thus 0111101 and 0111101.

The reader can check in Table 2.1 that these are the two orbits born in the saddle-node bifurcation 7_1 and that this bifurcation indeed occurs before 5_1 in the universal sequence. By repeating the above calculation for all low-period orbits of the logistic map, the whole universal sequence of Table 2.1 can be reproduced.

It is not a coincidence that the two period-7 orbits forced in the example are saddle-node partners and thus are associated with the same permutation. As was suggested by the simple example above, orbit forcing in one-dimensional maps is naturally expressed as an order on permutations rather than on periods. This is illustrated in Section 9.3.1. In unimodal maps, each bifurcation corresponds to a different permutation (Section 2.9.1) so that this order on permutations induces a total order on bifurcations, which corresponds to the universal sequence discussed in Section 2.7.5.3.

As is discussed in Chapter 9, there is no longer a total ordering of bifurcations in two-dimensional maps (and three-dimensional flows). The topological structure of a periodic orbit then is not specified by a permutation but by a braid type. There is a forcing relation on braid types. However, several saddle-node bifurcations can be associated with the same braid type, so that the induced order on bifurcations is only a partial order.

2.10 TWO-DIMENSIONAL DYNAMICS: SMALE'S HORSESHOE

2.10.1 Horseshoe Map

While one-dimensional maps display the most distinctive features of deterministic chaos, they lack a crucial property of time-one maps or Poincaré maps of flows: They are not invertible, as are the equations describing flows. Because many physical systems are described by differential equations, it is now time to turn to two-dimensional invertible maps to understand how the basic mechanisms studied in previous sections can be embedded in an invertible dynamics. This is not a trivial problem; many key points of our analysis of the tent and logistic map, such as property (2.31), rely heavily on noninvertibility. To help us in this task, our keystone will be the two-dimensional analog of the tent map, the paradigmatic Smale's horseshoe.

The tent map example makes it clear that a key ingredient of chaos is expansion, or *stretching*. Yet many experimental and numerical examples of chaotic systems show that for a large number of them (i.e., dissipative systems), the asymptotic dynamics is confined to a small region of state space. Thus, expansion must be balanced by contraction, or *squeezing*, so that trajectories remain forever in a bounded region. In continuous systems, these two antagonistic mechanisms can be combined by folding processes, as illustrated by Smale's horseshoe map f_S, defined as follows.

Take the unit square S, stretch it along one direction (the unstable direction) by approximately a factor of 2, while squeezing it in the transverse direction (the stable direction). Then fold the deformed rectangle so that it occupies roughly the same region as the original square, intersecting the latter in two disjoint strips [Fig. 2.19(a)]. Note that points located in the middle and the ends of S do not return to it.

The key topological property of the horseshoe map is that $f(S) \cap S$ consists of two disjoint components:

$$f(S) \cap S = H_0 \cup H_1 \tag{2.101}$$

which are the two horizontal strips H_0 and H_1 shown in Fig. 2.19(a).

To specify the action of the inverse map f^{-1}, we only need to determine $f^{-1}(H_0 \cup H_1)$ since $f^{-1}(f(S) \cap S) = S \cap f^{-1}(S)$. As shown in Fig. 2.19(b), all the points sent by f to a given horizontal strip H_i come from a vertical strip $V_i = f^{-1}(H_i)$ and thus

$$f^{-1}(S) \cap S = V_0 \cup V_1 \tag{2.102}$$

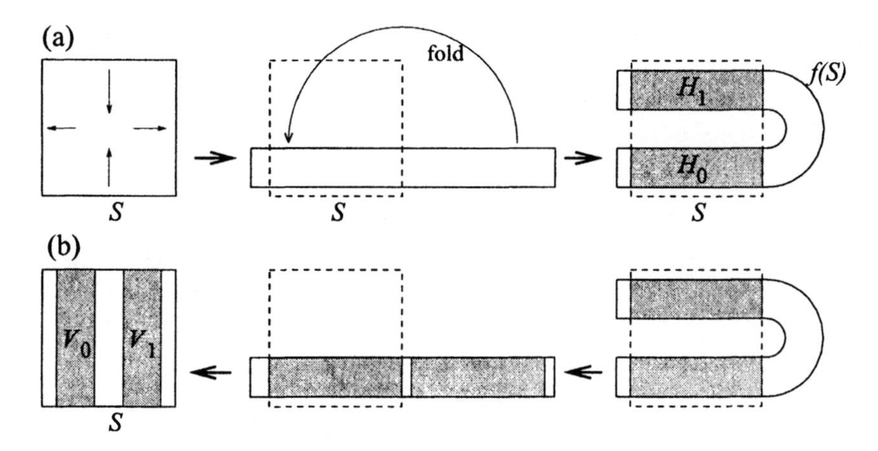

Fig. 2.19 Graphical representation of Smale's horseshoe map. (a) The unit square S is stretched horizontally and squeezed vertically, then folded so that $f(S)$ intersects S in two horizontal strips H_i. (b) The map is iterated backward to obtain the preimages V_i of the strips H_i.

2.10.2 Symbolic Dynamics of the Invariant Set

We want to study the structure of the invariant set $\Lambda = \cap_{-\infty}^{+\infty} f^n(S)$, which contains points whose orbits remain in S forever. As a first approximation, $\Lambda \subset f^{-1}(S) \cap S \cap f(S)$ or, by (2.101) and (2.102),

$$\Lambda \subset (V_0 \cup V_1) \cap (H_0 \cup H_1) \tag{2.103}$$

As illustrated in Fig. 2.20, Λ is partitioned into four components, $H_i \cap V_j$, by the horizontal rectangles H_i and the vertical rectangles V_j. Since $H_i = f(V_i)$, whether a point X belongs to a square $H_i \cap V_j$ depends only on the vertical rectangles V_j its orbit successively visits:

$$X \in H_i \cap V_j \Leftrightarrow X \in f(V_i) \cap V_j \Leftrightarrow f^{-1}(X) \in V_i \text{ and } X \in V_j \tag{2.104}$$

that is, on the symbolic sequence $\Sigma(X)$ obtained by encoding the orbit of X with the coding function

$$\forall X \in \Lambda, \quad s(X) = \begin{cases} 0 & \text{if } X \in V_0 \\ 1 & \text{if } X \in V_1 \end{cases} \tag{2.105}$$

corresponding to the partition $\Lambda = (\Lambda \cap V_0) \cup (\Lambda \cap V_1)$. Because f is invertible, the action of the shift operator on symbolic sequences must not discard information, unlike in the one-dimensional case. Thus, the sequence $\Sigma(X)$ associated to X must be bi-infinite:

$$\Sigma(X) = \ldots s_{-3}s_{-2}s_{-1}.s_0s_1s_2\ldots \qquad s_i = s(f^i(X)), i \in \mathbb{Z} \tag{2.106}$$

with the dot separating the *forward sequence* $\Sigma_+(X) = s_0 s_1 \ldots$ from the *backward sequence* $\Sigma_-(X) = s_{-1}s_{-2}\ldots$. These two sequences describe the future and the past of the point, respectively. Applying the shift operator to a sequence simply amounts to shifting the dot: $\sigma(\ldots s_{-1}.s_0 s_1 \ldots) = \ldots s_{-1}s_0.s_1 \ldots$.

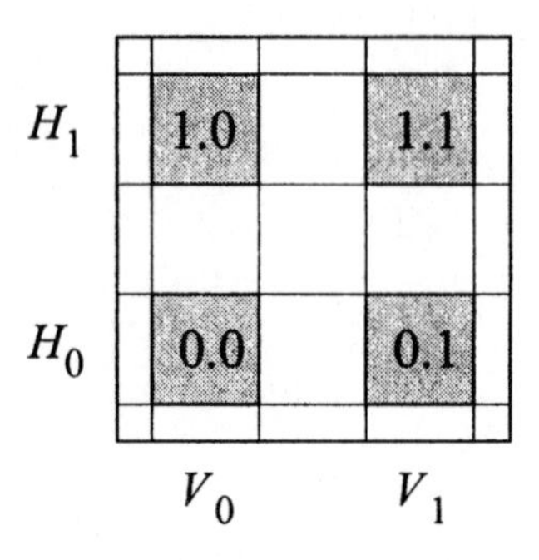

Fig. 2.20 The invariant set Λ is contained in the intersections of the vertical rectangles V_i with the horizontal rectangles H_i. As explained in the text, $H_i \cap V_j$ is labeled $i.j$.

As in (2.46), let us define cylinders as sets of points whose symbolic sequences contain a common substring, such as

$$C[s_{-2}s_{-1}.s_0 s_1] \equiv \{X \in \Lambda;\ \Sigma(X) = \ldots s_{-2}s_{-1}.s_0 s_1 \ldots\} \tag{2.107}$$

It is easily seen from (2.104) that the squares in Fig. 2.20 are cylinders:

$$(H_i \cap V_j) \cap \Lambda = C[i.j] \tag{2.108}$$

which explains the labeling of the squares in Fig. 2.20. To show that the partition based on (2.105) is generating, we now detail the structure of Λ and of higher-order cylinders.

Start from the cylinders $C[i.j] = H_i \cap V_j$ shown in Figs. 2.20 and 2.21(a) and compute their images $C[ij.] = f(C[i.j])$. These are the horizontal rectangles $H_{ij} = f(H_i) \cap H_j$ shown in Fig. 2.21(b), which are seen to be the components of $f^2(S) \cap S$. A dual decomposition of Λ is based on the preimages $C[.ij] = f^{-1}(C[i.j])$, which are the vertical rectangles $V_{ij} = V_i \cap f^{-1}(V_j)$ covering $f^{-2}(S) \cap S$ [Fig. 2.21(c)].

Figure 2.21(d) illustrates the fact that the four-symbol cylinders $C[ij.kl]$ can be obtained as the intersections of the vertical and the horizontal two-symbol rectangles:

$$C[ij.kl] = C[ij.] \cap C[.kl] = H_{ij} \cap V_{kl} = f^2(V_i) \cap f(V_j) \cap V_k \cap f^{-1}(V_l) \tag{2.109}$$

which follows directly from definition (2.108). The forward (resp., backward) sequence $\Sigma_+(X)$ [resp., $\Sigma_-(X)$] specifies in which of the vertical (resp., horizontal) rectangles V_{ij} (resp., H_{ij}) the point X lies.

Note that if one takes into account that $\Sigma_-(X) = s_{-1}s_{-2}\ldots$ should be read from the labels of Fig. 2.21(b) from right to left, forward and backward sequences

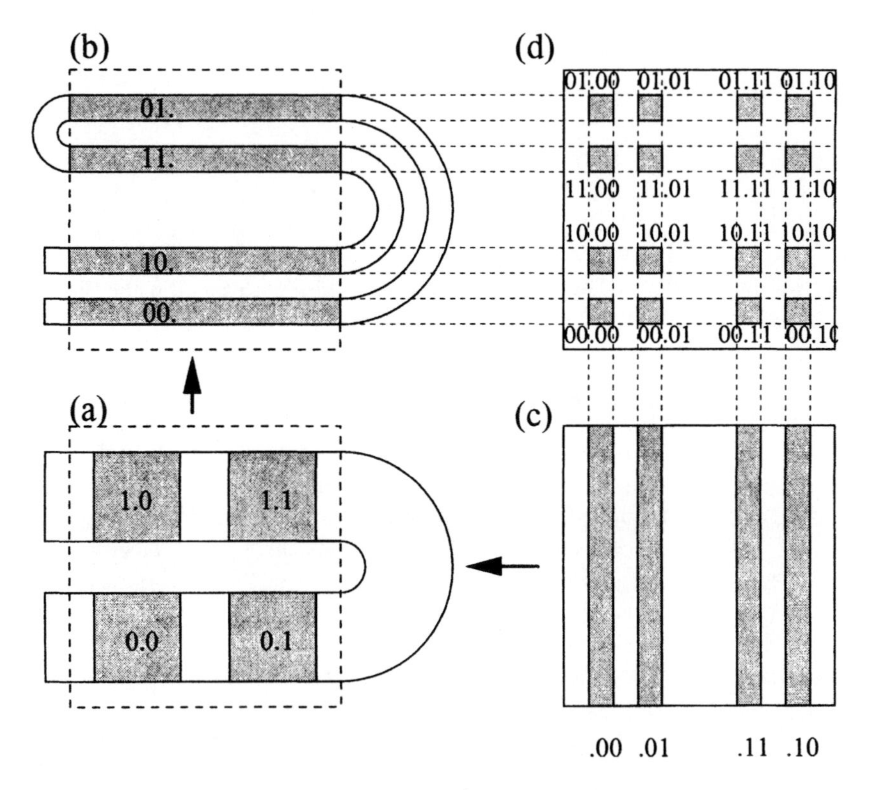

Fig. 2.21 (a) Decomposition of $f(S) \cap S \cap f^{-1}(S)$ in squares $H_i \cap V_j$. These are mapped by the horseshoe map f to (b) the horizontal rectangles $H_{ij} = f(H_i) \cap H_j$, which are the components of $f^2(S) \cap S$, and by f^{-1} to (c) the vertical rectangles $V_{ij} = V_i \cap f^{-1}(V_j)$, which cover $f^{-2}(S) \cap S$. Arrows indicate the action of the horseshoe map. (d) By intersecting the horizontal with the vertical rectangles, one obtains a cover of the invariant set Λ by cylinders $C[ij.kl]$.

are ordered in exactly the same way along the horizontal and vertical sides of S, respectively. In fact, they follow the unimodal order that governs the tent and logistic maps: This is a signature of the folding process (compare with Fig. 2.14).

In general, cylinders of the type $C[s_{-N} \dots s_{-1}.s_0 \dots s_{N-1}]$ will be contained in the 2^{2N} intersections of 2^N vertical rectangles with 2^N horizontal rectangles. Without loss of generality, the action of the horseshoe map can be chosen to be piecewise-linear in the regions of interest, with expansion and contraction rates $\lambda_u > 2$ and $\lambda_s < \frac{1}{2}$. In that case, the vertical rectangles $V_{s_0 \dots s_{N-1}}$ will be of width $1/\lambda_u^N$ while the height of the horizontal rectangles $H_{s_{-N} \dots s_{-1}}$ will be λ_s^N. It is easy to see that both sizes shrink to zero in the limit $N \to \infty$, and thus that the cylinder $C[s_{-N} \dots s_{-1}.s_0 \dots s_{N-1}]$ converges to a unique point, as required for a generating partition.

This shows that the invariant set Λ is the product of a Cantor set of vertical lines with a Cantor set of horizontal lines, and that it is in 1:1 correspondence with the set of bi-infinite binary sequences. It can even be shown that this correspondence is a homeomorphism if the space of symbolic sequences is equipped with the following metric:

$$d(\{s_k\}_{k\in\mathbb{Z}}, \{s'_k\}_{k\in\mathbb{Z}}) = \sum_{i=-\infty}^{\infty} \frac{\delta_i}{2^{|i|}} \qquad \delta_i = \begin{cases} 0 & \text{if } s_i = s'_i \\ 1 & \text{if } s_i \neq s'_i \end{cases} \tag{2.110}$$

which means that the closer two sequences are according to this metric, the closer the associated points are. This property will be useful in the following.

2.10.3 Dynamical Properties

A few dynamical properties of the horseshoe map can be deduced from the discussion of previous sections. Some of them follow directly from the correspondence with a full shift, as discussed in Section 2.6.5: sensitivity to initial conditions, existence of a dense set of periodic orbits, existence of dense orbits, and so on. Others are linked to the coexistence of two complementary mechanisms, stretching and squeezing, and to the horseshoe map being two-dimensional. In particular, a foliation of Λ can be constructed in the following way.

Consider a point X and the vertical line $\mathcal{V}(X)$ passing through X. From the results of Section 2.10.2, it is easy to see that X and any point $Y \in \mathcal{V}(X)$ have the same forward sequence $\Sigma_+(X) = \Sigma_+(Y)$ (the vertical line is the limit of a sequence of vertical rectangles). This implies that X and Y have the same future, more precisely that their orbits converge to each other in forward time: $d(f^n(X), f^n(Y)) \to 0$ as $n \to \infty$. Indeed, the two corresponding symbolic sequences converge to each other according to metric (2.110) as the discrepancies between them are pushed farther and farther in the backward sequence.

Let us define the stable and unstable manifolds $W_s(X)$ and $W_u(X)$ of a point X as the sets of points whose orbits converge to that of X in the future and in the past, respectively:

$$W_s(X) = \{Y \in \Lambda, \lim_{n\to\infty} d(f^n(X), f^n(Y)) = 0\} \tag{2.111a}$$

$$W_u(X) = \{Y \in \Lambda, \lim_{n\to\infty} d(f^{-n}(X), f^{-n}(Y)) = 0\} \tag{2.111b}$$

According to this definition, the vertical line $\mathcal{V}(X)$ is a segment of the stable manifold of X: $\mathcal{V}(X) \subset W_s(X)$.

Similarly, two points X' and Y' located along the same horizontal line $\mathcal{H}(X')$ have the same backward sequence, hence the same past. Their orbits converge to each other backward in time: $d(f^{-n}(X'), f^{-n}(Y')) \to 0$ as $n \to \infty$. Thus, they belong to the same segment of unstable manifold $W_u(X') = W_u(Y')$.

Hence Λ is foliated by horizontal and vertical lines that are segments of stable and unstable manifolds, respectively. The astute reader may have noticed that this could have been obtained directly from the geometrical description of the horseshoe map. However, it is in general not straightforward to determine these segments for a general two-dimensional invertible map. Because the discussion above is based on symbolic dynamical properties only, it can be extended directly to a large class of dynamical systems: As soon as a symbolic encoding of the invariant set is available, segments of stable and unstable manifolds can be obtained simply by following points with identical forward or backward sequences, respectively.

We conclude this section with an important remark regarding how noninvertibility can be embedded in an invertible dynamics. As discussed above, each vertical line $\mathcal{V}$ is associated with a unique forward sequence $\Sigma_+(\mathcal{V})$ shared by all the points in the line. Obviously, the images of all these points will have the same forward sequence: The image of the line by the horseshoe map is thus another vertical line $f(\mathcal{V})$, whose forward sequence is $\Sigma_+(f(\mathcal{V})) = \sigma\Sigma_+(\mathcal{V})$.

If $\mathcal{V}(\Lambda)$ represents the complete set of vertical lines (or equivalently, of classes of points with identical forward sequences), the horseshoe map induces a map f_V : $\mathcal{V}(\Lambda) \to \mathcal{V}(\Lambda)$. Because forward sequences are one-sided, the map f_V is intrinsically noninvertible; in fact, it is conjugate to the tent map since we know that forward sequences are ordered according to the unimodal order. In the invertible horseshoe map, there is thus *an underlying noninvertible map that organizes the dynamics*.

The clue to this paradox is that the shift operator acts on symbol sequences by transferring one symbol from the forward sequence to the backward sequence. The forward sequence loses information (i.e., the associated dynamics is noninvertible), gained by the backward sequence (whose dynamics thus depends on the forward sequence), while the net information flow is zero for the complete sequence (i.e., the global dynamics is invertible). Again, we see that this observation applies to a large class of dynamical systems, and it will be one of the building blocks of this book.

To conclude, let us emphasize that the two-sidedness of symbolic sequences for the horseshoe map is not due directly to the dynamics being two-dimensional. Rather, it originates in the distinction between stable and unstable subspaces, regardless of their dimension. For example, one could imagine that forward (resp., backward) sequences parameterize a two (resp., one)-dimensional subspace.

2.10.4 Variations on the Horseshoe Map: Baker Maps

The tent map is the simplest 1D map that is conjugate to a full one-sided 2-shift. To illustrate horseshoe dynamics, we now present examples of piecewise-linear maps that are conjugate to *two-sided* full shifts. These maps are the two-dimensional counterparts of the tent map. As we shall see, the stretching and squeezing mechanisms at play in the horseshoe map can be organized in a few inequivalent ways.

For definiteness, we restrict ourselves to bijections of the unit square into itself. Our first example, shown in Fig. 2.22(a), is a deformation of the horseshoe map of Fig. 2.19. It is easily seen that this map can be described by two linear maps defined on the domains $x < \frac{1}{2}$ and $x > \frac{1}{2}$. More precisely, the coordinates (x', y') of the image of a point (x, y) are given by

$$x' = 1 - 2|x - \frac{1}{2}|, \quad y' = \begin{cases} \frac{y'}{2} & \text{if } x \leq \frac{1}{2} \\ 1 - \frac{y'}{2} & \text{if } x > \frac{1}{2} \end{cases} \tag{2.112}$$

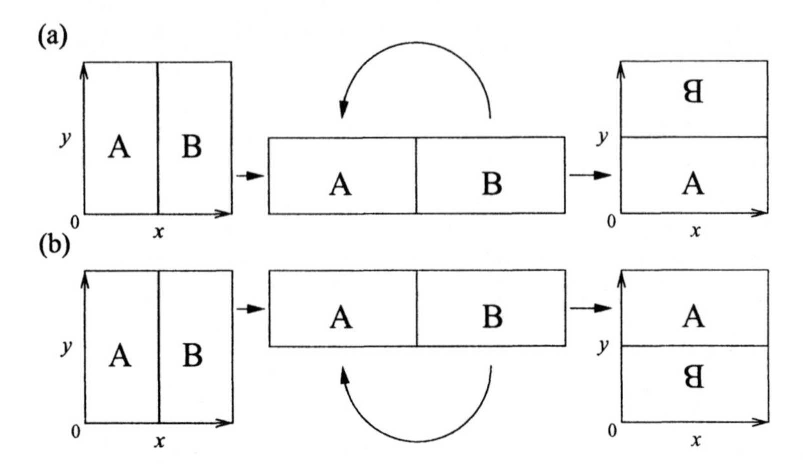

Fig. 2.22 Two topologically distinct horseshoe-like maps that differ in the folding process.

Note that while the 1D tent map is continuous but not differentiable at the critical point, the 2D piecewise-linear map (2.112) has necessarily a discontinuity, here at the line $x = \frac{1}{2}$. More importantly, Eqs. (2.112) are our first example of a *reducible* system. Indeed, they have the following structure:

$$x' = f(x) \tag{2.113a}$$

$$y' = g(x, y). \tag{2.113b}$$

The key property of (2.113) is that it has a lower-dimensional subsystem (2.113a) which can be iterated independently: The dynamics of the x variable depends only on x itself. From the discussion of previous section, this is not surprising. Indeed, the y direction is the stable direction and vertical lines $x = c$ are segments of the stable manifold. Since the vertical lines parameterized by x are mapped to other vertical lines, the original 2D map induces a 1D map $x' = f(x)$. As discussed before, this map is intrinsically non-invertible: An invertible chaotic dynamics in the global state space is associated with a non-invertible dynamics in the unstable space.

This structure allows one to represent the action of the map (2.112) by plotting the graphs (x, x') and (y, y'), as shown Fig. 2.23. We see that the chaotic two-dimensional map (2.112) is associated with two singular one-dimensional maps (f_u, f_s) such that $x' = f_u(x)$ and $y = f_s(y')$. The existence of f_u has been discussed previously and is a signature of the folding process. The existence of f_s follows from the fact that the map (2.112) is invertible and that for the inverse map, stable and unstable spaces are exchanged. Hence, the inverse map is associated with 1D maps (f_s, f_u). Note that f_s is not arbitrary: Each monotonic branch of f_s corresponds to a given branch of f_u. Therefore, both maps have the same number of branches.

Figures 2.22(b) and 2.23(b) display another map that is extremely similar to our first example, except that folding takes place in the other direction. We note that the two maps are associated with the same map in the unstable space, *but not in the stable*

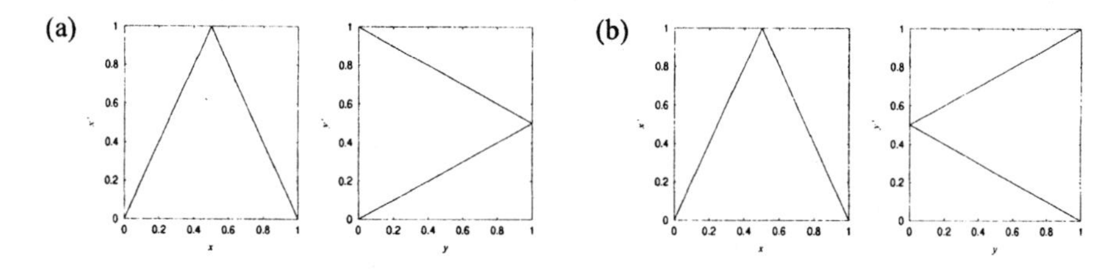

Fig. 2.23 Graphs of the noninvertible one-dimensional maps associated with the invertible maps of Fig. 2.22.

space. They are topologically distinct. These two topologically classes will later be referred to as the horseshoe and the *reverse* (or *twisted*) horseshoe. We shall see in Chapter 7 that both classes can be observed experimentally.

In these examples, reducibility is clearly apparent because the unstable and stable direction are fixed over the state space. This makes it easy to find coordinate systems whose axes are parallel to these directions. As the example of the Hénon map in the next section will show, this is not the case in general. However, it is tempting to conjecture that the equations describing a chaotic system can be always brought, at least approximately, to a reducible form (2.113) by a suitable change of coordinates. As noted previously, this should be the case if a good symbolic description of the system is available: Points which have identical forward (resp., backward) sequences should be on the same segment of the stable (unstable) manifold.

To elaborate on this classification, we display in Fig. 2.24 a few examples of piecewise-linear maps with three branches. The associated pairs of non-invertible maps are also shown. We note that all three maps have the same map in the unstable space. Maps in Figs. 2.24(a) and 2.24(c) have apparently the same map in the stable space. However, the way in which branches of the stable map are associated to branches of the unstable map differ. Thus, we see that a two-dimensional piecewise-linear map such as shown in Figs. 2.22 and 2.24 can be specified by: (1) the map f_u in the unstable space, and (2) the permutation indicating the order in which different branches are encountered along the stable direction (how they are stacked). This observation will help us to understand the construction of templates in Chapter 5.

In these examples, reducibility is clearly apparent because the unstable and stable direction are fixed over the state space. This makes it easy to find coordinate systems whose axes are parallel to these directions. As the example of the Hénon map shows (Section 2.11) shows, this is not the case in general. However, it is tempting to conjecture that the equations describing a chaotic system can be always brought, at least approximately, to a reducible form (2.113) by a suitable change of coordinates. The independent subsystem should then describe the dynamics in the unstable space. As noted previously, this should hold when a good symbolic description of the system is available: Points which have identical forward (resp., backward) sequences should be on the same segment of the stable (unstable) manifold. Again, this idea will return in Chapter 5.

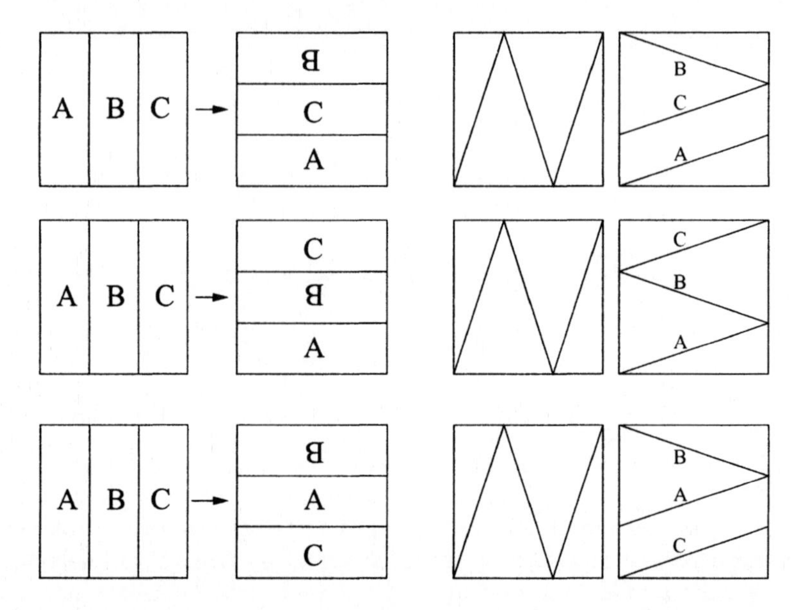

Fig. 2.24 Examples of piecewise-linear maps on three domains that are topologically distinct. The associated 1D maps are shown on the right hand side.

2.11 HÉNON MAP

2.11.1 A Once-Folding Map

The Smale's horseshoe is the classical example of a *structurally stable* chaotic system: Its dynamical properties do not change under small perturbations, such as changes in control parameters. This is due to the horseshoe map being *hyperbolic* (i.e., the stable and unstable manifolds are transverse at each point of the invariant set). Recall that the study of the symbolic dynamics of the horseshoe map involved horizontal and vertical rectangles used to define cylinders (Section 2.10.2). If these horizontal and vertical rectangles intersect transversely everywhere, intersections are preserved if the rectangles are slightly deformed: Thus, the symbolic dynamics remains described by a full shift. Consequently, every map sufficiently close to the standard horseshoe map has the same spectrum of periodic orbits.

This is usually not the case for real systems, where periodic orbits are created or destroyed as a control parameter is varied. This is well illustrated by the example of the logistic map studied in Section 2.7. As the control parameter a is increased from 0 to 2, this one-dimensional system displays a gradual transition from a perfectly ordered state to a chaotic behavior that is associated with a binary symbolic dynamics and thus is asymptotically as random as coin tossing.

Similarly, the Hénon map [48]

$$x_{n+1} = a - x_n^2 + y_n \tag{2.114a}$$
$$y_{n+1} = bx_n \tag{2.114b}$$

is undoubtly the most widely studied invertible two-dimensional map. It displays bifurcation sequences that lead to a complete horseshoe at large values of the parameter a. Since the Jacobian $\partial(x_{n+1}, y_{n+1})/\partial(x_n, y_n) = -b$, the parameter b admits of a simple interpretation: The area of an arbitrary region of the plane is reduced by a factor of $-b$ under the action of the map. When $|b| = 1$, areas are preserved and the map is said to be *conservative*. When $|b|$ is decreased to zero, the map becomes more and more dissipative. At $b = 0$, the y variable is forced to zero. The system then reduces to the logistic map, thereby becoming non invertible.

If Eqs. (2.114) are iterated at the usual parameter values $(a, b) = (1.4, 0.3)$ (which will be used throughout this section), and if successive points of a typical orbit are plotted in the (x, y) plane, the familiar picture shown in Fig. 2.25(a) is obtained. It has been widely used to illustrate the fractal structure of chaotic attractors. While few other chaotic systems have been more deeply studied, a number of questions about its dynamics have not received a final answer, such as: How to construct a symbolic encoding? Or: What are all the possible routes to a complete horseshoe? These questions are discussed in Section 2.11.2 and Chapter 9.

Interestingly, the Hénon map was originally introduced as a simplified model of a Poincaré map, but is orientation-reversing at the parameters given above (i.e., its Jacobian is negative). However, this is impossible for the Poincaré map of a flow, which can be viewed as resulting from a smooth deformation of the plane continuously connected to the identity. For most properties (fractal dimensions, Lyapunov exponents), this makes no difference. However, there are topological measures that distinguish the orientation-preserving Hénon map from the orientation-reversing one [49].

The action of the Hénon map on points of the plane is represented graphically in Fig. 2.25(b). The folded dark curve is the image of the enclosing square and shows that the Hénon map is a once-folding map. The lighter curve corresponds to the second iterate of this square and is accordingly folded twice. This figure illustrates the similarity of the Hénon map with the hyperbolic horseshoe but also shows a crucial difference. At the parameter values considered, the intersection of the square enclosing the invariant set and of its image is in one piece whereas it had two disjoint components in the case of the horseshoe map (Fig. 2.19).

Since this natural partition of state space in two regions was the keystone of our study of the symbolic dynamics of the horseshoe map, it is not obvious whether there is a symbolic coding for the Hénon map and how it can be constructed. We discuss this important question in Section 2.11.2. We shall see that it is directly connected to the nonhyperbolicity of the Hénon attractor (i.e., to the the lack of transversality of certain intersections between the stable and unstable manifolds).

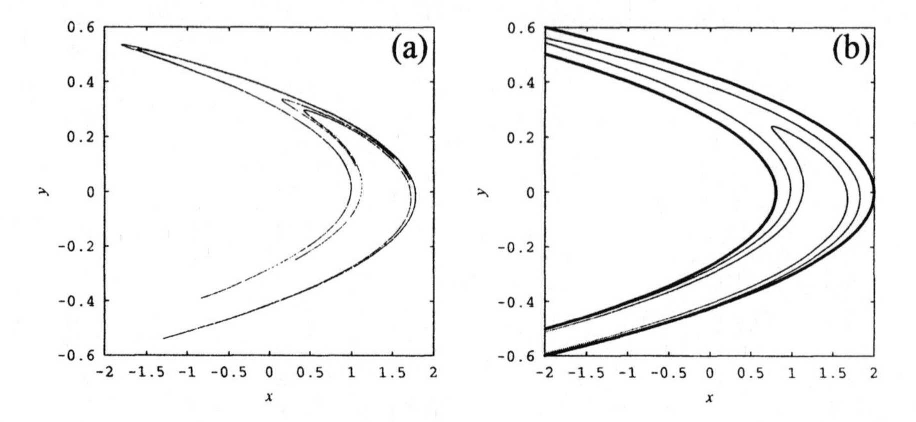

Fig. 2.25 (a) The Hénon attractor obtained with Eqs. (2.114) for $a = 1.4$, $b = 0.3$. (b) to illustrate the action of the map, the first and second iterate of the square $[-2, 2] \times [-0.6, 0.6]$ have been plotted, showing that the Hénon map is a once-folding map.

2.11.2 Symbolic Dynamics of the Hénon Map

The relevance of symbolic codings relies heavily on the mixing property of chaotic dynamics (Section 2.7.2). Assume that we have an arbitrary partition $M = M_1 \cup M_2$ of phase space. If a map f is mixing, there exists for any region $N \in M$ an integer $n_0(M)$ such that $f^{n_0}(N) \cap M_1 \neq \emptyset$ and $f^{n_0}(N) \cap M_2 \neq \emptyset$. However small a region N is, there exist points in N with different symbolic itineraries, and thus N can always be divided into smaller regions. Since this operation can in principle be repeated indefinitely, it could be tempting to believe that well-defined symbolic codings exist for any choice of the partition.

However the example of the logistic map in Section 2.7 has shown that this is not true. The weakness in the argument above is that mixing cannot be invoked if the orbits of two points are strongly correlated for arbitrarily long times. In particular, we noted in our study of the logistic map that the different preimages of a given point must be associated with different symbols. This rule is indeed necessary to prevent the symbolic itineraries of such points from being identical, since their orbits differ only in their initial conditions. If the map under study is expansive, separating preimages is actually sufficient for obtaining generating partitions, as was shown in Section 2.7.2.

However, this criterion cannot be applied to the Hénon map; since it is invertible, every point has one and only one preimage. Thus, we have to extend the simple rule used for the logistic map to handle the case of invertible chaotic maps. The extension should also be consistent with the coding of the hyperbolic horseshoe map.

The key point is that while the orbits of two different points cannot collapse onto each other in finite time if the map is invertible, they can become indiscernible after an infinite number of iterations: Two different points can have the same asymptotic future or the same asymptotic past (or both), as we see below.

In Fig. 2.26(a), we have plotted segments of the local stable and unstable manifolds of the fixed point $X^* \sim (0.884, 0.265)$. We denote these segments by W_s^l and W_u^l, respectively. Recall that the stable manifold $W_s(X^*)$ and unstable manifold $W_u(X^*)$ consist of the points whose orbits converge to the fixed point X^* under repeated action of the Hénon map f and of the inverse map f^{-1}, respectively:

$$x \in W_s(X^*) \quad \Leftrightarrow \quad \lim_{n \to +\infty} ||f^n(X) - X^*|| = 0, \tag{2.115a}$$

$$x \in W_u(X^*) \quad \Leftrightarrow \quad \lim_{n \to +\infty} ||f^{-n}(X) - X^*|| = 0 \tag{2.115b}$$

Obviously, the stable and unstable manifolds are invariant under both f and f^{-1}, hence the restrictions of these maps to the invariant manifolds behave like one-dimensional diffeomorphisms. Since the eigenvalues of the Jacobian at X^* along the stable and unstable directions are $0.156 > 0$ and $-1.92 < 0$, f and f^{-1} preserve (reverse) orientation along the stable (unstable) manifold. The action of f^{-1} on the stable segment W_s^l and that of f on the unstable segment W_u^l are illustrated in Figs. 2.26(b) to 2.26(f), which display $f^{-n}(W_s^l))$ and $f^n(W_u^l)$ for $n = 1, 2, 3, 4$, and 8, respectively. The two segments are stretched at each step and span a longer and longer part of the two invariant manifolds[5]. They are also folded at each iteration, a signature of the horseshoe-like process which organizes the Hénon attractor. For exemple, Fig. 2.26(c) can be compared with Fig. 2.21 (keeping in mind that, unlike the classical horseshoe map, the Hénon map is orientation-reversing along the unstable manifold).

An extremely important consequence of the folding mechanism is that the stable and unstable manifolds of the fixed point X^* intersect themselves not only at X^* but also at other locations, called *homoclinic points*. Since homoclinic points belong to both the unstable and the stable manifolds, their orbits converge to the fixed point both forward and backward in time. The stable and unstable manifolds being invariant, the set of homoclinic points is itself invariant: All images and preimages of a homoclinic point are themselves homoclinic points. More generally, two points are said to be homoclinic to each other if their orbits converge to each other both forward and backward in time, with "homoclinic points" denoting points that are homoclinic to a fixed point. The loop formed by the pieces of the stable and unstable manifolds joining X^* and P is called a homoclinic loop.

Fig. 2.26 also shows that, since the segments $f^{-n}(W_s^l)$ and $f^n(W_u^l)$ are increasingly folded as the iteration number n increases, the number of intersections $N_i(n)$ between the two segments (and thus of apparent homoclinic points) increases exponentially with n. However, it does not grow as fast as could have been expected from the study of the horseshoe map: $N_i(1) = 4$, $N_i(2) = 12$ and $N_i(3) = 30$ instead of

[5]This is how the segments W_s^l and W_u^l of this example have themselves been obtained, by iterating repeatedly from extremely small linear segments located around the fixed points and aligned along the stable and unstable direction at that point.

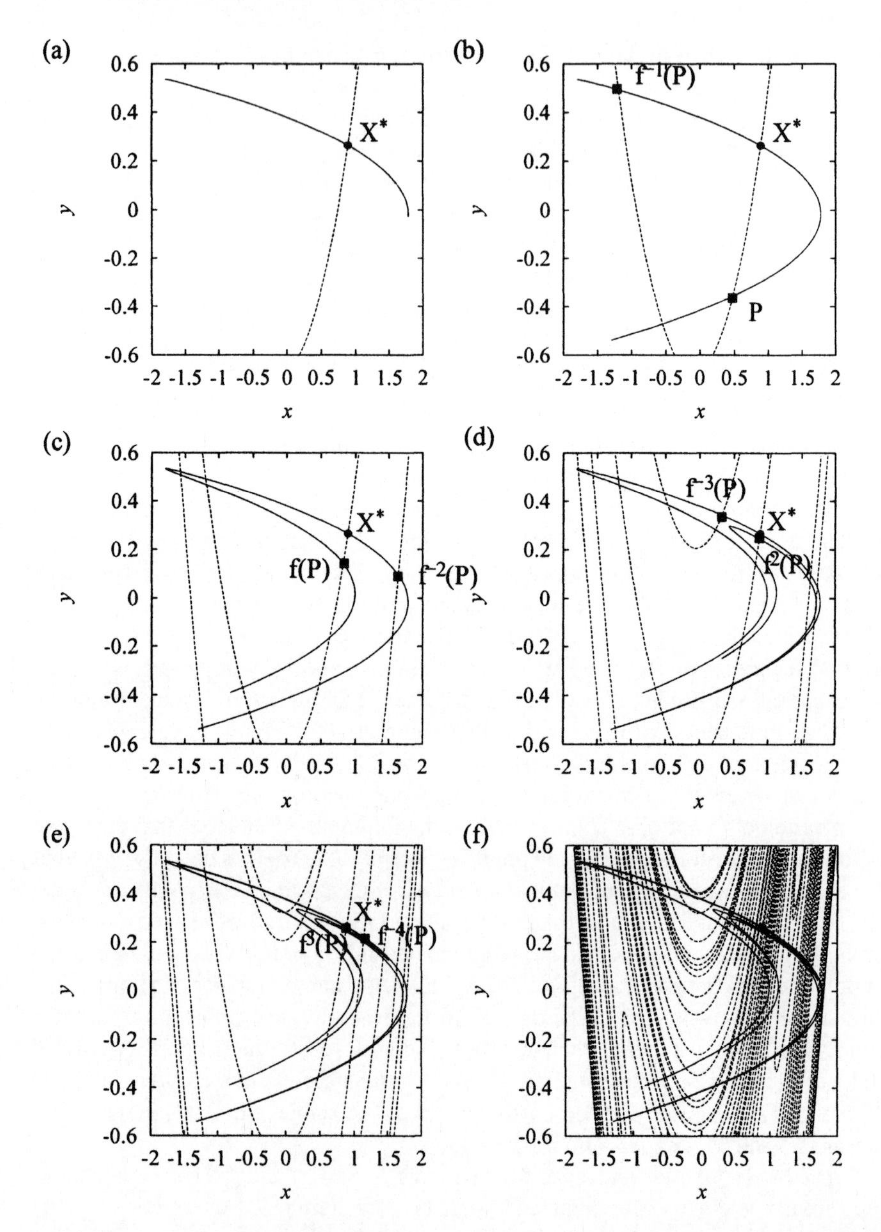

Fig. 2.26 (a) The solid (dotted) line represents a segment W_u^l (W_s^l) of the unstable (stable) manifold of the fixed point located at their intersection. The orbits of points belonging to this line converge to the fixed point backward (forward) in time. (b)-(e) Images $f^n(W_u^l)$ and preimages $f^{-n}(W_s^l)$ of the segments W_u^l and W_s^l for $n = 1, 2, 3, 4$. They are folded at each iteration and the number of intersections grows exponentially. (f) This is the same picture after 8 iterations. The image of the unstable segment becomes indiscernible from the strange attractor. One clearly notices locations where the stable and unstable manifolds are tangent. These degenerate intersections are called homoclinic tangencies.

$N_i(n) = 4^n$ for the horseshoe. This is an indication that the symbolic dynamics is not described by a full 2-shift.

In Fig. 2.26(b), we have distinguished two of the three homoclinic points present at that stage. By studying how f and f^{-1} act on the segments shown in Figs. 2.26(a) and 2.26(b), it is easy to see that one point (denoted by P) is the image of the other [indicated as $f^{-1}(P)$]. Images and preimages of P are easily located in Figs. 2.26(b)-(e). Since they are themselves homoclinic points, they can be found among the intersections that appear in the consecutive plots of Fig. 2.26. For example, $P = W_s^l \cap (f(W_u^l) \setminus W_u^l)$, hence $f(P) \in W_s^l \cap (f^2(W_u^l) \setminus f(W_u^l))$, which consists of two points. These two points are the images of P and of the unlabeled homoclinic point shown in Fig. 2.26(b). Since the latter is further from X^* than P and since f is continuous, $f(P)$ is the point of $W_s^l \cap (f^2(W_u^l) \setminus f(W_u^l))$ which is closest to X^* [Fig. 2.26(c)]. Determining the positions of all the images and preimages $f^i(P)$ shown in Figs. 2.26(b) to Figs. 2.26(e) is an interesting exercise left to the reader.

We see that the sequence $\{P, f(P), f^2(P), f^3(P), \ldots\}$ converges to the fixed point X^* along the stable manifold, approaching it from one side. Similarly, the sequence $\{P, f^{-1}(P), f^{-2}(P), f^{-3}(P), \ldots\}$ converges to the fixed point along the unstable manifold, alternating between the two sides of the fixed point and with a smaller convergence rate since the unstable eigenvalue is negative and has a modulus closer to one than the stable eigenvalue. Thus, the orbits of X^* and P are strongly correlated both in the past and in the future, and some care has to be taken in order to prevent the two points from being associated with the same symbolic itinerary.

Before addressing this problem, a few remarks are in order. First, we note that the existence of homoclinic points forces the unstable manifold to be infinitely folded. On the one hand, all the iterates $f^i(P)$ belong to the unstable manifold, by definition. On the other hand, it can be seen in Figs 2.26(b)-(e) that they all belong for $i > 0$ to the segment of the stable manifold joining P and X^* (f is contracting along the stable manifold). Thus, the unstable manifold intersects this segment infinitely many times and is infinitely folded, which is consistent with what Fig. 2.26 shows. The same argument can be applied to the stable manifold. This suggests that homoclinic points are associated with a complicated dynamics. As a matter of fact, it can be proven that if a dynamical system displays homoclinic points, then it has an invariant set on which the dynamics is chaotic and there are periodic orbits arbitrarily close to the homoclinic point [38].

Fig. 2.26(f) shows the segments $f^8(W_u^l)$ and $f^{-8}(W_s^l)$. At the resolution of the plot, they provide good approximations of the unstable and stable manifolds of the fixed point. Two important properties are illustrated by this picture. The first one is that the unstable segment $f^8(W_u^l)$ is indistinguishable from the strange attractor shown in Fig. 2.25. Indeed, it is believed that the strange attractor is nothing but the closure of the unstable manifold $W_u(X^*)$. Figure 2.26 thus provides an illustration of how strange attractors are built and of their hierarchical structure. The second property is that although the intersections between the unstable and stable manifolds are mostly transverse, places where the two manifolds are tangent to each other can be clearly discerned. These points, where it is not possible to define distinct stable and unstable directions, are called *homoclinic tangencies*. Their existence is a crucial difference between the horseshoe map and the Hénon map: The former is hyperbolic, the latter is not. Note that images or preimages of a homoclinic tangency are themselves homoclinic tangencies. Homoclinic tangencies play a foremost role in the problem of coding the Hénon attractor, which we now return to.

Assume that the fixed point X^* lies in the region $\mathcal{P}_k$ of the partition. Its symbolic itinerary $\Sigma(X^*)$ then consists of the symbol s_k infinitely repeated. A necessary and sufficient condition for the points P and X^* to have different symbolic itineraries is thus that at least one iterate $f^i(P)$ is separated from X^* by the border of the partition, so that it is associated with a symbol $s_{k'} \neq s_k$. Note also that if $\Sigma(P) \neq \Sigma(X^*)$, then obviously $\Sigma(f^i(P)) \neq \Sigma(X^*)$ so that the problem is solved at once for the whole orbit of homoclinic points based on P.

For the sake of simplicity, assume that we search for a partition where P itself and X^* belong to different regions. This is a natural choice, because P is more distant from X^* than other points of its orbit, most of them being extremely close to X^*. Furthermore, P is one of the three intersections in Fig. 2.26(b), a first-generation homoclinic point so to speak. To design a simple criterion for placing the border of the partition, we need a quantity that distinguishes P from X^* (and more generally two points that are homoclinic to each other), that is related in a natural way to the geometry of the invariant manifolds, and that makes it easy to define a "middle point".

We note that since X^* and P in Fig. 2.26(b) form an homoclinic loop, the unstable manifold crosses the stable manifold in different directions at X^* and P. This can be used to distinguish the two points in a robust way. To this end, we consider at each point X the angle $\beta(X)$ between the unstable and stable manifolds. This is a signed quantity because the two manifolds can be oriented. For example, assume that the two stable and unstable segments going from X^* to P are positively oriented. Then, an oriented frame $(\vec{v}_s(X), \vec{v}_u(X))$ can be defined at each point, with the two vectors being tangent to the stable and unstable manifold, and positively oriented. The quantity $\beta(X)$ is then defined as the oriented angle between $\vec{v}_s(X)$ and $\vec{v}_u(X)$, with $\beta(X) \in [-\pi, \pi]$.

The key observation is that $\beta(X)$ varies continuously in the plane, as do the unstable and stable directions[6], and that $\beta(P)$ and $\beta(X^*)$ have necessarily opposite signs because X^* and P are connected by a homoclinic loop [Fig. 2.27(a)]. Thus, any path going from X^* to P has a point where $\beta(X) = 0$. Such a point is nothing but one of the homoclinic tangencies we have mentioned earlier. This observation is a two-dimensional version of the theorem stating that two zeroes of a continuous function $g : \mathbb{R} \to \mathbb{R}$ must be separated by a zero of its derivative, i.e., by a singularity.

Now, we note that around a homoclinic tangency, there are arbitrarily small homoclinic loops (Fig. 2.27). Thus, all the corresponding pairs of homoclinic points will be encoded with unique itineraries if and only if the partition border passes exactly through the homoclinic tangency.

[6] In the pictures shown in this section, we have shown only the unstable manifold of the fixed point X^*. This manifold is confined to the strange attractor. However, a stable and an unstable direction can be determined at each point of the plane using definitions (2.115). Thus, $\beta(X)$ is well-defined everywhere, inside or outside of the strange attractor.

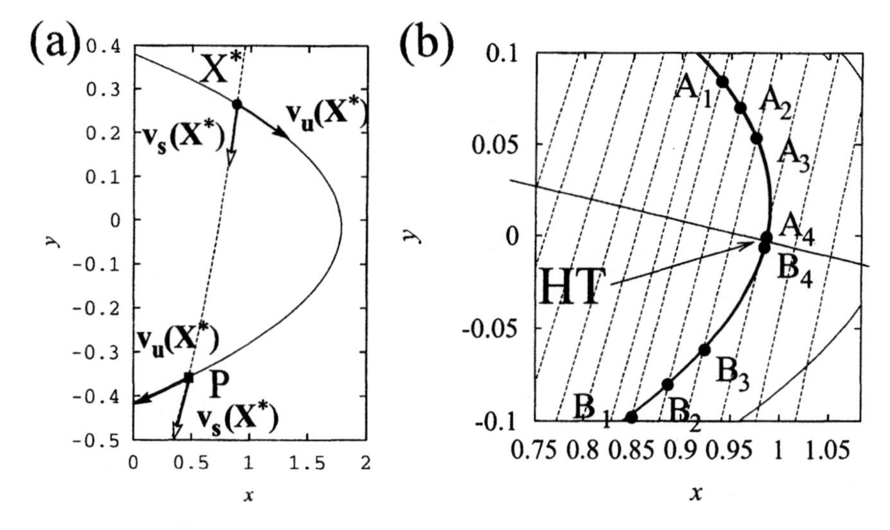

Fig. 2.27 (a) Illustration of the definition of the oriented angle $\beta(X)$. A two-dimensional frame $v_s(X), v_u(X)$ is attached to each point of the plane, each vector being tangent with one of the two invariant manifolds and positively oriented. The angles at two points connected by a homoclinic loop have necessarily opposite signs. (b) Magnified view of the neighborhood of a homoclinic tangency HT. One can find pairs (A_i, B_i) of points homoclinic to each other arbitrarily close to the homoclinic tangency. Consequently, the border of the partition (shown here as a straight line) must pass through the homoclinic tangency in order to separate each pair.

Accordingly, it has been conjectured by Grassberger and Kantz that a generating partition for a nonhyperbolic system can be obtained by connecting homoclinic tangencies [50, 51]. This generalizes the procedure for coding one-dimensional maps, with the singularity located at the homoclinic tangency replacing the singularity located at the critical point, and pairs of homoclinic points replacing preimages of a given point. This conjecture has not been proven yet, but extensive numerical evidence that it yields generating partitions has been given over the years [51–57]. The simplest partition that can be obtained in this way for the Hénon attractor is shown in Fig. 2.28. Note that with this partition, the itinerary of P is $\Sigma(P) = \cdots 111.0111 \cdots$ and differs from $\Sigma(X^*)$ by only one symbol.

However, there are some difficulties with this approach. Individual lines defined by the equation $\beta(X) = 0$ usually not make a satisfactory partition border. In fact, one generally finds that several such lines have to be followed, the connection between two different lines occurring outside the attractor. As a result, it is not always obvious which homoclinic tangencies to connect. Various criteria have been proposed to overcome this ambiguity (see, e.g., [54, 57]), but no definitive solution has been found so far, to our knowledge. Another problem is that there is a dramatic noise amplification precisely at homoclinic tangencies [56], which can make their extraction from experimental time series difficult. However, the folding and squeezing processes

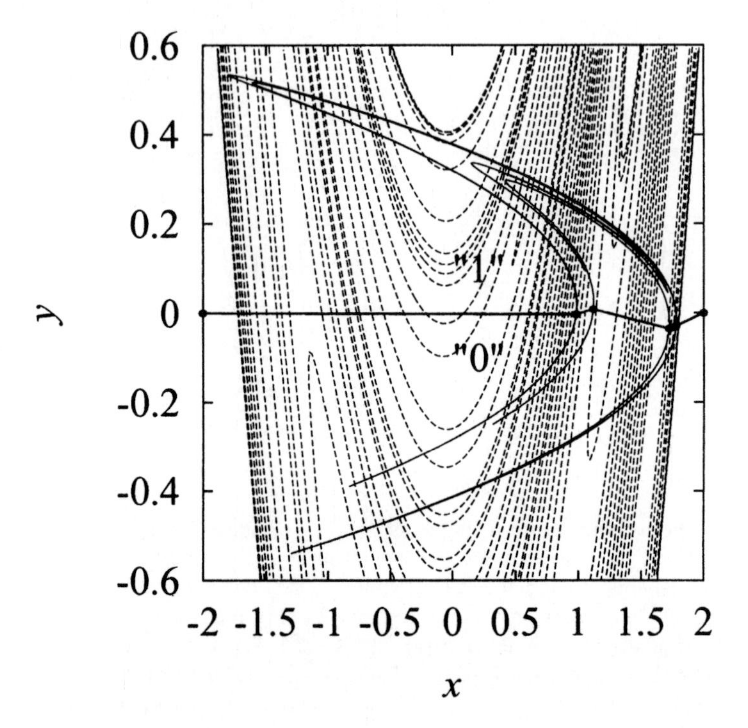

Fig. 2.28 A generating partition for the Hénon map constructed by connecting homoclinic tangencies.

which build the strange attractor not only create localized singularities (homoclinic tangencies) but also determine the global structure of the attractor. We shall see later that analyzing the topological organization of a chaotic attractor provides detailed information about its symbolic dynamics. In particular, the symbolic name of many low-period orbits can be determined directly from its topological invariants. This has been used to design methods for contructing generating partitions that are based on topological analysis and do not rely on the structure of homoclinic tangencies [49,58]. These methods will be outlined in Section A.4.

2.12 CIRCLE MAPS

2.12.1 A New Global Topology

In Section 2.9.4 we saw that the global topology of phase space can have dramatic consequences: A period-3 orbit forces orbits of every period if it belongs to a map of an interval into itself, none if the state space is two-dimensional or is the unit circle. This indicates that qualitatively different behaviors can appear when phase space topology is changed. Accordingly, this section is devoted to a brief review of dynamical properties of maps from the unit circle S^1 into itself. If S^1 is parameterized with an angular variable $\theta \in [0,1]$, these maps can be written as $\theta_{n+1} = f(\theta_n) \pmod 1$.

Physically, the study of circle maps is motivated by the problem of coupled oscillators. Assume that we have two systems oscillating on periodic cycles at frequencies ν and ν', respectively. The state of each oscillator can be described by an angular variable $\theta(t) = \nu t \pmod 1$. In the spirit of Poincaré sections, let us sample these angles stroboscopically at the frequency ν' so that we need only measure the successive samples $\theta_n = \theta(t_0 + n/\nu')$ of the first angle, given by

$$\theta_{n+1} = (\theta_n + w) \pmod 1 \tag{2.116}$$

where $w = \nu/\nu'$. The map (2.116) describes a rotation by a fraction w of a full turn per sampling period and is denoted $R(w)$ in the following.

Two different qualitative behaviors can occur depending on the value of w. If w is a rational p/q with $p, q \in \mathbb{Z}$, we have that $\theta_{n+q} = \theta_n + qw \pmod 1 = \theta_n$: The dynamical regime is a periodic orbit, and θ_n takes only a finite number of different values. If w is irrational, the sequence $\{\theta_n\}$ fills densely the interval $[0,1]$. This is a *quasiperiodic* regime and corresponds to the superposition of two incommensurate frequencies.

2.12.2 Frequency Locking and Arnol'd Tongues

It is known that the set of rational numbers is dense in $[0,1]$ but that it has zero measure: The frequency ratio of two uncoupled oscillators is irrational with a probability of 1, even if one can find rational values arbitrarily close. However, this changes as soon as a coupling is introduced. One then observes *frequency locking*: The frequency ratio of the two oscillators remains fixed at a rational value p/q in a finite range $w \in [p/q - \Delta\rho_1, p/q + \Delta\rho_2]$.

To study this phenomenon, the following circle map was introduced by Arnol'd [59]:

$$\theta_{n+1} = [\theta_n + w + \frac{K}{2\pi}\sin(2\pi\theta_n)] \pmod 1 \tag{2.117}$$

which features a nonlinear coupling characterized by its strength K. Figure 2.29 displays the graph of the map obtained for $(w, K) = (0.47, 0.8)$.

To describe the asymptotic regimes of (2.117), one introduces the *rotation number* [38,40,60]

$$\rho = \lim_{N\to\infty} \frac{1}{N} \sum_{i=0}^{N-1} \Delta\theta_n \quad \text{with } \Delta\theta_n = [w + K/2\pi \sin(2\pi\theta_n)] \tag{2.118}$$

Note that $\rho = w$ in the limit $K = 0$. The structure of the function $\rho(w, K)$ thus provides insight into the phenomenon of frequency locking as the nonlinear coupling is increased.

When the circle map (2.117) is a homeomorphism of S^1 into itself (i.e., for $K \leq 1$), the following properties of the rotation number $\rho(w, K)$ can be established:

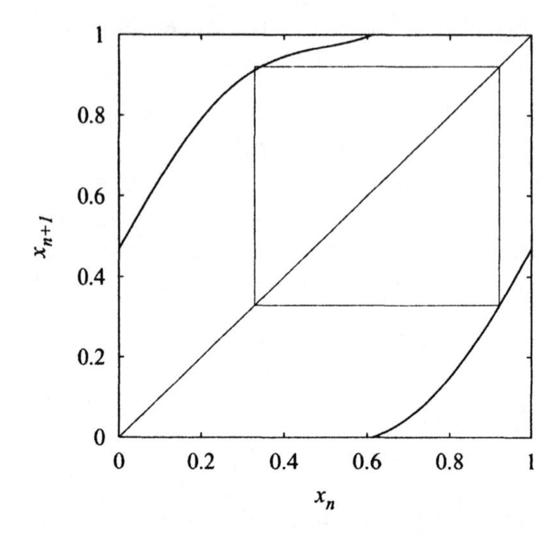

Fig. 2.29 Graph of the circle map (2.117) for $w = 0.47$ and $K = 0.8$. A period-2 orbit is also represented.

- The rotation number (2.118) does not depend on the orbit used to compute it.
- If $\rho(w, K)$ is irrational, the circle map is equivalent to the pure rotation $R(\rho)$; the motion is quasiperiodic.
- If $\rho(w, K) = p/q$ with p and q relatively prime integers, the asymptotic regime is a periodic orbit of period q. The periodic points of this orbit are ordered along the unit circle as with the pure rotation $R(p/q)$.

Thus, the classification of dynamical behaviors of the Arnol'd map for $K \leq 1$ amounts to determining the parameter regions in the (w, K) plane where $\rho(w, K)$ is rational.

As a simple example, let us consider the region $\rho(w, K) = 0$, where the oscillator frequencies are locked to each other in a 1:1 ratio. The corresponding asymptotic regime is a fixed point $\theta_{n+1} = \theta_n$ whose location, according to (2.117), is given by the equation $w = -(K/2\pi)\sin\theta$. It is easy to see that for $w \in [-K/2\pi, +K/2\pi]$, there are two solutions, one of which is stable in the whole domain $0 \leq K \leq 1$, at least (Fig. 2.30). Indeed, the slope of the graph at the two intersections is positive (the function is monotonic) and must be lower than 1 at one of the intersections. Hence, there is a periodic orbit having multiplier $0 \leq \mu \leq 1$. For $w = \pm K/2\pi$, the graph of the map is tangent to the diagonal, indicating that the stable and unstable periodic orbits are created and destroyed through saddle-node bifurcations. Note that the width of the frequency-locking interval $\rho = 0$ increases linearly with K and corresponds at $K = 1$ to almost one-third of the possible values of w.

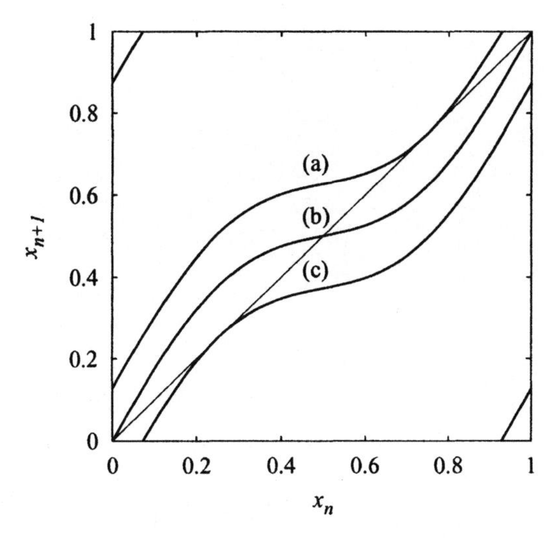

Fig. 2.30 Graph of the circle map for $K = 0.8$ and (a) $w = 0.8/2\pi$; (b) $w = 0$; (c) $w = -0.8/2\pi$.

By determining which regions of the (w, K) plane correspond to rotation numbers $\rho(w, K) = p/q$ with a small denominator q, the diagram shown in Fig. 2.31 is obtained for $q \leq 8$. The regions of frequency locking are called the *Arnol'd tongues*. Each of them corresponds to a different rational p/q, which governs the order in which they are encountered as w is increased at fixed K, since $\rho(w, K)$ is a monotonic function of w. As discussed above, tongues are bounded on both sides by saddle-node bifurcations where periodic orbits of the corresponding rotation number are created or destroyed.

It is interesting to note that the rotation numbers corresponding to the most important tongues can be classified according to a hierarchy based on an arithmetic operation on fractions. Indeed, it turns out that the principal tongue located between two tongues of rotation numbers p_1/q_1 and p_2/q_2 that satisfy $p_1q_2 - p_2q_1 = \pm 1$ is the one associated with the *Farey sum* of these two fractions, defined as follows: $p_1/q_1 \oplus p_2/q_2 = (p_1 + p_2)/(q_1 + q_2)$. Starting from the fundamental tongues $0/1$ and $1/1$, one first obtains the $1/2$ tongue. The latter is then separated from $0/1$ by $1/3$ and from $1/1$ by $2/3$. At the third level, one obtains $1/4, 2/5, 3/5, 3/4$, and so on. Tongues at a given level are wider than those at the next levels, as can be checked by visual inspection of Fig. 2.31.

As K is increased from 0 to 1, the relative proportions of the quasiperiodic and periodic regimes are exchanged. At $K = 0$, quasiperiodic regimes have a probability of 1, as mentioned above. Since there are an infinite number of tongues, it might not be obvious that the total length in w of the frequency-locked intervals goes to zero as $K \to 0$. That this is the case is due to the width $\Delta w(p/q)$ of the $\rho = p/q$ tongue decreasing sufficiently fast as $K \to 0$, more precisely as $\Delta w(p/q) \sim K^q$ or $\Delta w(p/q) \sim K^{q-1}$, depending on p/q.

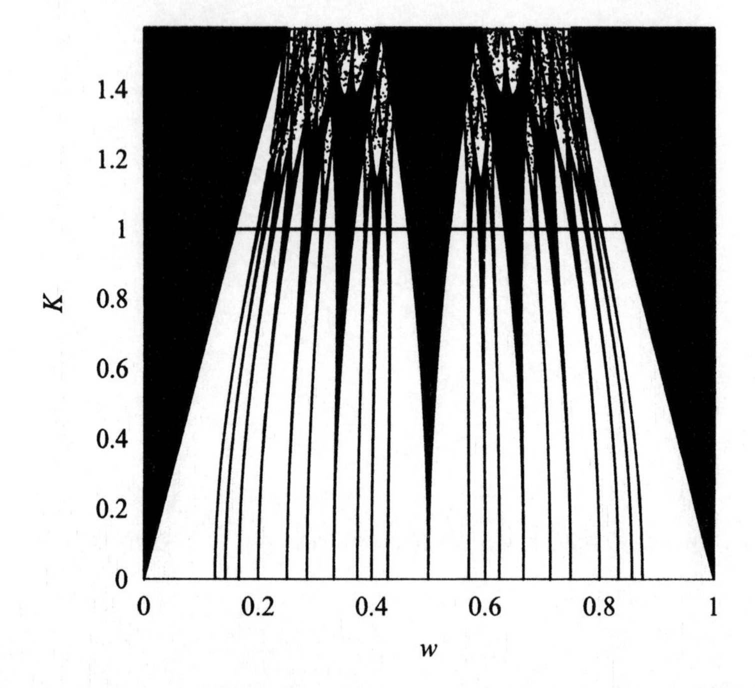

Fig. 2.31 Arnol'd tongues for the circle map (2.117) corresponding to rational rotation numbers $\rho(w, K) = p/q$ with $q \leq 8$.

At $K = 1$, values of w yielding quasiperiodic regimes are confined to a Cantor set of measure 0 and of fractal dimension $D \sim 0.87$; frequency-locked regimes have measure 1. The graph of the function $\rho(w, K = 1)$, shown in Fig. 2.32, has a very peculiar structure, known as a *devil's staircase*. It is continuous and monotonic but increases only where ρ is irrational: Each rational value occurs on a finite interval. Moreover, it is self-similar: Any part of the graph contains a reduced copy of the entire graph. Incomplete devil's staircases are observed for $K < 1$ (i.e., the set of parameters yielding irrational rotation numbers then has positive measure).

2.12.3 Chaotic Circle Maps and Annulus Maps

The $K = 1$ line in the phase diagram of Fig. 2.31 is called the *critical line*. Beyond it, the circle map (2.117) has a point with zero derivative and hence is no longer invertible, which has dramatic consequences on the dynamics. On the one hand, there are no longer quasiperiodic regimes. Indeed, the latter are equivalent to a pure rotation with an irrational rotation number, which cannot be conjugate to a noninvertible map. On the other hand, more complex behavior can then appear, including chaos. Since the map (2.117) has branches with negative slope, the stable periodic orbit can now have a negative multiplier and undergo a period doubling when the latter crosses -1. Most

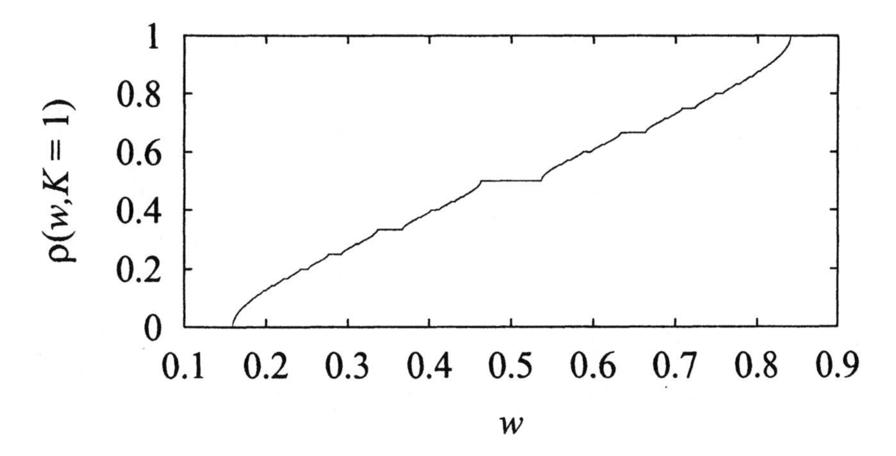

Fig. 2.32 Graph of the rotation number $\rho(w, K = 1)$ is a devil's staircase.

of the analysis carried out for the logistic map applies here: One observes cascades of period doubling leading to chaos. The white zones in the $K > 1$ part of Fig. 2.31 correspond to chaotic regimes or to periodic regimes of high period.

Another consequence of noninvertibility is that the rotation number (2.118) now depends on the initial condition. Accordingly, a given regime is characterized by a *rotation interval* $[\rho^-, \rho^+]$ rather than by a single number. This corresponds to Arnol'd tongues gradually overlapping as K is increased above 1, as can be seen in Fig. 2.31.

As discussed in the introduction to this section, invertible circle maps can be obtained rigorously as a first-return map when the dynamics is confined to a two-dimensional torus T^2. Obviously, this interpretation is not valid for noninvertible circle maps. However, exactly as the one-dimensional logistic map can be viewed as the infinitely dissipative of a two-dimensional horseshoe-like invertible map, noninvertible circle maps can be thought as limits of maps of an annulus into itself. Not that this interpretation is limited to circle maps having a degree of 1 (the image of the annulus winds once around the center). This is illustrated in Section 10.8 with the important example of the forced van der Pol oscillator.

2.13 SUMMARY

Although maps are very simple dynamical systems, they display most of the key features of chaos. This has allowed us to become familiar with concepts that will appear throughout this book, without excessive mathematical difficulty.

Even the simplest dynamical system that one can think of, the logistic map, is able to reproduce surprisingly well qualitative behaviors that are observed experimentally in the laser system described in Chapter 1. As a control parameter is varied, it experiences bifurcations, in particular a period-doubling cascade leading to chaos, and a variety of chaotic regimes.

The simple structure of the logistic map makes it possible to study one of the basic mechanisms responsible for chaotic behavior, namely stretching. In its most chaotic regime, the logistic map is basically a "multiply by two" machine. This has far-reaching consequences: Sensitivity to initial conditions, existence of an infinite number of periodic orbits which are dense in the invariant set, and so on.

Stretching is at the root of an extremely powerful tool for unfolding chaos, symbolic dynamics. Thanks to the unlimited magnification provided by sensitivity to initial conditions, a series of coarse-grained measurements of the system state suffice to specify it with arbitrary accuracy. Symbolic dynamics not only allows us to classify orbits (e.g., how many periodic orbits of period p?) but also to study their genealogies (e.g., in which order do orbits appear? Which orbit is a period-doubled orbit from another?). By studying the grammar of a chaotic system, we can classify regimes and compute quantitative invariants such as topological entropy. Not all of the results obtained (e.g., the universal sequence) can be directly extended to higher dimensions. That there are topological invariants (e.g., permutations) that both are deeply related to symbolic dynamics and play a major role in forcing relations will later prove to be a key property.

The logistic map is a noninvertible system. Many physical systems are described by ODEs and thus are associated with invertible maps, such as the Hénon map. A chaotic invertible map shares many properties with noninvertible ones. In particular, the dynamics in the unstable space is associated with a noninvertible map, as the example of the horseshoe map shows. There are also new problems, such as constructing relevant symbolic encodings in that case. Finally, global phase space topology can have a deep influence of phenomena observed, as exemplified by circle maps.

3

Continuous Dynamical Systems: Flows

All of the physical systems we deal with in this book are described by sets of coupled nonlinear ordinary differential equations. Such systems share many common features. In this chapter we survey the properties of such systems. The discussion ranges from the fundamental theorem of dynamical systems theory, the existence and uniqueness theorem, to the fundamental idea behind the classification of strange attractors. This idea is that most of the dynamics described by a set of ordinary differential equations is understood once the two basic mechanisms that operate in phase space have been identified. These mechanisms are stretching and squeezing.

3.1 DEFINITION OF DYNAMICAL SYSTEMS

Definition: A *dynamical system* is a set of ordinary differential equations of the form

$$\frac{dx_i}{dt} = F_i(x; c) \tag{3.1}$$

Here $x = (x_1, x_2, \ldots, x_n) \in R^n$ are called *state variables* and $(c_1, c_2, \ldots, c_k) \in R^k$ are called *control parameters*. The functions $F_i(x; c)$ are assumed to be sufficiently smooth.

Remark 1: The state variables and control parameters are usually considered to be in subspaces of Euclidean spaces, but they may more generally be in n- and k-dimensional manifolds.

Remark 2: The space of state variables is often called the *phase space*.

Remark 3: Initial conditions for any solution of the dynamical system equations belong in the phase space.

Remark 4: The functions $F_i(x; c)$ are usually assumed to be differentiable over the region of interest, but a Lipschitz condition $|F(x; c) - F(x'; c)| < K(c)|x - x'|$ is sufficient for the most important properties of dynamical systems which we will exploit: the existence and uniqueness theorem (cf. Section 3.2). In most cases of interest, the functions $F_i(x; c)$ are polynomials in the x_j whose coefficients depend on the control parameters c. Such functions are Lipschitz on bounded domains in R^n.

Remark 5: If the functions $F_i(x; c)$ are not explicitly dependent on time (i.e., $\partial F_i / \partial t = 0$, all i) the system is said to be *autonomous*. Otherwise, it is said to be *nonautonomous*. It is often possible to replace a nonautonomous system of equations in n dimensions by an autonomous system of equations in higher dimensions. We will see how this can be done in the two examples of periodically driven systems discussed below.

Remark 6: The laser equations (1.1) and (1.2) are nonautonomous since one of the source functions is periodically driven. For periodically driven dynamical systems the phase space is a torus $R^n \times S^1$. When the phase space is a torus, it is a simple matter to construct a Poincaré section by choosing ϕ = constant, $\phi \in S^1$. Mappings of the Poincaré section into itself are themselves discrete dynamical systems of the type studied in Chapter 2.

3.2 EXISTENCE AND UNIQUENESS THEOREM

The fundamental theorem for dynamical systems is the *existence and uniqueness theorem*, which we state here without proof [61]:

Theorem: If the dynamical system (3.1) is Lipschitz in the neighborhood of a point x_0, there is a positive number s, and

Existence: There is a function $\phi(t) = (\phi_1(t), \phi_2(t), \ldots, \phi_n(t))$ which satisfies the dynamical system equations $d\phi_i(t)/dt = F_i(\phi(t); c)$ in the interval $-s \leq t \leq +s$, with $\phi(0) = x_0$.

Uniqueness: The function $\phi(t)$ is unique.

This is a *local* theorem. It guarantees that there is a unique nontrivial trajectory through each point at which a dynamical system satisfies the Lipschitz property. If the dynamical system is Lipschitz throughout the domain of interest, this local theorem is easily extended to a global theorem.

Global Theorem: If the dynamical system (3.1) is Lipschitz, there is a unique trajectory through every point, and that trajectory can be extended asymptotically to $t \to \pm\infty$.

We cannot emphasize strongly enough how important the existence and uniqueness theorem is for the topological discussion that forms the core of this work. Much of our discussion deals with closed periodic orbits in dynamical systems which exhibit chaotic behavior. To be explicit, we depend heavily on the fact that in three dimensions, the topological organization of such orbits cannot change under any deformation. For the organization to change, one orbit would have to pass through another (cf. Fig. 4.3). This means that at some stage of the deformation, two orbits would pass through the same point in phase space. The existence and uniqueness theorem guarantees that this cannot occur. More specifically, if two closed orbits share a single point, the two orbits must be identical.

3.3 EXAMPLES OF DYNAMICAL SYSTEMS

The archetypical dynamical system is the set of equations representing the Newtonian motion of N independent point particles under their mutual interactions, gravitational or not. In three dimensions, this consists of $n = 6N$ first-order coupled ordinary differential equations: three for the coordinates and three for the momenta of each particle. Under the gravitational interaction this system is non-Lipschitz: The gravitational force diverges when any pair of particles becomes arbitrarily close.

In the following four subsections we give four examples of dynamical systems. These are four of the most commonly studied dynamical systems. The examples are presented in the historical order in which they were introduced. In the fifth subsection we give examples of some differential equations that are not equivalent to finite-dimensional dynamical systems.

3.3.1 Duffing Equation

The Duffing equation [62] was introduced to study the mechanical effects of nonlinearities in nonideal springs. The force due to such a spring was assumed to be a perturbation of the general spring force law $F(x) = -kx$, which retained the symmetry $F(-x) = -F(x)$. The simplest perturbation of this form is $F(x) = -kx - \alpha x^3$ with nonlinear spring 'constant' $k + \alpha x^2$. The spring is "harder" than an ideal spring if $\alpha > 0$, "softer" if $\alpha < 0$. The potential for this nonideal spring force is even: $U(x) = \frac{1}{2}kx^2 + \frac{1}{4}\alpha x^4$. The equations of motion for a mass m attached to a massless nonideal spring, with damping constant γ, are

$$\begin{aligned} \frac{dx}{dt} &= v \\ m\frac{dv}{dt} &= F = -\frac{dU}{dx} - \gamma v \end{aligned} \tag{3.2}$$

These two equations can be combined to a single second-order equation in the variable x:

$$\ddot{x} + \frac{\gamma}{m}\dot{x} + \frac{1}{m}\frac{dU}{dx} = 0 \tag{3.3}$$

In what follows we set $m = 1$ for convenience.

The potential $U(x)$ leads to exciting behavior if we relax the condition that it represents a small perturbation of an ideal spring. That is, we assume that the potential may be *un*stable near the origin but is globally stable. A family of potentials with these properties is

$$A_3: \qquad U(x) = -\frac{1}{2}\lambda x^2 + \frac{1}{4}x^4 \tag{3.4}$$

Here the parameter λ defines the strength of the instability at the origin. For $\lambda > 0$, the potential has an unstable fixed point at $x = 0$ and two symmetrically distributed stable fixed points at $x = \pm\sqrt{\lambda}$. This potential describes many physical systems of interest: for example, a column under sufficient stress that it buckles (cf. Fig. 3.1). The dynamics of the system (3.3) are relatively simple: Any initial condition eventually decays to one of the two ground states. However, if the oscillator is driven periodically by including a forcing term of the form $f\cos(\omega t)$, the two coupled equations of motion become

$$\begin{aligned} \frac{dx}{dt} &= v \\ \frac{dv}{dt} &= -\frac{dU}{dx} - \gamma\dot{x} + f\cos(\omega t) \end{aligned} \tag{3.5}$$

These two equations can be combined to a second-order equation in the single variable x:

$$\ddot{x} + \gamma\dot{x} + x^3 - \lambda x = f\cos(\omega t) \tag{3.6}$$

This equation for the driven damped Duffing oscillator leads to behavior that is still not completely understood. The Duffing oscillator is treated in more detail in Section 10.7.

3.3.2 van der Pol Equation

The van der Pol equation [63] arises in a natural way when we consider the effect of a nonlinear resistor in a series RLC circuit. A simple RLC circuit is shown in Fig. 3.2(a). If we label the current flowing through the resistor as I_R and the voltage across the resistor as V_R, and similarly for the condensor ($_C$) and inductor ($_L$), the

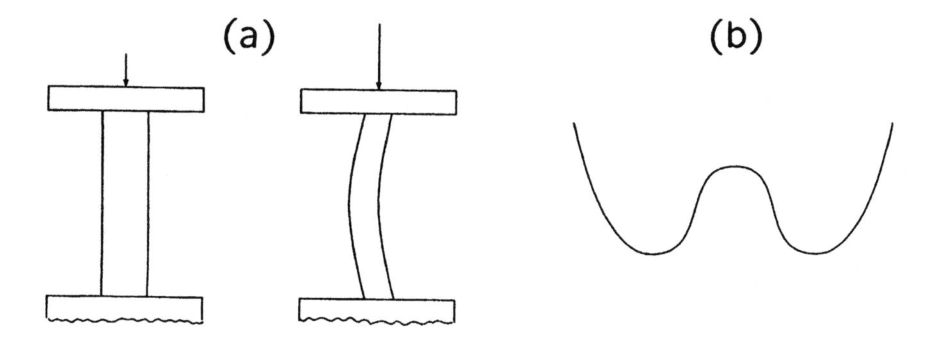

Fig. 3.1 (a) A vertical beam under excessive compression deforms to one of two stable equilibrium states. (b) The motion of a mass in a double well potential driven by a periodic symmetry-breaking term is described by the Duffing equation (3.6).

Kirchhoff current and voltage laws impose the following constraints on the currents and voltages in this simple series system:

$$\begin{array}{ccccccc} I_R & = & I_L & = & I_C & & \\ V_R & + & V_L & + & V_C & = & 0 \end{array} \tag{3.7}$$

The dynamics of the circuit is governed by the equations of motion relating currents and voltages in the condensor and inductor:

$$\begin{aligned} L\frac{dI_L}{dt} &= V_L \\ C\frac{dV_C}{dt} &= I_C \end{aligned} \tag{3.8}$$

and the constitutive relation for the resistor

$$V_R = f(I_R) \tag{3.9}$$

A linear relation ($V_R = RI_R$, R is the *resistance* of the resistor) is usually assumed. However, interesting dynamics occur if we can create a resistor that is nonlinear, especially if its linearization in the region of small currents has a negative value: $df(x)/dx < 0$ for $x \simeq 0$. It is convenient to retain the symmetry of $f(x)$, that is, $f(-x) = -f(x)$. A convenient choice for the functional form of f is

$$A_2: \qquad f(x) = \frac{1}{3}x^3 - \lambda x \tag{3.10}$$

The dynamical equations of motion are then easily obtained from (3.8). We simplify by setting $L = C = 1$, $I_L = I_C = I_R = x$ and $V_C = y$. Then

$$\begin{aligned} \frac{dx}{dt} &= V_L = -V_C - V_R = -y - f(x) \\ \frac{dy}{dt} &= I_C = x \end{aligned} \tag{3.11}$$

These two equations can be combined to a second-order equation in the single variable x:

$$\ddot{x} + (x^2 - \lambda)\dot{x} + x = 0 \tag{3.12}$$

This is one standard form of the time-independent van der Pol equation.

Driving terms can be introduced in two different ways, as shown in Fig. 3.2(b). We can place a periodic voltage source $-v\sin(\omega_1 t)$ in the circuit or put a periodic current $-i\cos(\omega_2 t)$ through the capacitor. The dynamical equations are

$$\begin{aligned} \frac{dI_L}{dt} &= V_L = -V_C - V_R + v\sin(\omega_1 t) \\ \frac{dV_C}{dt} &= I_C - i\cos(\omega_2 t) \end{aligned} \tag{3.13}$$

Employing the same substitutions as before ($x = I_L = I_R = I_C$ and $y = V_C$) we find that

$$\begin{aligned} \frac{dx}{dt} &= -y - f(x) + v\sin(\omega_1 t) \\ \frac{dy}{dt} &= x - i\cos(\omega_2 t) \end{aligned} \tag{3.14}$$

Once again, these two first-order equations in the dynamical variables x and y can be expressed as a single second-order equation in one of them:

$$\ddot{x} + (x^2 - \lambda)\dot{x} + x = v\omega_1\cos(\omega_1 t) + i\cos(\omega_2 t) \tag{3.15}$$

This equation for the periodically driven van der Pol oscillator also leads to behavior that is still not completely understood. This oscillator is treated in more detail in Section 10.8.

3.3.3 Lorenz Equations

The Lorenz equations [64] arise as a suitable truncation of the Navier–Stokes equations subject to appropriate boundary conditions. The physical problem that leads to this set of equations is summarized in Fig. 3.3. This cartoon shows a fluid that is heated from below. When the rate of heating of the fluid is low (a), heat is transferred from the bottom to the top by heat conduction alone. The fluid remains at rest. If the heating rate is increased, heat conduction alone is not sufficient to transport the heat rapidly enough from the bottom to the top: The fluid helps out by organizing itself

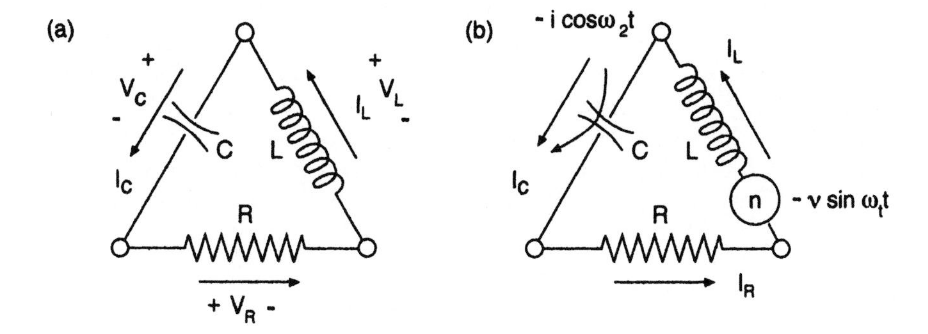

Fig. 3.2 (a) Triangular RLC series circuit; (b) triangular RLC series circuit with periodic forcing terms in the C and L legs.

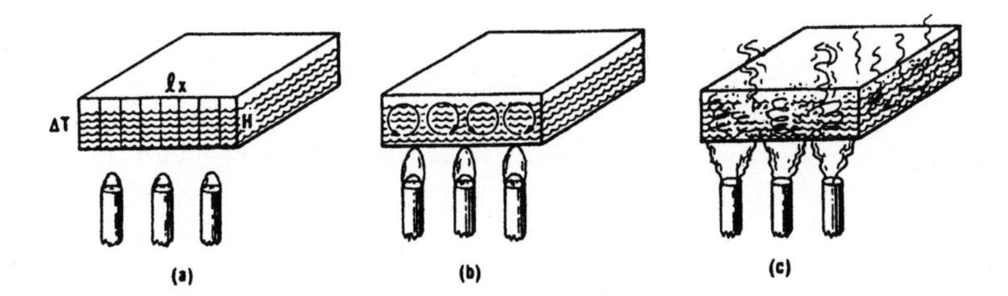

Fig. 3.3 (a) At low temperatures, heat is transported by thermal conduction alone. (b) At higher temperatures, the fluid helps out by moving in orderly, self-organized rolls. The rolls may be unchanging in time, may vary periodically, or may even vary chaotically. (c) At very high temperatures the spatially organized roll structures break up into turbulent motion.

into rolls which help to convey heat from the lower surface to the upper surface (b). When the heating rate is increased further, the simple organized fluid rolls break up and the behavior appears turbulent (c).

The physics of incompressible fluid motion under thermal stress is described by the Navier–Stokes equations:

$$
\begin{aligned}
\left(\frac{\partial}{\partial t} + \mathbf{u}\cdot\nabla\right)\mathbf{u} &= \epsilon \mathbf{g}\,\Delta T - \frac{1}{\rho}\nabla P + \nu\,\nabla^2\mathbf{u} \\
\left(\frac{\partial}{\partial t} + \mathbf{u}\cdot\nabla\right)T &= \kappa\nabla^2 T \\
\nabla\cdot\mathbf{u} &= 0
\end{aligned}
\tag{3.16}
$$

The fluid velocity field is $\mathbf{u} = \mathbf{u}(\mathbf{x}) = \mathbf{u}(x, y, z)$, and $T = T(x, y, z)$ is the temperature field. The first equation describes the evolution of the fluid velocity field. The second equation describes the evolution of the temperature field. The third equation is the incompressibility condition. In these equations, ϵ is the coefficient of thermal expansion, $\mathbf{g}$ describes the gravitational force, ΔT is the constant temperature difference imposed between the lower and upper surfaces of the fluid, P is the fluid pressure field, ν is the fluid viscosity, and κ is the thermal conduction coefficient.

This problem is simplified by making some assumptions and introducing new variables. First, we introduce a scalar stream function $\psi(x, y, z)$ whose gradient is the velocity field: $\nabla\psi = \mathbf{u}$. We also introduce a function θ which is the difference between the temperature field in the fluid and the temperature field in the static case [Fig. 3.3(a)]: $\theta(x, y, z) = T(x, y, z) - T_{av}$, where T_{av} decreases linearly between the bottom and top surfaces. We also assume homogeneity in the y direction.

The auxiliary functions obey the equations

$$\begin{aligned} \frac{\partial}{\partial t}\nabla^2\psi &= -\frac{\partial(\psi, \nabla^2\psi)}{\partial(x, z)} + \nu\,\nabla^4\psi + \epsilon g\frac{\partial\theta}{\partial x} \\ \frac{\partial}{\partial t}\theta &= -\frac{\partial(\psi, \theta)}{\partial(x, z)} + \frac{\Delta T}{H}\frac{\partial\psi}{\partial x} + \kappa\,\nabla^2\theta \\ \nabla^2\psi &= 0 \end{aligned} \tag{3.17}$$

The boundary conditions on the auxiliary functions $\psi(x, z, t)$ and $\theta(x, z, t)$ for a container of width L (in the x direction) and height H (in the z direction) are that both ψ and $(\partial\theta/\partial x)$ vanish on the boundaries. As a result, Fourier representations of these functions must assume the form

$$\begin{aligned} \psi(x, z, t) &= \sum_{m_1 m_2} A_{m_1 m_2}(t)\sin\left(\frac{m_1\pi x}{L}\right)\sin\left(\frac{m_2\pi z}{H}\right) \\ \theta(x, z, t) &= \sum_{n_1 n_2} B_{n_1 n_2}(t)\cos\left(\frac{n_1\pi x}{L}\right)\sin\left(\frac{n_2\pi z}{H}\right) \end{aligned} \tag{3.18}$$

Here m, n are positive integers, and n_1 can also be zero. These expressions can be substituted into the partial differential equations (3.17) for the auxiliary variables. The time derivatives act on the Fourier coefficients $A_{m_1 m_2}(t)$, $B_{n_1 n_2}(t)$ on the left-hand side of the equations (3.17). All spatial derivatives that appear on the right-hand side of these equations are easily computed. Integrating out the spatial functions leads to a set of coupled nonlinear ordinary differential equations in which the forcing terms on the right-hand side are polynomial functions (at most quadratic) of the Fourier coefficients. The set contains an infinite number (countable) of time-dependent Fourier coefficients.

The usual procedure is to truncate this set of equations by considering only a finite number of terms in the expansion for the two auxiliary functions. Saltzman [65] studied truncations containing 52 and seven Fourier coefficients. He found that in many instances all but three coefficients relaxed to zero after a sufficiently long time. Lorenz reduced the complexity of the problem by studying a very severe truncation of this problem containing only the three Fourier modes which persisted in Saltzman's simulations. These correspond to the expansions

$$\begin{aligned} \frac{a}{\kappa(1+a^2)}\psi &= \sqrt{2}\,X \sin\left(\frac{\pi x}{L}\right)\sin\left(\frac{\pi z}{H}\right) \\ \pi\frac{R_a}{R_c}\frac{\theta}{\Delta T} &= \sqrt{2}\,Y \cos\left(\frac{\pi x}{L}\right)\sin\left(\frac{\pi z}{H}\right) - Z\sin\left(\frac{2\pi z}{H}\right) \end{aligned} \tag{3.19}$$

Here the three real variables (X, Y, Z) do *not* represent spatial coordinates – they represent Fourier amplitudes as follows: $X \sim A_{11}$, $Y \sim B_{11}$, and $Z \sim B_{02}$. Substituting this ansatz into (3.17) and integrating out the spatial dependence leads to the Lorenz equations for the three Fourier amplitudes X, Y, Z:

$$\begin{aligned} \frac{d}{d\tau}X &= -\sigma X + \sigma Y \\ \frac{d}{d\tau}Y &= rX - Y - XZ \\ \frac{d}{d\tau}Z &= -bZ + XY \end{aligned} \tag{3.20}$$

The parameters that appear in this set of equations are $a = H/L$, $R_c = \pi^4(1 + a^2)^3/a^2$, $R_a = \epsilon g H^4(\Delta T/H)/\kappa\nu$, $r = R_a/R_c$, $\sigma = \nu/\kappa$, $b = 4/(1 + a^2)$, and $\tau = (\pi/H)^2(1 + a^2)\kappa t$.

The equations (3.20) are known as the *Lorenz equations*. These are the equations in which the property of sensitivity to initial conditions was first studied in detail.

Remark: We should emphasize here that even if a dynamical system depends on many variables (e.g., 52), often only a small number (e.g., 3) are important. In this sense, if we model a system correctly, its behavior is ultimately independent of the number of variables used, as long as enough variables are used. This observation will take on added importance in one of the modifications we will make of the Birman–Williams theorem (Chapter 5).

3.3.4 Rössler Equations

The Lorenz equations possess two nonlinearities ($-XZ$ in the second equation and $+XY$ in the third) and exhibit three fixed points. Rössler [66] felt that there should be a simpler set of equations that exhibits the same generic behavior: sensitivity to initial conditions and chaotic behavior. *Simpler* meant fewer nonlinear terms and a smaller number of fixed points. Accordingly, he set out to construct a simpler set, in the spirit of Dirac setting out to find a simpler version of the equation for an electron by factoring the Klein–Gordon equation into two simpler equations.

Rössler began by considering motion in the neighborhood of an unstable focus in the x–y plane. Such motion can occur in a two-dimensional dynamical system of the form

$$\begin{aligned} \dot{x} &= -y \\ \dot{y} &= x + ay \end{aligned} \tag{3.21}$$

He then added a term that forces the motion out of the x–y plane when it gets too far away from the origin. This involves an equation of motion for the z direction. He chose the simple form $\dot{z} = b + z(x - c)$. Here b is small and positive, causing the value of z to increase slowly when z and x are small. When x becomes larger than c, $\dot{z}$ becomes large and the z coordinate suddenly "blasts off" from the x–y plane. Finally, one more term was added to the original two-dimensional dynamical system, which forces x to decrease suddenly when z becomes large. The larger the value of $x - c$, the higher the phase-space point travels before being forced down to the $z \simeq 0$ plane, and the closer to the origin it returns. This produces the half twist that is observed in one part of the cartoon, which represents the flow dynamics (Fig. 3.4).

In final form, the Rössler equations are

$$\begin{aligned} \dot{x} &= -y - z \\ \dot{y} &= x + ay \\ \dot{z} &= b + z(x - c) \end{aligned} \tag{3.22}$$

These equations are simpler than the Lorenz equations in the sense that they contain only one nonlinearity (zx in the third equation) and possess only two fixed points.

The Rössler equations have been widely studied over a broad range of control parameter values (a, b, c). Although these equations were not motivated through consideration of some physical problem (as opposed to the Duffing, van der Pol, and Lorenz equations), the behavior of these equations serves to model a wide variety of real physical systems. An analysis of many such systems is provided in Chapter 7.

3.3.5 Examples of Nondynamical Systems

Many physical systems can be described, or at least reasonably well approximated, by sets of ordinary differential equations. Lest we leave the impression that all of physics is described by dynamical systems, we present three examples that do not fall under this rubric.

3.3.5.1 Equation with Non-Lipschitz Forcing Terms

The equation

$$\frac{dx}{dt} = \sqrt{2x} \tag{3.23}$$

has a forcing term ($\sqrt{2x}$) that is non-Lipschitz. Its derivative, $1/\sqrt{2x}$, is not bounded in the neighborhood of the origin. As a result, the existence and uniqueness theorem is not applicable. To emphasize this point, this equation has two distinct solutions through $x_0 = 0$. One solution is $x(t) = 0, -\infty \leq t \leq +\infty$. The other solution is $x(t) = \frac{1}{2}t^2, -\infty \leq t \leq +\infty$.

3.3.5.2 Delay Differential Equations

Populations of biological species are governed by the balance between birth and death. A simple model for the adult population, $x(t)$, of a single species can be constructed as follows. The rate of

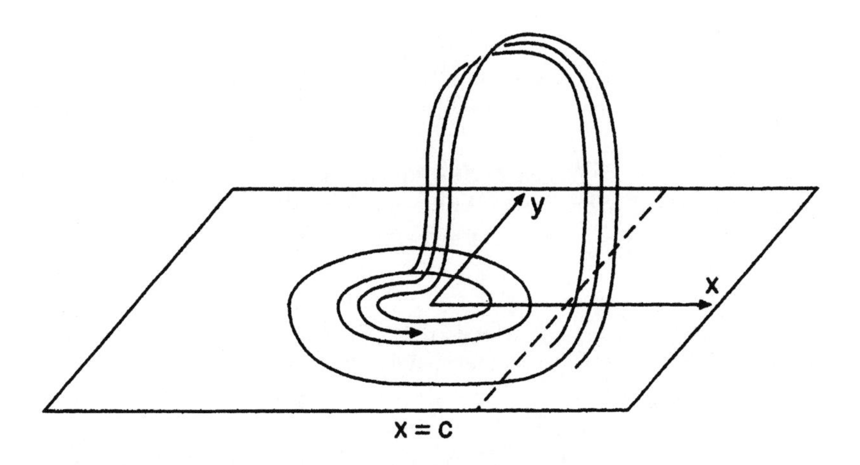

Fig. 3.4 Flow dynamics generated by the Rössler equations.

change of population at time t is the difference between the rate at which adults enter and leave the species pool. The number of adults that leave the pool is a function of the present population: $D(x(t))$. On the other hand, the number of adults that enter the pool at time t is a function of the number of adults at some previous time: $B(x(t-\tau))$, where τ is the time scale of one generation. As a result, the population dynamics equation is

$$\frac{dx}{dt} = B(x(t-\tau)) - D(x(t)) \tag{3.24}$$

The death function can be expanded as $D(x) = d_0 + d_1 x + d_2 x^2 + \cdots$. In this expansion $d_0 = 0$ and the nonlinear terms $d_2, d_3, \ldots$ model ecological effects. For many purposes it is sufficient to represent $D(x(t))$ as a linear function: $D(x(t)) = d_1 x(t)$. By similar simple reasoning we can assume that $B(x(t-\tau))$ is small and linearly increasing when $x(t-\tau)$ is small. This describes exponential growth of the population when the population is small. It is not unreasonable to assume that $B(x(t-\tau))$ is also small when $x(t-\tau)$ is large. This describes diminishing population growth when populations are large due to competition for diminishing resources. Thus, $B(x(t-\tau))$ is often assumed to be a single humped distribution of the type discussed extensively in Chapter 2.

The delay-differential equation (3.24) with B quadratic (or unimodal) and D linear has been studied extensively and exhibits a wide spectrum of exciting behavior, including chaotic behavior with Lyapunov dimension $d_L > 2$ [67].

The single scalar equation (3.24) is equivalent to an infinite-dimensional dynamical system. This is most easily seen as follows. To integrate this equation, an initial condition must first be specified. An initial condition takes the form of a function $x(t), 0 \leq t < \tau$. A *function* can be represented as a *point* in an infinite-dimensional space. Thus, the appropriate phase space for the single delay equation (3.24) is an infinite-dimensional space, such as a Hilbert space or a Banach space. Chaotic behavior in such spaces can generate strange attractors with $d_L \gg 2$.

3.3.5.3 Stochastic Differential Equations Partial differential equations are generally more difficult to handle than ordinary differential equations. For this reason a large industry has grown up around the problem of reducing partial to ordinary differential equations.

We have illustrated one approach to this reduction in our discussion of the Lorenz equations. This procedure is useful for partial differential equations of evolution type: $\partial u/\partial t = F(u, \nabla u, \ldots, t)$. The dependent fields are expanded in terms of a complete set of orthogonal spatial modes, $\Phi_\mu(x)$, which satisfy the appropriate boundary conditions. These expressions are inserted into the partial differential equations, and the spatial dependence is integrated out. This results in a set of coupled ordinary differential equations. There is one equation for each amplitude. The source terms consist of polynomial products involving the amplitudes and numerical coefficients which are Clebsch–Gordan coefficients for the orthogonal modes. If the function $F(*)$ contains quadratic, cubic, ... nonlinearities in the dependent fields, the source terms in the projected coupled ordinary differential equations will contain only quadratic, cubic, ... products of the amplitudes. For example, the Navier–Stokes equations contain only quadratic nonlinearities. Therefore, any Galerkin projection of these equations will contain only quadratic couplings involving the mode amplitudes, no matter what complete set of modes is used for expansion of the dependent functions.

Suppose, to be specific, that a partial differential equation contains only quadratic nonlinearities. We assume that the dependent fields have an expansion

$$u(x,t) = \sum_i x_i(t)\Phi_i(x) + \sum_\alpha y_\alpha(t)\Phi_\alpha(x) \tag{3.25}$$

where $i = 1, 2, \ldots, n$ and $\alpha = n+1, \ldots$. We assume that the modes numbered 1 through n are the most important modes under the conditions specified. Then the projection procedure outlined above will lead to a set of equations for the mode amplitudes x_i of the form

$$\frac{dx_i}{dt} = A_i + A_{i,j}x_j + A_{i,\beta}y_\beta + A_{i,jk}x_j x_k + A_{i,j\gamma}x_j y_\gamma + A_{i,\beta\gamma}y_\beta y_\gamma \tag{3.26}$$

A similar equation can be written for the dy_α/dt.

This set of equations is not closed. Evolution of the x_i depends on the time dependence of the y_α, which is not known. The $y_\alpha(t)$ dependence can be removed by writing these equations in the form

$$\frac{dx_i}{dt} = A_i + A_{i,j}x_j + A_{i,jk}x_j x_k + \epsilon_i(t) + \epsilon_{i,j}(t)x_j \tag{3.27}$$

The two time-dependent functions are

$$\begin{aligned} \epsilon_i(t) &= \sum_\beta A_{i,\beta} y_\beta + \sum_{\beta\gamma} A_{i,\beta\gamma} y_\beta y_\gamma \\ \epsilon_{i,j}(t) &= \sum_\gamma A_{i,j\gamma} y_\gamma \end{aligned}$$

If we are careful (or lucky), the time-dependent functions $\epsilon_i(t)$ and $\epsilon_{i,j}(t)$ will not have large amplitudes, and we might even be able to say something about their properties (mean values, moments, correlation functions).

In many cases, the function $\epsilon_i(t)$ is treated as an additive noise term and the functions $\epsilon_{i,j}(t)$ are treated as multiplicative noise terms in the equation for x_i. Equations of the type (3.27) are called *stochastic differential equations.*

Remark: Two distinct approaches exist for introducing normal modes $\Phi_\mu(x)$ into decompositions of type (3.25). For purely theoretical analyses, the normal modes $\Phi_\mu(x)$ are usually taken as some convenient set in which to analyze the appropriate equations: for example, Fourier modes. For analyses that involve observational data, empirical modes are very often constructed by processing the data. A much used procedure involves constructing a set of orthogonal modes using a singular-value decomposition (cf. Section 6.4.7).

3.3.6 Additional Observations

We can draw a number of general observations from the specific properties exhibited by the examples discussed above.

Nonautonomous → Autonomous Form: The Lorenz equations (3.20) and the Rössler equations (3.22) are autonomous. The periodically driven Duffing equations (3.5) and the periodically driven van der Pol equations (3.14) are nonautonomous.

All periodically driven dynamical systems are nonautonomous. If the driving terms are of the form $\sin(\omega t)$, $\cos(\omega t)$, the nonautonomous set of equations can be replaced by an autonomous dynamical system of higher dimension by adjoining to the original set the equations

$$\begin{aligned} \dot{r}_1 &= -\omega r_2 \\ \dot{r}_2 &= +\omega r_1 \end{aligned} \tag{3.28}$$

and then replacing $\sin(\omega t)$, $\cos(\omega t)$ by r_1, r_2 wherever they occur. When this is done, the source terms for the Duffing and van der Pol equations are low-degree polynomials in the appropriate variables.

Remark: Adjoining equations (3.28) to a dynamical system increases its dimension by only one, since there is a conserved quantity: $r_1\dot{r}_1 + r_2\dot{r}_2 = 0$. This corresponds to adding a single term of the form $\dot{\phi} =$ constant, where ϕ is a phase angle, to the original equations.

Structure of Phase Space: The phase spaces for the Lorenz and Rössler systems are R^3. The phase spaces for the periodically driven Duffing and van der Pol oscillators are also three-dimensional but topologically different: They are $R^2 \times S^1$. The S^1 part of the phase space refers to the periodic driving term.

Symmetry: Dynamical systems sometimes exhibit symmetries. When this occurs, the symmetries impose constraints on the behavior. These constraints often simplify the discussion of the properties of these systems.

To be more explicit, the scalar function A_2: $f(x) = \frac{1}{3}x^3 - \lambda x$, which appears in the van der Pol equations, and the scalar function A_3: $U(x) = \frac{1}{4}x^4 - \frac{1}{2}\lambda x^2$, which appears in the Duffing equations, both have a symmetry under the group action $x \to -x$. These symmetries propagate to the periodically driven equations, as follows. If either i or v is zero in the driven van der Pol equation (3.14), the equations are unchanged ("equivariant") under $(x, y, t) \to (-x, -y, t + \frac{1}{2}T)$, where T is the period of the driving term. Similarly, the Duffing equations (3.5) are equivariant under the same transformation $(x, y, t) \to (-x, -y, t + \frac{1}{2}T)$. In both cases, the symmetry is easily visualized as a symmetry in the phase space $R^2(x, y) \times S^1(\theta) \to R^2(-x, -y) \times S^1(\theta + \pi)$.

The Lorenz equations also exhibit symmetry. They are equivariant under $(X, Y, Z) \to (-X, -Y, +Z)$. This symmetry is easily visualized as a rotation through 180° about the Z axis in the phase space. The origin of this symmetry lies in the Navier–Stokes equations. For the boundary conditions chosen, these equations are unchanged under the real physical space coordinate transformation $(x, z) \to (L - x, z)$. Under this transformation the potentials satisfy

$$\begin{aligned} \psi(x, z, t) &= -\psi(L - x, z, t) \\ \theta(x, z, t) &= +\theta(L - x, z, t) \end{aligned} \tag{3.29}$$

This symmetry requires the following symmetries on the Fourier coefficients: $A_{m_1 m_2} \to (-)^{m_1} A_{m_1 m_2}$ and $B_{n_1 n_2} \to (-)^{n_1} B_{n_1 n_2}$. All truncations of the Navier–Stokes equations for this geometry will exhibit this twofold symmetry.

Singularities: In the work that follows, we show that singularities play an important role in the creation and classification of strange attractors. We have already encountered two of the simplest singularities: the fold and the cusp. The function $f(x, \lambda) = \frac{1}{3}x^3 - \lambda x$ is a one-parameter family of functions that contains a singularity ($d^2 f/dx^2 = 0$ when $df/dx = 0$) at $\lambda = 0$. This family of functions contains the fold singularity x^3 and is denoted A_2. The function $U(x, \lambda) = \frac{1}{4}x^4 - \frac{1}{2}\lambda x^2$ belongs to a two-parameter family of functions that contains the cusp singularity x^4, which is denoted A_3. The second parameter (b) in this family, $U(x, \lambda) = \frac{1}{4}x^4 - \frac{1}{2}\lambda x^2 + bx$, is restricted to be zero by symmetry considerations.

Singularities will be encountered in several different ways in our studies of dynamical systems that can exhibit chaotic behavior. For example, we will see in Section 3.5.3 that the fixed points of the Rössler equations can experience a fold singularity (A_2) and those of the Lorenz equations can experience a symmetry-restricted cusp singularity (A_3) as control parameters are varied.

***n*th-Order Equations:** Ordinary differential equations have also been represented by a scalar constraint on the first n derivatives of a single variable:

$$G(y, y^{(1)}, y^{(2)}, \ldots, y^{(n)}) = 0 \tag{3.30}$$

In this expression, $y = y^{(0)}$ and $y^{(i+1)} = dy^{(i)}/dt$. The scalar equation (3.30) represents a surface in the $(n+1)$-dimensional space whose coordinates are the state variable y and its first n derivatives. This is an nth order equation.

If the derivative $y^{(n)}$ occurs linearly in (3.30), so that $G(*) = y^{(n)} - g(y, y^{(1)}, \ldots y^{(n-1)})$, this equation can be written as a dynamical system by defining n variables $x_i = y^{(i-1)}$, with $x_1 = y$, so that

$$\begin{aligned} \frac{dx_1}{dt} &= x_2 \\ \frac{dx_2}{dt} &= x_3 \\ &\vdots \\ \frac{dx_{n-1}}{dt} &= x_n \\ \frac{dx_n}{dt} &= g(x_1, x_2, \ldots, x_n) \end{aligned} \tag{3.31}$$

The equations (3.3) and (3.12) for the undriven Duffing and van der Pol oscillators have the form (3.30) of second-order equations in a single variable. All fixed points of dynamical systems of the type (3.31) occur on the x_1 axis at the zeros of the function $fr(x_1, 0, 0, \ldots, 0)$.

If the function G in (3.30) is explicitly time dependent, the corresponding dynamical system is nonautonomous. The first remark in this subsection can then be used to construct a simple time-independent surface of the form (3.30) in a higher-dimensional space. This is possible for periodically driven dynamical systems.

Vector Fields: A dynamical system defines a vector field on its phase space. The following intuitive argument describes one way to make this association.

Assume that some function $P(x,t)$, $x \in R^n$, is defined on the phase space. This could represent some probability distribution, for example. If the evolution is due entirely to the dynamical evolution of the points in the phase space, $P(x,t) = P(x(t))$. The time evolution is then

$$\begin{aligned} P(x', t+dt) &= P(x(t+dt)) \\ &= P(x_i(t) + dt\dot{x}_i(t)) \\ &= P(x(t)) + dt\ \dot{x}_i(t)\frac{\partial}{\partial x_i}P(x) \end{aligned} \tag{3.32}$$

The differential operator

$$V = \frac{dx_i}{dt}\frac{\partial}{\partial x_i}$$

is the vector field associated with the dynamical system. This can be expressed in terms of the source functions using $\dot{x}_i = F_i(x)$ as follows:

$$V = F_i(x)\frac{\partial}{\partial x_i}$$

3.4 CHANGE OF VARIABLES

Physical results should be independent of the mathematics used to describe these results. To put this another way: our results should be independent of the coordinate system we choose to describe these results. This means that we should be free to choose different coordinate sytems without the risk of altering our predictions. In particular, one of the virtues of having many coordinate systems in which to perform calculations is that we are free to choose the coordinate system which simplifies, to the greatest extent possible, the computational difficulties.

3.4.1 Diffeomorphisms

Different but equivalent coordinate systems for sets of ordinary differential equations are related by diffeomorphisms.

Definition: A diffeomorphism from a space U to a space V is a $1 \leftrightarrow 1$ map that is onto and differentiable and whose inverse is also differentiable.

Example: A nonsingular linear transformation $x_i \to x'_i = \sum_j A^j_i x_j$ is a diffeomorphism.

Remark 1: Two n-dimensional dynamical systems that are related by a diffeomorphism are equivalent. That is, any property of one has a counterpart in the other dynamical system.

Remark 2: For most of our purposes, $U = V = R^3$.

Remark 3: It is sometimes useful to consider mappings which are *local* but not *global* diffeomorphisms. An example of a local diffeomorphism is the map $(x, y, z) \to (u, v, w) = (x^2 - y^2, 2xy, z)$. This mapping is not 1:1, but 2:1 for all points off the z axis. The Jacobian of this transformation is singular on the z-axis. Open subsets $U \subset R^3$ and $V \subset R^3$ exist on which this map is a local diffeomorphism. Dynamical systems on two subsets that are related by this map are *locally* equivalent but not *globally* equivalent. We discuss the relation between dynamical systems which are everywhere locally equivalent almost everywhere, but not globally equivalent, in Chapter 10.

3.4.2 Examples

In this subsection we consider three examples of changes of variables. The first two involve transformations from the original set of variables $(x_1, x_2, \ldots, x_n)$ to a new set of variables $(u_1, u_2, \ldots, u_n)$ which are differentially related to each other: $du_i/dt = u_{i+1}, i = 1, 2, \ldots, n-1$. In the first example the change of variables is a diffeomorphism; in the second it is not. In the third example, the change of variables is a local but not a global diffeomorphism.

Example 1: The general procedure for transforming a dynamical system $\dot{x}_i = F_i(x)$ from the original coordinates to a new set of differential coordinates $u_i = u^{(i-1)}$ is as follows.

1. Choose a function $\zeta = \zeta(x)$ of the original set of coordinates.

2. Define the new coordinate by $u_1 = \zeta(x)$.
3. Construct the next component using

$$u_2 = \frac{du_1}{dt} = \frac{d\zeta(x)}{dt} = \frac{dx_i}{dt}\frac{\partial}{\partial x_i}\zeta(x) = \left(F_i \frac{\partial}{\partial x_i}\right)\zeta(x) \tag{3.33}$$

4. Iterate this procedure, using the operator $F_i \partial_i$, to construct $u_3, \ldots, u_n$, and finally, $du_n/dt = g(x_1, \ldots, x_n)$.
5. Express the old coordinates x_i in terms of the new coordinates u_j.
6. Substitute the new coordinates into the last equation, to find

$$\begin{aligned} \dot{u}_1 &= u_2 \\ \dot{u}_2 &= u_3 \\ &\vdots \\ \dot{u}_{n-1} &= u_n \\ \dot{u}_n &= g(u_1, \ldots, u_n) \end{aligned} \tag{3.34}$$

We provide a simple illustration using the Rössler equations (3.22). We define $(x_1, x_2, x_3) = (x, y, z)$ and $(u_1, u_2, u_3) = (u, v, w)$, and choose $\zeta = y$. Then

$$\begin{aligned} u &= y \\ v &= x + ay \\ w &= ax + (a^2 - 1)y - z \\ \frac{dw}{dt} &= a(-y - z) + (a^2 - 1)(x + ay) - [b + z(x - c)] \end{aligned} \tag{3.35}$$

The first three equations are easily inverted:

$$\begin{aligned} y &= u \\ x &= v - au \\ z &= -u + av - w \end{aligned} \tag{3.36}$$

Since the transformation between the original and new set of coordinates is linear with determinant $+1$, it is a global diffeomorphism. In the new coordinate system the dynamical system equations are

$$\begin{aligned} \frac{du}{dt} &= v \\ \frac{dv}{dt} &= w \\ \frac{dw}{dt} &= g(u, v, w) = -\,cu + (ac - 1)v + (a - c)w - b \\ &\quad -\,(v - au)(-u + av - w) \end{aligned} \tag{3.37}$$

Example 2: We perform the same calculation now for the Lorenz set of equations (3.20), choosing $\zeta(x, y, z) = x$. Proceeding as above, we find that

$$
\begin{aligned}
x &= u \\
y &= u + (1/\sigma)v \\
z &= r - 1 - [w + (\sigma + 1)v]/\sigma u
\end{aligned}
\tag{3.38}
$$

This transformation is singular, with determinant $1/\sigma^2 u$. The equations of motion have the form (3.37), where the function $g(u, v, w)$ is now singular:

$$
\begin{aligned}
g(u, v, w) &= g_0(u, v, w) + g_{-1}(u, v, w)/u \\
g_0(u, v, w) &= \sigma b(r - 1)u - (\sigma + 1)bv - (\sigma + b + 1)w - u^2 v - \sigma u^3 \\
g_{-1}(u, v, w) &= (\sigma + 1 + w)v
\end{aligned}
\tag{3.39}
$$

In this form, the source term on the right-hand side does not obey a Lipschitz condition, so that the fundamental theorem does not apply to trajectories that approach the $u = 0$ plane.

Example 3: In this example we again carry out a change of variables on the Lorenz equations (3.20). These equations are equivariant (unchanged) under a group consisting of two elements: the identity and a rotation by π radians about the z axis: $R_z(\pi)$. We choose a change of variables that maps each pair of symmetry-related points (x, y, z) and $(-x, -y, +z)$ with $x^2 + y^2 \neq 0$ into the same point: $(u, v, w) = (x^2 - y^2, 2xy, z)$. The dynamical system equations in the new coordinate system are easily expressed in terms of the Jacobian of the transformation: $du_i/dt = (\partial u_i/\partial x_j)(dx_j/dt)$. This is explicitly

$$
\begin{aligned}
\frac{d}{dt}\begin{bmatrix} u \\ v \\ w \end{bmatrix} &= \begin{bmatrix} 2x & -2y & 0 \\ 2y & 2x & 0 \\ 0 & 0 & 1 \end{bmatrix} \begin{bmatrix} -\sigma x + \sigma y \\ rx - y - xz \\ -bz + xy \end{bmatrix} \\
&= \begin{bmatrix} -(\sigma + 1)u + (\sigma - r + w)v + (1 - \sigma)\rho \\ (r - \sigma - w)u - (1 + \sigma)v + (r + \sigma - w)\rho \\ -bw + v/2 \end{bmatrix}
\end{aligned}
\tag{3.40}
$$

where $\rho = \sqrt{u^2 + v^2}$. This last equation no longer possesses a symmetry. It is a $2 \to 1$ image of the original equations, which have twofold symmetry.

Remark: There is a general theory that describes the relation between dynamical systems with symmetry and closely related systems for which the symmetry has been eliminated (cover and image dynamical systems). This theory depends in a crucial way on transformations that are local but not global diffeomorphisms. The simplest applications of this theory, involving symmetry groups of order 2, are presented in Chapter 10.

3.4.3 Structure Theory

As the examples above show, the form of a dynamical system can change remarkably under a diffeomorphism. It would be useful to have a method for simplifying dynamical systems. Suppose that it is possible to find coordinates $y_1, y_2, \ldots, y_{n_1}$ and $z_1, z_2, \ldots, z_{n_2}$, $n_1 + n_2 = n$, so that the n-dimensional dynamical system (3.1) can be written in the form

$$\begin{aligned} \frac{dy_i}{dt} &= G_i(y, z) \qquad & i = 1, 2, \ldots, n_1 \\ \frac{dz_j}{dt} &= H_j(-, z) \qquad & j = 1, 2, \ldots, n_2 \end{aligned} \tag{3.41}$$

In this expression, $H_j(-, z)$ means that the functions H_j do not depend on the variables y. Then the second set of n_2 equations depends on the n_2 variables z_j alone. This dynamical subsystem can be integrated independent of the y variables. The explicit solutions can be used as time-dependent driving terms in the first set of n_1 equations which depend on the n_1 variables y: $dy_i/dt = G_i(y, z(t)) \to G_i(y, t)$. The phase space for the z subsystem is R^{n_2}; the phase space for the full dynamical system is $R^n \sim R^{n_1} \times R^{n_2}$.

A dynamical system with the form described in (3.41) is called *reducible.*

If reducibility of a dynamical system simplifies its treatment, complete reducibility simplifies the treatment even further. A dynamical system (3.1) is *fully reducible* (or *completely reducible*) if it is possible to find a diffeomorphism $x \to (y, z)$ so that the equations of motion assume the form

$$\begin{aligned} \frac{dy_i}{dt} &= G_i(y, -) \qquad & i = 1, 2, \ldots, n_1 \\ \frac{dz_j}{dt} &= H_j(-, z) \qquad & j = 1, 2, \ldots, n_2 \end{aligned} \tag{3.42}$$

In this case, each of the two dynamical subsystems evolves independent of the other.

A dynamical system is *irreducible* if there is no diffeomorphism that maps it to the form (3.41).

To date, the full power of transformation theory has not been exploited to simplify dynamical systems theory. Specifically, it would be very nice to have some kind of algorithm (analogous to Ado's algorithm in the theory of Lie algebras [68]) which identifies irreducible dynamical systems. Specifically, the algorithm should do three things. First, answer the question: Is this dynamical system reducible? If the answer to the first question is "yes," the algorithm should provide a method for constructing the diffeomorphism that effects the transformation to reduced form. Finally, the algorithm should carry out this transformation and display the reduced dynamical system equations.

Example: The driven Duffing equations can be written in the reducible form

$$\begin{aligned}\frac{dx}{dt} &= v \\ \frac{dv}{dt} &= \lambda x - x^3 - \gamma v + f r_1 \\ \frac{dr_1}{dt} &= -\omega r_2 \\ \frac{dr_2}{dt} &= +\omega r_1\end{aligned} \tag{3.43}$$

More generally, all periodically driven dynamical systems are equivalent to reducible autonomous systems.

Remark: Chaotically driven nonautonomous dynamical systems can be treated similarly. If the chaotic driving terms $z_1, z_2, \ldots, z_{n_2}$ appear in polynomial coupling terms in the original n_1-dimensional dynamical system, an n_2-dimensional dynamical system generating these terms can be adjoined to the original nonautonomous dynamical system and the time-dependent terms replaced by functions of the variables z. The resulting $n = (n_1 + n_2)$-dimensional dynamical system is autonomous and reducible.

Remark: The algorithm for reducibility should have a topological component. For example, the phase space for the Rössler attractor [the dynamics generated by equations (3.22) for certain ranges of the control parameter values] is $R^2 \times S^1$. This strongly suggests that it might be possible to find a coordinate system (a diffeomorphism) in which the Rössler equations assume reducible form. This is currently an open question.

3.5 FIXED POINTS

The global organization of a dynamical system $\dot{x}_i = F_i(x; c)$ is governed to a large extent by the number, distribution, and stability of its fixed points. The same is true of a map $x_i' = f_i(x)$.

3.5.1 Dependence on Topology of Phase Space

There is a strong relation between the fixed points of a dynamical system, the topological structure of its phase space, and the symmetry of the dynamical system. If the phase space of an n-dimensional dynamical system has the form $R^{n-1} \times S^1$, the dynamical system has no fixed points. This is the case for the periodically driven Duffing and van der Pol oscillators. In such cases two strategies are useful: (1) look for critical points of particular functions [e.g., $U(x)$ in Eq. (3.4), $f(x)$ in Eq. (3.10)] in the constant θ slices of S^1; (2) look for fixed points in the constant θ section of S^1 (a Poincaré section). These fixed points locate the period-1 orbits of the dynamical system. These closed orbits organize the dynamics in $R^2 \times S^1$ in much the same way as fixed points organize the dynamics in R^3.

If the dynamical system has symmetry, fixed points fall into classes. A simple example is provided by the Lorenz system, which possesses a twofold rotation symmetry about the Z axis. In this case, fixed points can exist on the symmetry axis or off the axis. Those that occur off the symmetry axis must occur in symmetry-related pairs. In general, if G is the symmetry group of the dynamical system and p is a fixed point that is invariant under the action of the subgroup $H \subset G$, then p will occur in a multiplet of order $|G|/|H|$, where $|G|$ is the order of G. The *partners* of p are the points in the *orbit* of p under G, or, more economically, under the quotient (*coset*) G/H. For the Lorenz equations there is one fixed point on the symmetry axis and a symmetry related pair of fixed points off the symmetry axis.

When the phase space is R^n and the dynamical system possesses no particular symmetry, algorithms exist for locating the fixed points of autonomous dynamical systems whose source terms are polynomials in the state variables.

3.5.2 How to Find Fixed Points in R^n

The fixed points of a dynamical system are located by solving the simultaneous nonlinear equations $F_i(x; c) = 0$. Typically, n equations in n phase-space variables will have only isolated solutions. The number of solutions will be finite if each of the functions $F_i(x; c)$ is a polynomial of finite degree in the n state variables x_i.

Often, the fixed points of a dynamical system can be located by the brute-strength approach favored by physicists. For example, the fixed points of the Lorenz system can be obtained using lower algebra. However, it would be comforting to know that more sophisticated methods ("higher" algebra) are available.

The study of solution sets of algebraic equations is the province of algebraic geometry [69]. A number of powerful computer programs have been developed to implement algorithms for locating the zeros of sets of algebraic equations. We illustrate in Fig. 3.5 how one of these algorithms is used in the widely available package Maple.

First, the Maple package *grobner* is loaded with the command $with(grobner)$;. The output list ($[finduni, \ldots, spoly]$) indicates the programs that are implemented in this package. Then the three polynomial equations that define (:=) the Lorenz equations are input without echoes(:), and collected as an array $F1$ (Maple is case sensitive) with echo(;). The variables $[x, y, z]$ are also collected into an array X. The two arrays of functions and coordinates, $F1$ and X, are arguments for the *gbasis* routine, which computes the Gröbner basis for these functions. Finally, the routine *gsolve* is called. The output is the set of roots of the simultaneous polynomial equations. Equivalently, it is the set of fixed points of the Lorenz flow.

Here is how to read the output. The roots consist of two sets. One set is $[x, y, z]$, which is to be read $x = 0, y = 0, z = 0$. The other set is $[x-y, y^2+b-br, -r+1+z]$. It is convenient to read these equations from right to left. Thus, in the second set, $z + 1 - r = 0$, then $y^2 + b - br = 0$, which has two solutions, and then $x - y = 0$, which means that for each of the two solutions $y_\pm$, $x = y$.

Although the fixed points of the Lorenz equations can easily be found by hand, such is not the case for more formidable sets of nonlinear polynomial equations. The

```
with(grobner);
        [finduni, finite, gbasis, gsolve, leadmon, normalf, solvable, spoly]
> f1 := -sigma * x + sigma * y :
> f2 := r * x - y - x * z :
> f3 := -b * z + x * y :
> F1 := [f1, f2, f3];
        F1 := [-σx + σy, rx - y - xz, -bz + xy]
> X := [x, y, z] :
> gbasis(F1, X);
        [x - y, y^2 - bz, yz + (1 - r)y, z^2 + (1 - r)z]
> gsolve(F1, X);
        [[x, y, z], [x - y, y^2 + b - br, -r + 1 + z]]
```

Fig. 3.5 This Maple worksheet computes the fixed points of the Lorenz equations.

procedure described here will *always* provide *all* the fixed points of such a dynamical system or map. However, this procedure may take awhile for large sets of polynomial equations.

3.5.3 Bifurcations of Fixed Points

The number of fixed points that a dynamical system has is an invariant. This number remains fixed as the control parameters are varied, if one counts their degeneracy. However, the number of *real* fixed points generally depends on the control parameter values, and is not invariant.

The values of the control parameters at which the number of real critical points changes is called the *bifurcation set*. As one passes transversally through the bifurcation set, two critical points with complex conjugate coordinate values scatter off each other to become real critical points.

We illustrate these ideas with two examples.

Example 1: The critical points of the Rössler system (3.22) are given by $z = -y$, $x = -ay$, where y satisfies $ay^2 + cy + b = 0$. The two solutions are

$$y_{\pm} = -\frac{c}{2a} \pm \sqrt{\left(\frac{c}{2a}\right)^2 - \frac{b}{a}}$$

The solutions are

Real			> 0
Degenerate	if	$c^2 - 4ab$	$= 0$
Complex			< 0

The bifurcation set in control parameter space is the two-dimensional surface $c^2 - 4ab = 0$. This result is also easily obtained from the canonical rank-1 form (3.37) for the Rössler equations by locating the doubly degenerate fixed point along the u axis $v = w = 0$.

Example 2: The three critical points for the Lorenz system were computed in Fig. 3.5. There is always one real critical point at $(x, y, z) = (0, 0, 0)$. The other two critical points have real coordinate $z = r - 1$. The x and y coordinates are imaginary if $r < 1$, real if $r > 1$. There is a threefold degeneracy of the critical points for $r = 1$. The bifurcation from one to three real critical points occurs as the two-dimensional surface $r = 1$ in the control parameter space (σ, b, r) is traversed.

Remark 1: The real fixed points of the Rössler and Lorenz equations lie on one-dimensional curves in phase space. These curves are given parametrically as follows:

$$\begin{array}{llcl} \text{Rössler} & (x, y, z) & = & \left(\frac{c}{2} - as, -\frac{c}{2a} + s, +\frac{c}{2a} - s\right) \\ \text{Lorenz} & (x, y, z) & = & (s, s, s^2/b) \end{array}$$

with $-\infty < s < +\infty$. If all the real fixed points of a dynamical system lie on a one-dimensional curve, the dynamical system has rank 1. More generally, if all the real fixed points of a dynamical system lie on a surface of minimal dimension r, the dynamical system has *rank* r. The rank of a dynamical system can be determined by identifying the bifurcation that takes place when all the critical points are degenerate. Specifically, rank is defined as for singularities and catastrophes [37]. The fixed-point rank of a dynamical system is the number of zero eigenvalues at the most degenerate critical point. The rank and number of real critical points are important invariants of a dynamical system. They provide a partial classification for dynamical systems. The Rössler and Lorenz systems are of type A_2 and A_3, since their real critical points arise through fold and cusp bifurcations. Both are of rank 1.

Remark 2: At least r independent time series are required for a topologically accurate embedding of a rank r dynamical system. For systems of type A_n ($n = 2$ for Rössler, $n = 3$ for Lorenz) a single time series suffices.

Remark 3: It is useful to choose a coordinate system that emphasizes the rank of a dynamical system. For the Lorenz system, such a coordinate system (u, v, w) has the property that all critical points lie on the u axis $(u, v = 0, w = 0)$. The diffeomorphism

$$\begin{array}{rcll} u & = & \alpha x + \beta y & \qquad \alpha + \beta = 1 \\ v & = & x - y & \\ w & = & bz - xy & \end{array} \tag{3.44}$$

brings the Lorenz system to the following form:

$$\begin{array}{rcl} \dot{u} & = & \beta(r-1)u - [\alpha(\alpha\sigma - \beta) + \beta(\alpha\sigma - \beta r)]\, v \\ & & -\ (\beta/b)(u + \beta v)\, [w + (u + \beta v)(u - \alpha v)] \\ \dot{v} & = & -(r-1)u - (\sigma + \alpha + \beta r)v \\ & & +(1/b)(u + \beta v)\, [w + (u + \beta v)(u - \alpha v)] \\ \dot{w} & = & -bw - r(u + \beta v)^2 - \sigma(u - \alpha v)^2 + (\sigma + 1)(u + \beta v)(u - \alpha v) \\ & & +(u + \beta v)^2\, [w + (u + \beta v)(u - \alpha v)] \end{array} \tag{3.45}$$

This set of equations remains invariant under the twofold rotation $R_z(\pi)$: $(u, v, w) \to (-u, -v, +w)$. Its three fixed points lie on the u axis $v = w = 0$ and satisfy $(r-1)u - (1/b)u^3 = 0$. If we choose $\alpha = 1/(\sigma+1)$, $\beta = \sigma/(\sigma+1)$, the linear part of this equation is

$$\frac{d}{dt}\begin{bmatrix} u \\ v \\ w \end{bmatrix} = \begin{bmatrix} \dfrac{\sigma(r-1)}{\sigma+1} & \dfrac{\sigma^2(r-1)}{\sigma+1} & 0 \\ -(r-1) & -\sigma - \dfrac{\sigma r+1}{\sigma+1} & 0 \\ 0 & 0 & -b \end{bmatrix}\begin{bmatrix} u \\ v \\ w \end{bmatrix} \tag{3.46}$$

On the bifurcation set $r - 1 = 0$ this matrix is diagonal, with two negative and one vanishing eigenvalue. The two new fixed points [real nonzero solutions of $(r-1)u - (1/b)u^3 = 0$ for $r > 1$] bifurcate along the u axis. The transformation (3.44) with $\alpha = 1/(\sigma+1)$, $\beta = \sigma/(\sigma+1)$ maps to the center manifold $v = w = 0$ at the bifurcation $r - 1 = 0$.

3.5.4 Stability of Fixed Points

Once the fixed points of a flow have been located, it is a relatively simple matter to determine their stability. The flow in the neighborhood of a fixed point can be determined by linearizing the system in the usual way. We write $x(t) = x^* + \delta x$. Then

$$\begin{aligned} \frac{d}{dt}(x^* + \delta x)_i &= F_i(x^* + \delta x, c) \\ \frac{d\delta x_i}{dt} &= \left.\frac{\partial F_i(x,c)}{\partial x_j}\right|_{x^*} \delta x_j \end{aligned} \tag{3.47}$$

Stability of motion in the neighborhood of the fixed point is determined by the eigenvalues $\lambda_i(c)$ of the stability matrix $\partial F_i(x^*, c)/\partial x_j$. Motion relaxes to the fixed point if all the eigenvalues of this matrix have negative real part. The motion is unstable in k directions if k eigenvalues have positive real part. If one or more eigenvalues have zero real part, then under weak conditions, a bifurcation is about to take place as the control parameters c are swept from $c - \delta c$ to $c + \delta c$.

Example: The Jacobian of the Lorenz flow is

$$J = \begin{bmatrix} -\sigma & \sigma & 0 \\ r - z & -1 & -x \\ y & x & -b \end{bmatrix} \tag{3.48}$$

The Jacobian matrix at the fixed point $(x, y, z) = (0, 0, 0)$ on the symmetry axis is simple to evaluate. The eigenvalues and eigenvectors are

Eigenvalue	Eigenvector
λ_+	$(\sigma, \sigma + \lambda_+, 0)$
λ_-	$(\sigma, \sigma + \lambda_-, 0)$
$-b$	$(0, 0, 1)$

(3.49)

where $\lambda_{\pm} = -\frac{1}{2}(\sigma+1) \pm \sqrt{(\frac{1}{2}(\sigma+1)^2 + \sigma(r-1)}$. For $\sigma, b > 0$, the two eigenvalues λ_- and $-b$ are negative. The third eigenvalue, λ_+, is negative for $r < 1$ and positive for $r > 1$. For $\sigma, b > 0$, this fixed point is stable for $r < 1$ and unstable in one direction for $r > 1$. At $r = 1$ its stablity is determined by including higher-order terms in the linearized equations of motion. However, stability is of less interest than the bifurcation which occurs as r passes through $r = 1$. The bifurcation is of type A_3 (cusp). It takes place along the eigenvector whose eigenvalue is λ_+. These two new real fixed points are both stable, by general arguments of topological type. Similar calculations can be carried out on the symmetric pair of fixed points $(x, y, z) = (\pm\sqrt{b(r-1)}, \pm\sqrt{b(r-1)}, r-1)$.

Remark: Imaginary fixed points (sometimes called *ghosts*) can have a significant influence on the dynamics, particularly as they approach a bifurcation.

3.6 PERIODIC ORBITS

Locating periodic orbits in a flow is far more difficult than finding the fixed points of a flow. However, the periodic orbits in a flow are a key ingredient in the description of the chaotic behavior of a dynamical system. In fact, the classification of low-dimensional chaotic dynamical systems depends in a fundamental way on the unstable periodic orbits which are all but invisible in such systems. It is therefore useful to describe methods for finding periodic orbits and assessing their stability.

3.6.1 Locating Periodic Orbits in $R^{n-1} \times S^1$

The problem of finding periodic orbits is simplified to some extent if the phase space for the flow has the topology of a torus: $R^{n-1} \times S^1$. It is then often possible to construct a first return map $R^{n-1} \to R^{n-1}$ which describes how a Poincaré section R^{n-1} at $S^1(\theta)$ is mapped into itself as θ increases to $\theta + 2\pi$. If we represent this map by

$$\begin{aligned} x_1' &= f_1(x_1, x_2, \ldots, x_{n-1}) \\ x_2' &= f_2(x_1, x_2, \ldots, x_{n-1}) \\ &\vdots \\ x_{n-1}' &= f_{n-1}(x_1, x_2, \ldots, x_{n-1}) \end{aligned} \tag{3.50}$$

then the problem of finding the period-1 orbits becomes equivalent to the problem of finding the zeros of the functions $x_i - f_i(x_1, x_2, \ldots, x_{n-1})$. If these functions are polynomials, the *grobner* package described above can be used to find the zeros of the difference map, or the period-1 orbits of the flow.

Higher-period orbits are located using a standard procedure. The pth iterate of the first return map is constructed, and its fixed points are located.

Example 1: The Hénon map $R^2 \rightarrow R^2$ is invertible if $s \neq 0$:

$$\begin{bmatrix} x \\ y \end{bmatrix}' = \begin{bmatrix} a - (x^2 + y) \\ sx \end{bmatrix} \tag{3.51}$$

The Gröbner package identifies the two fixed points as

$$[sx - y, y^2 + (s^2 + s)y - as^2] \tag{3.52}$$

The second factor is to be solved for the two values of the y coordinate. Then the linear equation $sx - y = 0$ is used to solve for each corresponding x coordinate.

Example 2: The cusp map $R^2 \rightarrow R^2$ is not invertible:

$$\begin{bmatrix} x \\ y \end{bmatrix}' = \begin{bmatrix} a_1 - y \\ a_2 - (x^3 + xy) \end{bmatrix} \tag{3.53}$$

The Gröbner package identifies the three fixed points as

$$[-a_1 + x + y, y^3 + (-3a_1 + 1)y^2 + (-1 - a_1 + 3a_1^2)y + (a_2 - a_1^3)] \tag{3.54}$$

The second factor is to be solved for the three values of the y coordinate. Then the linear equation $x + y - a_1 = 0$ is used to solve for each corresponding x coordinate.

3.6.2 Bifurcations of Fixed Points

Bifurcation sets for maps are described in exactly the same way as bifurcation sets for flows. We illustrate with two examples.

Example 1: The y coordinates for the fixed points of the invertible ($s \neq 0$) Hénon map (3.51) are obtained from (3.52):

$$y_{\pm} = -\frac{1}{2}(s^2 + s) \pm \sqrt{\left(\frac{1}{2}(s^2 + s)\right)^2 + as^2} \tag{3.55}$$

There are two real fixed points when $(s+1)^2 + 4a > 0$. The bifurcation set is defined by $(s+1)^2 + 4a = 0$.

Example 2: The bifurcation set for the noninvertible cusp map (3.53) is defined by the double degeneracy of the roots of the cubic equation that appears in (3.54). The condition for root degeneracy is imposed as follows:

$$\begin{array}{lcccccccc} \text{Roots of cubic}: & y^3 & + & Ay^2 & + & By & + & C & = 0 \\ \text{Degeneracy condition}: & 3y^2 & + & 2Ay & + & B & & & = 0 \end{array} \tag{3.56}$$

These two equations provide a parametric representation for the bifurcation set $(a_1(y), a_2(y))$ in terms of a dummy variable y. The bifurcation set is shown in Fig. 3.6. It consists of two components (*fold lines*). In the cusp-shaped region there are three real fixed points with y coordinate values $y_1 < y_2 < y_3$. The interior fixed point becomes degenerate with one of the two outside fixed points when either of the fold lines is crossed. The dummy variable is the y coordinate of the doubly degenerate fixed point.

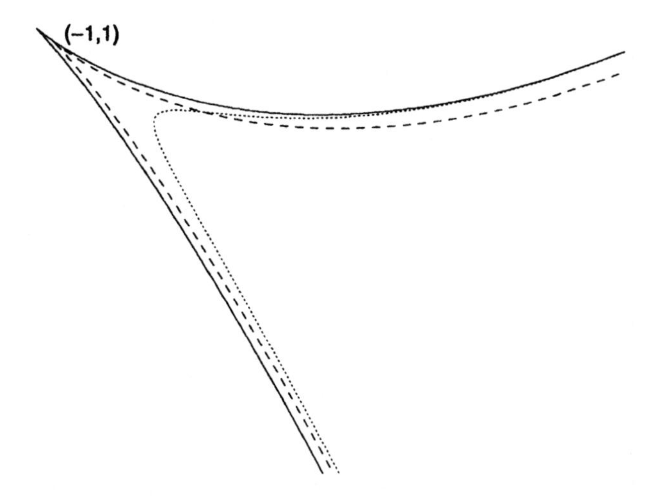

Fig. 3.6 Two fold lines (solid) form the bifurcation set for the noninvertible cusp map (3.53). On crossing a fold line from inside the cusp-shaped region to outside, the interior and one of the two exterior real fixed points collide, become degenerate, and scatter off each other to become a complex-conjugate pair. The dashed curve is the locus of the first period-doubling bifurcation. The dotted curve is the locus of Hopf bifurcations.

3.6.3 Stability of Fixed Points

The stability of periodic orbits is determined by linearizing the first return map in the neighborhood of fixed points. If x^* is a fixed point of a first return map, then

$$(x^* + \delta x)'_i = f_i(x^* + \delta x) = f_i(x^*) + \left.\frac{\partial f_i}{\partial x_j}\right|_{x^*} \delta x_j \tag{3.57}$$

In particular, the stability is determined from the eigenvalues of the Jacobian matrix

$$J = \left[\frac{\partial f_i(x^*)}{\partial x_j}\right] \tag{3.58}$$

The fixed point is stable if all eigenvalues $\lambda_i(c)$ are smaller than 1 in absolute value: $|\lambda_i(c)| < 1$. The fixed point of the map is unstable in k directions if k eigenvalues have magnitude larger than 1. If one or more eigenvalues have magnitude equal to 1, bifurcations can be expected as the control parameters are swept from $c - \delta c$ to $c + \delta c$.

Example: The Jacobian of the invertible Hénon map (3.51) is

$$J = \begin{bmatrix} -2x & -1 \\ s & 0 \end{bmatrix} \tag{3.59}$$

The two eigenvalues of this matrix are $\lambda_{\pm} = -x \pm \sqrt{x^2 - s}$. The x coordinates of the fixed points are $x_{\pm} = -\frac{1}{2}(s+1) \pm \sqrt{(\frac{1}{2}(s+1))^2 + a}$. At the A_2 bifurcation to two real fixed points, one of the two is stable and the other is unstable in one direction.

3.7 FLOWS NEAR NONSINGULAR POINTS

In the neighborhood of a nonsingular point $\bar{x}$ it is possible to find a local diffeomorphism that brings the flow to a simple canonical form. The canonical form is

$$\begin{aligned} \dot{y}_1 &= 1 \\ \dot{y}_j &= 0 \qquad\qquad j = 2, 3, \ldots, n \end{aligned} \tag{3.60}$$

The transformation to this canonical form is shown in Fig. 3.7.

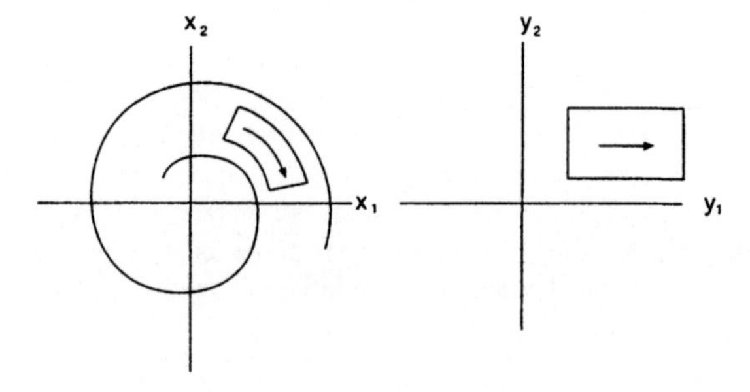

Fig. 3.7 A smooth transformation reduces a dynamical system to the very simple local normal form (3.60) in the neighborhood of a nonsingular point.

The local normal form (3.60) tells us nothing about how the phase space is stretched and squeezed locally by the flow. To provide this information, at least locally, we present another version of this local normal form theorem that is much more useful for our purposes. If $\bar{x}$ is not a singular point, there is an orthogonal (volume-preserving) transformation, centered at $\bar{x}$, to a new coordinate system in which the dynamical equations assume the following local canonical form in the neighborhood of $\bar{x}$:

$$\begin{aligned} \dot{y}_1 &= |F(x,c)| = \left(\sum_{k=1}^{n} F_k(\bar{x},c)^2 \right)^{1/2} \\ \dot{y}_j &= \lambda_j(\bar{x},c) y_j \qquad\qquad j = 2, 3, \ldots, n \end{aligned} \tag{3.61}$$

where the forcing terms $F_k(\bar{x}, c)$ are defined in Eq. (3.1). The local eigenvalues $\lambda_j(\bar{x}, c)$ describe how the flow deforms phase space in the neighborhood of $\bar{x}$. This deformation is illustrated in Fig. 3.8. The constant associated with the y_1 direction shows how a small volume is displaced by the flow in a short time Δt. If $\lambda_2 > 0$ and $\lambda_3 < 0$, the flow stretches the initial volume in the y_2 direction and shrinks it in the y_3 direction. The exponents $\lambda_j(\bar{x}, c)$ are called *local* Lyapunov exponents.

Remark: One eigenvalue of a flow at a nonsingular point always vanishes, and the associated eigenvector is the flow direction. The remaining eigenvectors describe stretching or shrinking in the directions transverse to the flow. The idea is as follows. The properties of the flow are determined by the leading terms in a Taylor series expansion of the forcing terms at any point x_0 in phase space. The leading (constant) terms define the flow direction. A coordinate system change $y = y(x)$ $(y(x_0) = 0)$ is introduced so that the flow *throughout* the neighborhood of x_0 has the form $dy_1/dt = k$, with k some constant. The remaining equations for $\dot{y}_i$ $(i > 1)$ have no constant term: Their leading terms are generically linear in the displacements: $dy_i/dt = F_{ij}y_j$. The eigenvalues of this $(n-1) \times (n-1)$ matrix are the local Lyapunov exponents. The eigenvectors are transverse to the flow direction and define the local directions in phase space in which the flow departs from or approaches the reference filament through x_0 at the rate $y_i(t) \sim y_i(0)e^{\lambda_i t}$.

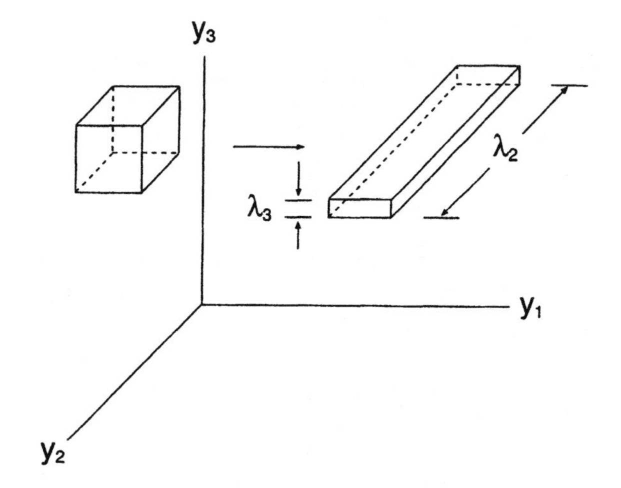

Fig. 3.8 Orthogonal transformation reduces a dynamical system to the very simple local normal form (3.61) in the neighborhood of a nonsingular point. The local Lyapunov exponents describe how the edges of the cube of initial conditions expand or contract.

3.8 VOLUME EXPANSION AND CONTRACTION

Under the flow, a small region in phase space will change shape and size, as can be seen in Fig. 3.8. The time rate of change of the volume of the region is determined by the divergence theorem.

We assume that a small volume V in phase space is bounded by a surface $S = \partial V$ at time t and ask how the volume changes during a short time dt. The volume will change because the flow deforms the surface. The change in the volume is equivalent to the flow through the surface. The volume change can be expressed as

$$V(t + dt) - V(t) = \oint_{\partial V} dx_i \wedge dS_i \tag{3.62}$$

Here dS_i is an element of surface area that is orthogonal to the displacement dx_i and $\wedge$ is the standard mathematical generalization to R^n of the vector cross product in R^3. The time rate of change of volume is

$$\frac{dV}{dt} = \oint_{\partial V} \frac{dx_i}{dt} \wedge dS_i = \oint_{\partial V} F_i \wedge dS_i \tag{3.63}$$

The surface integral is related to a volume integral by

$$\frac{1}{V}\frac{dV}{dt} = \lim_{V \to 0} \frac{1}{V} \oint_{\partial V} F_i \wedge dS_i \stackrel{\text{def}}{=} \text{div}F = \nabla \cdot F \tag{3.64}$$

In a locally Cartesian coordinate system, $\text{div}F = \sum_{i=1}^{n} \partial F_i / \partial x_i$. The divergence can also be expressed in terms of the local Lyapunov exponents:

$$\text{div}F = \sum_{j=1}^{n} \lambda_j(\bar{x}, c) \tag{3.65}$$

where $\lambda_1 = 0$ (flow direction) and λ_j ($j > 1$) are the local Lyapunov exponents in the directions transverse to the flow direction. This is a direct consequence of the local normal form result (3.61).

3.9 STRETCHING AND SQUEEZING

In this work we are interested principally in chaotic motion. Roughly speaking, this is motion that is recurrent but nonperiodic in a bounded region in phase space. Chaotic motion is defined by two properties: (1) sensitivity to initial conditions, and (2) recurrent nonperiodic behavior.

Sensitivity to initial conditions means that two nearby points in phase space "repel" each other. That is, the distance between two initially nearby points increases exponentially in time, as least for sufficiently small time:

$$d(t) = d(0)e^{\lambda t} \tag{3.66}$$

Here $d(t)$ is the distance separating two points at time t, and $d(0)$ is the initial distance separating them at time $t = 0$, t is sufficiently small, $d(0)$ is sufficiently small, and the *Lyapunov exponent* λ is positive. To put it graphically, two nearby initial conditions are *stretched apart*.

If two nearby initial conditions diverged from each other exponentially in time for all times, they would wind up at opposite ends of their universe (phase space). If the motion in phase space is bounded, the two points will eventually reach a maximum separation and then begin to approach each other again. To put it graphically again, the two points are *squeezed together*.

Chaotic motion is generated by the repetitive action of these two processes. Specifically, a strange attractor is built up by the repetition of the stretching and squeezing processes. It is this repetitiveness that is responsible for the self-similar geometry of strange attractors. More to the point, if we understand how to choose some appropriate piece of phase space, and if we know the nature of the stretching and squeezing processes, we have an algorithm for constructing strange attractors from the given input and the two processes acting together.

We illustrate these ideas in Fig. 3.9 for a process that develops a strange attractor in R^3. We begin with a set of initial conditions in the form of a cube. As time increases, the cube stretches in directions with positive local Lyapunov exponents and shrinks in directions with negative local Lyapunov exponents. Two typical nearby initial conditions (a) separate at a rate determined by the largest local Lyapunov exponent (b). Eventually, these points reach a maximum separation (c), and thereafter are squeezed to closer proximity (d). If the initial neighborhood is chosen properly, the image of the neighborhood may intersect the original neighborhood after a finite time (e),(a). This stretching and squeezing is then repeated, over and over again, to build up a strange attractor.

Remark 1: We make a distinction between shrinking, which must occur in a dissipative system since some eigenvalues must be negative [cf. Fig. 3.9(b); $\sum_{i=1}^{n} \lambda_i < 0$], and squeezing, which forces distant parts of phase space together.

Remark 2: When squeezing occurs, the two distant parts of phase space that are being squeezed together must be separated by a boundary layer. This boundary layer is indicated in Fig. 3.9(d). Boundary layers in dynamical systems are important but have not been studied extensively.

Remark 3: The intersection of the original neighborhood with its image under the flow creates the possibility of having fixed points in the first-, second-, ..., pth-return maps. These fixed points identify period-1, period-2, ..., period-p orbits in the flow.

3.10 THE FUNDAMENTAL IDEA

The stretching and squeezing processes, repeated over and over again in phase space, act to generate chaotic behavior and to build up strange attractors. Many different stretching and squeezing processes exist. For example, the strange attractors that are generated by the Duffing, van der Pol, Lorenz, and Rössler systems are built up by four inequivalent stretching and squeezing processes (cf. Fig. 4.1). This means, in particular, that no diffeomorphism exists that maps any one of these sets of equations into any of the others. They are distinct and inequivalent at a fundamental (topological) level.

Embedded in each strange attractor is a large collection of periodic orbits. In three dimensions these orbits are rigidly organized with respect to each other. The organization is determined by the particular stretching and squeezing mechanism which acts to build up the strange attractor. Each different stretching and squeezing

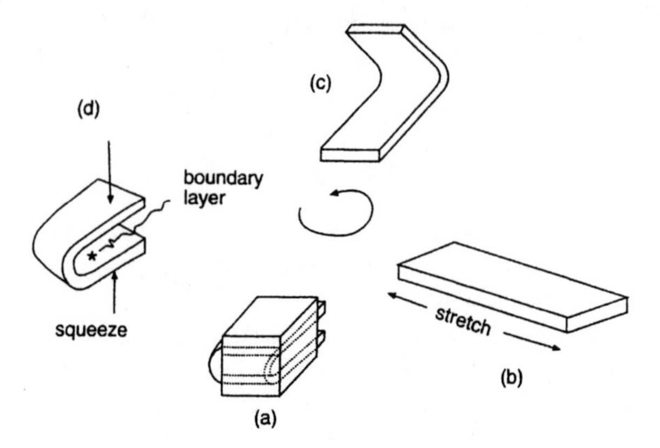

Fig. 3.9 A cube of initial conditions (a) evolves under the flow. The cube moves in the direction of the flow. The sides stretch in the direction of the positive local Lyapunov exponents and shrink in the direction of the negative local Lyapunov exponents (b). Eventually, two initial conditions reach a maximum separation (c) and begin to get squeezed back together (d). A boundary layer separates distant parts of phase space that are squeezed together. The image of the cube, under the flow, may intersect the original neighborhood. Repetition of the stretching and squeezing processes builds up a layered structure that is self-similar. This geometric object is called a *strange attractor*. The stretching and squeezing mechanisms also create large numbers of periodic orbits which are organized among themselves in a unique way. This organization is the Achilles' Heel for understanding strange attractors.

mechanism generates a different organization of periodic orbits in the host strange attractor.

Fundamental Idea: We will identify strange attractors, as well as the stretching and squeezing mechanisms that generate them, by identifying the organization of the periodic orbits "in" the strange attractor. The organization of the orbits in turn is determined by a spectrum of names for the orbits (symbolic dynamics) together with a set of topological indices for each orbit and each pair of orbits. These indices, linking numbers and relative rotation rates, are discussed in Chapter 4. In short, periodic orbits provide the fingerprints for identifying strange attractors and the stretching and squeezing mechanisms that generate them.

3.11 SUMMARY

Physical systems evolved continuously in time. They are described by sets of ordinary or partial differential equations. Sets of coupled first-order ordinary (nonlinear) differential equations are called *(continuous) dynamical systems*. We introduced the fundamental theorem, the existence and uniqueness theorem, for such systems. A number of examples of such systems were discussed. We return to these examples (Duffing, van der Pol, Lorenz, Rössler) again and again throughout the book, for properties of these four sets of equations are similar to the properties of a very large number of physical systems.

Much of the behavior of a dynamical system is governed by the number, distribution, and stability of its fixed points, and their evolution as control parameters are varied. We presented algorithms for locating the fixed points of flows, determining their stability, and studying their bifurcations. These algorithms apply also to the periodic orbits in flows.

The stability properties of a flow are summarized by its spectrum of Lyapunov exponents. These characterize local expansion and contraction directions in phase space. To generate chaotic behavior, which is recurrent nonperiodic motion that exhibits sensitivity to initial conditions, two mechanisms must operate in phase space: stretching and squeezing. Stretching is responsible for sensitivity to initial conditions. Squeezing is responsible for recurrence. The fundamental idea which we pursue is that these two mechanisms can be identified by analyzing chaotic data, and they can be used to classify strange attractors.

4

Topological Invariants

The objectives of the present work are to:

- Identify the stretching and squeezing mechanisms that act to build up strange attractors.
- Classify strange attractors.
- Understand the perestroikas of strange attractors: that is, how strange attractors change as control parameters change.

Many different stretch and squeeze mechanisms exist and occur in physical systems and their mathematical models. We survey several such mechanisms at the beginning of this chapter. The primary means of differentiating among these different mechanisms is through the periodic orbits which they generate. In particular, the topological organization of the unstable periodic orbits associated with a strange attractor distinguishes different strange attractors from each other. The bulk of this chapter is devoted to a description of the two tools that have proven most useful in determining the topological organization of periodic orbits. These tools, linking numbers and relative rotation rates, are useful not only for distinguishing among strange attractors, but also for distinguishing among classes of periodic orbits with different properties within a single strange attractor.

4.1 STRETCHING AND SQUEEZING MECHANISMS

To make these ideas concrete, in Fig. 4.1 we show four different mechanisms that generate strange attractors. These mechanisms are illustrated by cartoons of the flows which they describe. These cartoons, or caricatures of the flows, are described more fully in Chapter 5. For now, they can simply be considered as representations of the flows when the dissipation is very high.

The first mechanism, shown in Fig. 4.1(a), is the simple stretch and fold mechanism which generates a Smale horseshoe return map. The map related to this flow (the orientation-preserving Hénon map) was discussed in Section 2.11. This simple mechanism occurs frequently when the phase space for the flow is $R^2 \times S^1$. It is this mechanism that is responsible for creating the strange attractor generated by the Rössler equations (3.22).

The second mechanism, shown in Fig. 4.1(b), is an iterated stretch-and-fold mechanism. During the first half of a full cycle an initial neighborhood in phase space is stretched and folded over as in the Smale horseshoe mechanism. The deformation is repeated during the second half of the cycle. This mechanism occurs naturally when the phase space for the dynamical system is $R^2 \times S^1$ and the flow is unchanged (equivariant) under the symmetry $R^2 \times S^1 \rightarrow R^2 \times S^1$ given explicitly by $(x, y, t) \rightarrow (-x, -y, t + \frac{1}{2}T)$, where T is the period of the driving term. It is this mechanism that is responsible for creating the strange attractor generated by the Duffing equations (3.5).

The third mechanism, shown in Fig. 4.1(c), involves a shear in the angular direction of a mapping of an annulus to itself. This mechanism occurs frequently when the phase space for the flow is $(R^1 \times S^1) \times S^1$. It occurs for driven dynamical systems (the second S^1) whose constant θ (or t) slice is not R^2 but rather $R^1 \times S^1$. It is this mechanism that acts to generate the strange attractor of the van der Pol oscillator (3.14).

The fourth mechanism, shown in Fig. 4.1(d), tears the flow apart in phase space. This type of mechanism can occur when two or more focus-type fixed points, around which the flow takes place, are separated by a saddle. It is this mechanism that acts to generate the strange attractor of the Lorenz equations (3.20).

How is it possible to tease out information about the stretching and squeezing mechanisms which generate a strange attractor from the strange attractor itself? The following observations suggest a method.

1. Each stretching and squeezing mechanism that builds up a strange attractor also creates periodic orbits in the attractor. In Fig. 4.2 we show one such mechanism. The intersections of the strange attractor with a series of Poincaré sections is shown, along with a closed orbit of period 4. Many other closed orbits are generated by this mechanism. The knot and braid structure of these

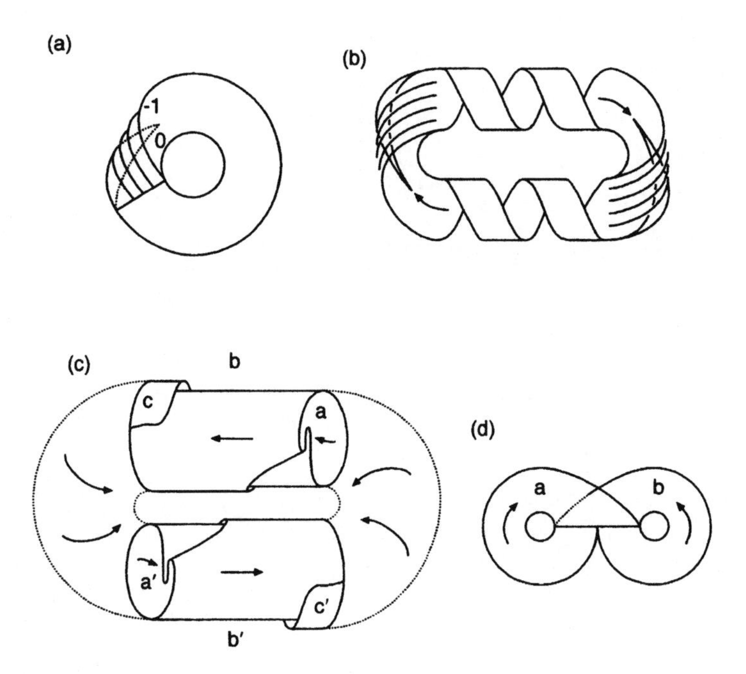

Fig. 4.1 Four distinct stretch and squeeze mechanisms generate well-studied strange attractors. (a) The simple stretch and fold mechanism acts on flows in $R^2 \times S^1$ and generates a Smale horseshoe return map. The Rössler equations generate such a mechanism. (b) The iterated stretch and fold mechanism also acts on flows in $R^2 \times S^1$ with a twofold symmetry under $(x, y, t) \to (-x, -y, t + \frac{1}{2}T)$. The Duffing equations generate such a mechanism. (c) The stretch and fold mechanism acting on an annulus in phase space $(R^1 \times S^1) \times S^1$ must generate a double reverse fold in the angular direction (the first S^1) by global topological boundary conditions. The van der Pol equations generate such a mechanism. (d) This mechanism tears the flow apart in the neighborhood of a saddle in R^3, forcing part of the flow to the influence of one unstable focus, the remainder to the influence of another unstable focus. The Lorenz equations generate such a mechanism.

orbits is determined uniquely by the mechanism that builds up the strange attractor.

2. A well-known theorem states that unstable periodic orbits are *dense* in a *hyperbolic* strange attractor. However, even a nonhyperbolic strange attractor is accompanied by an infinite number of periodic orbits, most of which are unstable.

3. It is sometimes possible to extract some unstable periodic orbits from chaotic time-series data. In fact, the lower the (Lyapunov) dimension of the strange attractor, the easier it is. In Chapter 6 we describe how to use the method of

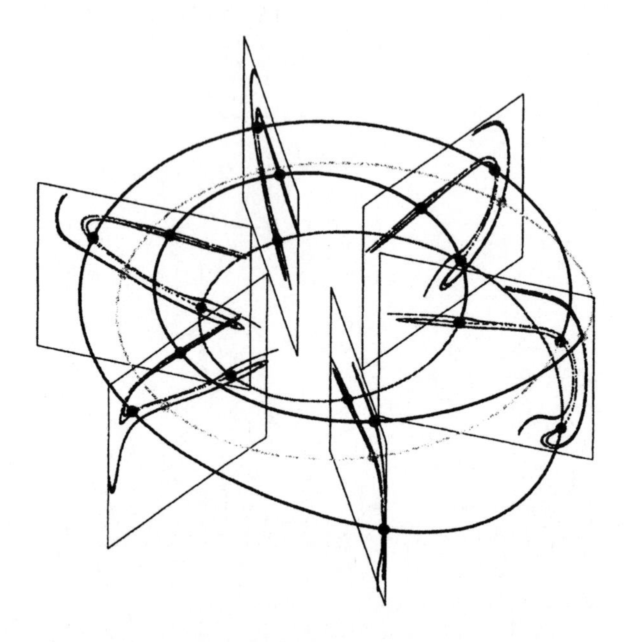

Fig. 4.2 The stretching and squeezing mechanisms that build up a strange attractor also create the unstable periodic orbits in the attractor. Here a period-4 orbit is shown for the mechanism illustrated in Fig. 3.9, which builds up the attractor shown in Figs. 7.20 and 7.21.

close returns to locate (surrogates of) unstable periodic orbits in chaotic time series.

4. The unstable periodic orbits are rigidly organized with respect to each other if they occur in a three-dimensional space: for example, R^3, $R^2 \times S^1$, or $(R^1 \times S^1) \times S^1$.

The idea of organizational rigidity is illustrated in Fig. 4.3. In this figure we show two orbits, A and B, which occur in a strange attractor in a three-dimensional space. Now suppose that we tune some control parameters in an attempt to displace orbit A to a new position, A'. If the orbit A continues to exist through this displacement (homotopy), it will carve out a two-dimensional surface in R^3, shown shaded in Fig. 4.3. If the original pair of orbits A, B and the displaced orbits A', B are organized differently, the orbit B will intersect the surface bounded by the orbit pair A, A'. At the point of intersection the fundamental theorem of dynamical systems is violated: two different orbits, B and some orbit interpolating between A and A', have the same initial condition but different futures. This means that orbit A cannot be deformed to orbit A' continuously if A and A' are organized differently with respect to orbit B.

The idea of orbit organization can be made quantitative. Intuitively, in Fig. 4.3 the orbits A and A' *link* the orbit B in different ways. In the following two sections we

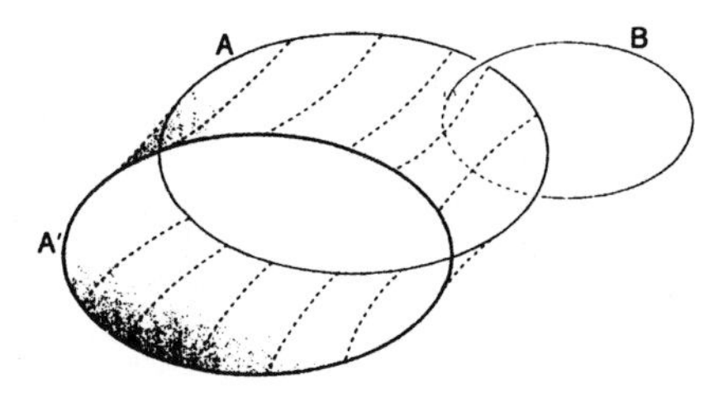

Fig. 4.3 Orbit A cannot be deformed to orbit A' continuously without intersecting orbit B if A and A' are organized differently with respect to B in R^3. At the intersection point the fundamental theorem of dynamical systems is violated. The organization of A and A' with respect to B is determined in part by the Gaussian linking numbers $L(A, B)$ and $L(A', B)$.

introduce topological invariants which are useful for describing the organization of unstable periodic orbits in strange attractors. The two invariants are linking numbers and relative rotation rates. These invariants, computed for a relatively small number of orbits, provide information that is sufficient to identify the stretching and squeezing mechanisms that generate a strange attractor while simultaneously organizing all the unstable periodic orbits in the attractor in a unique way.

Remark 1: There are many more periodic orbits in a strange attractor than there are stretching and squeezing mechanisms for generating strange attractors, speaking roughly but accurately. This is why it is possible to determine the topological organization of a limited number of periodic orbits extracted from a strange attractor and then use that information to classify the stretching and squeezing mechanisms which generate a strange attractor, and therefore the strange attractor itself.

Remark 2: The description of strange attractors involves at the same time a certain amount of rigidity and looseness. The rigidity comes from the spectrum of unstable periodic orbits associated with the strange attractor and the topological organization of these unstable periodic orbits among themselves. The looseness is associated with the ability to carry out smooth changes of coordinates. These deform the attractor while leaving unchanged both its spectrum of unstable periodic orbits as well as their rigid topological organization. Such deformations change the geometry of the attractor without changing its topology.

Remark 3: The rigid topological organization of unstable periodic orbits breaks down in higher dimensions. One way to understand this is through the concept of transversality. Two manifolds of dimensions n_1 and n_2 embedded in an N-dimensional space may or may not intersect each other. If they intersect, they intersect transversally in a manifold of dimension $n_1 + n_2 - N$ if $n_1 + n_2 - N \geq 0$; otherwise, they do not intersect [37]. In Fig. 4.3, n_1 is the dimension of the closed orbit B, n_2 is the dimension of the surface bounded by A and A', and these two manifolds either do not intersect or intersect at a point in R^3, since $1 + 2 - 3 = 0$. In R^N, $N > 3, 1+2-N < 0$, these two manifolds typically do not intersect, no violation of the fundamental theorem occurs, and the rigid organization of the unstable periodic orbits is lost.

Remark 4: When the closed orbits are dressed by their stable and unstable manifolds, the rigid organization is not lost.

4.2 LINKING NUMBERS

As with much of modern mathematics, the concept of linking number was first formulated quantitatively by Gauss, [70, 71].

There was a time when knots were thought to underlie the atomic nature of matter. Kelvin proposed that atoms were knotted vortex tubes of ether. Different knots corresponded to different atoms. This idea was taken seriously for awhile, since it provided explanations for some atomic properties. Specifically, physicists of the period felt that this idea could explain three important properties of atoms—their variety, stability, and spectra—as well as many molecular properties, as follows:

Variety: Each chemical element corresponded to a different ether knot.

Stability: Stable knots explained the stability of matter.

Spectrum: It was assumed that the ether knots could vibrate and that their normal modes corresponded to the spectral lines emitted and absorbed by the corresponding atoms.

Molecules: Bonds between atoms were caused by links between their ether knots. The variety, stability, and spectra of molecules arose from the properties of these ether links, as suggested above.

It is amusing to point out that this quaint view of atomic physics has returned in a renormalized form to potentially 'explain' the variety, stability, and spectra of the elementary particles.

4.2.1 Definitions

Gauss defined the linking number of two loops in R^3 in terms of an Ampère type of integral. Loops A and B are described by two three-vectors $\mathbf{x}_A(s)$ and $\mathbf{x}_B(s')$. Then the linking number of A and B is defined as the integral

$$L(A,B) = \frac{1}{4\pi} \oint_A \oint_B \frac{(\mathbf{x}_A - \mathbf{x}_B) \cdot (d\mathbf{x}_A \times d\mathbf{x}_B)}{|\mathbf{x}_A - \mathbf{x}_B|^3} \tag{4.1}$$

Gauss showed that this integral is an integer and is invariant, as the loops A and B are deformed as long as they do not intersect each other.

This integral is difficult to compute. A much simpler method for computing the linking number of two loops makes use of the fact that knots can be represented by plane diagrams. These are projections of the knots onto a plane that preserves

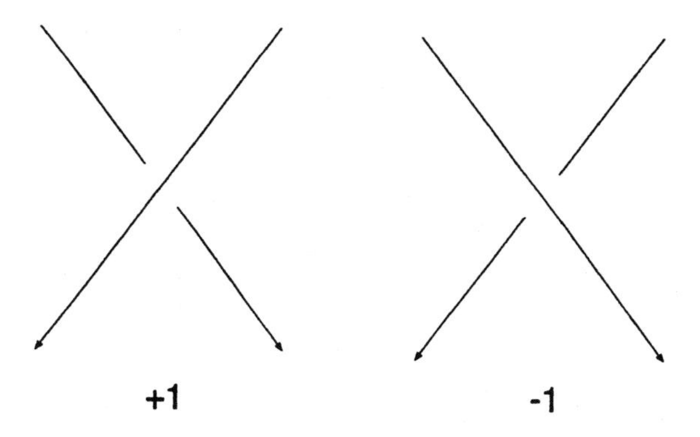

Fig. 4.4 Projections of curves in R^3 into a two-dimensional subspace can have crossings. A sign is associated with each nondegenerate crossing. The sign is $+1$ for right-handed crossings, -1 for left-handed crossings.

crossing information. The order of crossing (above, below) is clearly indicated in the projection. In the projection, it is typical for nondegenerate crossings to occur (Fig. 4.4). Degenerate crossings can be removed by a perturbation of either the loops or the projection plane.

Linking numbers are computed from projections as follows:

1. An integer, $\sigma_i(A,B) = \sigma_i(B,A) = \pm 1$, is assigned to each crossing.

2. At each crossing, tangent vectors are drawn in the direction of flow in the two loops.

3. The tangent vector to the upper segment (in the projection) is rotated into the tangent vector in the lower segment through the smaller angle.

4. If the rotation is right-handed, the crossing is assigned a value $+1$. If the rotation is left-handed, the crossing is assigned a value -1.

5. The linking number is half the sum of the signed crossings:

$$L(A,B) = L(B,A) = \frac{1}{2}\sum_i \sigma_i(A,B) \tag{4.2}$$

A loop can link itself. The self-linking number of an orbit with itself can be defined. It is

$$SL(A) = \sum_j \sigma_j(A,A) \tag{4.3}$$

Note that the factor of $\frac{1}{2}$ is missing in this definition.

Remark: The self-linking number of a closed orbit is not generally a topological invariant. However, if the orbit occurs in a cyclic phase space of the type $R^2 \times S^1$, it is a topological invariant.

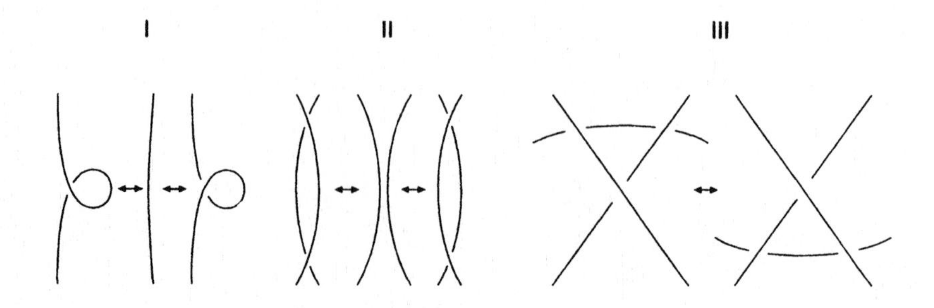

Fig. 4.5 The three Reidemeister moves describe deformations involving one, two, or three segments of a link projected onto a plane. I: One segment is unwound. II: Two segments are pulled apart. III: One segment is slid past the crossing point of two other segments.

4.2.2 Reidemeister Moves

Counting integers for a two-dimensional projection of a link of two or more knots, as in (4.2), is much simpler than computing the double integral (4.1) for the three-dimensional embedding of a pair of knots. Just as the integral in (4.1) is invariant under deformations of the knots which do not involve self-intersections, so also is the sum (4.2) unchanged under such deformations. Deformations of the projection of a link to a plane can arise through two sources:

1. Deformations of the knots which do not involve self-intersections

2. Changes of the plane onto which the projection is made

Reidemeister [72] showed that the sum (4.2) is invariant under three basic deformations in the plane of projection. These are now known as *Reidemeister moves*. All deformations of a projected link can be described by these three moves. They are illustrated in Fig. 4.5. The Reidemeister moves involve one, two, or three segments of a projected knot. They are:

I: Unwinding. A crossing of a single segment is unwound.

II: Separation. Two segments are pulled apart.

III: Sliding. One segment is moved past a crossing of two other segments.

Reidemeister showed that the sum (4.2) for the linking number of two knots is invariant under these three moves. He also showed that the sum (4.3) for self-linking numbers is invariant under *ambient isotopy*, which involves only the second and third moves.

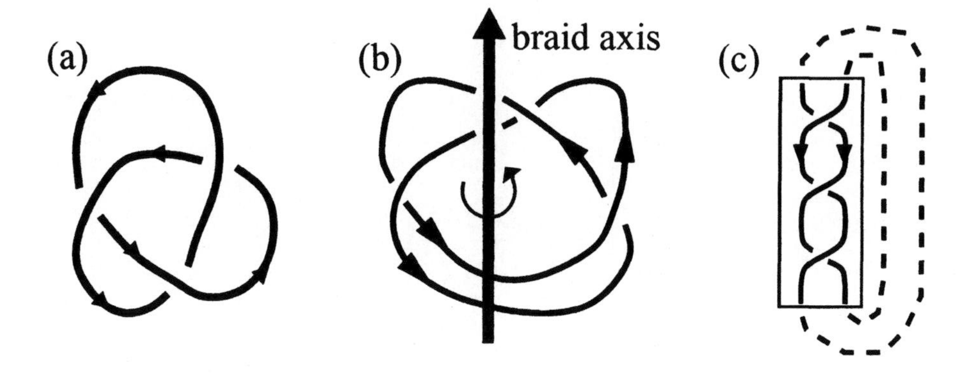

Fig. 4.6 Every oriented knot or link can be deformed into a braid. For the trefoil knot (a) an axis is found that maximizes the linking number (b). The link is then smoothly transformed into the standard braid representation (c). In this representation all crossings occur in the area localized within the rectangle.

4.2.3 Braids

The computation of the sums involved in the expressions for linking numbers (4.2) and self-linking numbers (4.3) can be simplified if it is possible to move all of the crossings to a limited range of the link's projection. Alexander [73] showed that it was always possible to transform an *oriented* link to a closed braid in which all of the crossings could be localized. An oriented link is a link in which there is a sense of travel, or flow, in each of the knots comprising the link.

This works for us. Periodic orbits in a strange attractor are closed, so they are knots. They also possess a flow direction, so they are oriented. Therefore, Alexander's theorem is applicable to links composed of periodic orbits in a dynamical system, in particular, in a strange attractor. Figure 4.6 illustrates how Alexander's construction can be used to transform a trefoil knot [Fig. 4.6(a)] into a braid. An axis that links the knot (or link) a maximum number of times is identified [Fig. 4.6(b)]. The knots are then deformed into a standard representation, shown in Fig. 4.6(c). All crossings are then rotated around the hole in the middle of the flow into a localized region. This interesting region is shown between two horizontal line intervals (subsets of R^1). The flow emerges from the lower line segment and returns to the upper segment without any crossings. As a result, only the part of the braid shown inside the rectangle in Fig. 4.6 is needed to determine the linking numbers of the knots.

In general, a braid contains n strands. It is only possible for two adjacent strands to cross. The right-handed crossing of strands i and $i+1$ can be represented algebraically by the operator σ_i. The left-handed crossing of these two strands is represented by its inverse σ_i^{-1}. When the upper and lower braid boundaries are intervals, $1 \leq i \leq n-1$, but when they have the topology S^1, $1 \leq i \leq n$. These operators are illustrated in Fig. 4.7.

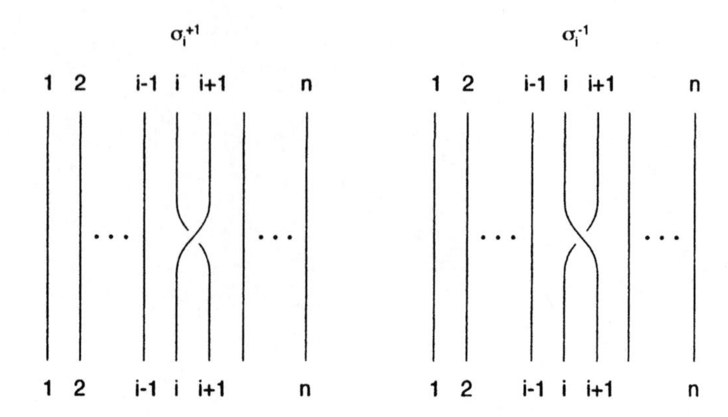

Fig. 4.7 The crossing between the ith and $(i + 1)$st strands in a braid with n strands is represented by the operator $\sigma_i^{\pm 1}$, depending on whether the crossing is right-handed (left) or left-handed (right).

A braid with four crossings on five strands is shown in Fig. 4.8. Since the crossings are discrete, they can be time ordered. Thus, each braid can be coded by a symbol sequence [74–77]. The sequence for the braid shown in Fig. 4.8 is $\sigma_4 \sigma_3^{-1} \sigma_1^{-1} \sigma_2^{-1}$ from top to bottom.

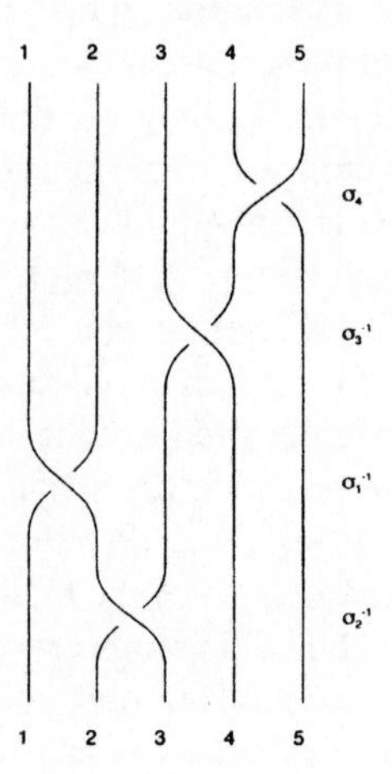

Fig. 4.8 Four crossings in a braid on five strands. The flow is from top to bottom (a convention). The sequence of operators describing this braid is $\sigma_4 \sigma_3^{-1} \sigma_1^{-1} \sigma_2^{-1}$.

Many different symbol sequences can code the same link. There are, in fact, braid moves analogous to the Reidemeister moves, derived from the Reidemeister moves. The braid moves (or braid relations) are shown in Fig. 4.9. The first, shown in Fig. 4.9(a), involves simply changing the time ordering of crossings on independent pairs of strands. The second involves changing the order in which the middle strand of three crosses its two neighbors. These operations correspond to the *braid relations*

$$\begin{aligned} \sigma_i\sigma_j &= \sigma_j\sigma_i \qquad\qquad |i-j|>1 \\ \sigma_i\sigma_{i+1}\sigma_i &= \sigma_{i+1}\sigma_i\sigma_{i+1} \end{aligned} \tag{4.4}$$

The operators σ_i and their inverses generate a group called the *braid group*. The braid group is defined by the braid relations (4.4). It is often very convenient to represent geometric operations by algebraic operations, and vice versa. The braid group provides exactly this capability.

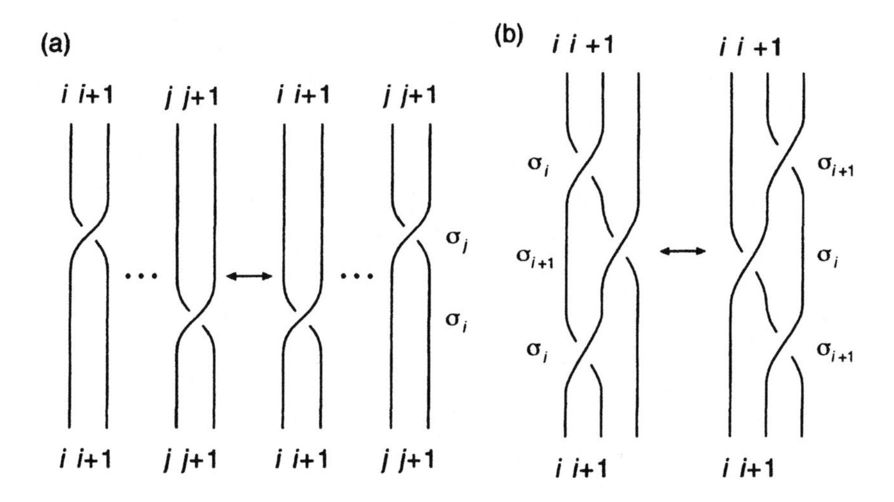

Fig. 4.9 Two braid relations leave a link invariant. (a) This relation describes the commutativity of sequential crossings on independent pairs of strands: $\sigma_i\sigma_j = \sigma_j\sigma_i$ for $|i-j|>1$. (b) This relation describes the interchange of the crossing order of the middle strand with its adjacent partners on either side: $\sigma_i\sigma_{i+1}\sigma_i = \sigma_{i+1}\sigma_i\sigma_{i+1}$.

Remark: Alexander's theorem guarantees that every oriented link can be deformed so that it has a hole in the middle. We exploit this capability when we introduce branched manifolds in Chapter 5. The basic idea is the following. We reorganize the period-1 orbits in a strange attractor into a braid. This braid then forms the skeleton of the branched manifold that describes the strange manifold. This idea will be elaborated on later.

Important Result: Katok [75] and Boyland [76] proved an important result relating braids to chaos. A period-n orbit is represented by a braid with n strands. The topological organization of the orbit is represented by a braid word, W. This is a product of the crossing operators $\sigma_i^{\pm 1}$. The word is not unique: Other words related by conjugation $W' = \sigma_j^{\pm 1} W \sigma_j^{\mp 1}$ represent the same orbit. If the sum of the exponents of the crossing operators in W is not an integer multiple of $n-1$, the orbit has positive topological entropy. If the exponent sum is a multiple of $n-1$ but W^n is not equivalent to a rotation, the orbit has positive topological entropy.

As an example, the period-3 orbit 001 in Rössler dynamics can be represented by the word $W = \sigma_2^{-1}\sigma_1^{-1}$. A conjugate representation for this orbit is $W' = \sigma_1^{-1}\sigma_2^{-1}$. The exponent sum -2 is divisible by $n-1$. The braid word $W^3 = (\sigma_2^{-1}\sigma_1^{-1})^3 = (\sigma_2^{-1}\sigma_1^{-1}\sigma_2^{-1})(\sigma_1^{-1}\sigma_2^{-1}\sigma_1^{-1}) = (\sigma_2^{-1}\sigma_1^{-1}\sigma_2^{-1})(\sigma_2^{-1}\sigma_1^{-1}\sigma_2^{-1})$. Since $(\sigma_2^{-1}\sigma_1^{-1}\sigma_2^{-1})$ is a half-twist, W^3 is equivalent to a rotation. As a result, the period-3 orbit 001 forces no other periodic orbits when it occurs in a map of a disk to itself.

4.2.4 Examples

Figure 4.10 shows two unstable periodic orbits which were extracted from a strange attractor generated by a Belousov–Zhabotinskii chemical reaction. The orbits A and B have periods 2 and 3, respectively. The self-linking numbers of these two orbits are $SL(A) = -1$ and $SL(B) = -2$.

In Fig. 4.11 the two orbits are superposed, all mutual intersections are identified, and an integer ± 1 is assigned to each. The linking number of these two orbits is half the sum of these signed crossings: $L(A, B) = \frac{1}{2}(+2 - 6) = -2$.

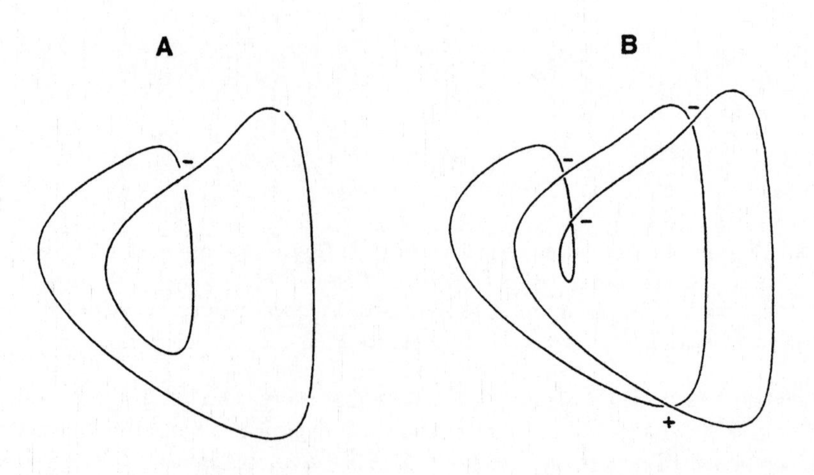

Fig. 4.10 The self-linking number of the period-2 orbit A is $SL(A) = -1$ and that of the period-3 orbit B is $SL(B) = -2$.

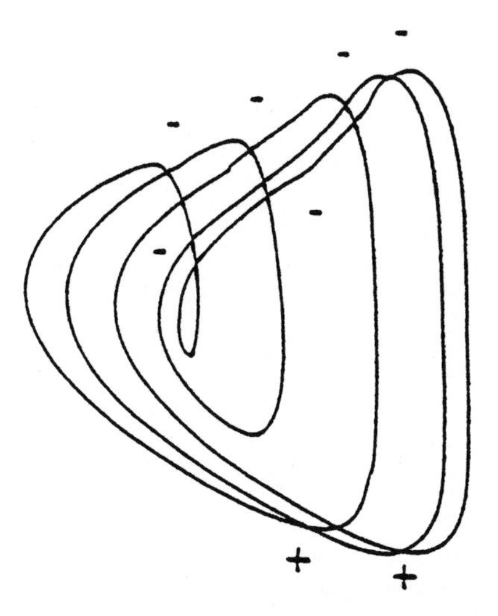

Fig. 4.11 The linking number of the period-2 orbit A with the period-3 orbit B is $L(A, B) = \frac{1}{2}(+2 - 6) = -2$.

In Fig. 4.12 we show three orbits extracted from a strange attractor that exhibits Lorenz-like dynamics. These orbits have periods 3, 3, and 4. All crossings are positive. The linking and self-linking numbers for these three orbits are

	A	A'	B
A	1	1	2
A'	1	1	2
B	2	2	2

(4.5)

The self-linking numbers are on the diagonal.

4.2.5 Linking Numbers for the Horseshoe

In this section we begin with a mechanism, locate the unstable periodic orbits in the strange attractor generated by the mechanism, and then compute the linking and self-linking numbers of all orbits to period 3.

The mechanism is the simple stretch and fold that generates a Smale horseshoe return map. This mechanism is shown in Figs. 3.4, 3.9, and 4.1(a). We illustrate it again in Fig. 4.13, with a view as seen from above. In this figure the flow spirals outward. When the radius exceeds some threshold, the flow is bent back under the outward-flowing spiral and inserted near the focus, which organizes the flow. This process is repeated indefinitely. The return map is a standard horseshoe. Orbits of

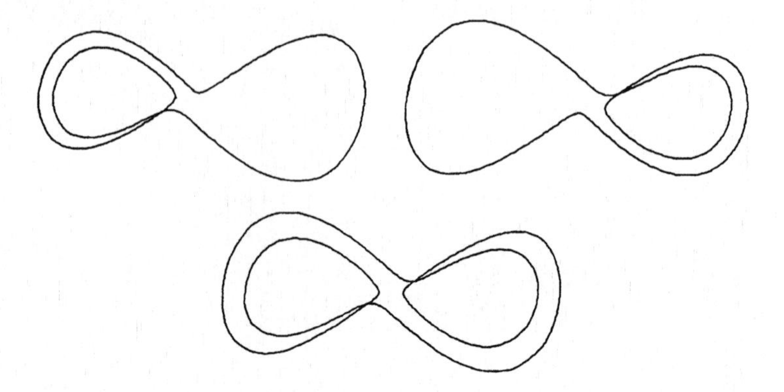

Fig. 4.12 These three periodic orbits have been extracted from a Lorenz-like strange attractor. All crossings are positive. The linking numbers and self-linking numbers are computed by inspection and presented in Eq. (4.5).

period p are located by finding fixed points of the pth return map. Each periodic orbit can be accorded a symbolic encoding that is unique up to cyclic permutation. The symbols are 0 and 1. The symbol 0 indicates that the flow is through the flat part of the attractor, the symbol 1 indicates that the flow is through the folded part of the attractor. For this mechanism the fold is in the clockwise direction, as seen from the direction of the flow, and all crossings are positive. The linking and self-linking numbers for all orbits, up to period 3, are listed in Table 4.1.

Table 4.1 Linking numbers for all orbits to period 3 generated by the right-handed Smale horseshoe mechanism (left) and the right-handed Lorenz mechanism (right)

		1_1	1_1	2_1	3_1	3_1
1_1	0	0	0	0	0	0
1_1	1	0	0	1	1	1
2_1	01	0	1	1	2	2
3_1	011	0	1	2	2	3
3_1	001	0	1	2	3	2

		1_1	1_1	2_1	3_1	3_1
1_1	L	0	0	0	0	0
1_1	R	0	0	0	0	0
2_1	LR	0	0	1	1	1
3_1	LRR	0	0	1	2	1
3_1	RLL	0	0	1	1	2

4.2.6 Linking Numbers for the Lorenz Attractor

Linking numbers for the low-period orbits generated by the Lorenz mechanism [Fig. 4.1(d)] have also been computed. The procedure followed for locating periodic orbits

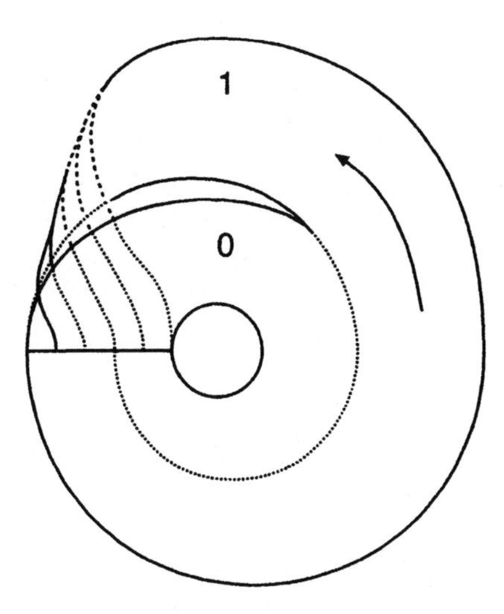

Fig. 4.13 The simple stretch and fold mechanism is seen from above. All crossings are positive. All periodic orbits have an encoding in terms of the symbols 0 and 1.

Table 4.2 Linking and self-linking numbers for orbits of period 2^k generated by the Smale horseshoe mechanism

k		0	1	2	3	4	5
	Period	1	2	4	8	16	32
0	1	0	1	2	4	8	16
1	2	1	1	3	6	12	24
2	4	2	3	5	13	26	52
3	8	4	6	13	23	51	102
4	16	8	12	26	51	97	205
5	32	16	24	52	102	205	399

for this mechanism is the same as that followed for the Smale horseshoe mechanism. The spectrum of orbits is the same in both cases: There are 2,1,2,3,6,9,18,30, ... orbits of period 1,2,3,4,5,6,7,8, ..., respectively. A useful unique symbolic encoding for each orbit involves the letters L and R, depending on whether the orbit segment goes through the left or right half of the strange attractor. The linking and self-linking numbers for all Lorenz orbits up to period 3 are also given in Table 4.1.

The tables of linking numbers for the (right-handed) Smale horseshoe mechanism and the Lorenz mechanism have been presented side by side to emphasize that knowledge of the linking numbers of only a small number of low-period orbits is sufficient to distinguish among the two stretch and squeeze mechanisms.

4.2.7 Linking Numbers for the Period-Doubling Cascade

The linking and self-linking numbers for the orbits of period 2^k that occur in the period doubling-cascade in the Smale horshoe are summarized in Table 4.2. These invariants obey some systematic properties. If the orbit of period 2^k is simply represented by k, these systematics are summarized as follows:

$$\begin{aligned} L(k,l) &= 2^{l-(k+1)}L(k,k+1) \qquad l > k \\ L(k,k+1) &= 2 \times SL(k) + T(k+1) \\ SL(k+1) &= 4 \times SL(k) + T(k+1) \end{aligned} \tag{4.6}$$

The auxiliary function $T(k+1)$ has a geometric interpretation which is described in Section 4.3.3. It obeys the Fibonacci-type equation

$$T(k+1) = T(k) + 2 \times T(k-1) \tag{4.7}$$

For this period-doubling sequence, the initial conditions are $T(0) = 0$, $T(1) = 1$ and $SL(0) = 0$. Other period-doubling cascades, for example, those based on the period-3 orbit 001, the period-4 orbit 0001, the period-5 orbits 01101, 00111, and 00001, exhibit their own systematics in the behavior of $SL(k)$, $L(k,l)$, and $T(k+1)$. In fact, the only differences from the relations above are for the initializations $T(1)$ and $SL(0)$. The mysteries of these relations will be explained in terms of relative rotation rates.

4.2.8 Local Torsion

Local torsion is an important geometric property of periodic orbits. It is defined roughly as follows. An arrow is defined at some point of an orbit, in a direction perpendicular to the flow. As the flow progresses, the base of the arrow moves along the orbit, and the tip rotates around the direction of the flow. When the initial condition returns to its starting point, the arrow has rotated through some angle θ. This angle θ, or more accurately θ/π, is defined as the local torsion of the flow around the periodic orbit.

In general, local torsion is not an integer. However, integer values of the local torsion force significant topological consequences. Specifically, if the local torsion about a closed period-n orbit is an even integer, it is possible for two orbits of period n to entwine the original period-n orbit. On the other hand, if the local torsion is an odd integer, the original period-n orbit can be entwined by an orbit that closes after two cycles: that is, an orbit of period $2n$.

When a saddle-node bifurcation creates a pair of orbits in R^3, both the saddle and the node have local torsion which is an even integer. The local torsion of the saddle cannot change. It is fixed because the two Lyapunov exponents are different. They are different because the saddle has both a stable and an unstable manifold. The local torsion of the node is initially an even integer: Both Lyapunov exponents are negative but different. As a control parameter is varied, these two negative numbers become equal. The Lyapunov exponents become complex-conjugate pairs. The complex phase factor typically increases from 0 to π radians, after which the two eigenvalues scatter off each other along the negative real axis. Thereafter, one decreases below -1, at which point a period-doubling bifurcation takes place. The other remains in the interval -1 to 0. At this point the local torsion of the node has increased or decreased to the the nearest odd-integer value. The local torsions of the original saddle and the node, now unstable, differ by ± 1.

The index $T(k)$ in Eqs. (4.6) and (4.7) is the local torsion of the orbit of period 2^k in the period-doubling cascade based on the original period-1 orbit.

4.2.9 Writhe and Twist

In Figure 4.14 we show a pair of closed orbits A and A' in several configurations. On the left the two orbits are shown in large loops. On the right the loops have been pulled apart, and the two orbits are shown to twist about each other.

The orbits, as shown on the left of Fig. 4.14, are said to *writhe* about each other. A writhing number can be associated with each of the configurations shown. The writhing number, $Wr(A, A')$, associated with each pair can be computed as follows. Treat the pair of orbits as the boundary of a strip, or a rubber band RB, so that $\partial RB = A \cup A'$. Then this strip has a self-linking number with itself which is simply the sum of signed crossings of the band with itself, under the usual right-handed convention. Then

$$Wr(A, A') = SL(RB) \tag{4.8}$$

This number is also the self-linking number of either of the two orbits: $Wr(A, A') = SL(A) = SL(A')$.

Tension destroys writhe by converting it to twist, as shown in Fig. 4.14. On the right-hand side of this figure, the closed orbits exhibit no writhe, but do twist around each other. No twist is exhibited on the left-hand side of this figure. The twist is quantified by a twisting number: $Tw(A, A')$.

For the orbits shown in Fig. 4.14, neither writhe nor twist is a topological invariant. Writhe can be converted to twist by pulling on the pair or even by changing the projection of the knot from R^3 to R^2. The sum of these two quantities is an invariant, the familiar linking number:

$$L(A, A') = Tw(A, A') + Wr(A, A') \tag{4.9}$$

Remark 1: Twist and writhe are geometric quantities. They can be expressed in terms of integrals which are not necessarily integers. Linking number is a topological quantity. It is *always* an integer. It is remarkable that the sum of two geometric quantities is a topological index.

Remark 2: For some dynamical systems, the flow in phase space exhibits large loops that can be measured by a nonzero writhing number. We have observed this

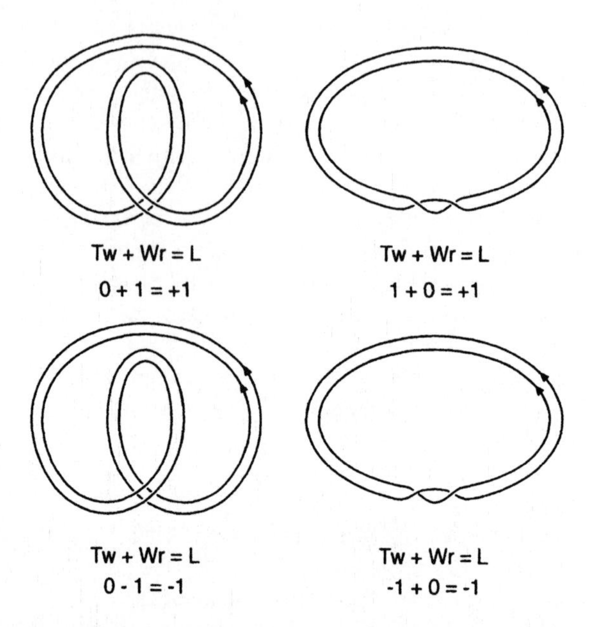

Fig. 4.14 Twist + writhe = link. Top: Two orbits have a linking number $+1$. On the left the link is exhibited as writhe, on the right it is exhibited as twist. Bottom: Two orbits have a linking number -1. On the left the link is exhibited as writhe, on the right it is exhibited as twist. Below each configuration we provide information on Tw, Wr, and L.

behavior in models of sensory neurons that exhibit subthreshold oscillations (cf. Figs. 7.37 and 7.40). The presence of nonzero writhing number is clearly evident in the time series observed. The fingerprint of writhing is the presence of spikes in the output (cf. Fig. 7.38).

4.2.10 Additional Properties

Linking numbers provide two different types of information about strange attractors.

First, they allow us to distinguish among different mechanisms that generate strange attractors. This can be seen by comparing tables of linking numbers for the Smale horseshoe mechanism and the Lorenz mechanism (Table 4.1).

Second, they provide information about the order in which different periodic orbits can be created in the formation of a strange attractor. This topic is treated in more detail in Section 9.3.2. The basic idea is this. Assume that two pairs of orbits, A and B, are created in saddle node bifurcations in the horseshoe mechanism. The regular saddle is labeled with the subscript R while the node is labeled with the subscript F. (cf. Fig. 9.5). In Fig. 9.5(a) the orbits B_R, B_F link A_R differently from A_F. Before orbits A_R and A_F can annihilate each other in an inverse saddle-node bifurcation, orbits B_R and B_F must first disappear. Otherwise, one of the orbits A_R or A_F must intersect both orbits B_R and B_F on the way to the inverse saddle-node bifurcation, violating the fundamental theorem. Running this scenario backwards, the orbit pair A_R, A_F must be created before the orbit pair B_R, B_F is created. This topological situation is reflected in the table of linking numbers. That is, for the situation shown in Fig. 9.5(a), $L(A_R, B_R) = L(A_R, B_F) \neq L(A_F, B_R) = L(A_F, B_F)$. Thus, it is possible to use information contained in the table of linking numbers to infer information about the order of creation of periodic orbits as control parameters are varied. As an example, the period-3 orbits 001, 011 and period-5 orbits 00111, 00101 are both created in saddle-node bifurcations. These orbits have the following linking numbers:

	00111	00101
001	4	4
011	5	5

(4.10)

As a result, the period-3 orbit pair must be created before this period-5 orbit pair is created on the way to creating the fully hyperbolic strange attractor generated by the horseshoe mechanism.

Similar considerations apply to the orbit pairs shown in Fig. 9.5(c), where orbit pair B_R, B_F must be created before orbit pair A_R, A_F. When all four linking numbers are equal, as in Fig. 9.5(b), the linking numbers do not indicate that either pair must be created before the other pair is created during the formation of the strange attractor.

These arguments can be extended to mother–daughter pairs of orbits involved in period-doubling cascades. Arguments of this type have been used to construct a forcing diagram (Fig. 9.8) for the periodic orbits in a horseshoe attractor, up to period 8. This is discussed more fully in Chapter 9.

4.3 RELATIVE ROTATION RATES

Linking numbers provide a powerful tool for determining the topological organization of the unstable periodic orbits in a typical strange attractor. There is an even more powerful tool for determining the organization of unstable periodic orbits in strange attractors embedded in phase spaces of the type $R^2 \times S^1$ or any variant of the form $D^2 \times S^1$, where D^2 is some compact domain in R^2: $D^2 \subset R^2$ (such as the annulus, which occurs in the van der Pol mechanism). This topological tool is called the *relative rotation rate*.

In Fig. 4.15 we show two period-1 orbits in a toroidal phase space $R^2 \times S^1$. In the generic perspective shown at the top of this figure it may be difficult to count crossings to compute linking numbers. This difficulty can be removed by viewing the dynamics directly from above (looking straight down into the hole in the middle). This difficulty can also be overcome by splitting the torus along a Poincaré section (middle of Fig. 4.15) and opening it up. Orbits in the phase space then propagate from $\phi = 0$ to $\phi = 2\pi$, which is identified with the plane $\phi = 0$ by periodic boundary

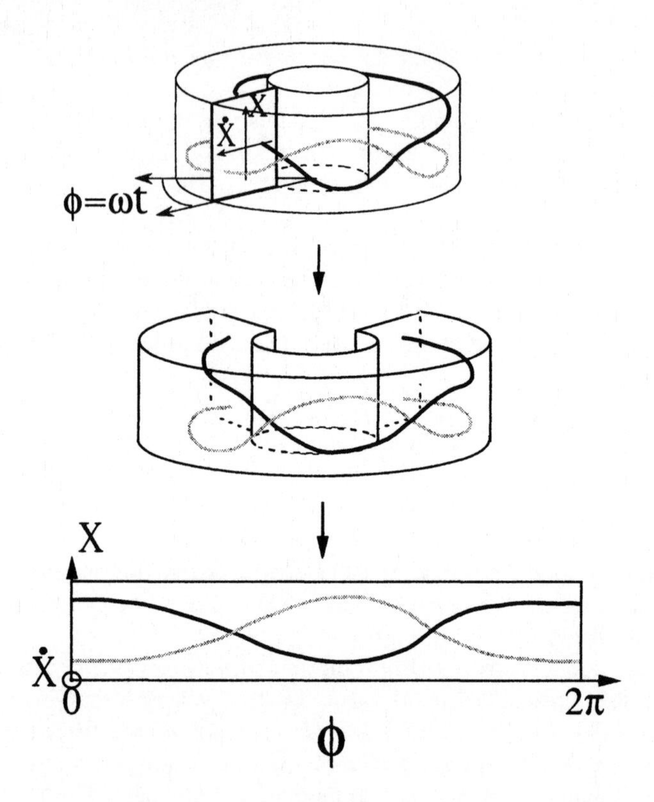

Fig. 4.15 Topological organization of orbits in a torus is easy to compute. The torus is opened up by splitting along a Poincaré section and straightening out. A point on an orbit flows from $\phi = 0$ to $\phi = 2\pi$, which is identified with the section at $\phi = 0$ (periodic boundary conditions). If the embedding coordinates are differentially related, so that $y = dx/dt$, all crossings in the projection at the bottom are left-handed.

conditions. This operation does for dynamics in a torus $R^2 \times S^1$ what Alexander's theorem does in transforming oriented links to braids.

If the coordinates (x, y, t) in the torus $R^2 \times S^1$ are differentially related, so that $y = dx/dt$, all crossings in the projection shown at the bottom of Fig. 4.15 are negative. When two segments cross, the one with the larger slope is closer to the observer with the conventions presented in the figure.

4.3.1 Definition

Suppose that A and B are two periodic orbits in the phase space $R^2 \times S^1$. Assume that they have periods p_A and p_B. Then orbit A intersects a Poincaré section $S^1(\theta)$, $\theta = \text{constant}$, in p_A distinct points $a_1, a_2, \ldots, a_{p_A}$, and B intersects the Poincaré section $S^1(\theta)$ in p_B distinct points $b_1, b_2, \ldots, b_{p_B}$. Under the flow $a_i \to a_{i+1}$, with $a_{p_A} \to a_1$, and similarly for B. Now connect two initial conditions, a_i and b_j, by an arrow. As θ increases by 2π, this arrow will rotate into a new orientation, connecting a_{i+1} and b_{j+1} (mod p_A and p_B, respectively). After $p_A \times p_B$ periods the arrow will return to its initial orientation, having rotated through some integer multiple of 2π radians [12].

The relative rotation rate, $R_{i,j}(A, B)$, of these two orbits is defined as the average number of rotations, per period, of these two orbits around each other, starting from initial conditions a_i and b_j. The relative rotation rates for two orbits can be expressed in terms of an integral, derived from the Gaussian expression for linking numbers.

$$R_{i,j}(A,B) = \frac{1}{2\pi p_A p_B} \oint_{i,j} \frac{\mathbf{n}\cdot(\mathbf{\Delta r}\times d\mathbf{\Delta r})}{\mathbf{\Delta r}\cdot\mathbf{\Delta r}} \tag{4.11}$$

For periodically driven flows, the unit normal $\mathbf{n}$ is automatically orthogonal to the Poincaré section. However, for autonomous flows in $R^2 \times S^1 \subset R^3$, this is not necessarily the case. This expression can also be used for flows in R^3.

4.3.2 How to Compute Relative Rotation Rates

As in the case of linking numbers, the integral expression for relative rotation rates of two orbits is not particularly easy to evaluate. Several simpler methods are available. We present four such methods [2].

In the first method, p_B copies of the orbit A are laid out end to end, beginning with initial condition a_i. This long orbit is overlaid with p_A copies of orbit B, starting from initial condition b_j. The linking number of these two orbits is computed as half the sum of their signed crossings, as usual. Then $RRR_{ij}(A,B) = L(\text{“}p_B A\text{”}, \text{“}p_A B\text{”})/(p_A \times p_B)$. This calculation must be repeated for each of the $p_A \times p_B$ initial conditions (a_i, b_j).

The second method is somewhat simpler. The periodic orbit A is broken up into p_A segments A_i, $i = 1, \ldots, p_A$, with segment A_i propagating between a_i and a_{i+1} in the Poincaré section. Orbit B is treated similarly. Then

$$R_{i,j}(A,B) = R_{j,i}(B,A) = \frac{1}{p_A \times p_B} \sum_{k=1}^{p_A \times p_B} \frac{1}{2}\sigma(A_{i+k}, B_{j+k}) \tag{4.12}$$

As usual, if $i + k > p_A$, it should be replaced by $1 + \text{mod}(i + k - 1, p_A)$ (e.g., for $p_A = 3$, then $i + k = 1, 2, 3, 4 \rightarrow 1, 5 \rightarrow 2, 6 \rightarrow 3, 7 \rightarrow 1$, etc.). Once again, this computation must be repeated for all $p_A \times p_B$ initial conditions. In fact, this method is simply an implementation of the first method.

Relative rotation rates are particularly easy to compute for closed orbits obtained from periodically driven dynamical systems. It happens often that a time series $x(t)$ is available. The series can be embedded in $R^2 \times S^1$ using an $x, \dot{x}, \theta$ embedding, shown in Fig. 4.15. In this embedding all crossings are left-handed. As a result, it is sufficient to count crossings and not worry about the signs of these crossings, as all signs are the same (-1).

To illustrate the computation of relative rotation rates by this method, we compute the self-relative rotation rates of a period-4 orbit extracted from chaotic laser data [78].

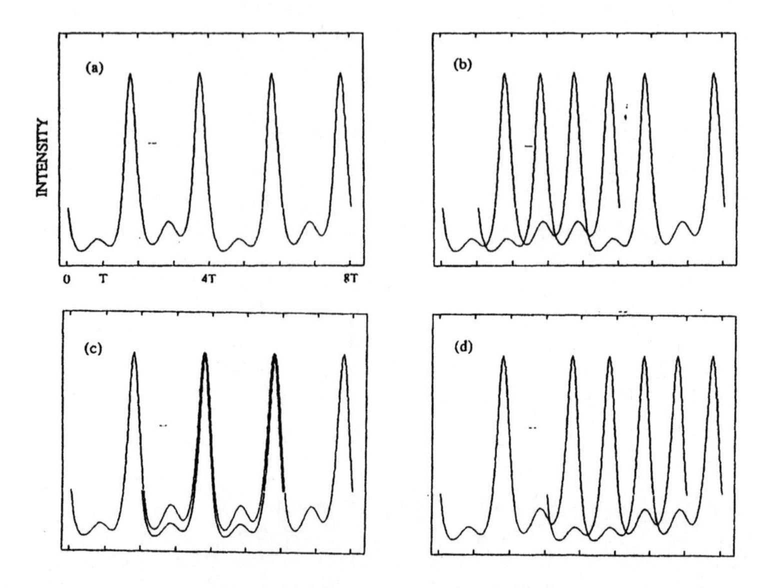

Fig. 4.16 Relative rotation rates can be computed by simple counting. (a) A period-4 orbit is repeated once. (b) The same period-4 orbit overlays the original in (a), shifted by one period; it is shifted two and three periods in (c) and (d). Arithmetic gymnastics leads immediately to the spectrum of self-relative rotation rates $(-\frac{1}{2})^8, (-\frac{1}{4})^4, 0^4$.

The orbit is shown in Fig. 4.16(a). It has been extended out to eight periods. In Fig. 4.16(b), (c), and (d) we have superposed the period-4 orbit on the lengthened orbit shown in Fig 4.16(a), shifting it one period in each of the three frames. The total number of crossings seen in Fig. 4.16(b) is 4. Extending the second period-4 orbit to eight periods would double the number of crossings to eight. Extending both to 16 periods would generate 16 crossings. The self-relative rotation rate for this period-4 orbit with the initial conditions shown is $-16/(2 \times 4 \times 4) = -\frac{1}{2}$. Similar arguments for Fig. 4.16(c) and (d) give self-relative rotation rates of $-\frac{1}{4}$ and $-\frac{1}{2}$, respectively. For convenience and other ulterior motives (cf. the remark at the conclusion of Section 4.4), we define the self-relative rotation rates of an orbit starting from the same initial condition to be zero.

In principle, we should repeat this computation three more times, shifting the original lengthened orbit one period each time. However, it is simple to see that each such computation generates some permutation of the same results. Thus, the full spectrum of 16 self-relative rotation rates for this period-4 orbit is $(-\frac{1}{2})^8, (-\frac{1}{4})^4, 0^4$. For convenience, the ratios of these values are displayed in tables of relative rotation rates.

The fourth method is a simple computer implementation of the previous three methods. It is simple, straightforward, can be done without much eyestrain, and leads to the full spectrum of both self- and mutual-relative rotation rates [1]. It is particularly convenient for computing these topological indices for orbits extracted from experimental data. This algorithm is illustrated in Fig. 4.17.

The orbit A of period p consists of p segments $A_1, A_2, \ldots, A_p$, and similarly for orbit B of period q. This decomposition is shown in Fig. 4.17(a) for the two orbits of periods 3 and 2 shown in Figs. 4.10 and 4.11.

Under forward iteration, each orbit segment evolves to the next mod p : $A_1 \rightarrow A_2 \rightarrow \cdots \rightarrow A_p \rightarrow A_1$. Therefore, forward iteration by one period is described by a cyclic permutation matrix P_p. Cyclic permutation matrices for $p = 2, 3, 4$ are

$$P_2 = \begin{bmatrix} 0 & 1 \\ 1 & 0 \end{bmatrix} \quad P_3 = \begin{bmatrix} 0 & 1 & 0 \\ 0 & 0 & 1 \\ 1 & 0 & 0 \end{bmatrix} \quad P_4 = \begin{bmatrix} 0 & 1 & 0 & 0 \\ 0 & 0 & 1 & 0 \\ 0 & 0 & 0 & 1 \\ 1 & 0 & 0 & 0 \end{bmatrix} \tag{4.13}$$

The permutation matrices for the period 3 and 2 orbits A and B are shown in Fig. 4.17(b). The direct sum of these permutation matrices describes the time evolution of these five orbit segments.

A $(p+q) \times (p+q)$ crossing matrix C is constructed for the two orbits A and B. The (i, j) matrix element, C_{ij}, is the crossing number of the two segments A_i and A_j, for $1 \leq i, j \leq p$: $C_{ij}(A, A) = \sigma(A_i, A_j)$. For either, or both, (i, j) in the range $p+1 \leq i, j \leq p+q$, the segment B_r or B_s of orbit B are used, $r = i - p$, $s = j - p$. The crosssing matrix C is shown in Fig. 4.17(c) for the two orbits of periods 3 and 2.

Finally, construct the $(p+q) \times (p+q)$ matrix RRR, defined by the sum

$$RRR = \begin{bmatrix} RRR(A, A) & RRR(A, B) \\ RRR(B, A) & RRR(B, B) \end{bmatrix} =$$

$$\frac{1}{p \times q} \sum_{k=1}^{p \times q} \begin{bmatrix} P_p & 0 \\ 0 & P_q \end{bmatrix}^k \begin{bmatrix} C_{ij} & C_{is} \\ C_{rj} & C_{rs} \end{bmatrix} \begin{bmatrix} P_p & 0 \\ 0 & P_q \end{bmatrix}^{-k} \tag{4.14}$$

The inverse of the matrix on the right is easy to compute, since it is block diagonal and the inverse of a permutation matrix is its transpose: $P_p^{-1} = P_p^t$. The diagonal $p \times p$ submatrix $RRR(A, A)$ contains the p^2 relative rotation rates of the period-p orbit A, as does $RRR(B, B)$ for the period-q orbit B. The $p \times q$ relative rotation rates for the orbit pair (A, B) are contained in either of the off-diagonal submatrices $RRR(A, B) = RRR(B, A)^t$. This algorithm is a simple implementation of the computation shown in Eq. (4.12).

The 5×5 matrix RRR is shown in Fig. 4.17(d) for the two orbits of periods 3 and 2. The two matrices on the diagonal contain the self-relative rotation rates of the two orbits. These are $(-\frac{1}{3})^6, 0^3$ for the period-3 orbit and $(-\frac{1}{2})^2, 0^2$ for the period-2 orbit. All six relative rotation rates for the two orbits are equal to $-\frac{1}{3}$: $RRR(01, 011) = (-\frac{1}{3})^6$.

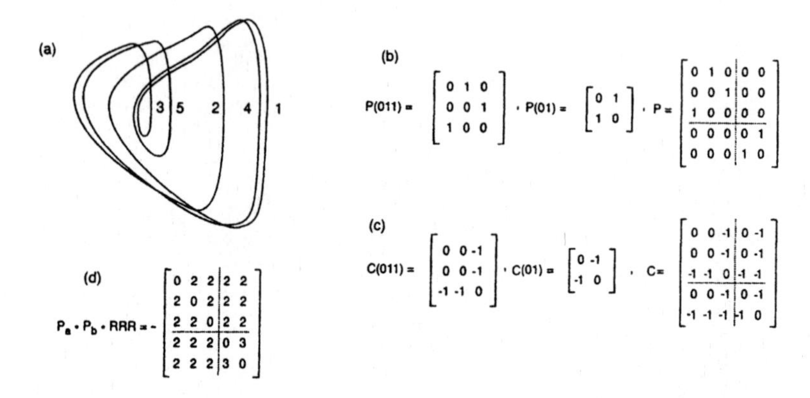

Fig. 4.17 All relative rotation rates for a pair of orbits can easily be computed. (a) Two orbits are superposed. These are the two orbits shown in Figs. 4.10 and 4.11. The orbit segments between Poincaré sections are numbered. Orbit segments 1, 2, 3 belong to the period-3 orbit, and segments 4 and 5 belong to the period-2 orbit. (b) The permutation matrices for the period-3 and period-2 orbits, as well as their direct sum, are given. (c) The self-crossing matrices, and the mutual-crossing matrix, for the two orbits, are used to construct a 5×5 crossing matrix. (d) The crossing and permutation matrices are used to compute the matrix of relative rotation rates, as shown in Eq. (4.14).

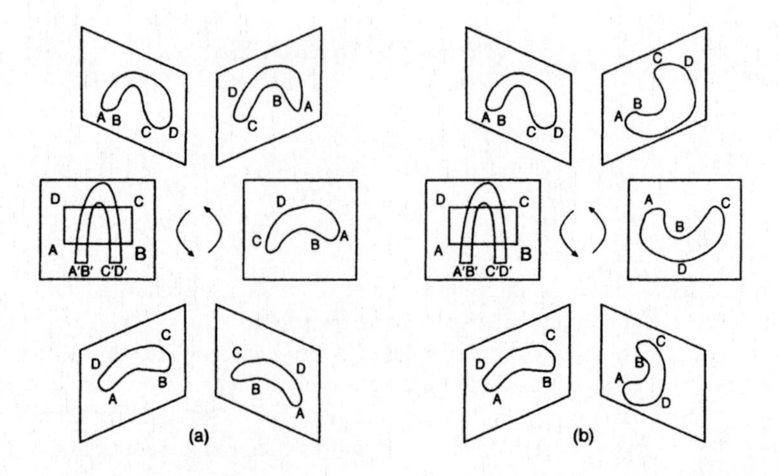

Fig. 4.18 Two inequivalent flows can share the same return map. (a) A zero-torsion lift of the horseshoe has intersections with the Poincaré sections θ = constant, as shown. (b) This lift of the horseshoe has global torsion -1. One complete counterclockwise rotation (looking in the direction of the flow) is made per period.

4.3.3 Horseshoe Mechanism

The spectrum of relative rotation rates can be computed for all periodic orbits created by the Smale horseshoe mechanism. It is sufficient to use the same orbits as those used for computing linking numbers.

One subtlety of this computation is shown in Fig. 4.18. This figure shows two distinct mechanisms which generate the same horseshoe return map. In one mechanism (left) an initial rectangular neighborhood is simply deformed into a horseshoe. In the second mechanism (right), in addition to the deformation, there is a full counterclockwise rotation through 2π radians, facing in the direction of the flow. These two flows have *global torsion* 0 and -1. More generally, many different flows can generate the same horseshoe return map. These differ by their global torsion, T, which is a homotopy index.

In Table 4.3 we present the relative rotation rates for all orbits up to period 5 in the right-handed Smale horseshoe with zero global torsion. In the left-handed horseshoe the signs of all fractions are reversed. The global torsion, T, of the lift is the linking number, or relative rotation rate, of the two period-1 orbits. If T is nonzero, it must be added to all fractions in this table.

Orbits in period-doubling cascades are organized in a rigid way with respect to each other. This organization is a very useful fingerprint for identifying orbits in a cascade. The relative rotation rates of the orbits in the basic period-doubling cascade in the right-handed zero-torsion Smale horseshoe are summarized in Table 4.4.

On the way to creating a fully hyperbolic horseshoe, two types of events take place in a systematic way: saddle-node and period-doubling bifurcations. Periodic orbits are created in saddle-node bifurcations. The saddle is unstable when created and remains unstable during the formation of the horseshoe. The node is stable when created, but at some point loses its stability, giving rise to a period-doubling cascade.

Relative rotation rates display a systematic behavior in both saddle-node bifurcations and period-doubling cascades.

When two orbits, A_S and A_N, of period p are created in a saddle-node bifurcation, these two orbits are localized in the phase space. This localization implies that the two orbits are isotopic. In short, this implies that all their knot invariants are identical when they are created. In particular, their linking numbers and relative rotation rates with other already existing knots are the same.

Just before the saddle-node bifurcation occurs, the local twist, or local torsion, in the neighborhood of the imminent pair creation approaches some fraction $r_p = n/p$ (n integer). Just after the saddle-node bifurcation, the relative rotation rates of the two orbits are the same, with one minor exception. They are:

	A_S	A_N
A_S	$(r_1, r_2, \ldots, r_{p-1}, 0)^p$	$(r_1, r_2, \ldots, r_{p-1}, r_p)^p$
A_N	$(r_1, r_2, \ldots, r_{p-1}, r_p)^p$	$(r_1, r_2, \ldots, r_{p-1}, 0)^p$

(4.15)

Table 4.3 Relative rotation rates for all orbits to period 5 generated by the right-handed Smale horseshoe mechanism

		1_1	1_1	2_1	3_1	3_1	4_1	4_2	4_2	5_1	5_1	5_2	5_2	5_3	5_3
1_1	0	T	0	0	0	0	0	0	0	0	0	0	0	0	0
1_1	1	T	0	$\frac{1}{2}$	$\frac{1}{3}$	$\frac{1}{3}$	$\frac{1}{2}$	$\frac{1}{4}$	$\frac{1}{4}$	$\frac{2}{5}$	$\frac{2}{5}$	$\frac{2}{5}$	$\frac{2}{5}$	$\frac{1}{5}$	$\frac{1}{5}$
2_1	01	0	$\frac{1}{2}$	$\frac{1}{2}0$	$\frac{1}{3}$	$\frac{1}{3}$	$\frac{1}{2}\frac{1}{4}$	$\frac{1}{4}$	$\frac{1}{4}$	$\frac{2}{5}$	$\frac{2}{5}$	$\frac{3}{10}$	$\frac{3}{10}$	$\frac{1}{5}$	$\frac{1}{5}$
3_1	011	0	$\frac{1}{3}$	$\frac{1}{3}$	$(\frac{1}{3})^2 0$	$\frac{1}{3}$	$\frac{1}{3}$	$\frac{1}{4}$	$\frac{1}{4}$	$\frac{1}{3}$	$\frac{1}{3}$	$\frac{1}{3}$	$\frac{1}{3}$	$\frac{1}{5}$	$\frac{1}{5}$
3_1	001	0	$\frac{1}{3}$	$\frac{1}{3}$	$\frac{1}{3}$	$(\frac{1}{3})^2 0$	$\frac{1}{3}$	$\frac{1}{4}$	$\frac{1}{4}$	$\frac{1}{3}$	$\frac{1}{3}$	$\frac{4}{15}$	$\frac{4}{15}$	$\frac{1}{5}$	$\frac{1}{5}$
4_1	0111	0	$\frac{1}{2}$	$\frac{1}{2}\frac{1}{4}$	$\frac{1}{3}$	$\frac{1}{3}$	$(\frac{1}{2})^2 \frac{1}{4} 0$	$\frac{1}{4}$	$\frac{1}{4}$	$\frac{2}{5}$	$\frac{2}{5}$	$\frac{7}{20}$	$\frac{7}{20}$	$\frac{1}{5}$	$\frac{1}{5}$
4_2	0011	0	$\frac{1}{4}$	$\frac{1}{4}$	$\frac{1}{4}$	$\frac{1}{4}$	$\frac{1}{4}$	$(\frac{1}{4})^3 0$	$\frac{1}{4}$	$\frac{1}{4}$	$\frac{1}{4}$	$\frac{1}{4}$	$\frac{1}{4}$	$\frac{1}{5}$	$\frac{1}{5}$
4_2	0001	0	$\frac{1}{4}$	$\frac{1}{4}$	$\frac{1}{4}$	$\frac{1}{4}$	$\frac{1}{4}$	$\frac{1}{4}$	$(\frac{1}{4})^3 0$	$\frac{1}{4}$	$\frac{1}{4}$	$\frac{1}{4}$	$\frac{1}{4}$	$\frac{1}{5}$	$\frac{1}{5}$
5_1	01111	0	$\frac{2}{5}$	$\frac{2}{5}$	$\frac{1}{3}$	$\frac{1}{3}$	$\frac{2}{5}$	$\frac{1}{4}$	$\frac{1}{4}$	$(\frac{2}{5})^4 0$	$\frac{2}{5}$	$(\frac{2}{5})^4 \frac{1}{5}$	$(\frac{2}{5})^4 \frac{1}{5}$	$\frac{1}{5}$	$\frac{1}{5}$
5_1	01101	0	$\frac{2}{5}$	$\frac{2}{5}$	$\frac{1}{3}$	$\frac{1}{3}$	$\frac{2}{5}$	$\frac{1}{4}$	$\frac{1}{4}$	$\frac{2}{5}$	$(\frac{2}{5})^4 0$	$(\frac{2}{5})^3(\frac{1}{5})^2$	$(\frac{2}{5})^3(\frac{1}{5})^2$	$\frac{1}{5}$	$\frac{1}{5}$
5_2	00111	0	$\frac{2}{5}$	$\frac{3}{10}$	$\frac{1}{3}$	$\frac{4}{15}$	$\frac{7}{20}$	$\frac{1}{4}$	$\frac{1}{4}$	$(\frac{2}{5})^4 \frac{1}{5}$	$(\frac{2}{5})^3(\frac{1}{5})^2$	$(\frac{2}{5})^2(\frac{1}{5})^2 0$	$(\frac{2}{5})^2(\frac{1}{5})^3$	$\frac{1}{5}$	$\frac{1}{5}$
5_2	00101	0	$\frac{2}{5}$	$\frac{3}{10}$	$\frac{1}{3}$	$\frac{4}{15}$	$\frac{7}{20}$	$\frac{1}{4}$	$\frac{1}{4}$	$(\frac{2}{5})^4 \frac{1}{5}$	$(\frac{2}{5})^3(\frac{1}{5})^2$	$(\frac{2}{5})^2(\frac{1}{5})^2 0$	$(\frac{2}{5})^2(\frac{1}{5})^2 0$	$\frac{1}{5}$	$\frac{1}{5}$
5_3	00011	0	$\frac{1}{5}$	$\frac{1}{5}$	$\frac{1}{5}$	$\frac{1}{5}$	$\frac{1}{5}$	$\frac{1}{5}$	$\frac{1}{5}$	$\frac{1}{5}$	$\frac{1}{5}$	$\frac{1}{5}$	$\frac{1}{5}$	$(\frac{1}{5})^4 0$	$\frac{1}{5}$
5_3	00001	0	$\frac{1}{5}$	$\frac{1}{5}$	$\frac{1}{5}$	$\frac{1}{5}$	$\frac{1}{5}$	$\frac{1}{5}$	$\frac{1}{5}$	$\frac{1}{5}$	$\frac{1}{5}$	$\frac{1}{5}$	$\frac{1}{5}$	$\frac{1}{5}$	$(\frac{1}{5})^4 0$

Table 4.4 Relative rotation rates for orbits of period 2^k in the period-doubling cascade[a]

k	Period	0 1	1 2	2 4	3 8	4 16
0	1	0	$(t_1)^2$	$(t_1)^4$	$(t_1)^8$	$(t_1)^{16}$
1	2	$(t_1)^2$	$(t_1 0)^2$	$(t_1 t_2)^4$	$(t_1 t_2)^8$	$(t_1 t_2)^{16}$
2	4	$(t_1)^4$	$(t_1 t_2)^4$	$(t_1^2 t_2 0)^4$	$(t_1^2 t_2 t_3)^8$	$(t_1^2 t_2 t_3)^{16}$
3	8	$(t_1)^8$	$(t_1 t_2)^8$	$(t_1^2 t_2 t_3)^8$	$(t_1^4 t_2^2 t_3 0)^8$	$(t_1^4 t_2^2 t_3 t_4)^{16}$
4	16	$(t_1)^{16}$	$(t_1 t_2)^{16}$	$(t_1^2 t_2 t_3)^{16}$	$(t_1^4 t_2^2 t_3 t_4)^{16}$	$(t_1^8 t_2^4 t_3^2 t_4 0)^{16}$

[a] In this cascade: $t_1 = 1/2^1, t_2 = 1/2^2$. $t_3 = 3/2^3, t_4 = 5/2^4, t_5 = 11/2^5$, and so on.

For example, as seen from Table 4.3, the local torsions in the neighborhoods of the three saddle-node pairs 5_1, 5_2, and 5_3 at the time of their creation are $\frac{2}{5}$, $\frac{1}{5}$, and $\frac{1}{5}$, respectively.

Eventually, as control parameters are changed, the node A_N will become unstable and become the progenitor of a period-doubling cascade. This occurs when the daughter orbit of period $2p$ is able to wind an odd number of times around the mother orbit A_N. An equivalent way to look at this is that the flow in the neighborhood of A_N becomes one-sided (i.e., on a Möbius band). At the time of the saddle node bifurcation, the local torsion about both orbits A_S and A_N is $r_p = n/p$. The total torsion that a nearby orbit makes during p periods is an integer n. The total number of times a nearby orbit winds around either member of the saddle-node pair in $2p$ periods is an *even* integer $2n$. As control parameters are varied, this even number of rotations around $2A_N$ either increases or decreases to the nearest odd integer. At this point the first of the period-doubling bifurcations takes place. The mother orbit $M = A_N$ of period p and the daughter orbit D of period $2p$ are localized in space at the period-doubling bifurcation. As in the case of saddle-node bifurcations, this means that $2M$ and D are isotopic with each other. Their linking numbers and relative rotation rates with other already existing knots are the same. In particular, the relative rotation rates of M, $2M$ and D are the same (modulo multiplicity). The mother and daughter orbits have the following spectrum of relative rotation rates:

	M	D
M	$(r_1, r_2, \ldots, r_{p-1}, 0)^p$	$(r_1, r_2, \ldots, r_{p-1}, t_1)^{2p}$
D	$(r_1, r_2, \ldots, r_{p-1}, t_1)^{2p}$	$(r_1^2, r_2^2, \ldots, r_{p-1}^2, t_1, 0)^{2p}$

(4.16)

The new fraction, t_1, is determined by the *quantization condition* for the creation of the daughter orbit:

$$t_1 = \frac{2n \pm 1}{2p} = \frac{2pr_p \pm 1}{2p} = r_p \pm \frac{1}{2p} \tag{4.17}$$

For the basic period-doubling cascade $t_1 = \frac{1}{2}$ [more generally $(2T \pm 1)/2$]. For the cascade based on 5_1, $t_1 = \frac{1}{2}$ or $\frac{3}{10}$, and for the cascades based on 5_2 and 5_3 the only possible values for t_1 are $\frac{3}{10}$ and $\frac{1}{10}$.

As the period-doubling cascade continues, a granddaughter orbit bifurcates from the daughter orbit. Daughter and granddaughter wind in the same way around the mother orbit. This means that the relative rotation rates of the daughter and granddaughter orbits about the mother orbit are the same, except for multiplicity. More generally, in a period-doubling cascade based on a period p orbit, the orbits of period $p \times 2^{k+1}, p \times 2^{k+2}, \ldots$ wind the same way around the orbit of period $p \times 2^k$. This means that the table of relative rotation rates for a period-doubling cascade is essentially trivial, except along the diagonal. They are

$$\begin{aligned} R(k,k) &= [2^k(r_1, r_2, \ldots, r_{p-1}), 2^{k-1}(t_1), 2^{k-2}(t_2), \ldots, 2(t_{k-1}), t_k, 0]^{p\times 2^k} \\ R(k,l) &= [2^k(r_1, r_2, \ldots, r_{p-1}), 2^{k-1}(t_1), 2^{k-2}(t_2), \ldots, 2(t_{k-1}), t_k, t_{k+1}]^{p\times 2^l} \\ & \qquad l > k \end{aligned} \tag{4.18}$$

The fractions t_k are given by

$$t_k = \frac{T(k)}{p \times 2^k} \tag{4.19}$$

where the integers $T(k)$ obey the Fibonacci-type series

$$T(k+1) = T(k) + 2 \times T(k-1) \tag{4.20}$$

The initialization is $T(0) = 0$, $T(1) = 2n \pm 1$, and $t_1 = (2n \pm 1)/2p$. For the period-doubling cascade, $T(1), T(2), T(3), \ldots$ are $1, 1, 3, 5, 11, 21, 43, \ldots$ so that $t_1, t_2, t_3, \ldots$ are $1/2, 1/2^2, 3/2^3, 5/2^4, 11/2^5, 21/2^6, 43/2^7, \ldots$, as given in the footnote to Table 4.4.

Example: The period-3 saddle 011 and node 001 are born in a saddle-node bifurcation with relative rotation rates

	011	001
011	$(\frac{1}{3}, \frac{1}{3}, 0)^3$	$(\frac{1}{3}, \frac{1}{3}, \frac{1}{3})^3$
001	$(\frac{1}{3}, \frac{1}{3}, \frac{1}{3})^3$	$(\frac{1}{3}, \frac{1}{3}, 0)^3$

(4.21)

In this case, $r_1 = r_2 = \frac{1}{3}$ and $r_3 = \frac{1}{3}$. The fraction t_1 is $t_1 = r_3 \pm \frac{1}{2\times 3} \rightarrow \frac{1}{3} - \frac{1}{6} = \frac{1}{6}$. The relative rotation rates for the mother–daughter pair are therefore

	011	001011
011	$(\frac{1}{3}, \frac{1}{3}, 0)^3$	$(\frac{1}{3}, \frac{1}{3}, \frac{1}{6})^6$
001011	$(\frac{1}{3}, \frac{1}{3}, \frac{1}{6})^6$	$(\frac{1}{3}^2, \frac{1}{3}^2, \frac{1}{6}, 0)^6$

(4.22)

In this cascade, $T(0) = 0$ and $T(1) = 1$, just as for the original period-doubling cascade. As a result, from the Fibonacci relation (4.20), the sequence $T(2), T(3), \ldots$ for the cascade based on the period-3 node 001 is the same as the sequence based on the period-1 node 1.

4.3.4 Additional Properties

The relative rotation rates have a number of additional useful properties which we now summarize.

Robustness: The integral (4.11) is invariant under small perturbations of initial conditions in the Poincaré section. It is invariant under changes in the Poincaré section. It is invariant under changes in the dynamical system, brought about by changes in control parameter values, provided that the orbits A and B exist at all times during these changes.

Symmetry: $R_{i,j}(A,B) = R_{j,i}(B,A)$.

Spectrum of Values: There are $p_A \times p_B$ relative rotation rates for orbits A and B of periods p_A and p_B, but no more than (p_A, p_B) of these fractions are distinct. Here (p_A, p_B) is the largest common divisor of p_A and p_B. A period-3 and period-5 orbit have only one distinct value for the 15 relative rotation rates.

Localization Property: When orbits A_S and A_N are created in a saddle-node bifurcation, their relative rotation rates with all orbits currently in existence at the time of the saddle-node bifurcation are the same. Similarly, when a daughter orbit D is created in a period-doubling bifurcation from a mother orbit M, their relative rotation rates with all orbits currently in existence at the time of the bifurcation are the same, except for doubled multiplicity.

4.4 RELATION BETWEEN LINKING NUMBERS AND RELATIVE ROTATION RATES

There is a simple relationship between linking numbers and relative rotation rates:

$$L(A,B) = \sum_{i=1}^{p_A} \sum_{j=1}^{p_B} R_{i,j}(A,B) \tag{4.23}$$

This equality is obtained immediately from (4.12) by summing over the two indices i and j:

$$\sum_{i=1}^{p_A} \sum_{j=1}^{p_B} \frac{1}{2 \times p_A \times p_B} \sum_{k=1}^{p_A \times p_B} \sigma(A_{i+k}, B_{j+k}) =$$

$$\sum_{k=1}^{p_A \times p_B} \frac{1}{p_A \times p_B} \left\{ \sum_{i=1}^{p_A} \sum_{j=1}^{p_B} \frac{1}{2} \sigma(A_{i+k}, B_{j+k}) \right\} \tag{4.24}$$

The double sum within the braces { }, is independent of k, by a simple combinatorial argument. It is equal to $L(A,B)$. The outer sum then becomes trivial, giving the desired result (4.23).

We give three simple examples of this relation.

Example 1: For the orbits 00101 and 0111, $R(00101, 0111) = \frac{7}{20}$, so that taking account of the multiplicity (20), we find $L(00101, 0111) = 7$.

Example 2: The self- and relative rotation rates of the saddle-node pair 5_2 with correct multiplicities are:

	00111	00101
00111	$\frac{2}{5}^{10}\frac{1}{5}^{10}0^5$	$\frac{2}{5}^{10}\frac{1}{5}^{15}$
00101	$\frac{2}{5}^{10}\frac{1}{5}^{15}$	$\frac{2}{5}^{10}\frac{1}{5}^{10}0^5$

$\xrightarrow{\text{sum}}$

	00111	00101
00111	6	7
00101	7	6

(4.25)

Example 3: By summing the relative rotation rates for the period-doubling cascade given in Table 4.4, we immediately obtain the linking and self-linking numbers for the orbits in the period-doubling cascade, given in Table 4.2.

When the global torsion is nonzero, its value T must be added to all relative rotation rates. This provides a simple relation between the linking numbers of orbits A and B in a zero-torsion lift of some map (horseshoe or otherwise) and a lift with nonzero global torsion:

$$L_T(A, B) = L_{T=0}(A, B) + p_A \times p_B \times T \tag{4.26}$$

The orbits A and B are generally determined by finding fixed points of the pth iterate of a return map. This procedure ignores global torsion. There seems to be no simpler way to derive the relation (4.26) than the derivation based on (4.24).

Remark: We have defined the self-relative rotation rate with the same initial conditions, $R_{ii}(A, A)$, to be zero, so that (4.24) is true for the case $B = A$ and self-relative rotation rates.

4.5 ADDITIONAL USES OF TOPOLOGICAL INVARIANTS

The relative rotation rates have a number of additional uses. We describe several of them here. They depend on the fact that self-relative rotation rates are topological invariants (as are self-linking numbers in the special case $R^2 \times S^1$).

4.5.1 Bifurcation Organization

Relative rotation rates provide information about the organization of periodic orbits, and in particular about the priority of saddle-node bifurcations on the road to chaos. They do this in the same way as linking numbers, except that they provide a more refined tool. This tool is more refined because entire sets of relative rotation rates must be equal, rather than single integers in the case of linking numbers, in order to carry out the tests shown in Figs. 4.3 and 9.5.

The relative rotation rates for the period-3 orbits 001 and 011 and the saddle-node pair 00111 and 00101, are shown below:

	00111	00101
001	$\frac{4}{15}$	$\frac{4}{15}$
011	$\frac{1}{3}$	$\frac{1}{3}$

(4.27)

In this simple case, we find no information not already provided in (4.10). However, for more complicated, higher-period orbits, the relative rotation rates provide information not provided by linking numbers.

Period-Doubling Cascades: The orbits belonging to a period-doubling cascade are simple to recognize from their self-relative rotation rates, which obey the systematic properties described in Section 4.3.3.

4.5.2 Torus Orbits

Some of the periodic orbits generated by the Smale horseshoe mechanism are *torus knots*. These knots can be placed on a torus, winding around the long direction of the torus p times while winding around the short direction m times. Such knots are very "tame": In mappings of the plane to itself, they can be created without any other knots having previously been created. In short, they have zero topological entropy.

Torus knots are easily recognized from their self-relative rotation rates. An m/p torus knot has self-relative rotation rates $\left[(m/p)^{p-1}, 0\right]^p$. Here p is the period of the orbit, the integers m and p are relatively prime, and in the zero-torsion lift of the right-handed horseshoe, $0 < m < p/2$. The horseshoe torus knots, up to period 5, are $3_1, 4_2, 5_1$, and 5_3 with $m/p = \frac{1}{3}, \frac{1}{4}, \frac{2}{5}$, and $\frac{1}{5}$, respectively. Proceeding beyond period 5, there are: one torus knot of period 6 with $m/p = \frac{1}{6}$ (2 and 6 are not relatively prime), three torus knots of period 7 with $m/p = \frac{3}{7}, \frac{2}{7}$, and $\frac{1}{7}$, and two torus knots of period 8 with $m/p = \frac{3}{8}$ and $\frac{1}{8}$.

4.5.3 Additional Remarks

A number of prejudices have developed regarding these maps, not all of which are true. These include:

1. There is a 1:1 correspondence between the unstable periodic orbits of the logistic map and the orientation-preserving Hénon maps in their respective hyperbolic limits.

2. Every orbit in the orientation-preserving Hénon map has a unique symbolic code, and this coding remains unchanged as ϵ is varied, in particular, as the logistic limit $\epsilon \to 0$ is approached.

3. The order in which orbits are created in saddle-node bifurcations is independent of the ratio, ϵ, of the image to source areas in the orientation-preserving Hénon map.

4. Saddle-node pairs created in the logistic limit are always created together, in all orientation-preserving Hénon maps.

The first prejudice is true.

It is not known if the second is true, but it holds up to period 6 [49].

The third is not. In fact, as ϵ is increased from 0 to 1, there is a general reversal in the order in which orbits are created in saddle-node bifurcations. This reversal was

Table 4.5 Self-relative rotation rates and properties of period-7 horseshoe orbits

		Self Relative					
	Symbolic	Rotation Rates					
Orbit	Dynamics	$\frac{3}{7}$	$\frac{2}{7}$	$\frac{1}{7}$	0	Remarks	
7_1	011 11$\underline{1}$1	6	0	0	1	Torus knot	$\frac{3}{7}$
7_2	011 01$\underline{1}$1	4	2	0	1	QOD	$\frac{2}{5}$
7_3	001 01$\underline{1}$1	2	4	0	1	Braid	7_A
7_4	001 11$\underline{1}$1	2	4	0	1	Braid	7_A
7_5	001 10$\underline{1}$1	0	6	0	1	Torus knot	$\frac{2}{7}$
7_6	000 10$\underline{1}$1	0	4	2	1	Braid	7_B
7_7	000 11$\underline{1}$1	0	4	2	1	Braid	7_B
7_8	000 01$\underline{1}$1	0	2	4	1	QOD	$\frac{1}{5}$
7_9	000 00$\underline{1}$1	0	0	6	1	Torus knot	$\frac{1}{7}$

discussed by Holmes [42]. In fact, there are severe topological constraints (*forcing*) on the order in which orbits can be created in the transition from a map with one simple fixed point to a Hénon-like map exhibiting a fully developed hyperbolic horseshoe. Forcing relations are discussed at length in Chapter 9.

The fourth prejudice is also false. To undergo a saddle-node bifurcation, two distinct orbits must be very similar topologically. Specifically, they must have the same braid type. Two orbits with the same braid type experience the same deformation under a flow [79–82]. Deformations under a flow are expressed by the self-relative rotation rates. This suggests that self-relative rotation rates might be an effective probe for the braid type of an orbit. Specifically, it might be conjectured that orbits with the same spectrum of self-relative rotation rates belong to the same braid type (and can therefore undergo saddle-node bifurcations with each other). This conjecture is also not true in general: It is true for horseshoe orbits up to period 10 but not for period 11.

In Table 4.5 we present the self-relative rotation rates for all horseshoe orbits of period 7. These orbits are listed by their order of appearance in unimodal maps of the interval—the universal-sequence order. For example, 7_5 = 001 10$\underline{1}$1 is the fifth orbit (pair) of period 7, which is created in the logistic map. This consists of a saddle-node pair. The symbol sequence 001 1011 represents the regular saddle (even parity); the partner node orbit has a parity change in the penultimate symbol: 001 1001. From the spectrum of self-relative rotation rates it is simple to characterize most of these orbits:

Torus Knots: The three orbit pairs $7_1, 7_5, 7_9$ are torus knots with $m/p = \frac{3}{7}, \frac{2}{7}, \frac{1}{7}$, respectively. For torus knots all self-relative rotation rates are m/p except for the diagonal one, $SR_{i,i}(A)$, which is 0 by definition.

Braids: The quartet $7_3, 7_4$ belongs to one braid type and the quartet $7_6, 7_7$ belongs to a different braid type. Since $p < 11$, the members of each quartet have the same spectrum of relative rotation rates.

QOD Orbits: The two remaining orbits, 7_2 and 7_8, are quasi-one-dimensional. These orbits in two-dimensional maps of the plane to itself force the full spectrum of orbits, which they force in unimodal maps of the interval. These orbits form singlets of braid type. They are described more fully in Chapter 9.

Table 4.6 Self-relative rotation rates and properties of period-8 horseshoe orbits

		Self – Relative						
	Symbolic	Rotation Rates						
Orbit	Dynamics	$\frac{4}{8}$	$\frac{3}{8}$	$\frac{2}{8}$	$\frac{1}{8}$	0	Remarks	
8_1	0111 0101	4	1	2	0	1	PD	4_1
8_2	011 111 $\underline{1}$1	4	3	0	0	1		
8_3	011 011 $\underline{1}$1	0	7	0	0	1	Torus knot	$\frac{3}{8}$
8_4	001 011 $\underline{1}$1	0	5	2	0	1	Braid	8_A
8_5	001 010 $\underline{1}$1	0	3	4	0	1	Braid	8_B
8_6	001 110 $\underline{1}$1	0	3	4	0	1	Braid	8_B
8_7	001 111 $\underline{1}$1	0	5	2	0	1	Braid	8_A
8_8	001 101 $\underline{1}$1	0	3	4	0	1	Braid	8_B
8_9	0001 0011	0	0	6	1	1	PD	4_2
8_{10}	000 101 $\underline{1}$1	0	2	4	1	1	Braid	8_C
8_{11}	000 111 $\underline{1}$1	0	2	4	1	1	Braid	8_C
8_{12}	000 110 $\underline{1}$1	0	0	6	1	1		
8_{13}	000 010 $\underline{1}$1	0	0	4	3	1	Braid	8_D
8_{14}	000 011 $\underline{1}$1	0	0	4	3	1	Braid	8_D
8_{15}	000 001 $\underline{1}$1	0	0	2	5	1	QOD	$\frac{1}{6}$
8_{16}	000 000 $\underline{1}$1	0	0	0	7	1	Torus knot	$\frac{1}{8}$

Similar information for horseshoe orbits of period 8 is provided in Table 4.6. This table shows that:

Torus Knots: The two orbit pairs 8_3 and 8_{16} are torus knots with $m/p = \frac{3}{8}$ and $\frac{1}{8}$, respectively.

Braids: The three saddle-node pairs of orbits $(8_4, 8_7)$, $(8_{10}, 8_{11})$, and $(8_{13}, 8_{14})$ belong to three different braid quartets, while the three saddle-node orbit pairs $8_5, 8_6, 8_8$ belong to a braid consisting of six orbits.

QOD Orbits: There is a single QOD orbit 8_{15} labeled by the fraction $\frac{1}{6}$.

In the case of period-8 orbits, the saddle 001 010 11 of 8_5 can be created with either of the three nodes 001 010 01 of 8_5 or 001 110 11 of 8_6 or 001 101 11 of 8_8, depending on the particular path the horseshoe deformation takes on the road to complete hyperbolicity.

4.6 SUMMARY

Our long-range objectives are to determine the stretching and squeezing mechanisms which occur repeatedly to build up a strange attractor. These mechanisms also organize all the unstable periodic orbits in a strange attractor in a unique way. As a result, the most effective way of identifying mechanisms and classifying strange attractors is to determine how the unstable periodic orbits in it are organized.

The organization is topological. Topological organization is determined quantitatively by two topological invariants: the Gaussian linking numbers and a newer topological invariant, the relative rotation rates. The latter are not as widely applicable as the former—they are useful in phase spaces with the topology of a torus ($R^2 \times S^1$). However, when they are applicable, they are more powerful than the linking numbers.

These two invariants have been defined and methods for their computation have been introduced. The methods described are primarily of use for the analysis of experimental data. The topological invariants have a number of uses. The linking numbers are able to distinguish among different stretching and squeezing mechanisms. Relative rotation rates can distinguish among many (but not all) different mechanisms. However, they are able to differentiate among different classes of orbits in a strange attractor. Both can be used to place constraints on the order in which bifurcations can occur as control parameters are changed.

5

Branched Manifolds

All of the unstable periodic orbits in a strange attractor can be placed on a single, simple geometric structure. This structure has been called variously a *knot holder*, an *orbit organizer*, and a *template*. Mathematically, it is a branched manifold. A branched manifold describes the topological organization of all the unstable periodic orbits in a strange attractor [83,84]. This means that the branched manifold for a strange attractor provides information about the stretching and squeezing mechanisms that generate the strange attractor. Branched manifolds can be classified discretely. This means that a discrete classification exists for low-dimensional strange attractors [85].

5.1 CLOSED LOOPS

Branched manifolds are central to the classification theory for strange attractors of low dimensional dynamical systems. In this section we introduce the idea of branched manifolds in an amusing but very nontrivial way.

5.1.1 Undergraduate Students

Many of us in physics or electrical engineering departments have had to teach Maxwell's equations in one form or another. One standard problem that we always give to undergraduates is to compute the magnetic field generated by a current carrying wire. Needless to say, the wire is straight and the current is constant. The standard approach is to find a closed loop, or magnetic field line, and perform a Gaussian type integral around it. As Fig. 5.1(a) shows, there is a two (continuous)-parameter family of closed field lines around the wire. These are parameterized by distance along the wire and radius of the loop.

5.1.2 Graduate Students

Through sheer perversity, we always make our students go through another round of electricity and magnetism in graduate school. The material is the same, but the problems have to be different—and harder. This time around, we bend the straight wire into a circular loop and then ask our students to compute the magnetic field in its vicinity. As Fig. 5.1(b) shows, there is still a real two-parameter family of closed loops around the current-carrying wire. One parameter is the angular distance around the wire; the other is the *perihelion* of the closed magnetic field line with respect to the current-carrying wire. The perihelion is the distance of closest approach. Some students (the smart ones) choose not to solve the problem with this approach.

5.1.3 The Ph.D. Candidate

When the poor student finally finishes his (her) thesis, you may be put on his committee. If you like the student, you give easy questions. If not, you give impossible questions.

Here is one. You take the current-carrying wire and tie it into a knot: a figure 8 knot, to be specific. Then ask: Are there still closed magnetic field lines? If so, what are they like?

It turns out that most of the closed field lines of undergraduate and graduate days break when the current-carrying wire is tied into the figure 8 knot. However, a few do not break. "Few" is, in fact, a countable infinity—significantly fewer than a continuous two-parameter family.

Not only are the closed magnetic field lines countable, they can also be named. More surprisingly, they are organized among themselves in a rigid and surprisingly simple way. The organizational mechanism is illustrated in Fig. 5.2. The current-carrying wire, tied into a figure 8 knot, is shown in Fig. 5.2(a). One of the closed

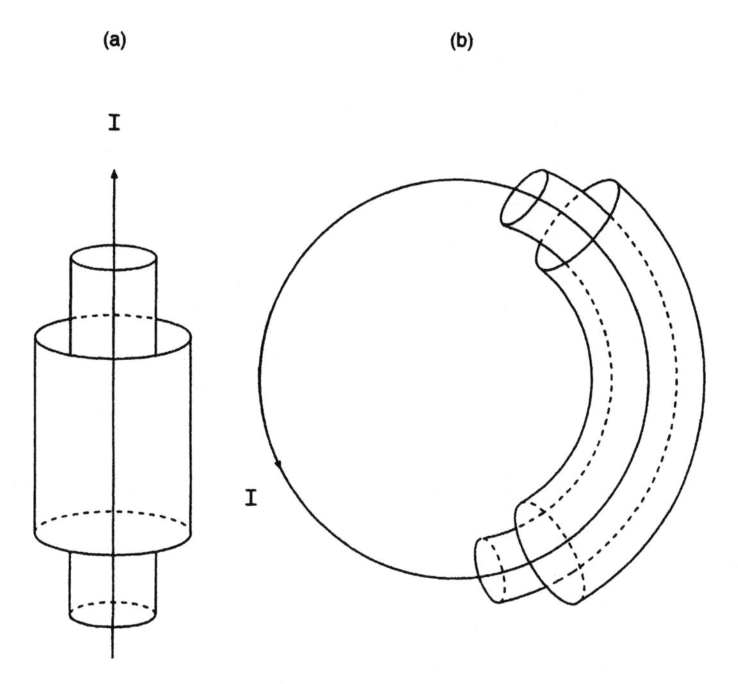

Fig. 5.1 (a) The closed magnetic field lines surrounding a straight wire carrying a uniform constant current can be identified by two real continuous parameters. (b) This remains true if the wire is deformed into a closed circular loop.

field lines generated by the current in this wire is shown in Fig. 5.2(b). In Fig. 5.2(c) we show a structure that contains all the information about the organization of all the closed magnetic field lines generated by the current in this wire. This structure is a branched manifold (or knot-holder, or template).

All of the closed field lines surrounding the Figure 8 knot can be deformed (*isotoped*) down to lie on this two-dimensional surface without undergoing any self-intersections. On this surface it is a relatively simple matter to compute the topological invariants of these closed field lines, their linking numbers. There is a 1:1 correspondence between symbol sequences for closed paths along the "one-way streets" (branches) in this branched manifold and the closed magnetic field lines surrounding the current-carrying wire. The symbols may identify either the branch lines, as encountered, or the branches of the branched manifold, as traversed. As a result, the closed field lines are clearly countable.

5.1.4 Important Observation

We point out here, forcefully, that this first encounter with branched manifolds has occurred for a "conservative" dynamical system.

In the past it seems that there has been a prejudice against the use of branched manifolds as a valuable tool for classifying strange attractors. This prejudice had been brought about by the incorrect assumption that this tool is useful only in the highly dissipative limit. The central tool for the classification theory, the Birman–Williams theorem, is applicable to dissipative three-dimensional dynamical systems ($\lambda_1 + \lambda_2 + \lambda_3 < 0$), but these systems need not be highly dissipative.

In the example just discussed, a branched manifold describes the organization of all the closed magnetic field lines around a current-carrying figure 8 knot. The analog dynamical system is *conservative*, not even *slightly dissipative*.

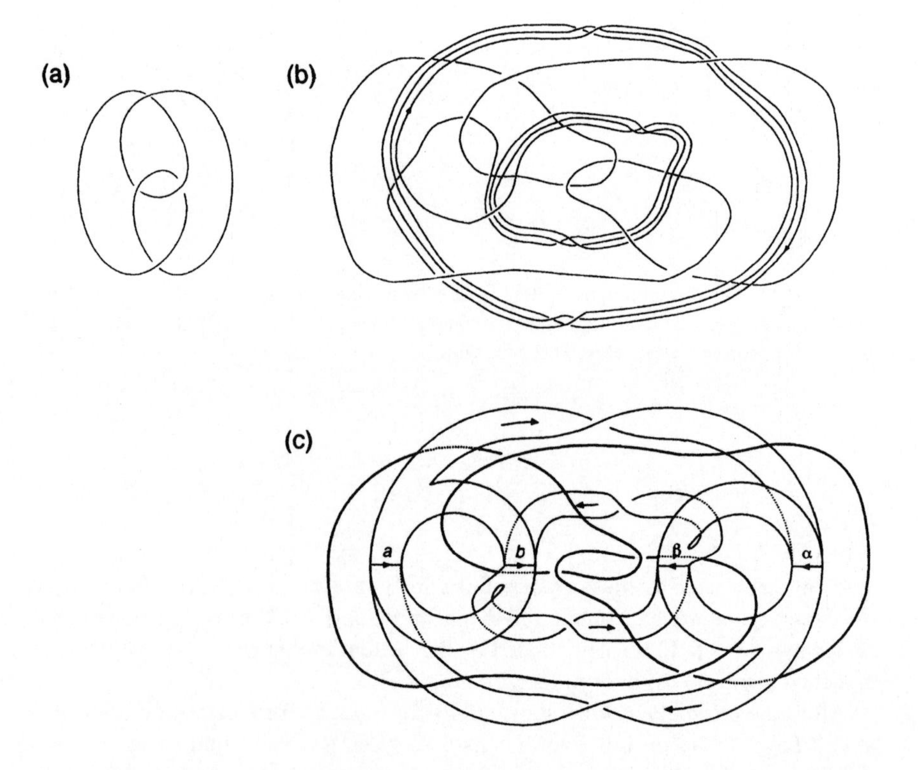

Fig. 5.2 A current-carrying wire tied into a figure 8 knot (a) generates a countable number of closed magnetic field lines, one of which is shown in part (b). (c) This branched manifold describes all of the closed field lines generated by the wire in (a). It can also be used to label all these orbits, and to compute their topological invariants, such as linking numbers. The closed loop shown in (b) can be identified as $(a\alpha)^3 a(b\beta)^3 b$. Adapted with permission from Birman and Williams [84].

5.2 WHAT HAS THIS GOT TO DO WITH DYNAMICAL SYSTEMS?

We are interested in classifying dynamical systems by the strange attractors that they generate. When a dynamical system in R^3 generates a chaotic signal, a large number of periodic orbits occur in the strange attractor. They are organized among themselves in a unique way. This organization reflects the stretching and squeezing mechanisms, which act to generate chaotic behavior.

There is a theorem, due to Birman and Williams, which guarantees that all these orbits can be isotoped down to a two-dimensional branched manifold, preserving their topological organization. As a result, we can identify a dynamical system by the branched manifold that describes the periodic orbits in its strange attractor.

The Birman–Williams theorem is valid for dissipative dynamical systems in R^3. The Lyapunov exponents for the strange attractor obey $\lambda_1 > 0$, $\lambda_2 = 0$, and $\lambda_3 < 0$, with $\lambda_1 + \lambda_2 + \lambda_3 < 0$. The Lyapunov dimension of such an attractor is $d_L = 2 + (\lambda_1 + \lambda_2)/|\lambda_3| = 2 + \epsilon$. When $\epsilon = \lambda_1/|\lambda_3|$ is small, it is easy to discern the shape of the branched manifold from the numerically computed strange attractor [cf. Figs. 5.7(c) and 5.8(c)]. However, when $\epsilon \simeq 1$ there is still a branched manifold which describes the dynamics, even though it may not be easy to identify from the strange attractor.

We emphasize once again that the branched manifold for the figure 8 knot describes the topological organization of the closed field lines in a system which has the properties of a strange attractor with $\epsilon = 1$: the conservative limit.

5.3 GENERAL PROPERTIES OF BRANCHED MANIFOLDS

The branched manifold shown in Fig. 5.2(c) consists of two kinds of structures. These describe stretching and squeezing. The origin of these structures is shown in Fig. 5.3. On the left we show a cube of initial conditions. Under the stretching process, the cube is deformed: It stretches in one direction and contracts in the other. Eventually, the flow goes off in two different directions in phase space. In the limit of very high dissipation, the three-dimensional structure becomes two-dimensional. This structure describes stretching.

On the right we show two neighborhoods in different parts of phase space. Under the flow they are squeezed together. Between the two deformed neighborhoods there is a boundary layer. In the limit of very high dissipation, the three-dimensional structure becomes two-dimensional. This structure describes squeezing.

Remark: The two-dimensional structures shown at the bottom of Fig. 5.3 do not depend on the dissipation being large. They, in fact, result from projecting the flow down along the stable direction. We emphasize again that the construction of branched manifolds does not depend on the dissipation being large.

The most general branched manifold is built up from just these two building blocks in Lego fashion. The simple rules are:

Out $\rightarrow$ in: Every outflow feeds into an inflow.

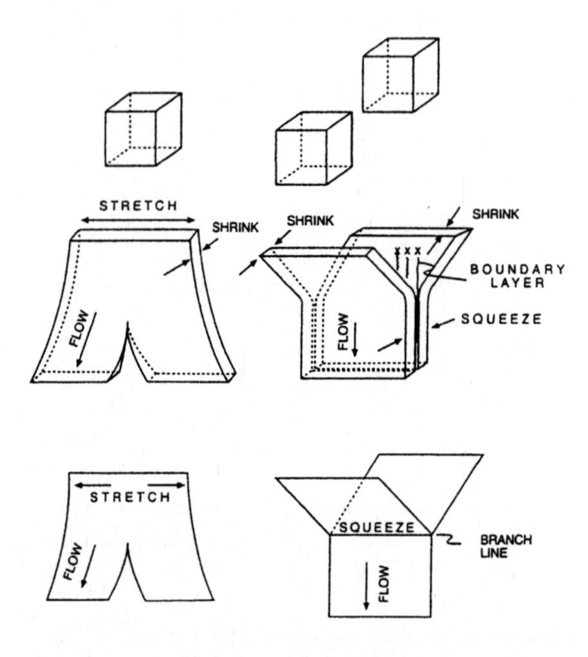

Fig. 5.3 Left: A cube of initial conditions (top) is deformed under the stretching part of the flow (middle). A gap begins to form for two parts of the flow heading to different parts of phase space. Under further shrinking (higher dissipation) a two-dimensional structure is formed which is not a manifold because of the tear point, which is an initial condition for a trajectory to a singular point. Right: Two cubes of initial conditions (top) in distant parts of phase space are squeezed together and deformed by the flow (middle). A boundary layer separates the deformed parallelepipeds at their junction. Under more dissipation the two inflow regions are joined to the outflow region by a branch line.

No free ends: There are no uncoupled outflow or inflow edges.

The two-dimensional branched manifolds that we use to classify dynamical systems are two-dimensional manifolds *almost* everywhere. Of the two dimensions: one dimension corresponds to the flow direction; the other corresponds to the unstable invariant manifold of a low-period orbit. The structure fails to be a manifold because of singularities. There are two types of singularities:

Zero-dimensional: The splitting points identify stretching mechanisms.

One-dimensional: The branch lines identify squeezing mechanisms.

It is possible to describe branched manifolds algebraically. The algebraic description for a branched manifold with n branches has three components:

Topological Matrix T: This is an $n \times n$ matrix that describes the topological organization of the branches. The diagonal element T_{ii} describes the local torsion of branch i. This is the signed number of crossings of the two edges of branch i with each other. The off-diagonal elements $T_{ij} = T_{ji}$ describe how branches i and j cross. The crossing convention adopted for segments of orbits (Fig. 4.4) is extended to projections of the branches in an obvious way.

Joining Array A: This is a $1 \times n$ array that describes the order in which branches are joined at branch lines. A simple convention is: The closer a branch is to the observer, the lower the branch number.

Incidence Matrix I: This is an $n \times n$ transition or incidence matrix. It describes which branches flow into which branches. If branch i flows into branch j, then $I_{ij} = 1$; it is zero otherwise.

The algebraic description for the branched manifold shown in Fig. 5.2(c) is given in Fig. 5.4.

We remark here that the algebraic description of a branched manifold is not unique. The branched manifold is embedded in R^3. As such, it can be rotated and projected to a variety of two-dimensional surfaces. Different projections have different algebraic representations. This nonuniqueness is the nonuniqueness of projections, discussed in Section 4.2.1. There are several other ways in which branched manifolds for an underlying dynamics may not be unique. However, there is one invariant: They all describe the same spectrum of periodic orbits with the same topological organization.

We also remark here that the algebraic description of branched manifolds is ideally suited for the computation of some topological invariants, such as linking numbers and relative rotation rates, but is not suitable for computing other invariants, such as knot polynomials.

5.4 BIRMAN–WILLIAMS THEOREM

We refer the interested reader to [83] and [84] for proof of the Birman–Williams theorem. In the first subsection we introduce the projection method (Birman–Williams projection) that is used to project a flow onto a branched manifold. In the second subsection we state the Birman–Williams theorem.

5.4.1 Birman–Williams Projection

Two points, x and y, are defined to be *equivalent* under a flow if they have the same asymptotic future:

$$x \sim y \qquad \text{if} \qquad |x(t) - y(t)| \stackrel{t \to \infty}{\longrightarrow} 0 \tag{5.1}$$

The Birman–Williams projection (5.1) has the effect of projecting the flow in a strange attractor down along the stable direction onto a two-dimensional branched manifold. The dimensions include the flow direction and part of the unstable direction. It is illustrated in Fig. 5.5.

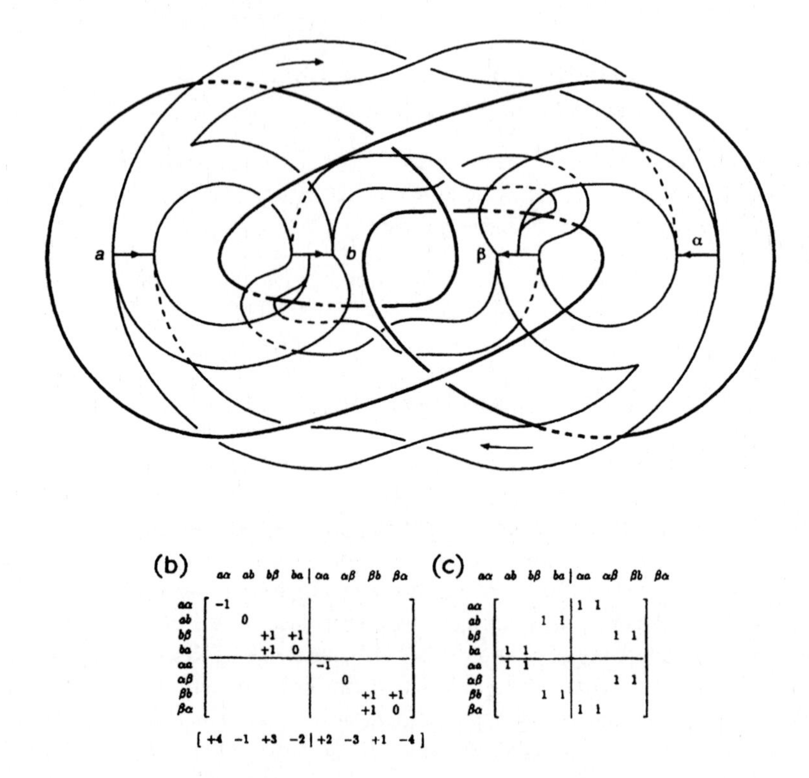

Fig. 5.4 Algebraic representation of a branched manifold with n branches (a) consists of: (b) an $n \times n$ matrix T that describes how the branches cross over or under each other (off-diagonal matrix elements) or how they twist about their flow axis (diagonal matrix elements), and underneath this matrix a $1 \times n$ array A that describes the order in which the branches join each other at branch lines, with the convention: The larger the number, the farther behind; and in addition, an $n \times n$ matrix I (incidence matrix) (c) describes how the branches are connected to each other. The branches may be labeled by numbers or by indicating which branch lines they connect. Adapted with permission from Birman and Williams [84].

We represent the flow in R^3 by Φ_t, so that for x in the basin of the strange attractor $\mathcal{SA}$, $\Phi_t(x(0)) = x(t)$. The flow has a unique future and past; that is, given $x(0)$, the points $x(t)$ are determined uniquely for all t in the range $-\infty < t < +\infty$.

The Birman–Williams projection maps every point x in the basin of $\mathcal{SA}$ into a point $\bar{x}$ in a branched manifold $\mathcal{BM}$. This projection is illustrated in Fig. 5.6. This figure shows how a flow that exhibits a stretch and fold mechanism [Fig. 5.6(a)] is transformed into a pair of branches that meet at a branch line [Fig. 5.6(b)]. The projection also maps the flow Φ_t in the basin of $\mathcal{SA}$ to a semiflow $\bar{\Phi}_t$ on $\mathcal{BM}$. Under the semiflow, every point $\bar{x} \in \mathcal{BM}$ has a unique future $\bar{x}(t)$. Every point $\bar{x}$ also has

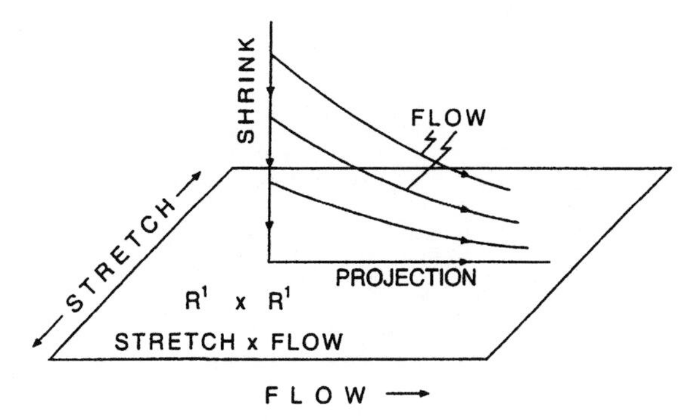

Fig. 5.5 The Birman–Williams projection identifies all points with the same asymptotic future. This has the effect of projecting the flow in a low-dimensional strange attractor down onto a two-dimensional manifold almost everywhere.

a unique past up to the first branch line in its past. At the branch line, information about its previous history is lost.

It is useful to extend each splitting point back to the nearest branch line in its past, as shown in Fig. 5.4. Then each branch line is split into a number of segments. Each branch of the branched manifold can then be labeled by the segments of the branches that it connects. These two symbols, the first the source, the second the sink, can be used to label the rows and columns of the transition matrix I.

Every point in a branched manifold has a unique future. In particular, every point on a branch line has a unique future. The future may be:

Aperiodic: A nonrepeating, chaotic orbit of infinite period.

Periodic: A periodic or ultimately orbit of finite period p.

Roughly speaking, each branch line can be considered to be like the closed interval $[0, 1]$. The points on a branch line that are initial conditions for aperiodic orbits are like the irrational numbers, and the points on a branch line that are initial conditions for periodic or ultimately periodic orbits are like the rational numbers. Both point sets are dense on the interval. We refine this classification slightly in Section 5.9.1.

5.4.2 Statement of the Theorem

The Birman–Williams theorem is as follows [83, 84]:

Theorem: Assume that a flow Φ_t:

- On $\underline{R^3}$ is dissipative ($\lambda_1 > 0, \lambda_2 = 0, \lambda_3 < 0$ and $\lambda_1 + \lambda_2 + \lambda_3 < 0$).
- Generates a <u>hyperbolic</u> strange attractor $\mathcal{SA}$.

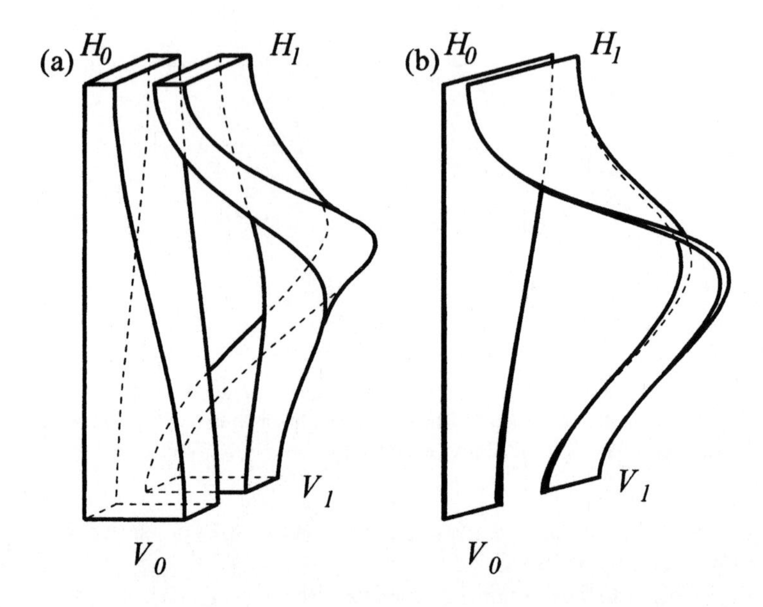

Fig. 5.6 (a) Suspension of the horseshoe map represented as a continous deformation of the two "vertical" rectangles V_0 and V_1 of Fig. 2.19 into the "horizontal" rectangles H_0 and H_1, with time flowing from bottom to top. Top and bottom should be identified. (b) When the flow is squeezed along the stable direction (i.e., dissipation is increased to infinity), the two prisms in (a) are transformed into a pair of two-dimensional branches that meet at a branch line. A complete branched manifold is obtained by connecting the branch line to the bottom with a flat ribbon.

The projection (5.1) maps the strange attractor $\mathcal{SA}$ to a branched manifold $\mathcal{BM}$ and the flow Φ_t on $\mathcal{SA}$ in R^3 to a semiflow $\bar{\Phi}_t$ on $\mathcal{BM}$ in R^3. The periodic orbits in $\mathcal{SA}$ under Φ_t correspond 1:1 with the periodic orbits in $\mathcal{BM}$ under $\bar{\Phi}_t$, with perhaps one or two specified exceptions. On any finite subset of periodic orbits the correspondence can be taken to be via isotopy.

This means, roughly but accurately, that the flow Φ_t on $\mathcal{SA}$ can be deformed continuously to the flow $\bar{\Phi}_t$ on $\mathcal{BM}$. During this deformation, periodic orbits are neither created nor destroyed, and orbit segments do not pass through each other (there are no crossings). In addition, their topological organization, as described by their linking numbers, remains invariant.

5.5 RELAXATION OF RESTRICTIONS

There are two serious restrictions on the Birman–Williams theorem. They have been underlined in the statement of the theorem. If they were unavoidable, they would render the theorem much less useful for experimental applications than it actually is. In this section we describe how these restrictions can be circumvented.

5.5.1 Strongly Contracting Restriction

The very first application of the Birman–Williams theorem to a physical system [1] ran into an unexpected and fortuitous problem. This involved the analysis of experimental data taken from a chemical system, the oscillating Belousov–Zhabotinskii reaction. Every theoretical description of this reaction involved more than three variables [86]. The Birman–Williams theorem is valid for three-dimensional systems. Knots fall apart in dimensions higher than 3. So, in principle, it appears that both the theorem and knowledge of the periodic orbits of this system are useless.

Despite this, we were able to carry out a successful analysis of the data and determine a branched manifold which described the organization of all the periodic orbits that we were able to extract from the experimental data.

Why?

This success in the face of inapplicable theorems leads to a deeper understanding of the Birman–Williams theorem, and more generally of low-dimensional strange attractors. First, the data do not care about the theoretical description (such descriptions are often incorrect, anyway). Suppose that the data are embedded in n dimensions and the Lyapunov exponents obey

$$\lambda_1 > \lambda_2 = 0 > \lambda_3 > \cdots > \lambda_n \tag{5.2}$$

Assume also that the attractor is *strongly* contracting. By definition, this means that

$$\lambda_1 + \lambda_2 + \lambda_3 < 0 \tag{5.3}$$

Then the Birman–Williams projection can be carried out in two steps. First, the projection is carried out along the strongly contracting directions corresponding to λ_4, $\lambda_5, \ldots, \lambda_n$. This has the effect of projecting the flow in R^n into a three-dimensional manifold, $\mathcal{IM}$. The manifold $\mathcal{IM}$ is called an *inertial manifold.* In this three-dimensional manifold:

- The conditions of the Birman–Williams theorem are met.
- The topological organization of periodic orbits is defined (knots don't fall apart).

The last projection along the least stable direction (λ_3) maps $\mathcal{SA} \subset \mathcal{IM}$ down to a two-dimensional branched manifold $\mathcal{BM} \subset \mathcal{IM}$ and preserves the topological organization of the unstable periodic orbits in the strange attractor.

For strongly contracting flows, the Lyapunov dimension

$$d_L = 2 + \frac{\lambda_1}{|\lambda_3|} < 3 \tag{5.4}$$

is less than 3. More specifically, if $d_L(x)$ is the local Lyapunov dimension at $x \in \mathcal{SA}$, and if $d_L(x) < 3$ everywhere on $\mathcal{SA}$, the Birman–Williams projection (5.1) provides a projection of $\mathcal{SA} \subset R^n$ down to a two-dimensional branched manifold $\mathcal{BM} \subset \mathcal{IM}^3 \subset R^n$, where $\mathcal{IM}^3$ is the three-dimensional manifold that results from the projection along the strongly stable directions λ_j, $j = 4, \ldots, n$.

5.5.2 Hyperbolic Restriction

We have never encountered a hyperbolic strange attractor, either in experimental data or in numerical simulations of ordinary differential equations.

Speaking roughly but accurately once again, the condition of hyperbolicity guarantees that the strange attractor is structurally stable under perturbations: Periodic orbits are neither created nor destroyed under perturbation of the control parameters. We get around this problem by assuming hyperbolicity for the strange attractor of interest. In doing so, we predict the existence of many more periodic orbits than actually exist in the strange attractor. Then we "unfold" the attractor. This means that we find a family of dynamical systems depending on one or (usually) more control parameters. The family contains the hyperbolic attractor for some control parameter value. Then we change the values of the control parameters. Under these changes many periodic orbits can be destroyed. However, the orbits that remain during the unfolding are organized in exactly the same way as in the hyperbolic attractor.

Unfolding comes in two forms. There is a global version and a local version. In the local version, as control parameters are changed, the branches in the branched manifold remain unchanged: It is the spectrum of periodic orbits on these branches that changes. In fact, the possible changes are restricted by topological considerations, as described in Sections 4.2 and 4.3. If we push the control parameters too far, new branches can come into existence and old branches can go out of existence. This is seen clearly in the perestroika of the Duffing oscillator and is visible in the experimental data described in Chapter 7. Unfoldings are discussed extensively in Chapter 9.

5.6 EXAMPLES OF BRANCHED MANIFOLDS

In this section we classify each of the four dynamical systems described in Section 3.3. This is done by integrating the dynamical equations for certain parameter values and then identifying the branched manifold which describes the strange attractor generated by each of these sets of equations. Precisely how the identification is made is discussed in detail in Chapter 6, which presents the topological analysis algorithm. We emphasize the fact that the branched manifold may change as the control parameters are varied. The possible changes are discussed more extensively in Chapter 9, which deals with unfoldings. A large number of branched manifolds are described in [87].

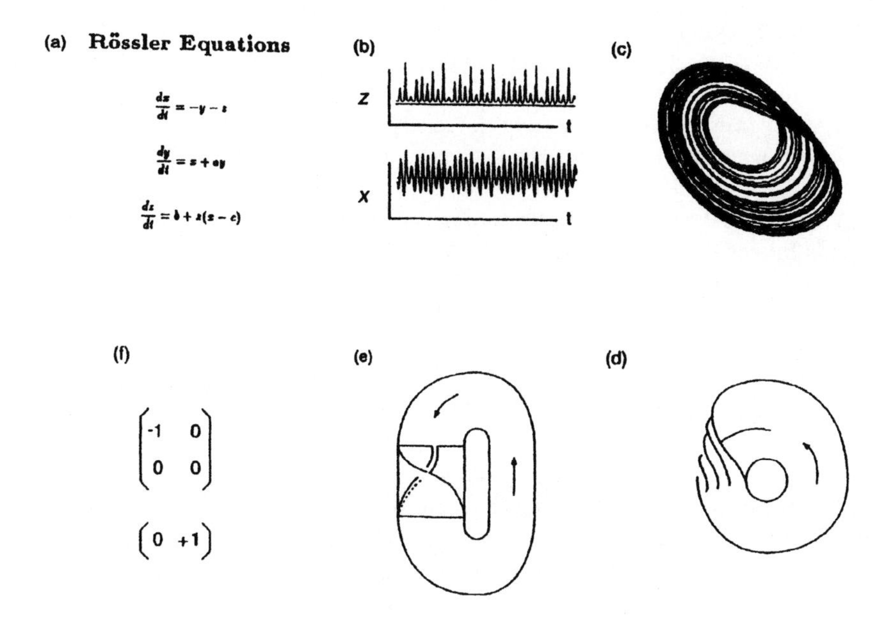

Fig. 5.7 (a) Rössler equations (3.22). (b) $x(t)$ and $z(t)$ after the transients have died out and the trajectory has relaxed to the strange attractor. Control parameter values: $(a, b, c) = (0.398, 2.0, 4.0)$. (c) Projection of the strange attractor onto the x–y plane. (d) Caricature of the flow on the attractor. (e) Branched manifold for this attractor. (f) Algebraic representation of this branched manifold. The topological matrix is shown at the top and the array at the bottom.

5.6.1 Smale–Rössler System

The Smale horseshoe mechanism consists of simple stretching and folding in phase space. It occurs very frequently in experiments that exhibit chaotic behavior [1, 78, 88–92]. This mechanism is exhibited by the Rössler equations.

The classification of the Rössler dynamical system is illustrated in Fig. 5.7 [2]. This figure consists of six parts. The equations of motion are shown in Fig. 5.7(a). These equations were integrated for control parameter values $(a, b, c) = (0.398, 2.0, 4.0)$. The traces $x(t)$ and $z(t)$ are recorded in Fig. 5.7(b). They were recorded after the transients died out. That is, an initial condition in the basin of attraction was chosen, and the integration was carried out beyond the point at which the trajectory relaxed to the strange attractor before the recording was begun. Figure 5.7(c) shows the attractor as projected onto the x–y plane $z = 0$. The flow is counterclockwise. The fold occurs at the 12 o'clock position (these comments are for analog people only). During the fold, the outer part of the attractor at the 5 o'clock position folds over the top of the inner part of the flow.

A schematic representation of this flow is presented in Fig. 5.7(d). It is clear that Fig. 5.7(d) is not a totally accurate representation of the dynamics shown in Fig. 5.7(c). In Fig. 5.7(d) the outer edge of the flow is reinjected to the inner edge of the flow, whereas in Fig. 5.7(c) the outer and inner edges of the flow at the 5 o'clock (or 9 o'clock) position are not squeezed together. As a result, the branched manifold shown in Fig. 5.7(d) contains more periodic orbits than actually exist in the flow shown in Fig. 5.7(c). However, the periodic orbits that exist in Fig. 5.7(c) are organized in exactly the same way as they are in Fig. 5.7(d).

The branched manifold is shown in a standard (*braid*) representation in Fig. 5.7(e). In this standard representation, all of the stretching and squeezing occurs between the two horizontal lines shown on the left-hand side. These two lines are branch lines—in fact, the same branch line. The flow emerging from the branch line at the bottom is returned to the branch line at the top without undergoing stretching and squeezing. It is no exaggeration to claim that all the nonlinear mechanisms responsible for chaotic motion are expressed between these two branch lines.

The algebraic representation for the branched manifolds in Fig. 5.7(d) and (e) is given in Fig. 5.7(f). There are two branches. Each branch contains one period-1 orbit. The 2×2 matrix T provides topological information. The diagonal elements describe the torsion of the two branches. The off-diagonal matrix elements T_{ij} are twice the linking number of the period-1 orbits in the branches i and j: $T_{ij} = 2L(i,j)$. The 1×2 array provides information about how the branches are ordered when they join at the branch line. In this case, the left-hand branch lies over the right-hand branch (in this projection). Its index is lower than the index for the right-hand branch, according to the convention adopted. The flow represented by this branched manifold is fully expansive. The incidence matrix I is therefore full: $\begin{bmatrix} 1 & 1 \\ 1 & 1 \end{bmatrix}$. When the incidence matrix is full, it is generally not presented explicitly.

Before leaving this dynamical system, we make a few observations about qualitative behavior. A small change in control parameter values will generally produce a small modification in Fig. 5.7(c); that is, there will be only a small change in how the two bands overlap. This results in only a small change in the spectrum of unstable periodic orbits in the attractor. If we continue to push the control parameters in an appropriate direction, the attractor will grow bigger. The outer edge will extend farther from the center, and when folded over, it will come closer to the center. One might easily believe that the folded-over region will never reach the center. If true, at some point the approach to the center will reverse itself. When this occurs, a second fold will occur at the inner edge of the attractor. In short, a third branch will be created. This third branch is connected to the second branch [-1 in Fig. 5.7(f)]. By continuity arguments, one might expect that its local torsion could only have values differing from -1 by ± 1. We could also expect that the local torsion value would place constraints on how this new branch could join with the two existing branches. These suspicions are true. It is in this way that the classification of strange attractors by branched manifolds allows us to make predictions about the behavior of nonlinear dynamical systems under perturbations both small and large.

5.6.2 Lorenz System

The Lorenz mechanism consists of a tear and a squeeze in phase space. It occurs in experiments that exhibit chaotic behavior and have some twofold symmetry.

The classification of the Lorenz dynamical system is illustrated in Fig. 5.8 [2]. This figure is identical in structure to Fig. 5.7 for the Rössler system. It consists of six parts. The Lorenz equations are shown in Fig. 5.8(a). These equations were integrated for control parameter values $(b, \sigma, r) = (\frac{8}{3}, 10.0, 30.0)$. The traces $x(t)$ and $z(t)$ are recorded in Fig. 5.8(b) after transients have died out. Figure 5.8(c) shows the attractor as projected onto the $x = y$ vs. z plane. The flow is clockwise on the left and counterclockwise on the right. The squeeze and tear occur in the middle.

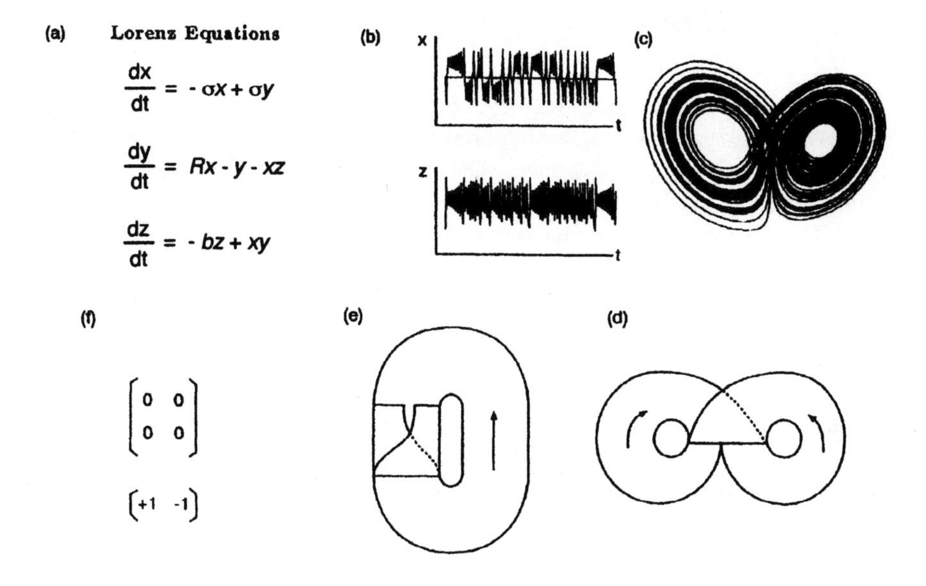

Fig. 5.8 (a) Lorenz equations (3.20). (b) $x(t)$ and $z(t)$ plotted after transients have died out and the trajectory has relaxed to the strange attractor. Control parameter values: $(b, \sigma, r) = (\frac{8}{3}, 10.0, 30.0)$. (c) Projection of the strange attractor onto the $x = y$–z plane. (d) Caricature of the flow on the attractor. (e) Branched manifold for this attractor. (f) Algebraic representation for this branched manifold. The topological matrix is shown at the top and the array at the bottom.

A schematic representation of this flow is presented in Fig. 5.8(d) and has been deformed into the branched manifold shown in Fig. 5.8(e). Once again, the stretching and squeezing mechanisms responsible for generating chaotic behavior are contained between the two horizontal lines.

The algebraic representation for the branched manifolds in Fig. 5.8(d) and (e) is given in Fig. 5.8(f). Neither branch exhibits twist, and the period-1 orbits in each branch correspond, in fact, to the two unstable foci. They clearly do not link.

The topological matrix appears trivial (all matrix elements are zero); nevertheless, it describes highly nontrivial dynamics. The array describes the order in which the two branches are connected. The incidence matrix is full, indicating that each branch flows into both.

Once again, the branched manifold description of this dynamics introduces the possibility of making educated guesses as to the behavior under control parameter variation. That is, one might expect the two outer edges to fold over when reinjected to the interior of the flow on the "opposite side." When new branches are visited by the flow (in symmetric pairs), they can only be related to the previously existing branches in a limited number of ways.

5.6.3 Duffing System

The dynamics of the Duffing oscillator are governed by a simple stretch and fold mechanism, in much the same way as in the Rössler system. However, unlike the Rössler oscillator, the Duffing oscillator has a twofold symmetry. As a result, the dynamics of the Duffing oscillator and its description by means of branched manifolds are much richer than those of the Rössler oscillator.

We describe the Duffing oscillator more thoroughly in Chapter 10, but in principle, what happens is simple. During the first half of a cycle, phase space undergoes a stretch and fold. The fold may be simple or may not be simple (i.e., multiple nondegenerate folds may occur). An identical stretch and fold occurs during the second half of the cycle. As a result, Duffing dynamics are the "square" of Rössler dynamics. More precisely, they are essentially Rössler dynamics twice iterated.

In Fig. 5.9 we present the branched manifold (to be accurate, only the central part, between the two horizontal lines in the standard representation) which describes an *extended fold.* This type of mechanism occurs for the Rössler equations for suitable control parameter values [2,93]. What we show in this figure is a branched manifold with four branches. This branched manifold is obtained as follows. A branch line, shown at the top right, is stretched out by a factor of 4 (i.e., $e^{\lambda_1} = 4$). This stretched branch line is then rolled up (right, middle) and then squeezed back down to the original interval (right, bottom). The four period-1 orbits in these dynamics are shown by symbols x in this figure. The four branches in this branched manifold are labeled by their local torsion, which varies systematically from 0 to 3 on going from left to right. It is relatively simple to verify that the linking numbers of the period-1 orbits in the four branches satisfy $L(i, j) = 1$ if $i \neq 0$ and $j \neq 0$, and are zero otherwise. These simple calculations define the 16 elements of the topological matrix. The array can be read off from the scrolling action shown on the right (see especially the middle figure on the right). The template and its algebraic representation are shown on the left in this figure.

In the Duffing oscillator this *scroll and squeeze mechanism* occurs twice. We illustrate this mechanism in Fig. 5.10 for the case where the stretch is by a factor of 3 in each half of a cycle. At the top of Fig. 5.10 we show a branch line. It is divided into three equal parts, labeled 1, 2, 3. These integers indicate twist during the first half cycle. Each part is further subdivided into three equal parts. During

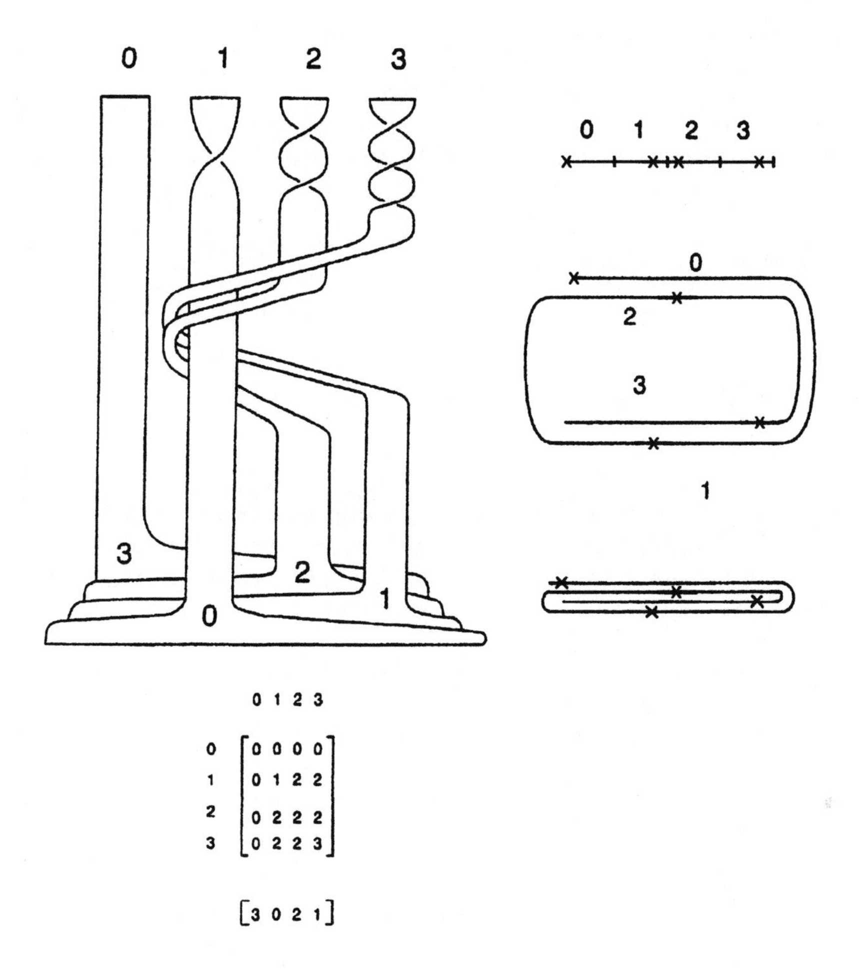

Fig. 5.9 Extended folding, as seen in the Rössler attractor for some control parameter values. Right: A branch line (top) is stretched by a factor of 4, rolled around like a jellyroll (middle), and then squeezed (bottom). The resulting branched manifold (left, top) and its algebraic representation (left, bottom) are then easily constructed.

the first half cycle this branch line is stretched out and rolled to a configuration as shown in Fig. 5.9, containing only branches 1, 2, and 3. This deformed branch line is then rotated through π radians, and then squeezed. The rotation accounts for the symmetry $(x, y, t) \to (-x, -y, t + \frac{1}{2}T)$ during half a cycle. This process is then repeated. The nine branches are conveniently labeled by two symbols: (1,1), (1,2), ..., (2,3), (2,2), ..., (3,3). The nine period-1 orbits for this iterated threefold stretch-and-roll mechanism can be located as indicated in Fig. 5.9, and their linking numbers computed. The ordering of the branches can be identified by inspection. This leads directly to the 9×9 topological matrix and the 1×9 array given at the bottom of Fig. 5.10.

Although the branched manifold and its algebraic description appear very complicated, both are generated by very simple rules. In particular, it is a simple matter to predict what happens when suitable control parameters are varied. New branches can be added and old branches removed only in a systematic and predictable way.

5.6.4 van der Pol System

We discuss a version of the van der Pol equations studied by Shaw [94,95]:

$$\begin{aligned} \dot{x} &= -0.7y + x(1 - 10y^2) \\ \dot{y} &= +x - 0.25\sin(2\pi t/T) \end{aligned} \tag{5.5}$$

The van der Pol oscillator exhibits the same half-period symmetry as the Duffing oscillator: The equations (5.5) are invariant under $(x, y, t) \to (-x, -y, t + \frac{1}{2}T)$, where T is the period of the driving term. The strange attractor generated by these equations must exhibit the same invariance. We therefore creep up on the description of this strange attractor in two steps, as we did for the branched manifold of the Duffing oscillator. We first describe what happens during half a period. Then we iterate.

Up to now, the branch lines we have encountered have been intervals—segments of R^1. However, it is only necessary that the branch line be one-dimensional. In the case of the van der Pol oscillator the branch line(s) is a segment with endpoints identified, a circle S^1. This comes about because the van der Pol oscillator undergoes a Hopf bifurcation on its way to chaos. The intersection of the strange attractor with a Poincaré section can be embedded in an annulus. Under the Birman–Williams projection, the annulus is mapped to S^1, and under the (semi)flow, S^1 is mapped to S^1.

Under the flow, stretching takes place. Stretching is followed by folding. However, only an even number of folds can occur, because of the global boundary conditions. In Fig. 5.11(a) we show both a branched manifold that describes the flow and a return map of S^1 to itself [Fig. 5.11(b)]. It is clear from this figure that two folds must occur (more generally, folds must be paired). The standard representation of a branched manifold is shown in Fig. 5.11(c). This is obtained by splitting open the flow shown in Fig. 5.11(a) and identifying the edges of the flow. The algebraic description for this flow is shown in Fig. 5.11(d). The discontinuity of local torsion for contiguous branches, as shown by the diagonal matrix elements of the topological matrix, is intimately related to the global boundary conditions (S^1 instead of R^1).

Figure 5.11 describes a strange attractor generated by a stretch and fold mechanism acting on an annulus but without the twofold symmetry exhibited by the van der Pol oscillator [2]. To construct a branched manifold for the van der Pol oscillator, the mechanism shown in Fig. 5.11 must be appropriately iterated. We illustrate what happens, in some range of control parameters, in Fig. 5.12. If the pinching during the first half cycle occurs at the top, the pinching during the second half cycle must occur at the bottom to account properly for the inversion part of the symmetry

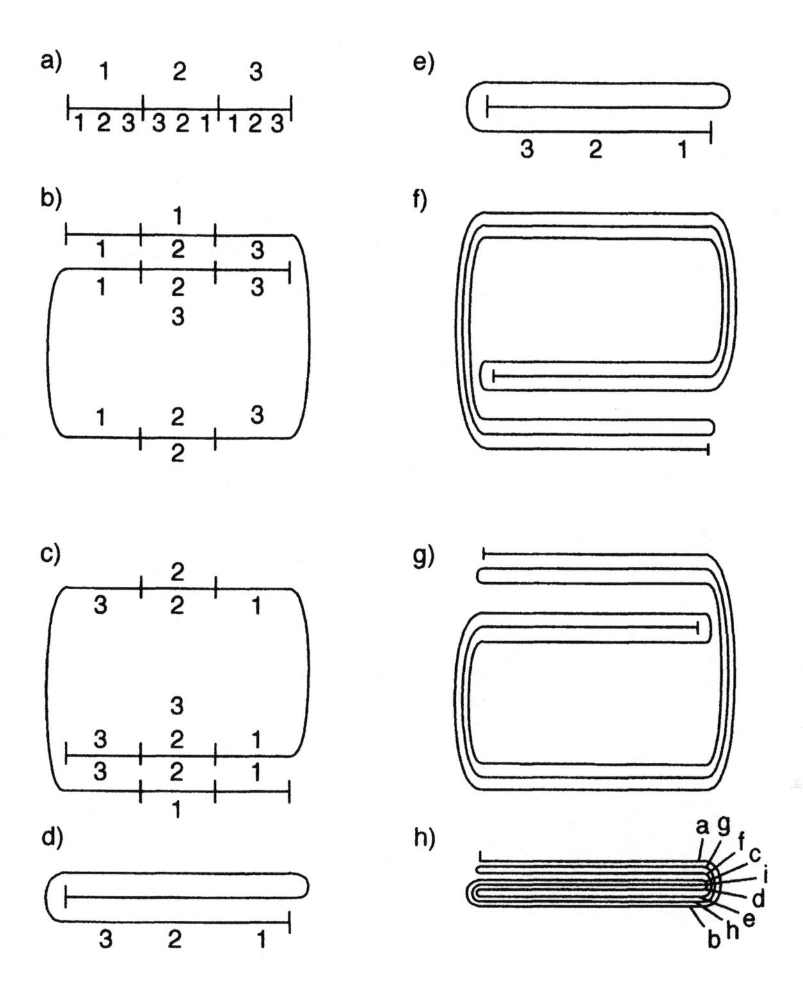

Fig. 5.10 The stretching and squeezing mechanism for the Duffing oscillator is essentially the second iterate of the stretch and fold mechanism for the extended Smale horseshoe, with suitable modifications. Stretching and squeezing during the first half cycle is shown on the left. The second half cycle is shown on the right. (a) A branch line with three large segments is shown. Each segment is divided into three smaller segments. All are labeled as shown. (b) The branch line is stretched by a factor of 3 and scrolled. (c) The stretched branch line is rotated by π radians in the direction of the scroll rotation and squeezed (d). (e) The squeezed branch line is again stretched by a factor of 3 and scrolled (f) and rotated by π radians (g) and squeezed (h). The period-1 orbits can be located by the method indicated in Fig. 5.9 and their linking numbers computed. The local torsion of the period-1 orbit through branch (i, j) is $i + j$. This information is sufficient to construct the topological matrix. Array information can be read directly from (g) or (h).

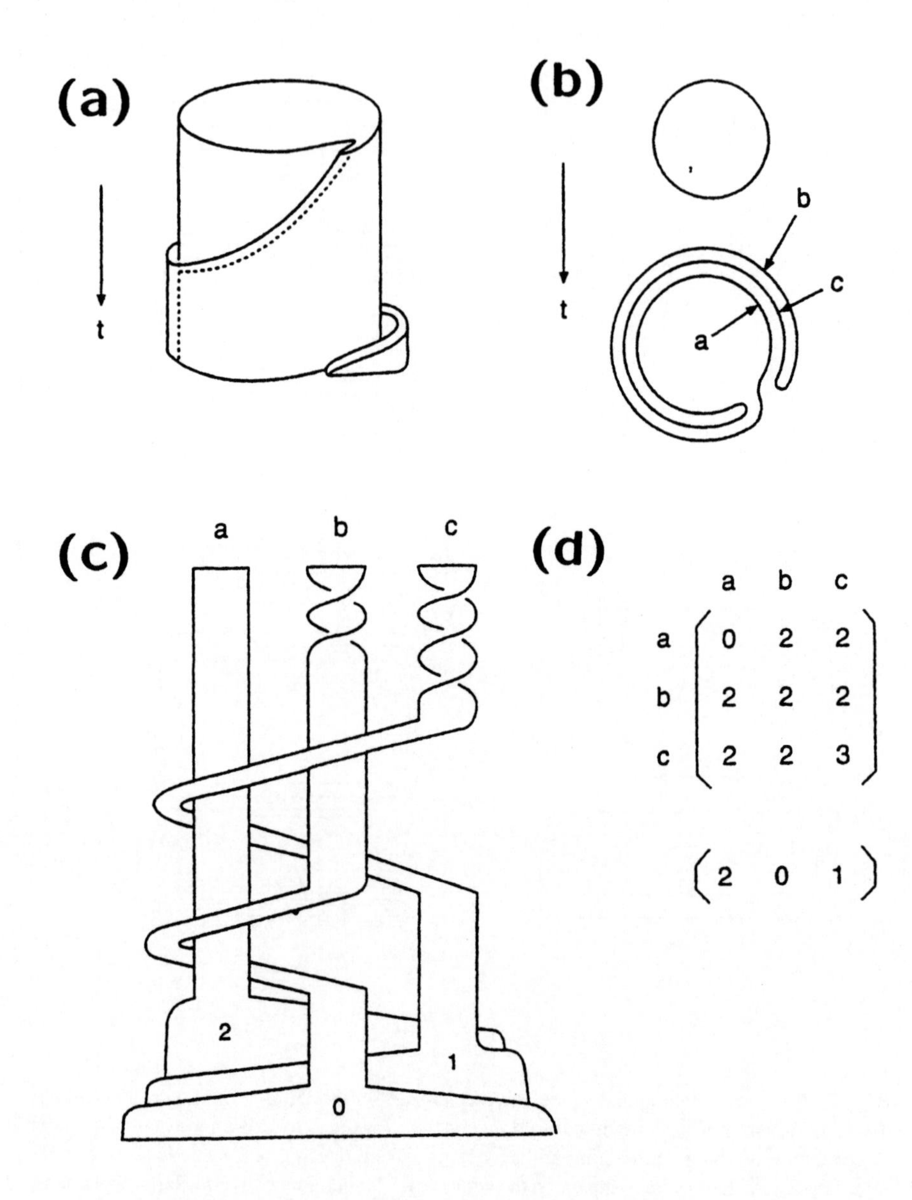

Fig. 5.11 What happens during half a cycle in the van der Pol oscillator. (a) The flow along the cylinder is pinched out, deformed, and folded back to the cylinder. The branched manifold is shown on the left. (b) The return map of the branch line S^1 is a circle map. (c) The flow in (a) is slit open, showing three branches for the branched manifold. (d) The topological matrix and array can be determined by inspection. The discontinuity of local torsions for contiguous branches is a signature that nonlocal boundary conditions must be imposed.

$[(x, y) \rightarrow -(x, y)]$. In this iteration, a total of nine branches is created. In Fig. 5.12 we present a caricature (cartoon) of this van der Pol mechanism which is similar to that presented in Fig. 5.10 for the mechanism at work in the Duffing oscillator.

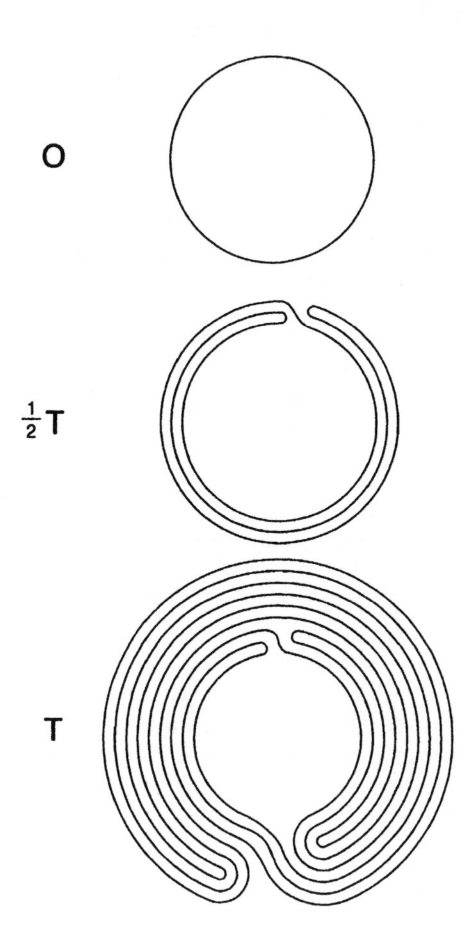

Fig. 5.12 What happens during each half of a full cycle in the van der Pol oscillator in terms of return maps. The flow along the cylinder is pinched out, deformed, and folded back to the cylinder during each half cycle. The deformations occur on opposite sides of the cylinder to respect the symmetry $(x, y, t) \rightarrow (-x, -y, t + \frac{1}{2}T)$. Each iterated stretch (by a factor of 3) and squeeze creates a total of nine branches.

As control parameters vary, the size of the pinched regions changes. It is possible to predict how new branches must be added to describe the chaotic dynamics as the pinched region becomes enlarged.

5.7 UNIQUENESS AND NONUNIQUENESS

Flow dynamics are conveniently represented by limits, or cartoons. This cartoon is a branched manifold. Branched manifolds were originally introduced to describe the unique organization of all the unstable periodic orbits in hyperbolic strange attractors. They also succinctly describe the stretching and squeezing mechanisms that generate strange attractors.

Many apparently different branched manifolds predict the same spectrum of orbits and the same topological organization (spectrum of linking numbers) of these orbits. Thus, there is not a 1:1 correspondence between branched manifolds and flow dynamics. This is somewhat analogous to the representation theory of groups. A single group can have many different inequivalent 1:1 (faithful) representations. In some sense, the group is fundamental and the matrix representations are simply a convenient means of performing computations. In the same way, the dynamics is fundamental and branched manifolds are convenient ways for doing calculations and classifying dynamics.

Definition: Two branched manifolds are **flow equivalent** if they predict the same spectrum of periodic orbits and these orbits have the same topological organization.

At the simplest level, a single branched manifold can have many different algebraic representations. An algebraic representation is obtained by projecting a branched manifold $\mathcal{BM} \subset R^3$ onto a plane $R^2 \subset R^3$. Different projections give different algebraic representations.

Definition: Two branched manifolds are **projection equivalent** if their algebraic representation differs only through their projection.

More generally, branched manifolds for the same flow can be geometrically different structures. The geometric differences can be due either to local moves or to global moves. A theory seems to exist to describe the equivalence of geometrically distinct branched manifolds under local moves. At present, there seems to be no theory to describe the equivalence of geometrically distinct branched manifolds under global moves.

In the first subsection we describe the local moves that can be used to transform one branched manifold into a geometrically different but flow-equivalent branched manifold. In the second subsection we describe three flow-equivalent branched manifolds that differ by global moves.

5.7.1 Local Moves

Knots and links remain invariant under a small number of Reidemeister moves. Braids remain invariant under the two types of braid relations that define braid groups. In the same way, branched manifolds remain invariant under a small number of local moves. These moves are:

- Branch line twists
- Writhe–twist exchange

- Branch line reversal
- Concatenation of inflows or outflows
- Branch line splitting

These moves are illustrated in Fig. 5.13.

In Fig. 5.13(a) we show two inflow branches joined at a branch line to an outflow branch. If the branch line is given half a twist in the clockwise direction, as shown in the middle of Fig. 5.13(a), the two inflow branches have their local torsion changed by $+1$ and the outflow branch will have its local torsion changed by -1. In addition, the order in which the two inflow branches are joined at the branch line is reversed. If an additional half twist is given to the branch line (right), the local torsion of the incoming and outgoing branches is again changed by ± 1, the order of joining is again reversed, and in addition the two incoming branches have their linking number increased by $+1$. Thus, the effect of a full twist on a branch line is that the local torsion of the incoming branches is changed by $+2$ and that of the outgoing branch is changed by -2, the linking number of the two incoming branches is changed by $+1$, and the order in which the incoming branches is joined at the branch line is unchanged. Twisting in the opposite direction changes all signs.

Figure 5.13(b) shows how writhe and twist can be exchanged; this was described earlier. In Fig. 5.13(c) we show how interchanging the spatial position of two branch lines will force a full twist into a branch connecting these branch lines. The direction of the twist depends on whether branch a moves in front of or behind branch b.

In Fig. 5.13(d) and (e) we show that orbit organization is unchanged by the concatenation of inflows with inflows or outflows with outflows. In Fig. 5.13(d) we show how the order of two branch lines can be exchanged. In fact, it is sometimes convenient to draw the branched manifold with degenerate branch lines, as shown. In Fig. 5.13(e) we show that splitting points can also be concatenated. This representation is convenient when the stretch in a local region of phase space is larger than a factor of 2. In fact, it is useful to show a branch line feeding $[\exp(\lambda_1)] + 1$ branches in regions of phase space where the maximum local Lyapunov exponent is λ_1 ($[x]$ is the integer part of x).

In Fig. 5.13(f) we extend the inflow to the splitting point back beyond the nearest branch line into the two inflows that join at the branch line. This does not affect any periodic orbits, since no inverse image of any splitting point lies on a periodic orbit.

5.7.2 Global Moves

In Fig. 5.14 we show three geometrically inequivalent branched manifolds that are flow equivalent. The first is the branched manifold that holds all the closed magnetic field lines produced by a current flowing in a wire tied into the shape of a figure 8 knot. This branched manifold holds aesthetic appeal since it manifestly exhibits a rotation symmetry. This branched manifold has eight branches, which may be labeled $a\alpha$, ab, and so on. The incidence matrix shows the connectivity of these branches; for example, $a\beta$ is not an allowed transition.

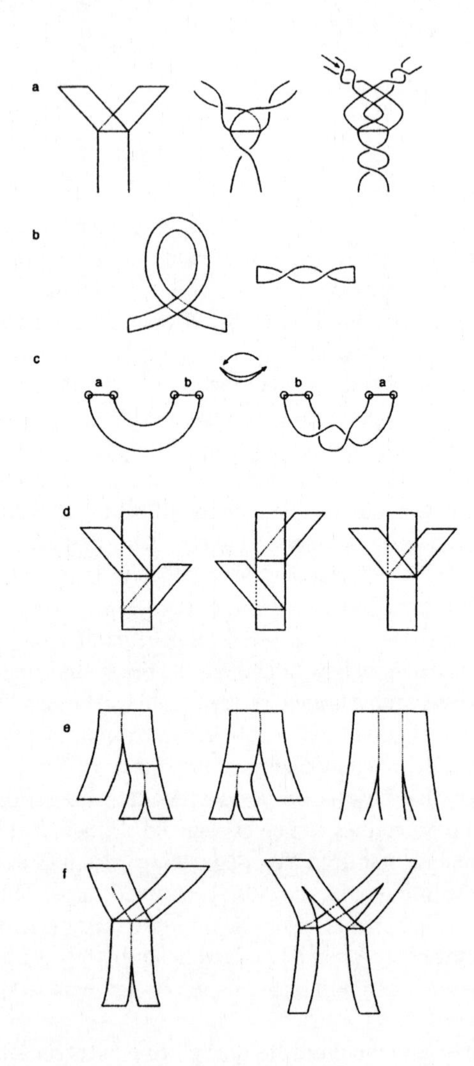

Fig. 5.13 Branched manifolds are flow equivalent under the following local moves. (a) The branch line is given a half twist. The inflow branches have their local torsion changed by $+1$, the outflow branch local torsion is changed by -1, and the order in which the inflow branches are joined at the branch line is reversed. When the branch line is given a full twist, the order of joining is unchanged. However, the local torsion of the inflow and outflow branches is changed by $+2$ and -2, and in addition the inflow branches link each other with a linking number $+1$. (b) Writhe and twist can be exchanged. (c) Interchanging the location of branch lines introduces a full twist into branches connecting them. (d) Topological organization is respected by interchanging the order of inflows. It is sometimes convenient to make the branch line degenerate. (e) Concatenated splitting charts can be treated the same way. This is convenient when the local Lyapunov exponent is larger than $\ln 2$. (f) The splitting point is extended backward beyond the nearest branch line in its past.

Table 5.1 Linking numbers for orbits to period 3 on the three representation of figure 8 flow dynamics

	$\alpha\beta$ ab x,y,z 0,1,2	αa xy 01	βb yz 12	$\alpha a\alpha\beta$ x^2y 0^21	αaba xy^2 01^2	$\alpha\beta ba$ $\beta\alpha ab$ xyz 012	$\alpha\beta b\beta$ y^2z 1^22	βbab yz^2 12^2
0,1,2	0	0	0	0	0	0	0	0
01	0	−1	0	−1	−1	−1	0	0
12	0	0	1	0	0	1	1	1
0^21	0	−1	0	−2	−1	−1	0	0
01^2	0	−1	0	−1	−2	0	0	0
012	0	−1	1	−1	−1	0	1	1
1^22	0	0	1	0	0	1	2	1
12^2	0	0	1	0	0	1	1	2

The second branched manifold shown in Fig. 5.14 was computed in [84]. There are again eight branches. The incidence matrix shows that the only transition not allowed is xz. The third branched manifold is flow equivalent to the second; both have the same spectrum of periodic orbits with the same topological organization. However, this third representation of the flow dynamics has a hole in the middle. This feature automates the computation of linking numbers.

With two exceptions, there is a 1:1 correspondence between the periodic orbits of the branched manifolds in Fig. 5.14(a) and (b). The two orbits $\alpha\beta$ and ab correspond to the three period-1 orbits x, y, and z, while the two orbits $\alpha\beta ba$ and $\beta\alpha ab$ correspond to the single period-3 orbit xyz. There is a 1:1 correspondence between the periodic orbits of the two branched manifolds shown in Fig. 5.14(b) and (c). Table 5.1 provides the linking numbers for the closed orbits up to period 3 on these three flow-equivalent branched manifolds.

For some purposes, it is convenient to simplify the description of the dynamics by expressing the branched manifold shown in Fig. 5.14(c) as a subbranched manifold of one showing a *full shift*. This idea is illustrated in Fig. 5.15. In the full-shift case, all periodic orbits based on three symbols are possible, including orbits containing the symbol sequence $\cdots 02 \cdots$. Such orbits do not occur in the branched manifold shown in Fig. 5.15(a).

For convenience, we show the return map for four branched manifolds in Fig. 5.16. Figure 5.16(a) provides the return map (tent map) for a Smale horseshoe template. Figure 5.16(b) and (c) provide the return maps for the two branched manifolds shown in Fig. 5.15. Figure 5.16(d) provides the return map for a branched manifold with four branches, of which branch 1 is orientation reversing. In each case the expansion is uniform. In three cases the branched manifold is fully expanding: case (b) is not.

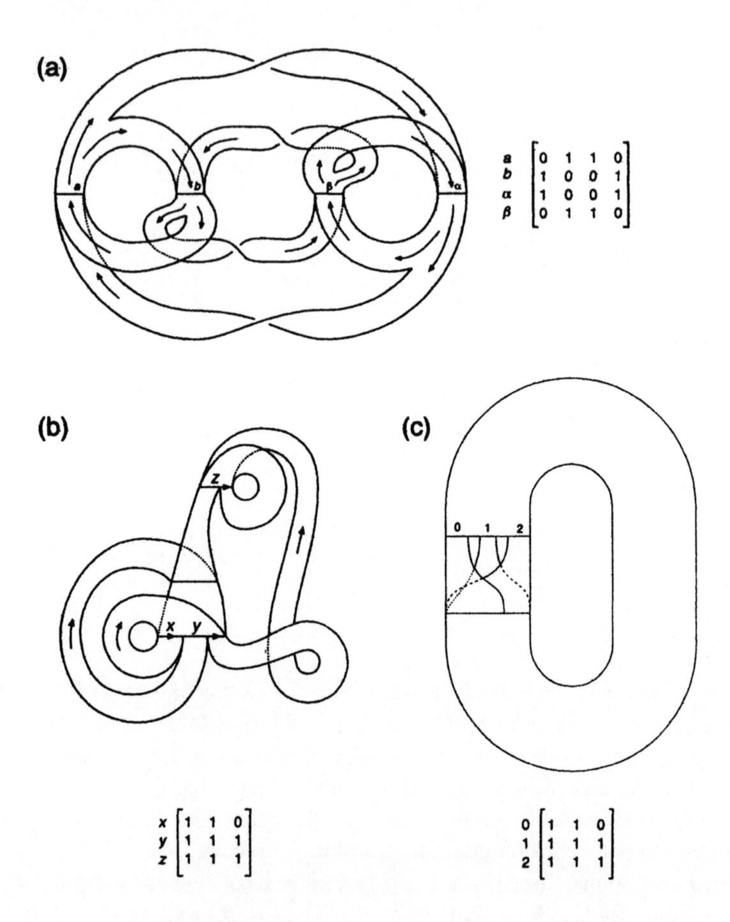

Fig. 5.14 (a) The branched manifold that describes the organization of all the closed magnetic field lines generated by a current-carrying wire tied into the shape of a figure 8 knot has eight branches. In this representation it manifests the rotational symmetry of the figure 8 knot. (b) The direct model of this branched manifold is simpler to deal with. The two branched manifolds are flow equivalent. With two exceptions, there is a 1:1 correspondence between periodic orbits on these two branched manifolds. (c) This third branched manifold is flow equivalent to the first two. It has a hole in the middle. This greatly simplifies the problem of computing the linking numbers for all the periodic orbits in the flow. The incidence matrices are given for each of these branched manifolds. The algebraic description of the third is given explicitly. Adapted with permission from Birman and Williams [84].

5.8 STANDARD FORM

By using the moves described in Section 5.3, any branched manifold can be transformed, after projection to R^2, into the standard form shown in Fig. 5.17 [96, 97].

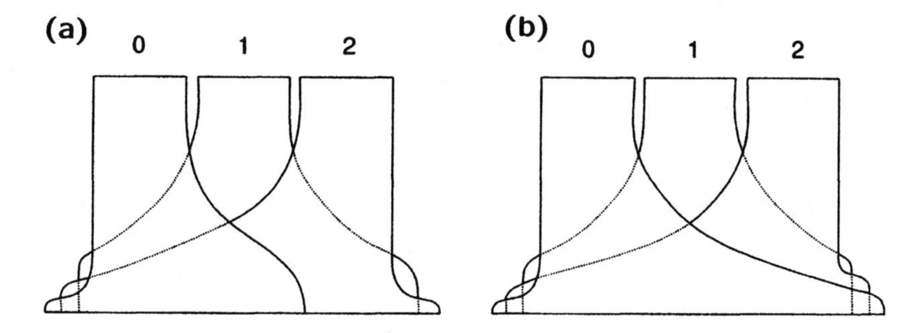

Fig. 5.15 The branched manifold (a) is a subtemplate of the one in (b). Some orbits on the right-hand branched manifold do not exist on the left-hand template. Those that exist on the subtemplate are organized in exactly the same way as their counterparts on the right.

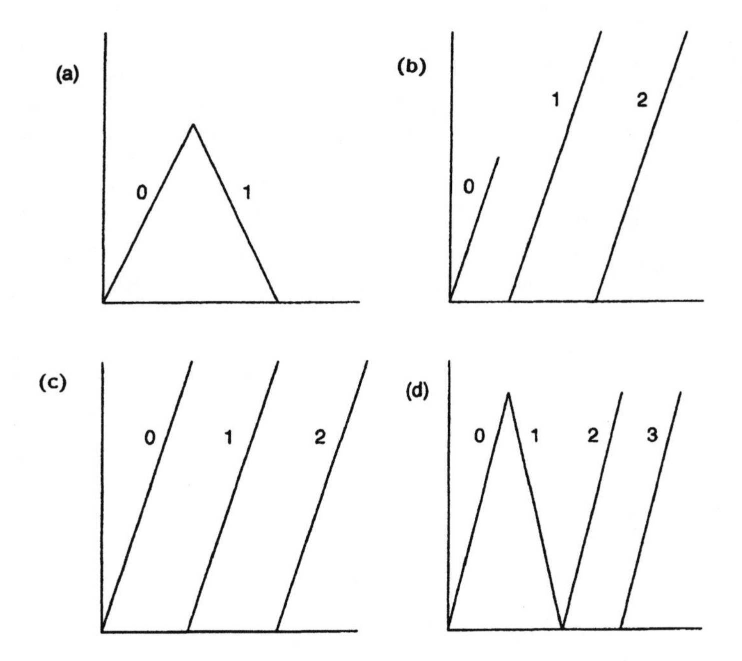

Fig. 5.16 Return maps for four branched manifolds. In each case the expansion is uniform. (a) The two branch Smale horseshoe template has a return map that is a tent map. (c) The three branch template shown in Fig. 5.15(b) is fully expanding, whereas (b) the subtemplate shown in Fig. 5.15(a) is not. (d) The four-branch template has one orientation-reversing branch.

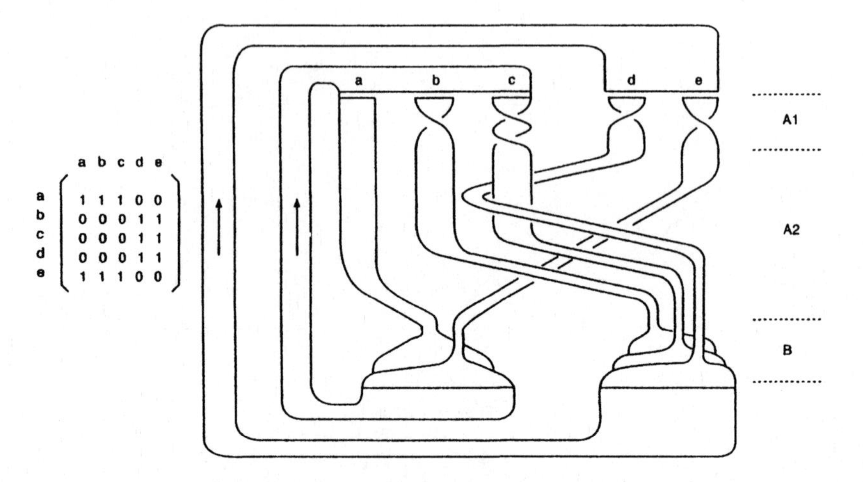

Fig. 5.17 Useful standard form for branched manifolds. All twisting occurs in region A1 and all crossing occurs in region A2. This information is summarized in a topological matrix T. All squeezing occurs in region B. This information is summarized in an array A. A Markov transition matrix (shown on the left for this template) shows how the branches are connected.

Each branch line is divided into segments by locating preimages of each tear point on the branch lines. The return flow from each branch line (bottom) feeds the segments of the branch lines (top). The stretch and squeeze mechanisms that generate chaos are described as follows:

- Branches twist but do not cross in the region labeled A1. The twists are assigned integer values $0, \pm 1, \pm 2, \ldots$ in the same way as for knots: The twist of a branch is the signed number of crossings of the edges of that branch.

- Branches cross but do not twist in the region labeled A2. The crossings are assigned integer values $0, \pm 1, \pm 2, \ldots$ in the same way as for knots by shrinking each branch down to a single curve. The information contained in regions A1 and A2 is summarized algebraically by a topological matrix T.

- Branches are squeezed together in the region labeled B. One convention is that the integers indicating ordering are larger the farther from the observer (i.e., increasing from top to bottom). The information contained in region B is summarized in array A.

- A Markov transition matrix is introduced to identify which branches are connected to which.

5.9 TOPOLOGICAL INVARIANTS

Linking numbers for any pair of periodic orbits on a branched manifold can be computed. The computation depends only on the algebraic description of the branched manifold. The computations simplify considerably when the branched manifold has a hole in the middle. That is, much simplification occurs when a series of local or global moves can be exploited to transform the branched manifold to a form in which all the stretching and squeezing is represented between two branch lines that are identified by a return flow which neither stretches nor squeezes. This representation for a branched manifold is particularly convenient because each trip around the hole in the middle corresponds to one period. Such a flow has one Poincaré section, which can be taken as the branch line. We describe the systematics of linking number computations for branched manifolds of this type below.

5.9.1 Kneading Theory

When only one branch line is present, it is possible to define an order along this branch line. We adopt the convention that the order increases from left to right. We assume that the branched manifold has n branches, labeled $0, 1, 2, \ldots, n-1$ from left to right, for lack of imagination. We also assume for convenience that the incidence matrix is full (cf. Fig. 5.15). This causes no problem: We can simply ignore periodic orbits that are forbidden by the original incidence matrix.

Under these conditions every orbit of minimal period p is represented by a symbol sequence $(\sigma_1\sigma_2\cdots\sigma_p)^{\text{“}\infty\text{”}}$, or

$$\sigma_1\sigma_2\cdots\sigma_p \;\; \sigma_1\sigma_2\cdots\sigma_p \;\; \cdots$$

After one period the symbol sequence advances to

$$\sigma_2\cdots\sigma_p\sigma_1 \;\; \sigma_2\cdots\sigma_p\sigma_1 \;\; \cdots$$

Advancing by a period amounts to cyclic permutation of the symbols (*symbolic dynamics*).

We now wish to locate periodic orbits on the branched manifold. We do this by computing the "address," or "zip code," along the branch, for each of the p initial conditions of a period p orbit.

The address along the branch line is computed as follows:

1. Write out the symbol code for one of the initial conditions. For example:

$$\sigma_1\sigma_2\cdots\sigma_p \;\; \sigma_1\sigma_2\cdots\sigma_p \;\; \cdots$$

2. Conjugate each symbol following passage through an orientation-reversing branch. Orientation-reversing branches are branches that twist through an odd multiple of π radians. These branches have negative parity, where the parity

of branch i is defined as $\mathcal{P}(i) = (-1)^{T_{ii}}$ and T_{ii} is the appropriate element of the topological matrix. The conjugate of σ_i is $\bar{\sigma}_i$, where

$$\sigma_i + \bar{\sigma}_i = n - 1 \tag{5.6}$$

3. This process produces a symbol sequence of period either p or $2p$, depending on whether the orbit goes through an even or odd number of orientation-reversing branches. The symbol sequence is the address, in normal numerical order, for the initial condition along the branch line.

4. This process is repeated for the remaining initial conditions of the period-p orbit.

Example: Assume that we have a template with four branches 0, 1, 2, 3, and branch 1 is orientation reversing [cf. Fig. 5.16(d)]. To find the address of 0213 along the branch line, we perform the following simple calculation:

$$\begin{aligned} 0213\,0213\,0213\,\cdots &\rightarrow 021\bar{3}\,\bar{0}\bar{2}\bar{1}\bar{\bar{3}}\,021\bar{3}\cdots \\ &= 0210\,3123\,0210\,\cdots \end{aligned} \tag{5.7}$$

This is repeated three more times for the additional three initial conditions. The four addresses for the passage of this period-4 orbit through the branchs of this four-branch manifold have period 8:

Initial Condition	Address	Fraction Base 10	Decimal
0213	0210 3123	9435/65535	0.143969
2130	2103 1230	37740/65535	0.575875
1302	1031 2302	19890/65535	0.303502
3021	3021 0312	51510/65535	0.785992

(5.8)

Every point on a branch line is the address for an initial condition for some orbit through the branched manifold. The address may be represented by a symbol string: $a_1a_2a_3\cdots$. Two possibilities arise:

Irrational: The symbol string is never repeating. Such symbol strings represent irrational numbers and nonrepeating (chaotic) orbits.

Rational: The symbol string is eventually repeating. Such symbol strings represent rational numbers and orbits that are either periodic, finite, or eventually periodic.

Periodic orbits of period p are described by a repeating sequence of p symbols $(\sigma_1\sigma_2\cdots\sigma_p)^\infty$, as described above. The address is a symbol sequence of period p or $2p$. We compute the rational fraction for a single case, then present the general result. On the Smale horseshoe template with orientation-preserving branch 0 and orientation-reversing branch 1, the period-3 orbit $(011)^{\infty}$ has a period-3 address $(010)^{\infty}$. The rational fraction for this address is

$$
\begin{aligned}
010\,010\,010\cdots &\rightarrow \tfrac{0}{2}+\tfrac{1}{2^2}+\tfrac{0}{2^3}+\tfrac{0}{2^4}+\tfrac{1}{2^5}+\tfrac{0}{2^6}+\tfrac{0}{2^7}+\tfrac{1}{2^8}+\tfrac{0}{2^9}+\cdots \\
&\rightarrow \left(0\times 2^2+1\times 2^1+0\times 2^0\right)\left(\tfrac{1}{2^3}+\tfrac{1}{2^6}+\tfrac{1}{2^9}+\cdots\right) \\
&= \frac{0\times 2^2+1\times 2^1+0\times 2^2}{1\times 2^2+1\times 2^1+1\times 2^2}=\frac{010}{111}
\end{aligned}
\tag{5.9}
$$

This calculation has been done in the binary system. To be strictly accurate, we should write the exponents $2, 3, 4, 5, \ldots$ as $10, 11, 100, 101, \ldots$. Conversion to the more familiar decimal system is simple:

$$
\frac{010}{111} \rightarrow \frac{0\times 2^2+1\times 2^1+0\times 2^2}{1\times 2^2+1\times 2^1+1\times 2^2}=\frac{2}{7}=0.285714\,285714\ldots \tag{5.10}
$$

For the more general case, of a periodic address on an n-branched manifold, the result proceeds in a similar fashion. The "n-imal" rational fraction address for the period-4 orbit 0213 described above is $(0210\,3123)/(3333\,3333)$, where $3333\,3333 = 4^8 - 1$. This fraction is easily converted to base 10 and its decimal equivalent:

$$
0213 \rightarrow 0210\,3123 \rightarrow \frac{0210\,3123}{3333\,3333} \rightarrow \frac{9435}{65535} \rightarrow 0.1439688716 \tag{5.11}
$$

These results are summarized for the four initial conditions of this orbit in Eq. (5.8).

Finite orbits are orbits that reach a splitting point on a branch after a finite number of periods. Splitting points are initial conditions for flows to a fixed point. The address for a finite orbit is a finite symbol sequence. Splitting points for templates with two, three, and four branches are shown in Fig. 5.16. For the three fully expansive branched manifolds shown in this figure, with two, three, and four branches, the addresses of the splitting points are $\frac{1}{2}$; $\frac{1}{3}$ and $\frac{2}{3}$; and $\frac{1}{4}$, $\frac{2}{4}$, and $\frac{3}{4}$; respectively. For the nonfully expanding template, the addresses are $\frac{2}{8}$ and $\frac{5}{8}$.

We illustrate the basic idea by computing the itinerary of the finite orbit 3212 on the four-branch manifold discussed above, whose return map is shown in Fig. 5.16(d).

Initial Condition	Address	Fraction
3212	3211	$\frac{3}{4^1}+\frac{2}{4^2}+\frac{1}{4^3}+\frac{1}{4^4}$
212	211	$\frac{2}{4^1}+\frac{1}{4^2}+\frac{1}{4^3}$
12	11	$\frac{1}{4^1}+\frac{1}{4^2}$
2	2	$\frac{2}{4^1}$

(5.12)

Eventually periodic orbits are represented by symbol sequences that eventually become periodic. As an example, the orbit with symbol sequence $(01)^2(011)^{\infty}$ on the Smale horseshoe template settles down to a period-3 orbit after four periods. The address for this initial condition is

$$
\begin{aligned}
01\,01\,011\,011\,\cdots &\rightarrow 01\,10\,010\,010\,010 \\
&\rightarrow \tfrac{0}{2^1} + \tfrac{1}{2^2} + \tfrac{1}{2^3} + \tfrac{0}{2^4} + \tfrac{0}{2^5} + \tfrac{1}{2^6} + \tfrac{0}{2^7} + \tfrac{0}{2^8} + \tfrac{1}{2^9} + \tfrac{0}{2^{10}} + \cdots \\
&= \frac{0110}{10000} + \frac{1}{10000}\frac{010}{111} \\
&= \tfrac{6}{2^4} + \tfrac{1}{2^4}\tfrac{2}{7} \rightarrow 0.39\overline{285714}^{\infty}
\end{aligned}
\tag{5.13}
$$

We summarize these results now for a branched manifold with n branches, a full incidence matrix, and uniform expansion along the branch. It is convenient to express results in a number system based on n and the corresponding n-imal fractions in the interval $[0, 1]$ of the branch line.

1. There is a 1:1 correspondence between irrational numbers and initial conditions for chaotic orbits.

2a. There is a 1:1 correspondence between n-mal fractions of the form

$$
\frac{\text{integer}}{n^p - 1} \quad \text{or} \quad \frac{\text{integer}}{n^{2p} - 1} \tag{5.14}
$$

and initial conditions for orbits of period p.

2b. There is a 1:1 correspondence between n-mal fractions of the form

$$
\frac{\text{integer}}{n^k} \tag{5.15}
$$

and initial conditions for finite orbits of k periods.

2c. There is a 1:1 correspondence between all other n-mal fractions, which have the form

$$
\frac{\text{integer}}{n^k} + \frac{1}{n^k} \times \frac{\text{integer}}{n^p - 1} \quad \text{or} \quad \frac{\text{integer}}{n^k} + \frac{1}{n^k} \times \frac{\text{integer}}{n^{2p} - 1} \tag{5.16}
$$

These fractions describe orbits that settle down to period-p orbits after k transient periods.

The irrationals are dense on the interval. So also are (separately) all fractions of the form (5.14), (5.15), and (5.16). As a result:

1. Chaotic orbits are dense on branched manifolds.

2a. Periodic orbits are dense on branched manifolds.

2b. Finite orbits are dense on branched manifolds.

2c. Eventually periodic orbits are dense on branched manifolds.

These denseness statements hold when the branched manifold is blown back up to the original strange attractor.

5.9.2 Linking Numbers

Once the locations of periodic orbits on a branched manifold have been determined, computation of the linking numbers is simply a matter of counting crossings. We illustrate by computing the linking number of the orbits 01 and 011 on a Smale horseshoe template. The results of the computation are shown in Fig. 5.18.

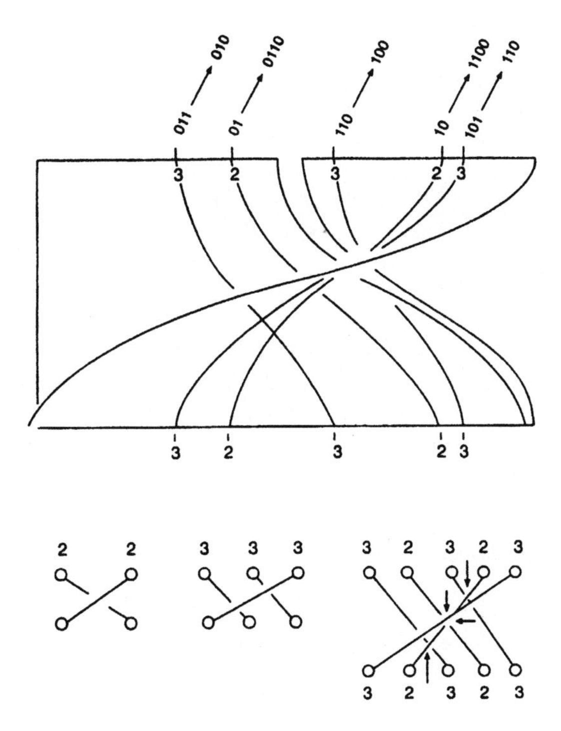

Fig. 5.18 Initial conditions for period-2 orbit 01 and period-3 orbit 011 on the Smale horseshoe template are computed in Eq. (5.18). These two orbits are draped over the interesting part of this branched manifold. The linking and self-linking numbers are computed by counting crossings. For these orbits, $SL(01) = +1$, $SL(011) = +2$, and $L(01, 011) = +2$.

The period-2 orbit 01 goes through the orientation reversing branch 1 once. Therefore the addresses of its two initial conditions have period 4. The period-3 saddle 011 goes through the orientation-reversing branch twice, so its three initial conditions have period 3. The addresses, and corresponding decimal fractions, are:

Initial Condition	Address	Decimal Fraction
01	01 10	6/15
10	11 00	12/15
011	010	2/7
110	100	4/7
101	110	6/7

(5.17)

These five addresses are shown along the branch line of the Smale horseshoe template. The order of these initial conditions is simple to read off from either the binary representation or the decimal fraction:

$$\begin{matrix} 010 \\ 2/7 \end{matrix} < \begin{matrix} 01\,10 \\ 2/5 \end{matrix} < \begin{matrix} 100 \\ 4/7 \end{matrix} < \begin{matrix} 11\,00 \\ 4/5 \end{matrix} < \begin{matrix} 110 \\ 6/7 \end{matrix} \tag{5.18}$$

Computing the linking numbers is now simply a matter of counting crossings. The self-linking numbers of the period-2 and period-3 orbits are $SL(01) = +1$ and $SL(011) = +2$. The computation is shown at the bottom of Fig. 5.17. The computation of the linking number $L(01, 011) = \frac{1}{2}(1 + 1 + 1 + 1) = +2$ is also shown at the bottom of Fig. 5.18. Computation of linking numbers on branched manifolds with a hole in the middle have been reduced to a FORTRAN code, which is available at the authors' Web site.

5.9.3 Relative Rotation Rates

Computation of relative rotation rates follows a very similar algorithm. Two orbits of periods p_A and p_B are draped over a branched manifold. Two initial conditions are joined by an oriented line segment, and the number of half twists that this segment undergoes as it evolves through $p_A \times p_B$ periods is counted. This integer is divided by $2 \times p_A \times p_B$. This calculation is repeated for all other initial conditions. This bookkeeping has also been reduced to a FORTRAN code, which is available at the Web site listed above. The inputs are the same as for the linking number computation. The output is a table of relative rotation rates.

The computation can be simplified using the procedure introduced in Section 4.3.2. That is, it is sufficient to construct a $(p_A + p_B) \times (p_A + p_B)$ crossing matrix C which summarizes the crossing information in the nontrivial part of the branched manifold. A cyclic permutation matrix P must also be constructed. This indicates how the initial conditions from one period are mapped to the initial conditions for the next. Then it is sufficient to construct the matrix [cf. Fig. 4.17].

$$\text{RRR} = \frac{1}{2 \times p_A \times p_B} \sum_{k=1}^{p_A \times p_B} P^{-k} C P^k \tag{5.19}$$

The p_a^2 matrix elements of the upper diagonal $p_A \times p_A$ matrix are the self-relative rotation rates for orbit A, and similarly for orbit B. The $p_A \times p_B$ matrix elements of either of the two off-diagonal matrices are the relative rotation rates of orbit A with orbit B.

5.10 ADDITIONAL PROPERTIES

Branched manifolds have a number of additional properties that we have not yet discussed. We describe a number of these properties below.

5.10.1 Period as Linking Number

It is often a simple matter to define the period of a closed orbit. For example, in driven dynamical systems the (dynamical) period of a closed orbit is an integer multiple of the driving period. Specifically, it is the number of distinct intersections of the orbit with a Poincaré section. This remains true when the phase space has the structure of a torus: $R^2 \times S^1$.

In more complicated cases it is not quite so obvious how to define the period of a closed orbit. For example, what is the period of a closed orbit on the Figure 8 branched manifold?

The period of a nontrivial closed orbit must always be a positive integer. There is a natural way to construct integers from closed orbits. This involves computation of linking numbers. Thus, it should be no surprise that it is always possible to define a topological period of a closed orbit on a branched manifold as the linking number of that orbit with some reference closed orbit which does not intersect the branched manifold. Then the topological period of a closed orbit A with respect to the reference loop is [84]

$$\text{topological period}(A) = LN(\text{Ref}, A) \tag{5.20}$$

where Ref is the reference loop. For the figure 8 branched manifold, one reference loop may be taken as the figure 8 knot itself. Other reference loops can also be used.

On the figure 8 branched manifold shown in Fig. 5.19, the closed orbits ab and $\alpha\beta$ each have topological period 1, while the closed loops $a\alpha$ and $b\beta$ each have period 2.

5.10.2 EBK–like Expression for Periods

Computing the topological period of a closed orbit on the figure 8 branched manifold can be simplified by deforming the reference loop. Such a deformation is shown in Fig. 5.19. The reference loop consists of the figure 8 knot itself. The deformed loop consists of a union of four loops. One each surrounds the branch lines b and β, one each surrounds the branches $a\alpha$ and αa. Then the period of any closed loop is the sum of the linking numbers of that loop with the four loops resulting from deformation of the figure 8 knot.

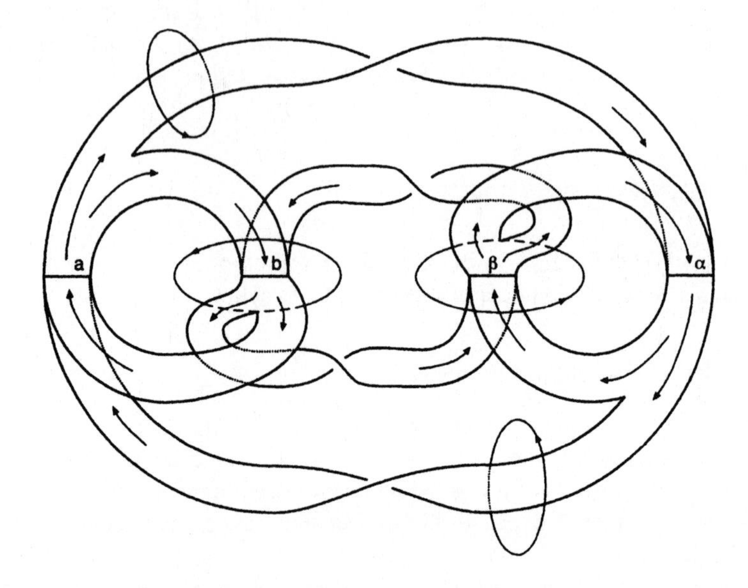

Fig. 5.19 The figure 8 knot is deformed to the union of four much simpler loops. This greatly simplifies computation of the topological period of any closed orbit on the branched manifold of the figure 8 knot. Adapted with permission from Birman and Williams [84].

This argument is general. We can deform the reference orbit for any branched manifold and write it as the union of "fundamental loops": Ref $\rightarrow \cup_i C_i$. We can then write the topological period of any closed orbit A as the sum of the linking numbers of A with each of the fundamental loops C_i:

$$\text{period}(A) = LN(\cup_{i=1} C_i, A) = \sum_{i=1} LN(C_i, A) \tag{5.21}$$

This result is very similar to the Einstein–Brillouin–Keller (EBK) quantization formula. The phase change of a single-valued wavefunction around any closed loop in its configuration space must be an integer multiple of 2π. The phase change is an action integral, so that

$$\oint_L p\, dq = \oint_{\sum_i m_i C_i} p\, dq = \sum_i m_i \oint_{C_i} p\, dq \tag{5.22}$$

The integrals around the closed loops C_i are themselves quantized:

$$\frac{1}{2\pi\hbar} \oint_{C_i} p\, dq = n_i + \frac{1}{4}\beta_i \tag{5.23}$$

Here both n_i and β_i are integers. The integer β_i is the Maslov index for the loop. The integers n_i have a natural interpretation as quantum numbers.

The loops C_i in (5.21) must be chosen as a basis set in the sense that every closed orbit A satisfies $LN(C_i, A) \geq 0$, at least one of these linking numbers is positive for each A, and inequivalent closed orbits A and B have inequivalent linking numbers with the basis loops C_i.

5.10.3 Poincaré Section

Poincaré introduced an ingenious idea to reduce the study of n-dimensional flows to the study of $(n-1)$-dimensional maps. The idea is to introduce $(n-1)$-dimensional surface(s) into the phase space. The surfaces are called variously the *Poincaré surface*, the *surface of section*, the *Poincaré section*, and so on. These surfaces have the property that the flow is transverse to the surface, always meets the surface from the same side, and almost all initial conditions meet the surface of section a countable number of times.

It is possible to define a Poincaré surface for any flow that satisfies the conditions of the Birman–Williams theorem. We first define the Poincaré section for the semiflow $\bar{\Phi}_t$ on the branched manifold $\mathcal{BM}$. The Poincaré section is simply the union of the branch lines:

$$\text{Poincaré section} = \cup \text{ branch lines} \tag{5.24}$$

For the figure 8 branched manifold the Poincaré section is the union of the four branch lines a, b, β, and α: Poincaré section $= a \cup b \cup \beta \cup \alpha$.

To construct the Poincaré section for the original flow, we "undo" the original Birman–Williams projection. This blowing-up process is described in more detail below. Briefly, each branch line is expanded against the stable direction to form a disk or rectangle. The expansion must be sufficient to ensure that the flow under Φ_t in the neighborhood of the branch line intersects the disk or rectangle. Then the Poincaré section for the original flow is the union of the disks obtained by blowing up the branch lines of $\mathcal{BM}$.

5.10.4 Blow-Up of Branched Manifolds

It is often useful to "inflate" or "blow up" a branched manifold in order to get a better approximation of the original dynamics. This is done by expanding the branch lines to "branch rectangles" against the strongly contracting direction. In a sense, this reverses the Birman–Williams projection.

We illustrate this process in Fig. 5.20. To do this, the negative Lyapunov exponent, whose limit is $-\infty$ in the construction of the branched manifold, is allowed to be finite. The stretch and squeeze factors for the map $R^2 \to R^2$ are then $\mu = \pm e^{\lambda_1}$ and $\nu = \pm e^{\lambda_3}$, with $|\mu| > 1 > |\nu| > 0$.

Periodic orbits in the map $R^2 \to R^2$ can be located by a method somewhat more involved than the kneading theory construction. Forward and backward iterates of the map are constructed. Their intersection defines a fractal in R^2. This construction is carried out explicitly for the two-branch mapping associated with a Smale horseshoe in Section 2.10. The intersections of the forward and backward iterates provide

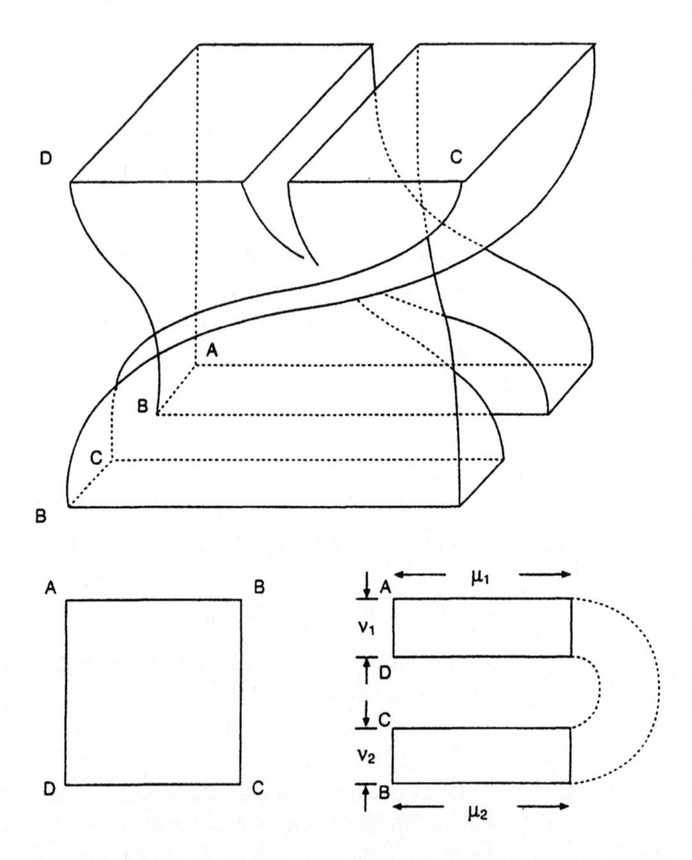

Fig. 5.20 A two-branch template is inflated by expanding against the contracting direction. The result provides an invertible map $R^2 \to R^2$.

addresses for all orbits in the flow and map. Needless to say, topological invariants remain invariant under inflation.

We should point out here, forcefully, that the fractal structure of a strange attractor comes about from repetition of the stretching and and squeezing processes in phase space. If one knows the topological structure of the attractor, as exemplified by its branched manifold, it is possible to compute geometric quantities such as Lyapunov dimension and fractal dimensions, simply by inputting the positive and negative Lyapunov exponents λ_1 and λ_3, or else a distribution of values for the local Lyapunov exponent $\lambda_1(x)$ and $\lambda_3(x)$. In this sense, the topological structure (branched manifold) is fundamental, the Lyapunov exponents are inputs, and all geometric invariants are derived quantities. From the geometric quantities alone, it is not possible to make any statements about stretching and squeezing mechanisms, much less about the branched manifold, which defines and classifies the dynamics.

5.10.5 Branched-Manifold Singularities

A two-dimensional branched manifold has two types of singularities. These are of dimension 0 and 1. The zero-dimensional singularity is a splitting point. This point describes stretching mechanisms. The one-dimensional singularity is the branch line. This line describes squeezing mechanisms. Between them, these two singularities describe the processes that build up a strange attractor in R^3.

Things are not quite as simple in higher dimensions. For starters, there is no Birman–Williams theorem. However, it is still possible to carry out a Birman–Williams projection (5.1). In the simple case where $\lambda_1 > \lambda_2 > \lambda_3 = 0 > \lambda_4$ and $\lambda_1 + \lambda_2 + \lambda_3 + \lambda_4 < 0$, the identification (5.1) maps the flow to a three-dimensional structure. This is a manifold almost everywhere. Its three directions correspond to the two stretching directions and the flow direction. However, it is not a manifold because it contains singularities. Singularities occur with dimension 0, 1, and 2. We do not have a clearcut identification of singularities with the stretching and squeezing processes.

5.10.6 Constructing a Branched Manifold from a Map

It is sometimes necessary to reconstruct properties of a flow simply from a return map. We illustrate how this can be done, using a simple example.

We consider here a map that has a period-3 orbit. The intersections of this orbit with a Poincaré section are s_1, s_2, s_3, and under the flow $s_1 \rightarrow s_3 \rightarrow s_2 \rightarrow s_1$. The map, and the folding that it forces, are shown in Fig. 5.21(a).

Other representations of this flow are possible. A second is shown in Fig. 5.21(b). Here the cyclic permutation is expressed as the composition of two interchanges: $(s_1, s_2, s_3) \rightarrow (s_2, s_1, s_3) \rightarrow (s_2, s_3, s_1)$. The branched manifold that describes this process is shown in Fig. 5.21(c). A return map on the interval $(s_1 s_2 s_3)$ is given in Fig. 5.21(d).

5.10.7 Topological Entropy

The incidence matrix provides information about the connectivity of a branched manifold. Equivalently, it provides information about which paths through the branched manifold are allowed and which are not. It is very useful to label the rows and columns of the incidence matrix by branch lines. This was done to describe the connectivity of the branched manifold for the figure 8 knot in terms of a 4×4 matrix [cf. Fig. 5.14(a)]. Although this is sometimes useful, it is often inadequate. For example, the Smale horseshoe template has one branch line, which would yield a 1×1 incidence matrix. What *is* sufficient is the following. Extend each splitting point back to the nearest branch line in its past. Then each branch line is the union of a small number of pieces. Under this construction, the branch line for the Smale

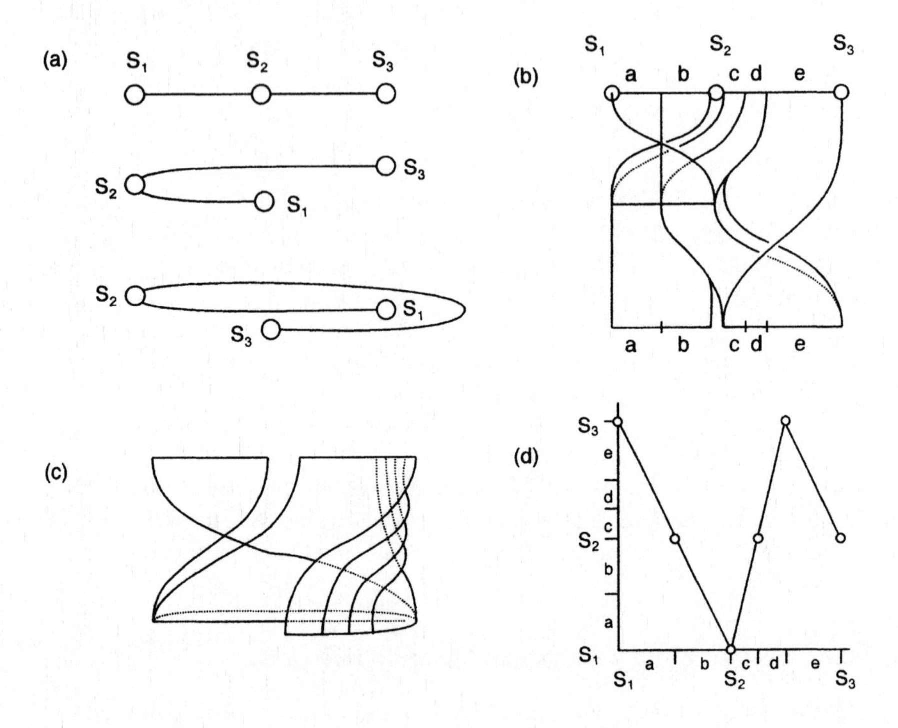

Fig. 5.21 Two ways to construct a branched manifold from a map. (a) The deformation of the line joining the three points of the period-3 orbit is shown. (b) The deformation of the line segment is carried out in two steps. The first interchanges s_1 and s_2, the second interchanges s_1 and s_3. (c) This branched manifold describes the deformation. (d) The return map for the branch line onto itself is shown. Adapted with permission from Birman and Williams [84].

horseshoe template has two pieces, while each of the four branch lines in Fig. 5.14(a) has two components. The incidence matrix for the Smale horseshoe is the full 2×2 matrix $\begin{pmatrix} 1 & 1 \\ 1 & 1 \end{pmatrix}$, and that for the template of the figure 8 knot is a sparse 8×8 matrix. In general, we adopt the smallest suitable incidence matrix.

The matrix element $(I^2)_{ik}$ describes the number of distinct ways it is possible to travel from branch i to branch k in two steps,

$$(I^2)_{ik} = \sum_j I_{ij} I_{jk} \tag{5.25}$$

For the branched manifold in Fig. 5.14(c),

$$\begin{pmatrix} 1 & 1 & 0 \\ 1 & 1 & 1 \\ 1 & 1 & 1 \end{pmatrix}^2 = \begin{pmatrix} 2 & 2 & 1 \\ 3 & 3 & 2 \\ 3 & 3 & 2 \end{pmatrix} \tag{5.26}$$

This shows, for example, that there are two ways to go from branch y to branch z:

$$\begin{array}{ccccl} y & \rightarrow & x & \rightarrow z & \text{No} \\ y & \rightarrow & y & \rightarrow z & \text{Yes} \\ y & \rightarrow & z & \rightarrow z & \text{Yes} \end{array} \tag{5.27}$$

More generally, $(I^p)_{ik}$ is the number of distinct ways of going from i to k in p distinct steps. If $i = k$, $(I^p)_{ii}$ is the number of distinct ways of starting and ending at branch i in p distinct steps.

If each step is one period, then $(I^p)_{ii}$ is the number of distinct ways of starting and ending at branch i in p periods. For the matrix I^2 above, we find seven ways of going around the branched manifold and getting back to the starting point in two periods:

$$\begin{array}{lll} x \rightarrow x \rightarrow x & y \rightarrow x \rightarrow y & \\ x \rightarrow y \rightarrow x & y \rightarrow y \rightarrow y & z \rightarrow y \rightarrow z \\ & y \rightarrow z \rightarrow y & z \rightarrow z \rightarrow z \end{array} \tag{5.28}$$

Three of these involve period-1 orbits iterated twice. The remaining four paths are

$$\begin{array}{lcl} x \rightarrow y \rightarrow x & \text{and} & y \rightarrow x \rightarrow y \\ y \rightarrow z \rightarrow y & \text{and} & z \rightarrow y \rightarrow z \end{array} \tag{5.29}$$

The two paths $x \rightarrow y \rightarrow x$ and $y \rightarrow x \rightarrow y$ belong to the single period-2 orbit $(xy)^\infty$. Similarly for the paths $y \rightarrow z \rightarrow y$ and $z \rightarrow y \rightarrow z$, which are the paths of $(yz)^\infty$ through the branched manifold starting from the two initial conditions. The number of distinct period-2 orbits on this branched manifold is

$$N(2) = \frac{1}{2}\left[\text{tr}(I^2) - N(1)\right] = \frac{1}{2}(7-3) = 2 \tag{5.30}$$

where $N(1) = 3$ is the number of period-1 orbits: $N(1) = \text{tr}(I)$.

The number of orbits of minimal period p, $N(p)$, is given by

$$pN(p) = \text{tr}(I^p) - \sum_{k \text{ divides } p} kN(k) \tag{5.31}$$

The subtraction removes orbits that are of period p but not of *minimal* period p. Such orbits are of period p which "go around" m times, where $p/k = m$ (integer, $m \geq 2$).

This algorithm is simple to implement. The only input is a suitable incidence matrix. In Table 5.2 we show the number of distinct closed orbits of minimal period p, $N(p)$, for the Smale horseshoe template and the two branched manifolds shown in Fig. 5.15.

Caution: This spectrum is sometimes presented differently. Many orbits of period p are saddle node pairs. Such pairs are often counted as a single orbit (type). In the Smale horseshoe template the period-3 saddle node pair 001 and 011 is counted as one orbit type. The period-4 saddle node pair 0001 and 0011 is counted as one, while the orbit 0111, period-doubled daughter of 01, is counted as another. Since 5 is prime, the six period-5 orbits are counted as three saddle-node pairs. At period 6, one orbit (001 011) is the period-doubled daughter of 001; the remaining eight comprise four saddle-node pairs. The number 7 is prime, while at period 8 there are the daughter of 0001 and the daughter of 0111 (itself the daughter of $01, \ldots$), giving $2+\frac{1}{2}(30-2) = 16$ orbit (types) of period 8. With this type of counting, the spectrum of orbit types of periods $1, 2, 3, 4, 5, 6, 7, 8, \ldots$ on the Smale horseshoe template is $1, 1, 1, 2, 3, 5, 9, 16, \ldots$.

Inspection of Table 5.2 reveals that the number of orbits of period p increases rapidly with p. In fact, the increase is exponential. We can write

$$N(p) \sim e^{ph_T} \tag{5.32}$$

where h_T is the topological entropy. This number can be estimated directly from the incidence matrix.

Here is how. If the eigenvalues of I are $\lambda_1 > \lambda_2 > \cdots > \lambda_n$, then

$$\begin{aligned} \mathrm{tr}(I) &= \lambda_1 + \lambda_2 + \cdots + \lambda_n \\ \mathrm{tr}(I^p) &= \lambda_1^p + \lambda_2^p + \cdots + \lambda_n^p \end{aligned} \tag{5.33}$$

For p large and $\lambda_1 > \lambda_2$, $\lambda_1^p \gg \lambda_2^p$ and, to a good approximation, $\mathrm{tr}(I^p) \sim \lambda_1^p$. In addition, the terms $kN(k)$ which are subtracted from $\mathrm{tr}(I^p)$ in (5.31) are of the order of $(\lambda_1)^k$, $k \leq p/2$. As a result, they can be neglected, and

$$N(p) \simeq \frac{1}{p}(\lambda_1)^p \simeq e^{ph_T} \tag{5.34}$$

Taking logarithms, we find that

$$h_T = \ln(\lambda_1) - \frac{1}{p}\ln(p) \overset{p\to\infty}{\longrightarrow} \ln(\lambda_1) \tag{5.35}$$

We provide the approximation $(\lambda)^p/p$ for orbits up to period 9 for the three cases discussed in Table 5.2. For the two fully expanding templates, the incidence matrix has rank 1, with eigenvalues 2, 0 and 3, 0, 0. In these cases, $N(p) \sim 2^p/p$ and $3^p/p$. In the third case, the incidence matrix has rank 2 with two nonzero eigenvalues $\frac{1}{2}(3 \pm \sqrt{5})$. Then $N(p) \sim (2.618)^p/p$. In this case it is seen that the correction of the second eigenvalue to $\mathrm{tr}(I^p) = \lambda_+^p + \lambda_-^p \to (2.618)^p + (0.382)^p$ becomes insignificant as p becomes large ($p = 2$, for example). It is also clear that the subtraction $-\sum kN(k)$ has the smallest effect for p prime (i.e., $p = 3, 5, 7$) and the largest effect when p is the smallest number with the largest number of prime factors. Thus, $N(p) \simeq \lambda^p/p$ is a worse approximation for $p = 8 = 2^3$ than for $p = 9 = 3^2$.

Table 5.2 Number of closed orbits up to period 9 in three different branched manifolds[a]

Incidence Matrix	1	2	3	4	5	6	7	8	9
$\begin{pmatrix} 1 & 1 \\ 1 & 1 \end{pmatrix}$	2	1	2	3	6	9	18	30	56
	2.0	2.0	2.7	4.0	6.4	10.7	18.3	32.0	56.9
$\begin{pmatrix} 1 & 1 & 0 \\ 1 & 1 & 1 \\ 1 & 1 & 1 \end{pmatrix}$	3	2	5	10	24	50	120	270	640
	2.6	3.4	6.0	11.7	24.6	53.7	120.4	275.9	642.0
$\begin{pmatrix} 1 & 1 & 1 \\ 1 & 1 & 1 \\ 1 & 1 & 1 \end{pmatrix}$	3	3	8	18	48	116	312	810	2184
	3.0	4.5	9.0	20.2	48.6	121.5	312.4	820.1	2187.0

[a] Top row, exact; bottom row, λ^p/p.

5.11 SUBTEMPLATES

Topological invariants of orbits and orbit pairs are unchanged under control parameter variation as long as the orbits exist. However, as control parameters are varied, periodic orbits are created and/or annihilated. Therefore, it is not obvious that the topological description of a strange attractor is invariant under control parameter variation.

5.11.1 Two Alternatives

In fact, there are two options, which will be illustrated with respect to both the Rössler and Lorenz attractors. Suppose that the Rössler equations are integrated for control parameter values for which there is a strange attractor, and that all the unstable periodic orbits in the attractor are constructed from an alphabet with two symbols, 0 and 1. If every possible symbol sequence is allowed, the attractor is hyperbolic. We have never encountered such an attractor, either in simulations of dissipative systems or in the analysis of experimental data. In our experience, it is always the case that some symbol sequences are forbidden.

For example, if the symbol sequence 00 is the only symbol sequence that is forbidden, every periodic orbit is constructed from a vocabulary with the two words $a = 01$ and $b = 1$. The flow, projected down onto a standard Smale horseshoe branched manifold, does not extend over the entire branched manifold, as can be seen in Fig. 5.22(a). The Markov transition matrix for the original two-letter alphabet consisting of 0 and 1 changes

$$\text{from} \quad M = \begin{bmatrix} 1 & 1 \\ 1 & 1 \end{bmatrix} \quad \text{to} \quad M = \begin{bmatrix} 0 & 1 \\ 1 & 1 \end{bmatrix} \tag{5.36}$$

The part of the Smale horseshoe template that is not traversed by the projection of the flow (the semiflow) is shown shaded in Fig. 5.22(a). It is constructed by observing that the flow must never enter the left quarter of the branch line shown at the top, for this encodes 00. Therefore, the two preimages of this part of the branch line must be removed, as well as all the preimages of their preimages. What remains is a fractal subset of the original branched manifold.

An alternative representation of this dynamics is given by the branched manifold shown in Fig. 5.22(b). This is a *subtemplate* of the original two-branch template shown in Fig. 5.22(a). The two branches $a = 01$ and $b = 1$ represent flows through (a) branches 0 followed by 1 in the Smale horseshoe template, and (b) through branch 1 in that template. All possible sequences involving the two words a and b are allowed. The Markov transition matrix for this subtemplate is full. However, constructing the subtemplate of Fig. 5.22(b) from the original shown in Fig. 5.22(a) is not easy—it borders on nightmarish even for this simple case.

The subtemplate of Fig. 5.22(b) describes dynamics at the creation of the period-3 orbit 3_1. For other parameter values other vocabularies and grammars describe the dynamics. In general, the number of words required grows with the wordlength. For example, to wordlength 4 the required words might be $01, 011$, and 0111. In general, as longer and longer symbol sequences occur, new inadmissible sequences appear. We can take this into account by increasing the number of words in the vocabulary. Then in this representation of the dynamics:

- The subtemplate can in principle be constructed from the original template.
- It typically has an infinite number of branches.
- The number of branches corresponding to words of finite length is finite.
- Every possible sequence of words is allowed.

We are faced with a similar choice with another branched manifold. The flow generated by the Shimizu–Morioka equations [2,98] is similar to the flow generated by the Lorenz equations [cf. Fig. 5.8(c) with Fig. 8.5]. However, the former occupies a subtemplate of the latter. The restriction of the Shimizu–Morioka flow on a Lorenz template is shown in Fig. 5.23(a). On the original Lorenz template, some periodic orbits are allowed and others forbidden. This corresponds to the fact that some symbol sequences are forbidden in the Shimizu–Morioka flow. One possibility is to restrict the projection of the flow to the part of the branched manifold that is shaded. Another is to construct a subtemplate representing a vocabulary of allowed words which can occur in arbitrary order. Such a subtemplate is shown in Fig. 5.23(b). Once again, constructing this simple subtemplate from the original Lorenz template borders on the nightmarish.

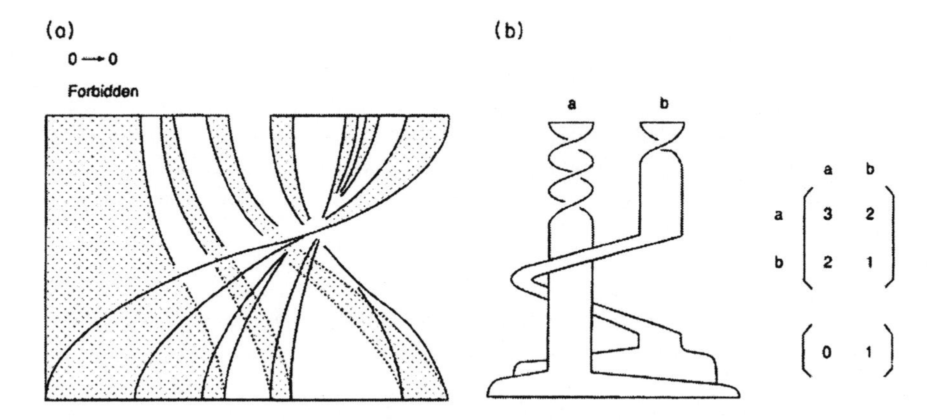

Fig. 5.22 (a) When the only forbidden symbol sequence 00, the flow is restricted to the unshaded part of the Smale horseshoe template. Some orbits on the original template are allowed; others are forbidden. The forbidden region consists of all preimages of the left quarter of the upper branch. (b) The flow can be represented by this subtemplate of the Smale horseshoe template when only the symbol sequence 00 is forbidden.

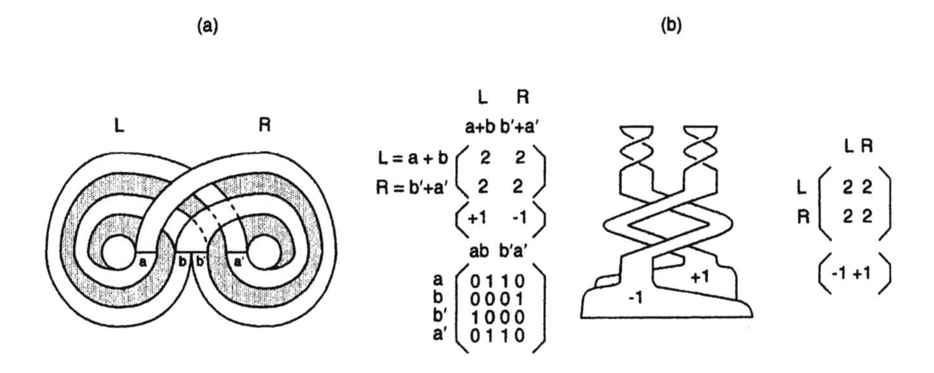

Fig. 5.23 (a) The flow generated by the Shimizu–Morioka equations is restricted to the shaded subset of the Lorenz template. (b) All orbits in the Shimizu–Morioka strange attractor can be represented by this subtemplate of the original Lorenz template.

5.11.2 A Choice

The two alternatives for representing dynamics that have been presented above are summarized as follows:

1. As control parameters are varied, the basic template remains and describes all the unstable periodic orbits in the strange attractor, and then some. Some of the orbits predicted to exist by the template do not exist—they have been "pruned away." All that remain are organized as they were in the hyperbolic limit.

2. As control parameters are varied, the dynamics is represented by a series of subtemplates. The vocabulary changes from one control parameter to another, as does the template. In general, the subtemplates have an infinite number of branches, but all possible word sequences are allowed.

Of these two alternatives, we adopt the first without hesitation, for the following reasons:

- The template is invariant, or at least robust, under control parameter variation.

- It is much easier to see how the flow gets "pushed around" on a template than to work out how one subtemplate metamorphoses into another as control parameters vary.

- With only one template to work with, the topological invariants of all orbits need to be computed only once. As long as those orbits remain in existence as the attractor changes with the control parameters, these quantities remain invariant.

- It makes no sense to force an interpretation in terms of subtemplates to preserve an idea of hyperbolicity or full-shift dynamics when this is nongeneric in dissipative physical systems in the first place.

- The global organization of a flow is largely determined by its fixed points and their insets and outsets, and by some low-period orbits and their stable and unstable manifolds. Since these are robust under large variations in parameter values, we also want the caricature (template) to be robust under these variations.

With this interpretation, templates are topological invariants under change of coordinates and initial conditions. They are robust under change of control parameter values. That is, they can remain unchanged under large changes of the control parameter values. However, under sufficiently large changes in control parameter values, they can change (cf. Chapter 9). They can change by adding new branches. They can change also if the flow ceases to visit branches. In any case, the template must change when the basic alphabet required for a symbolic encoding of the dynamics undergoes a change.

The changing nature of the dynamics over a fixed template can be described as follows, using the Lorenz template as an example. The two segments of the branch lines L and R are divided into n_1 and n_2 segments $L_1, L_2, \ldots, L_{n_1}$ and $R_1, R_2, \ldots, R_{n_2}$. Then the linking numbers (topology) depend only on the symbol name ($LRLL \cdots$), but the dynamics depend on the $(n_1 + n_2) \times (n_1 + n_2)$ Markov transition matrix. This matrix describes, to some extent (the better the larger n_1 and n_2) which orbits are allowed in the flow and which have been pruned from the flow.

5.11.3 Topological Entropy

The problem of computing topological entropy for a map or a (semi)flow over a branched manifold is simple when the Markov transition matrix describes allowed and forbidden period-1 processes. Specifically, the topological entropy is the logarithm of the largest real root of this matrix.

If time steps of varying length are the basic units, the problem of computing topological entropy becomes more interesting. Since chaotic dynamics is described in terms of letters, vocabularies, and grammars, it might be expected that there is some nontrivial relation between the concepts of chaos and those of communication. This hope is not in vain: There is a strong connection. Many of the major problems were formulated and answered by Shannon in his seminal contributions to communications theory [44,45]. We first present Shannon's results for communication channels. Then we map these results to dynamical systems theory.

The capacity of a transmission channel is

$$C = \lim_{T \to \infty} \frac{1}{T} \log N(T)$$

Here $N(T)$ is the number of allowed signals of duration T and log is to base e. First, assume that an alphabet contains n symbols $S_1, S_2, \ldots, S_n$ of lengths $t_1, t_2, \ldots, t_n$, and that every possible symbol sequence is allowed. The number of symbol sequences of length t is

$$N(t) = N(t - t_1) + N(t - t_2) + \cdots + N(t - t_n)$$

A well-known result from the theory of finite difference equations states that $N(t)$ is asymptotic to AX_0^t, where A is a constant and X_0 is the largest real solution of the characteristic equation

$$X^t = X^{t-t_1} + X^{t-t_2} + \cdots + X^{t-t_n}$$

or, equivalently,

$$1 = X^{-t_1} + X^{-t_2} + \cdots + X^{-t_n}$$

We assume that all words in the vocabulary have integer length and that there are $w(1)$ words of length 1, $w(2)$ words of length 2 (i.e., they are two symbols long in the original alphabet), and so on. Then the characteristic equation for this vocabulary and grammar is

$$1 = \sum_{p=1}^{\infty} \frac{w(p)}{X^p} \tag{5.37}$$

The topological entropy is the logarithm of the largest real root of this equation.

In many grammars, not all symbol sequences are allowed (qu is OK but qv is KO). In such cases, assume that there are m states $b_1, b_2, \ldots, b_m$. For each state only certain symbols from the set $S_1, S_2, \ldots, S_n$ can be transmitted (different subsets for different states). The transmission of symbol S_k from state b_i to state b_j (b_i may be the same as b_j) takes time $t_{ij}^{(k)}$. This process is illustrated by a graph such as that shown in Fig. 2.17.

Theorem: The channel capacity C is $\log X_0$, where X_0 is the largest real root of the $m \times m$ determinantal equation

$$\det \left| \sum_k X^{-t_{ij}^{(k)}} - \delta_{ij} \right| = 0 \tag{5.38}$$

We now translate these results into statements useful for computing the topological entropy for a dynamical system. The table that effects the isomorphism between topological entropy for dynamical systems and channel capacity for communication systems is

Communication Systems	Dynamical Systems
Graph	Branched manifold
S_i	Branch
t_i	Period
b_j	Branch line
Channel capacity	Topological entropy

Remark: Assume that a dynamical system is described by a branched manifold with m branches and incidence matrix I. Transit through each branch takes one period. Then (5.38) becomes

$$\det \left[\frac{1}{X} I_{ij} - \delta_{ij} \right] = X^{-m} \det \left[I_{ij} - X\delta_{ij} \right] = 0$$

As a result, in this case the topological entropy is the logarithm of the largest real eigenvalue of the incidence matrix I.

In the following two subsections we consider a series of applications of the expressions (5.37) and (5.38) for topological entropy to subtemplates of the Smale horseshoe template and subtemplates involving branches describing the dynamics seen in circle maps.

5.11.4 Subtemplates of the Smale Horseshoe

In the following three examples the alphabet has the two letters 0 and 1. The grammar is full. It is just the words that differ from one example to the next.

Example 1: There are two words: $S_1 = 0$ and $S_2 = 1$, $t_1 = t_2 = 1$, and (5.37) becomes

$$1 = \frac{1}{X} + \frac{1}{X}$$

The solution is $X_0 = 2$, $h_T = \log 2 = 0.693147$.

Example 2: There are again two words, $S_1 = 1$ and $S_2 = 01$. Then $t_1 = 1$ and $t_2 = 2$, so (5.37) becomes

$$1 = \frac{1}{X} + \frac{1}{X^2}$$

The solution is $X_0 = \frac{1}{2}(1 + \sqrt{5})$, $h_T = 0.481212$.

Example 3: There are four words: 01, 011, 0111, and 01111. All combinations of these symbol sequences are allowed, and (5.37) becomes

$$1 = \frac{1}{X^2} + \frac{1}{X^3} + \frac{1}{X^4} + \frac{1}{X^5}$$

The solution is $X_0 = 1.534158$, $h_T = 0.427982$.

5.11.5 Subtemplates Involving Tongues

Some dynamical systems do not follow a simple stretch and fold route to chaos as exhibited by the Rössler system. The best known of these is the van der Pol oscillator, but it is one of many dissipative systems that follow an alternative route. In this route a Hopf bifurcation occurs, followed eventually by some kind of transition to chaos. The inertial manifold has the topology of a hollow donut: $(I^1 \times S^1) \times S^1$ In this topology the second S^1 parameterizes a periodic driving term. A Poincaré section is easily defined. In a Poincaré section the intersection $I^1 \times S^1$ is topologically an annulus (I^1 is an interval). By the Birman–Williams theorem this projects down to a one-dimensional set that is topologically a circle (S^1). The return map is then a map of the circle to itself. The properties of the circle map were summarized in Section 2.12.

Invertibility is lost when the circle folds over on itself during the return map. Because of the boundary conditions (S^1 is topologically different from R^1), two folds must occur. The flow from S^1 to its folded over image is described by a three-branch manifold. Branches L and R are orientation preserving. On branch L the rotation angle increases by less than 2π, on branch R it increases by more than 2π. Branch C occurs between the two folds and is orientation reversing.

While the circle map is still invertible, mode locking occurs. Each mode-locked region is characterized by a rational fraction $\omega = p/q$, with $0 \leq \omega \leq 1$ for the case of zero global torsion. In the rational fraction, q is the number of times the orbit goes around the long circumference of a torus and p is the number of times it goes around the short circumference: q is the period and p is the winding number.

The symbol sequence of the saddle-node pair in the Arnol'd tongue p/q is $W(1)$ $W(2) \cdots W(q)$, where

$$W(i) = \left[i \times \frac{p}{q} \right] - \left[(i-1) \times \frac{p}{q} \right] = \begin{pmatrix} 0 \\ 1 \end{pmatrix} \longrightarrow \begin{pmatrix} W(i) = L \\ W(i) = R \end{pmatrix}$$

where $[x]$ is the integer part of x. For $p/q = \frac{3}{5}$, $W(1)W(2)W(3)W(4)W(5) \to LRLRR$. The partner orbit is obtained by replacing the penultimate symbol by C (e.g., $LRLCR$).

Chaotic behavior occurs when the map loses the invertibility property and the Arnol'd tongues begin to overlap. We describe the chaotic behavior when tongues described by rational fractions $\omega_1 = p_1/q_1$ and $\omega_2 = p_2/q_2$ just begin to overlap. We assume that $\omega_1 < \omega_2$ and

$$\det \begin{bmatrix} p_1 & p_2 \\ q_1 & q_2 \end{bmatrix} = \pm 1$$

At this point the behavior is chaotic and the vocabulary contains three words. These are:

A The symbol sequence for the left-hand tongue p_1/q_1
B The symbol sequence for the right-hand tongue p_2/q_2
$\overline{B}$ The partner of B

Not every symbol sequence is allowed, for $\overline{B}$ must be preceded by A. Each word labels a branch in a branched manifold. This is a subtemplate of the branched manifold that describes the dynamics in the fully expansive case (L, R, and C have a fully expansive incidence matrix). The incidence matrix for the three words A, B, and $\overline{B}$ is

$$\begin{matrix} A \\ B \\ \overline{B} \end{matrix} \begin{bmatrix} 1 & 1 & 1 \\ 1 & 1 & 0 \\ 1 & 1 & 0 \end{bmatrix}$$

Applying this information to Eq. (5.38), we find that

$$\det \begin{bmatrix} \frac{1}{X^{q_1}} - 1 & \frac{1}{X^{q_1}} & \frac{1}{X^{q_1}} \\ \frac{1}{X^{q_2}} & \frac{1}{X^{q_2}} - 1 & 0 \\ \frac{1}{X^{q_2}} & \frac{1}{X^{q_2}} & -1 \end{bmatrix} = 0$$

This reduces to

$$X^{q_1+q_2} - X^{q_1} - X^{q_2} - 1 = 0 \quad \text{or} \quad \frac{1}{X^{q_1}} + \frac{1}{X^{q_2}} + \frac{1}{X^{q_1+q_2}} = 1$$

Example 1: Compute the vocabulary and the topological entropy for the strange attractors that occur when the tongues p_1/q_1 and p_2/q_2 just overlap, for the pairs $(\frac{1}{2}, \frac{2}{3})$, $(\frac{1}{2}, \frac{3}{5})$, $(\frac{3}{5}, \frac{2}{3})$.

p_1/q_1	p_2/q_2	A	B / $\overline{B}$	X_0	h_T
$\frac{1}{2}$	$\frac{2}{3}$	LR	LRR	1.429108	0.357051
			LCR		
$\frac{1}{2}$	$\frac{3}{5}$	LR	$LRLRR$	1.307395	0.268037
			$LRLCR$		
$\frac{3}{5}$	$\frac{2}{3}$	$LRLRR$	LRR	1.252073	0.224801
			LCR		

Example 2: Compute the topological entropy for the low-period Arnol'd tongues for which $p_1 q_2 - q_1 p_2 = \pm 1$. *Solution*: The results depend only on the periods q_1 and q_2. To period 9, here they are. Entries for which the periods are not relatively prime have been left blank.

	3	4	5	6	7	8	9
2	0.357051		0.268037		0.219131		0.187366
3		0.253442	0.224801		0.186002	0.172048	
4			0.196620		0.164136		0.142458
5				0.160664	0.148188	0.137920	0.129277
6					0.135847		
7						0.117680	0.110713
8							0.103803

5.12 SUMMARY

Branched manifolds were introduced by Birman and Williams [83,84] as a simple tool to describe completely the organization of all the unstable periodic orbits in the Lorenz dynamical system [83]. Their theorem guarantees that branched manifolds can be used to describe the organization of unstable periodic orbits in any three-dimensional dissipative dynamical system with a (hyperbolic) strange attractor. However, one of the first branched manifolds discussed in detail by Birman and Williams, the figure 8 knot-holder, describes the topological organization of all the closed magnetic field lines generated by a constant current flowing in a wire knotted into a figure 8 shape [84]. This is a conservative dynamical system. It is not clear that the Birman–Williams theorem can be applied *only* to dissipative systems.

As stated, the Birman–Williams theorem is not immediately useful for the analysis of chaotic data. Two of the input assumptions are too restrictive. Both assumptions (hyperbolicity, three-dimensional flow) can be relaxed. Once these modifications were made, the Birman–Williams theorem became a key component in the topological analysis of chaotic data and the classification of strange attractors.

We have described the branched manifolds for the four standard testbeds of dynamical systems theory: the Duffing, van der Pol, Lorenz, and Rössler attractors. Each branched manifold has an algebraic representation in terms of three matrices. The topological matrix T determines how the various branches twist and cross each other. The joining array A identifies the order in which two or more branches are joined at a branch line. The transition or incidence matrix I determines the flow ordering: which branches flow into which other branches.

The location of periodic orbits on branched manifolds can be determined by kneading theory. Once orbits have been located, their linking numbers and relative rotation rates can be determined algorithmically. The inputs to the algorithm are the two matrices T and A. Conversely, a symbolic coding of the orbits in a flow determines I, and information about the linking numbers of these orbits can be used to construct the two matrices T and A. The result is that branched manifolds can be identified on the basis of properties of unstable periodic orbits identified in the flow.

6

Topological Analysis Program

The topological analysis program [1,2] was developed to extract the information required to identify and classify strange attractors from experimental data. It is summarized briefly in Fig. 6.1. It consists of the first five steps shown in the figure. The remaining two steps are natural extensions of the first five steps, which involve making models to explain physical processes and then testing these models. The vertical arrows in this figure are feedback loops which are absent in other dynamical and metric tests for chaotic behavior.

6.1 BRIEF SUMMARY OF THE TOPOLOGICAL ANALYSIS PROGRAM

Very briefly, the topological analysis program consists of the following steps:

Locate Periodic Orbits: Periodic orbits are extracted from chaotic data.

Embed Data: The experimental data are embedded in a three-dimensional space.

Compute Topological Invariants: The linking numbers and/or relative rotation rates of the periodic orbits extracted from the strange attractor are computed.

Identify Template: A template is proposed that is consistent with the topological invariants for a relatively small number of these periodic orbits.

Verify Template: Topological invariants for all orbits supported by the template are computed. These are compared with the topological invariants for all orbits and orbit pairs extracted from the data. If the two sets differ, either the proposed branched manifold is incorrect or some of the experimental orbits have been misidentified. If both sets are the same, the template proposed can be accepted as describing the experimental strange attractor. (More accurately: It is not possible to reject the null hypothesis that the template proposed is correct.)

The remaining two steps in Fig. 6.1 are not strictly of a topological nature. Nevertheless, they are natural extensions in this program for understanding and describing chaotic dynamical systems.

Model Dynamics: A mathematical model is developed that attempts to reproduce the experimental data. If the chaotic signal generated by the model and the experimental signal are described by inequivalent branched manifolds, the model can be rejected. Otherwise, the model is "qualitatively correct."

Validate Model: The goodness of fit of model output with chaotic data is tested. At present, there is no quantitative test for chaotic systems that is analogous to standard tests (e.g., the χ^2 test) for general linear models, although there is a good candidate that still needs to be made quantitative. If the model fails this quantitative test, it must be modified or rejected.

6.2 OVERVIEW OF THE TOPOLOGICAL ANALYSIS PROGRAM

We now describe in more detail the method developed for the topological analysis of strange attractors generated by dynamical systems operating in a chaotic regime. The method consists of the steps summarized in Fig. 6.1 and described briefly above. These steps are described in much more detail below. At present, the methods are applicable to low-dimensional dynamical systems—that is, to systems for which the strange attractor can be embedded in a three-dimensional manifold.

6.2.1 Find Periodic Orbits

Unstable periodic orbits are abundant in strange attractors and dense in hyperbolic strange attractors. If an initial condition enters the neighborhood of an unstable

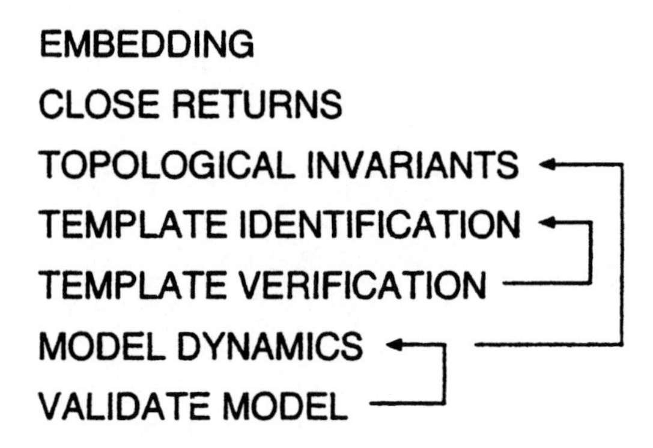

Fig. 6.1 Steps in the topological analysis program. Vertical arrows are feedback loops that are absent from other tests for chaotic behavior. The first five steps properly belong to the topological analysis program. The remaining two quantitative steps are extensions of this procedure.

periodic orbit, it will evolve in the neighborhood of this unstable periodic orbit for awhile. If this initial condition falls close enough to the unstable periodic orbit along its unstable manifold and its unstable Lyapunov exponent is not too large, the initial condition may evolve all the way around the attractor and return to a neighborhood of its starting point. If this happens, it will evolve in the neighborhood of phase space that it visited previously. When this occurs, the difference $|\mathbf{x}(t) - \mathbf{x}(t+\tau)|$ remains small for awhile. Here τ is the period of the closed orbit. This signature is used to locate segments in a chaotic data set that can be used as surrogates for unstable periodic orbits. That is, the segment lies in a neighborhood of the unstable periodic orbit and so behaves to some extent like the unstable periodic orbit. This method of finding unstable periodic orbits in data is called the *method of close returns*.

It is not sufficient simply to locate surrogates for unstable periodic orbits. The name of each orbit must be identified by a symbol sequence. This is necessary because we eventually need to identify orbits in the flow with orbits on a template in a 1:1 way. Identifying the symbolic dynamics of an orbit in a flow can often be done with a low error rate, which decreases as the dissipation rate increases. Identification of an orbit's symbolic dynamics is facilitated by constructing a return map on a Poincaré section. We have shown that a Poincaré section can always be constructed when the Birman–Williams theorem is applicable: It is the blowup of the branch lines of the underlying branched manifold. A data file of successive encounters of the flow with the Poincaré section is created. Then a pth return map is generated, and those intersections closest to the diagonal are interesting candidates for the endpoints of an unstable periodic orbit of period p (or $p/2, \dots$).

The first return map is often useful for suggesting the appropriate symbol sequence for an orbit. This is especially true when critical points are present in the return map. This is often the case for highly dissipative dynamical systems. However, even if symbolic dynamics is not available for orbits at this stage, the topological analysis can be carried out. The details are presented in Appendix A. At this stage, the identification of a symbol sequence with an unstable periodic orbit must be regarded as tentative, if available.

6.2.2 Embed in R^3

For a topological analysis to be carried out, the strange attractor must first be embedded in R^3, for it is only in R^3 that linking numbers can be computed. If the dynamical system is given analytically and is already three-dimensional, the problem is already solved. If the dynamical system is given analytically but is of dimension greater than 3, it is necessary to compute the local Lyapunov dimension, $d_L(x)$, on the attractor. If $d_L(x) < 3$ everywhere on the attractor, the Birman–Williams theorem is applicable. The projection of the attractor along the strong stable directions corresponding to $\lambda_4, \lambda_5, \ldots$ into an inertial manifold $\mathcal{IM}^3$ then provides the appropriate embedding.

When the chaotic dynamics is generated by a physical system, the analysis becomes more interesting. If three or more independent time series $x(t), y(t), z(t), \ldots$ are available, the situation is as described previously. In many cases only a single time series is available. This time series is always discretely sampled and may not have an optimum signal-to-noise ratio. In this case we must construct an embedding from this single time series. In other instances, we have an entire data field (e.g., CCD frames) sampled at each time and must reduce this to a small number of time series. These two situations occur in laser experiments. In one case only the integrated output intensity on a cross section may be available. In another, a succession of frames (e.g., 120×240 pixels of 16-bit data/pixel) may be available. In both cases we wish to generate a small number ($n = 3$) of time series so that a $d_L = 2 + \epsilon$ ($0 < \epsilon < 1$)-dimensional strange attractor can be embedded in R^3. We discuss embedding procedures in more detail in Section 6.4.

6.2.3 Compute Topological Invariants

Topological invariants for periodic orbits embedded in a strange attractor in a three-dimensional phase space include linking numbers. They also include relative rotation rates in case the strange attractor has a hole in the middle. We have described how to compute both linking numbers (in Section 4.2) and relative rotation rates (in Section 4.3). Topological invariants are computed for all surrogate periodic orbits extracted from chaotic data.

In higher dimensions, it is not possible to compute linking numbers and relative rotation rates. Knots and links fall apart in higher dimensions. The only impediment to extending the topological analysis program to higher-dimensional strange attractors occurs at this point: It is the construction of topological invariants for a strange attractor when $d_L > 3$.

6.2.4 Identify Template

The next stage in identifying an underlying stretch and squeeze mechanism for generating chaotic behavior involves providing a name for each surrogate periodic orbit. Naming orbits involves creating an alphabet and sometimes even a grammar to go along with the alphabet. If the alphabet has N letters $A, B, \ldots, N$, the template containing all orbits contains N branches. The *grammar*, or statement of allowed and forbidden transitions, defines the incidence matrix for the branched manifold.

A preliminary guess at the branched manifold is made as follows:

Topological Matrix T: The diagonal matrix elements T_{ii} are determined by the local torsion of the surrogates for the letter i. Specifically, if two or more surrogates for orbit segment i are available, T_{ii} is the signed number of times they cross. The off-diagonal matrix elements T_{ij} are the number of times surrogates i and j cross. This is unique unless the branches containing i and j meet at a common branch line. This nonuniqueness is then accounted for in the $N \times 1$ array A. Altogether, $\frac{1}{2}N(N+1)$ pieces of information are required to specify the symmetric matrix T.

Array A: The order in which branches meet at a branch line is contained in this array. If B_α branches meet at branch line α, the order in which they meet (viewed in projection from front to back) is specified by $B_\alpha - 1$ pieces of information. The amount of information required to specify the array A is then $\sum_\alpha (B_\alpha - 1)$. This information is not available from period-1 orbits but is available from period-2 orbits.

Incidence Matrix I: This information is provided by the grammar that the alphabet satisfies. If there are N branches, N^2 pieces of information are specified by the incidence matrix.

When the attractor is embedded in a torus $R^2 \times S^1$, the branched manifold that classifies the strange attractor has a hole in the middle. Extraction of topological information from surrogate periodic orbits then simplifies, since each letter $A, B, \ldots, N$ in the alphabet corresponds to a period-1 orbit. In this frequently occurring case:

Topological Matrix T: The linking numbers and local torsions are sufficient to construct the topological matrix according to

$$\begin{aligned} T(i,i) &= LT(i) \\ T(i,j) &= 2L(i,j) \end{aligned} \tag{6.1}$$

Array A: The linking numbers of the period-1 orbits with the period-2 orbits, and the linking and self-linking numbers of the period-2 orbits among themselves are sufficient to determine the order in which the N branches meet at a common branch line.

Incidence Matrix I: It is usually convenient to take this as full in the sense that all matrix elements are 1, even if there are some forbidden transitions. Then the missing orbits can be regarded as having been pruned away by an unfolding of the branched manifold, as described in Chapter 5. The invariants do not depend on whether or not the incidence matrix is full.

In most instances not all period-1 and period-2 orbits are available. Then other low-period orbits from the attractor can be used to extract the information required to make a first tentative identification of the branched manifold. As an example, in horseshoe dynamics the orbit 0 is often not available. The period-1 orbit 1, the period-2 orbit 01, and one of the period-3 orbits 011 or 001 can be used to extract the four necessary pieces of information from the chaotic signal.

6.2.5 Verify Template

A branched manifold is tentatively identified using topological information from a minimal set of low-period orbits. This identification must then be verified. This is done by using the template to construct a table of topological invariants for all the periodic orbits that it supports. If the original template identification was correct, these numbers must include as a subset the topological invariants determined for the unstable periodic orbits extracted from the strange attractor. If the two sets of numbers do not agree, either the original template identification was incorrect or the symbolic names attributed to some surrogate orbits were in error. We have usually found complete agreement between the topological invariants computed from surrogate data and from the corresponding orbits on a template. In a few cases there was not complete agreement. This has always been due to a questionable symbol assignment to some part of a surrogate orbit. In all cases this validation step has helped to refine the identification of a few surrogate orbits.

Orbit labeling and template identification are not isolated problems. They constitute one global problem, which must be resolved so that the table of topological invariants computed from the data is identical to that computed for corresponding orbits on the branched manifold.

We remark that the template identification step provides a loop closing step or self-consistency check. Loop-closing steps are represented by the return arrows in Fig. 6.1. This process is analogous to the process involved in the statistical evaluation of experimental data. For example, a least squares fit of *any* linear model will *always* converge to *some* result. The follow-up question, "Is this best-fit model any good?" must then be answered by additional tests (see Press et al. [99]). Such loop-closing tests are absent from the older metric methods for analyzing chaotic data.

Technical Remark: The threshold problem and its consequences are frequently encountered. This problem is as follows. In locating surrogates for unstable periodic orbits, a threshold for close returns must be established. Choosing a stringent threshold guarantees very good surrogates. It also guarantees very few surrogates. The more stringent the threshold, the fewer the orbits. The limiting case is obvious. There is therefore pressure to go in the other direction. Loosening the threshold increases the number of orbits. It also increases the probability that the invariants computed for these surrogates contain errors, in the sense that invariants computed for some of these surrogates are not the same as the invariants for the corresponding unstable periodic orbits. In this case the table of invariants computed for the surrogates will contain elements that differ from the table computed for the unstable periodic orbits. One response to this small error rate is to reject the initial template identification. This response is less constructive than the alternative response, which is to reject a small number of questionable surrogates and test whether this removes all the errors. If so, the initial template identification can be accepted, while the surrogates can be divided into a subset that is acceptable and another that is not.

6.2.6 Model Dynamics

In the end, a branched manifold provides only a caricature for a flow. It identifies the stretching and squeezing mechanisms responsible for generating chaos. However, it provides more unstable periodic orbits than actually exist in the strange attractor, and it is more dissipative than the actual dynamics. The branched manifold provides a qualitative model for the dynamics but not a quantitative model.

Once a qualitative model for some dynamics has been validated, it is natural to attempt to develop a quantitatively correct model for the dynamics. The art of model building falls into two broad categories: building analytical models and thinking the unthinkable.

Building Analytical Models: In the traditional approach, a system of ordinary differential equations (3.1) is proposed and one attempts to represent the forcing functions $F_i(x)$ as linear superpositions of basis functions $\Phi_j(x, c)$:

$$F_i(x, c) = \sum_{j=1}^{N} A_{ij}(c)\Phi_j(x, c) \tag{6.2}$$

Many criteria exist for choosing the basis functions $\Phi_j(x, c)$ and estimating the coefficients $A_{ij}(c)$. This field has been extensively discussed in a beautiful review [100] and a monograph [101]. We have only one contribution to add to the assorted wisdom presented there, which deals with estimating the coefficients $A_{ij}(c)$.

We have found it useful to consider the vector $(B_0, B_1, B_2, \ldots, B_N)$ coupled to the functions $(\Phi_0 = dx_i/dt, \Phi_1, \Phi_2, \ldots, \Phi_N)$ for each $i = 1, 2, 3$. A singular-value decomposition of the $(1 + N) \times T$ matrix $\Phi_j(x_t, c)$, $j = 0, 1, \ldots N$, and $t = 1, 2, \ldots, T$, where T is the number of measurements, or length of the data set, produces a series of eigenvectors $[B_0(\alpha), B_1(\alpha), \ldots, B_N(\alpha)]$ with eigenvalues $\lambda(\alpha)$, $\alpha = 0, 1, 2, \ldots, N$ [2]. The square of each eigenvalue provides a noise estimate. We search over α for the minimum value of $[\lambda(\alpha)/B_0(\alpha)]^2$ and identify the coefficients A_{ij} in (6.2) with $[-B_j(\alpha)/B_0(\alpha)]$. This is done for each of the three values of i. This eigenvalue analysis avoids the singularities that occur too frequently when one attempts to normalize the functions $\Phi_j(x, c)$ with respect to the measure on the strange attractor.

Thinking the Unthinkable: The second approach is based on the spirit that motivated a remarkable paper entitled "Computers and the theory of statistics: thinking the unthinkable" [102]. We ask the question: "Even if we have an analytic expression of the form (6.2), does that really help us to undersand the dynamics?" Usually: NO!

For this reason, we partition the phase space into a small number of flow tubes, which are essentially inflations of branches of a branched manifold. We then provide a numerical algorithm for the flow through each region. In this way the physical nature of the flow is apparent, and the lack of a (global) analytic expression for the dynamics is no great drawback.

6.2.7 Validate Model

Once again, an estimation step must be followed by a loop-closing (validation) step. Two ways exist to validate a model of a chaotic system: Compare invariants and test for entrainment.

In the first method, the topological, dynamical, and metric invariants of the experimental strange attractor are compared with those generated by the model. In particular, the branched manifolds must be identical and the spectrum of unstable periodic orbits, as represented by a basis set of orbits (cf. Section 9.3.4), should be close. In addition, the average Lyapunov exponents and Lyapunov dimensions should be close. If sufficient data are available (often not the case), metric properties determined from the data can be compared with those computed from the model.

The second idea is based on a beautiful idea due to Fujisaka and Yamada [103] and independently to Brown, Rulkov, and Tracy [104]. This idea in turn is based on one of the oldest observations in the field of nonlinear dynamics: The seventeenth-century observation by Huyghens that two pendulum clocks will synchronize when placed close together on a wall that provides coupling between them [105]. Synchronization between two physical systems has been studied in some detail (Pecora and Carroll [106]). The idea of Brown, Rulkov, and Tracy is that if a model is a sufficiently good representation of a physical process, the model can be entrained by the data. If y_i^m are the model variables and y_i^d are the data variables, one cannot expect the model

$$\frac{dy_i^m}{dt} = F_i(y^m; c) \tag{6.3}$$

to reproduce the data in the sense $y_i^m(t) = y_i^d(t)$ for all times, no matter how good the model is. However, one might expect a small coupling between the model and data variables of the form

$$\frac{dy_i^m}{dt} = F_i(y^m; c) - \sum_j \lambda_{ij}(y_j^m - y_j^d) \tag{6.4}$$

will entrain the model output to the data. In the case of entrainment, a plot of $y_j^m(t) - y_j^d(t)$ vs. t is zero. This test already provides a useful method for model validation, even though it has not yet been made quantitative.

6.3 DATA

Data sets generated by a number of physical systems have been subjected to a topological analysis. These include the Belousov–Zhabotinskii reaction, the laser with modulated losses, the laser with saturable absorber, the CO_2 laser, a dye laser, a catalytic reaction, a model of a collapsing globular cluster, and a musical instrument. In most cases the data collected consisted of a single (scalar) time series. However, some data sets consisted of entire data fields, an amplitude or intensity distribution in one, two, or three dimensions as well as time. In this section we describe the data characteristics and the processing methods that have been useful for implementing topological analyses.

Data processing can be carried out in the time domain, the frequency domain, or a combination of time and frequency domains. Frequency-domain processing for linear systems has a long history and is well understood. Reliable tools (Fast Fourier Transform, see [99] and [107]) are easily available for such processing. Time-domain processing of data generated by chaotic processes is a more recent development [108]. Some tools are robust; others are in the development stage. More recently, a combination of time- and frequency-domain methods have been developed for processing chaotic data [109]. For the most part, we have found that frequency-domain methods have been sufficient for processing chaotic data. In some instances, time-domain processing with singular-value methods has been useful [110].

We emphasize strongly that the topological analysis procedure is carried out in the time domain only. However, time-domain, frequency-domain, or singular-value methods may be used to construct the embeddings on which the topological analysis method is based. We assume throughout that the data sample the entire strange attractor: That is, all transients have died out and motion is not confined to a subset of the attractor during the data acquisition process.

6.3.1 Data Requirements

Data requirements for a topological analysis can conveniently be expressed in terms of cycles and cycle time. Roughly, a *cycle* is a trip around the strange attractor, and *cycle time* is the time it takes to make this trip. Usually, the meaning of *around* is clear once an embedding is available. This time scale can often be estimated by direct inspection of the data: It is the average peak-to-peak separation. If necessary, it can be estimated as the inverse of the highest-frequency peak in the power spectrum or the lowest time-delay peak in a close-returns plot.

6.3.1.1 $\sim$ 100 Cycles From experience, $\sim$ 100 cycles is more than enough for a topological analysis for a dissipative system. When data are plotted in a suitable embedding, the first dozen cycles outline the shape of the strange attractor, the next 20 to 50 cycles fill in the details, and beyond 100 cycles no additional detail is provided. To make this point, in Fig. 6.2 we show an embedding of data describing the Belousov–Zhabotinskii reaction. The number of cycles plotted increases from about 10 in Fig. 6.2(a) to about 25 in Fig. 6.2(b) and about 80 in Fig. 6.2(c). This figure suggests that too much data is not always a good thing. The number of cycles needed for a successful analysis increases as the dissipation decreases.

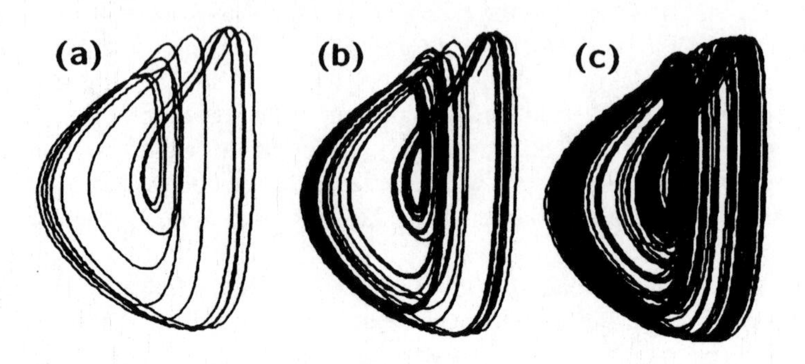

Fig. 6.2 Too much data is not useful for a topological analysis. (a) A dozen cycles outline the shape of the attractor; (b) 20 to 50 cycles fill in some details; (c) and more begin to obscure the details.

6.3.1.2 ~ 100 Samples/Cycle From experience, ~ 100 samples/cycle provides a convenient sampling rate. More than 100 samples/cycle provides redundant information. Fewer than 50 samples/cycle usually means that the data must be smoothed or interpolated in some way. We have analyzed data sets with as many as 200 samples/cycle and as few as 12 samples/cycle. In the former case we carried the overhead of larger-than-necessary data files. In the latter case we had relatively short data sets but paid the price of being forced to interpolate the data. We have found frequency-domain methods fast and efficient for data interpolation purposes.

Our preference has been to deal with shorter rather than longer data sets. Many of our analyses have been carried out on a subset (often, a small subset) of the available data. In our experience, an optimum file for a topological analysis contained 8K ($8192 = 2^{13}$) scalar measurements of 8- or 16-bit data, with the sampling parameters in the range recommended above. These modest requirements should be compared with the data requirements for fractal dimension estimation, which are usually obscenely large. The presence of strong higher-order harmonics requires a larger sampling rate.

The modest data requirements outlined above are strongly correlated with the development of these methods on a modest, old, and very reliable computer (Mac II) running at 16 MHz with a disk capacity of 10 MB using a very friendly MacTran compiler.

6.3.2 Processing in the Time Domain

Very clean data sets can sometimes be used directly for topological analysis, without any processing. This is not usual. Data processing can be carried out in the time domain or in the frequency domain. In the time domain, the principal tool for processing is rational function fitting and interpolation. This is recommended if the data are not equally sampled. If the data are equally sampled, frequency-domain methods are far more efficient and highly recommended, even when there are missing measurements.

6.3.2.1 *Interpolating with Rational Approximations* It is sometimes necessary to know a data value that has not been measured. Such a value can often justifiably be inferred from nearby data values. One useful way to do this is to fit a smooth curve through data points surrounding the value that is desired. The fitted function can then be used both to interpolate and extrapolate values. Interpolation is usually safe; extrapolation should be avoided.

The procedure is as follows. A set of $k+1$ successive data points (x_i, t_i), (x_{i+1}, t_{i+1}), ..., (x_{i+k}, t_{i+k}) is chosen that surround the point at which the data value is to be interpolated. A smooth curve through these $k+1$ points can be fitted by a rational function of the form [99]

$$x(t) = \frac{P_\mu(t)}{Q_\nu(t)} = \frac{p_0 + p_1 t + p_2 t^2 + \cdots + p_\mu t^\mu}{q_0 + q_1 t + q_2 t^2 + \cdots + q_\nu t^\nu} \tag{6.5}$$

when the number of independent parameters in the rational function is equal to the number of data points: $k+1 = (\mu+1) + (\nu+1) - 1$. This interpolation satisfies $x(t_i) = x_i$. It is then usually safe to evaluate x for values of t in the midrange of values t_i to t_{i+k}. In this way, data values $x(t)$ can be estimated which have not been observed.

The interpolation procedure is inefficient, in that the fit has to be carried out for each value of t at which the data are to be estimated. If the data are not equally spaced, this may be the only reasonable approach to reconstructing an equally spaced data set.

6.3.2.2 *Smoothing* If the number of degrees of freedom $(\mu + \nu + 1)$ is reduced, the rational function will no longer go through the $k+1$ data points. In this case the coefficients can be chosen to provide a best fit to the data. In other words, they can be used to smooth the data. This is a useful way to reduce noise in the data set.

Choosing the length of a data sample $k+1$ and the degrees μ and ν of the numerator and denominator polynomials to provide good smoothing characteristics for the data set is an art form. When the data are equally spaced, smoothing by rational interpolations is very inefficient compared to frequency-domain methods.

6.3.2.3 *Derivatives* For purposes of embedding data into R^3 it is often useful to construct derivatives and integrals of the time series. Derivatives can be constructed in the time domain directly by taking first differences, $dx_i/dt \sim x_{i+1} - x_{i-1}$. This is fast and efficient when the data are equally spaced and the signal-to-noise ratio is high.

If the data are not equally spaced, or are noisy, derivatives can be computed from the rational fractional interpolations or smoothing functions constructed in the neighborhood of each value of t. Once these functions have been constructed, it is a simple matter to compute the values dx/dt and d^2x/dt^2.

6.3.2.4 *Integrals* Integrating the time series is more problematic. Before integration, the average value of the data set must be removed. If it is not, the integrated time series will not be stationary: It will increase in time at a rate $\langle x \rangle t$. The mean value can be estimated from $\langle x \rangle = \frac{1}{T}\sum_{i=1}^{T} x_i$. However, this mean value may be sensitive to the endpoints of the observed time series. In short, it is also an art form to ensure that the time series

$$y(t) = \int_{-\text{``}\infty\text{''}}^{t} [x(t') - \langle x \rangle] \, dt' \tag{6.6}$$

is stationary. To finesse this problem, it is very useful to integrate backwards with a memory. That is, an integrated time series can be constructed in the form

$$y(t) = \int_{-\text{``}\infty\text{''}}^{t} [x(t') - \langle x \rangle] \, e^{-(t-t')/\tau} \, dt' \tag{6.7}$$

Here τ is a memory time. We have found it useful to use a multiple of the cycle time for τ ($\tau \sim 10$ cycles).

If the data are equally spaced, the integrated time series can be created on line as follows:

$$y_i = x_i + e^{-1/(10P)} y_{i-1} \tag{6.8}$$

Here P is the cycle time, measured appropriately. With this algorithm, it is not even necessary to remove the mean value $\langle x \rangle$ of the data set. However, it is necessary to discard roughly the first 10 periods of the integrated data set unless care has been taken in choosing the initial condition, y_1.

6.3.3 Processing in the Frequency Domain

Processing in the frequency domain is almost always more efficient than working in the time domain. This is particularly true when the data are equally spaced in time. Most of the procedures described below depend on the Fast Fourier Transform (FFT) [99].

6.3.3.1 *High-Frequency Filter* Experimental data sets often consist of two components: a signal on top of which is superposed noise. Even very clean data sets that have been recorded and stored in digital form have a noise component induced by round-off or truncation. An extensive industry has arisen to deal with the separation of signal from noise. For our purposes, it is usually sufficient to remove the high-frequency components in the Fourier transform of the data set. Typically, it is sufficient to filter out components in the Fourier spectrum whose frequency is more than a factor of 10 greater than the frequency corresponding to the cycle time.

6.3.3.2 *Low-Frequency Filter* Most of the important information in a chaotic signal is contained in the low frequencies. It is therefore usually a very bad idea to filter frequencies any smaller than those corresponding to the cycle time.

However, exceptions do exist. When performing an integral-differential embedding (cf. Section 6.4.3), one of the three variables created from the scalar data set is the integral of the data (6.18). In some cases, when plotting the two-dimensional projection $y(t)$ vs. $x(t)$, we were able to see the "attractor" drift in the x–y plane along the y axis (cf. Fig. 7.5). This drift can be traced to secular and long-term variability in the initial data set. This is not atypical of experiments that last longer than a substantial fraction of a day due to slow variations in the electrical grid. Such variations are not apparent in the initial data set x_i but are amplified by the integration procedure.

When the slow variations are removed by low-frequency filtering of the initial data set x_i, the attractor remains stationary (cf. Fig. 7.6). We can provide no hard-and-fast rules for the low-frequency cutoff. We can only suggest that the cutoff gradually be increased until the projection of the strange attractor ceases to wander in phase space.

6.3.3.3 Derivatives and Integrals To carry out a topological analysis, an embedding of data must first be constructed. One way to create an embedding is to use derivatives and/or integrals of the original scalar time series as components of the embedding vector. These can be constructed in the time domain, as described above.

They can also be constructed in the frequency domain. The time series and its derivative and integral can be represented in the form

$$\begin{aligned} x(t) &= \int \hat{x}(\omega) e^{i\omega t}\, d\omega \\ \frac{dx}{dt} &= \int i\omega \hat{x}(\omega) e^{i\omega t}\, d\omega \\ \int^t x(t')\, dt' &= \int \frac{\hat{x}(\omega)}{i\omega} e^{i\omega t}\, d\omega \end{aligned} \tag{6.9}$$

Figure 6.3 illustrates how to compute the derivative.

1. Compute the Fast Fourier Transform of the data set.

2. Interchange the real and imaginary components, with phase information. Multiply by $|\omega|$.

3. Compute the inverse FFT.

If the power spectrum is not small near the Nyquist frequency, frequency-domain processing is a bad idea and should be abandoned.

The integral is computed in much the same way. The differences are:

1. The zero-frequency component should be zeroed out. This is equivalent to removing the mean value $\langle x \rangle$ in the time domain.

2. The phase change is opposite that for the derivative, which is shown in Fig. 6.3.

3. The Fourier coefficients should be divided by $|\omega|$ rather than multiplied by $|\omega|$.

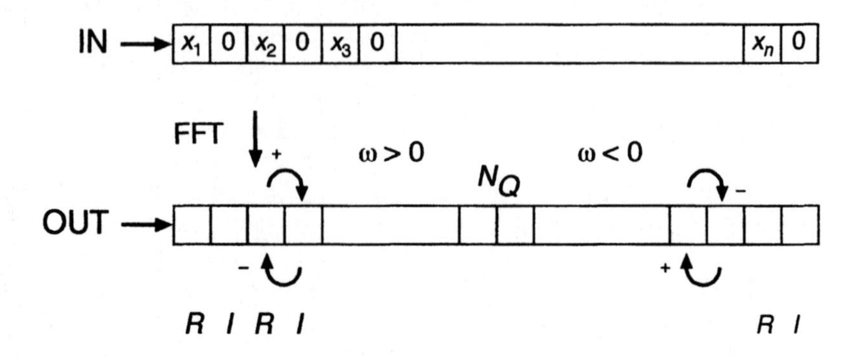

Fig. 6.3 Derivatives are computed in the frequency domain as shown. The input data set of length 2^k is place in the real part of an array of length 2×2^k. The Fast Fourier Transform is computed [99]. The real and imaginary parts are interchanged, with phase changes as shown. The Fourier coefficients are also multiplied by $|\omega|$. The inverse FFT is then performed. The output data file contains the derivatives dx_i/dt. Only slight modifications are needed to compute the integral of the data set, fractional derivatives and integrals, and the Hilbert transform.

Generalized derivatives and integrals of degree d can also be computed. They are constructed by the algorithms above, except that the Fourier coefficients are multiplied by $|\omega|^d$ instead of $|\omega|^{\pm 1}$. Generalized derivatives for experimental data are shown in Fig. 6.4 for the following cases:

$d = -\frac{1}{2}$ "square root of the integral"

$d = \ \ 0$ Hilbert transform

$d = +\frac{1}{2}$ "square root of the derivative"

6.3.3.4 *Hilbert Transforms* The scalar time series $x(t)$ can be regarded as the real part of a complex data set $z(t)$:

$$x(t) = \text{Re}[x(t) + iy(t)] = \text{Re}\ z(t) \tag{6.10}$$

If $z(t)$ is analytic in the upper half plane, the imaginary part $y(t)$ is the Hilbert transform of the real part $x(t)$. The Hilbert transform of $x(t)$ is given by an integral that is an immediate result of Cauchy's theorem:

$$y(t) = \mathcal{P}\frac{1}{\pi}\int_{-\infty}^{+\infty} \frac{x(t')}{t - t'}\, dt' \tag{6.11}$$

Here $\mathcal{P}$ means "take the principal part of this integral." It is a simple matter to construct $y(t)$ using the Fast Fourier Transform [107]. Essentially, $y(t)$ is the "noise-free" derivative of $x(t)$. Its construction, illustrated in Fig. 6.3, is as follows:

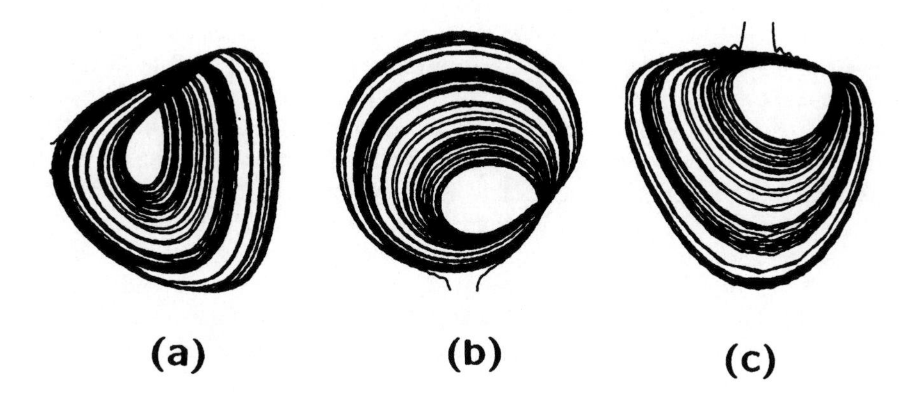

Fig. 6.4 Generalized derivatives computed for an experimental data set: (a) $d = -\frac{1}{2}$; (b) $d = 0$; (c) $d = +\frac{1}{2}$. Transients at the edges are expressions of Gibbs' phenomenon. They arise because the input data set is not periodic. They can be reduced substantially by choosing a segment of the data set in which the beginning and end values are approximately equal: $x_i = x_{i+2^k}$.

1. Compute the FFT of $x(t)$.
2. Interchange the real and imaginary parts of $\hat{x}(\omega)$, with phase as shown in Fig. 6.3. Multiply by $|\omega|^{d=0} = 1$ (optional).
3. Compute the inverse FFT.

The output is the real signal $y(t)$. It is the generalized derivative with $d = 0$. We call it the "noise-free" derivative because in the construction of the derivative, the Fourier components are multiplied by $|\omega|^1$, which emphasizes noise in the high-frequency components.

This algorithm can be implemented more efficiently by computing $y(t)$ as an imaginary signal. If the array that results in step 2 is multiplied by i, it is no longer necessary to interchange the real and imaginary parts of the Fourier coefficients. The positive-frequency components are multiplied by -1 and the negative-frequency components are multiplied by $+1$. If this array is now added to the original array representing $\hat{x}(\omega)$, all positive-frequency terms are zero and all negative-frequency terms are doubled in value. For bookkeeping purposes, it is simpler to construct $\hat{x}(\omega) - i\hat{y}(\omega)$. This is done by zeroing out all negative-frequency terms and leaving the positive-frequency terms unchanged. The inverse FFT is then $\frac{1}{2}[x(t) - iy(t)]$.

The Hilbert transform of some data from the Belousov–Zhabotinskii reaction is shown in Fig. 6.4(b). The beginning and end of the time series diverge from the projected attractor $[y(t)$ vs. $x(t)]$ because of the Gibbs phenomenon. We did not make an effort to match values at the beginning and end of the date segment that was used for this transform.

6.3.3.5 Fourier Interpolation In some cases data sets are undersampled. It is then useful to interpolate between data points. Time-domain interpolation schemes have been described above. We find these time-domain schemes time consuming since they require interpolation between successive small subsets of data all along a long data set.

We prefer a Fast Fourier Transform–based interpolation method, which is illustrated schematically in Fig. 6.5 [2]. This method can be used whenever the power spectrum drops to zero or an acceptable noise level at the Nyquist frequency. The data set of length $N = 2^l$ is placed in an array of length $2N$ in the usual way. The FFT is then performed. The output array is then extended to length $2 \times 2N$ by inserting $2N$ zeros at the Nyquist frequency N_Q. The inverse FFT is real and of length $4N$, so that $2N$ values are nonzero. These $2N$ values consist of the original N data values together with N additional values, which are interpolations between each of the original data values.

The interpolation of $1, 3, \ldots, 2^k - 1$ values between each original observation can be achieved using the same method, except that $(2^k - 1) \times 2N$ zeros must be inserted at the Nyquist frequency. In Fig. 6.6 we show how this method was used to clean up an undersampled time series.

Remark: Output amplitudes are reduced by a factor of $1/2^k$ in this interpolation scheme.

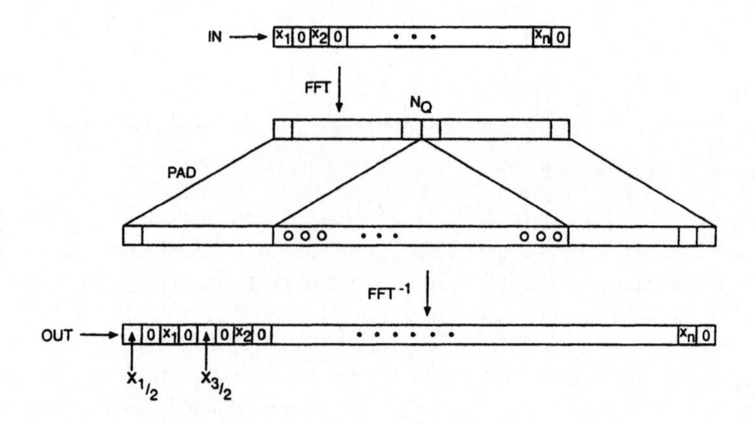

Fig. 6.5 Fourier interpolation: A real data set of length N is interpolated by inserting $2N$ zeros at the Nyquist frequency in the frequency domain.

6.3.3.6 Transform and Interpolation The algorithms described above for computing a Hilbert transform and for interpolating a data set can be combined into a single algorithm that does both and that is more efficient than either algorithm singly. Given a real data set of length $N = 2^l$:

1. Compute the FFT of $x(t)$, stored in the usual way as a data set of length $2N$.

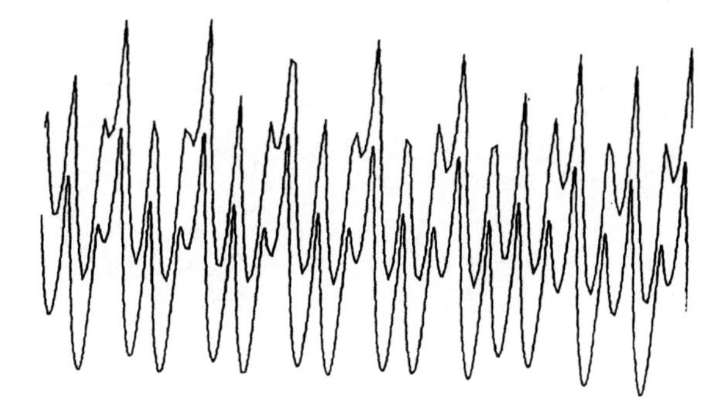

Fig. 6.6 An undersampled time series (top) is made smoother by an 8:1 interpolation. The two are offset for clarity. The output of the interpolation has been multiplied by 8, so the two data sets can be plotted on the same scale.

2. Zero out all terms except the positive-frequency Fourier coefficients between the low-frequency cutoff and the high-frequency cutoff.

3. Extend the length of the data file to $2N \times 2^k$ by padding the end of the file with zeros.

4. Compute FFT^{-1}.

The output data file contains a complex data set of length $N \times 2^k$. The real part contains the original signal with $2^k - 1$ points interpolated between each of the original 2^l observations. The imaginary part is the Hilbert transform of the interpolated real part.

Extending the data file in frequency space by a factor of 2^k reduces the amplitude of the output signal by the same factor. Zeroing out the negative-frequency components reduces the output amplitude by another factor of 2. However, it is generally not necessary to multiply the output signals by the lost factor of 2^{k+1} unless it is desired to plot the interpolated function right through the original observations on the same scale.

The effect of this algorithm on an undersampled data set is shown in Fig. 6.7. In Fig. 6.7(a) is a plot of $\dot{x}$ vs. x; in Fig. 6.17(b) is a plot of the Hilbert transform y vs. x using an 8:1 interpolation.

6.4 EMBEDDINGS

A strange attractor is a geometric structure embedded in a space of dimension $n \geq 3$. The problem of mapping observational data to a strange attractor in R^n is called the *embedding problem*. If the observations comprise n independent time series, in principle there is no problem. Unfortunately, this usually occurs only in numerical simulations of dynamical systems. In most experiments only a single scalar time

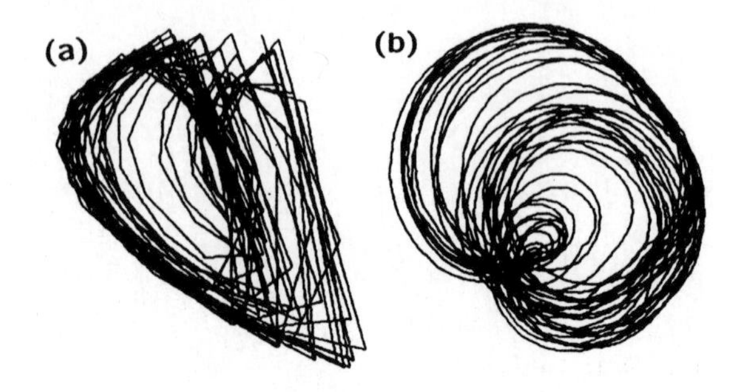

Fig. 6.7 (a) An undersampled data set is shown in an $\dot{x}$ vs. x embedding. The derivatives are simulated by first differences. (b) The data set has been treated to an 8:1 Hilbert transform-interpolation scheme, and y vs. x is plotted.

series is available. This is due to the high cost of recording even a single time series. It is then necessary to find ways of creating additional time series to carry out the embedding. Most of this section is devoted to this important case.

Under some conditions a surplus of data is available. This occurs when two-dimensional data are recorded by a camera or CCD. In these cases, methods must be used that reduce the data to a managable, small number of independent time series. This problem is also treated below. Finally, we describe the role that embedding theorems play in the topological analysis program.

6.4.1 Embeddings for Periodically Driven Systems

In many experiments scalar data $x_i = x(t_i)$ are generated by a dynamical system that is periodically driven with a period T, or P in discrete sample units. A very convenient embedding for this case is the map

$$x(t) \to (x(t), \dot{x}(t), t \bmod T) \quad \text{or} \quad x_i \to (x_i, x_i - x_{i-1}, i \bmod P) \qquad (6.12)$$

This maps the scalar time series into points in $R^2 \times S^1$. Since the phase space is a torus, both linking numbers and relative rotation rates can be computed for periodic orbits.

An $x, \dot{x}, t$ embedding is shown in Fig. 4.15. At any intersection of orbit segments in the x, t projection, the segment with the larger slope has a larger $\dot{x}$ value, and therefore lies closer to the observer. As a result, all crossings in this projection are negative under the usual crossing rules (cf. Section 4.2). This means that computing linking numbers and relative rotation rates reduces simply to counting crossings, since all crossings have the same sign.

In fact, computation of these invariants can be made even simpler. If $x_A(t)$ is an orbit of period p_A and $x_B(t)$ is an orbit of period p_B, their linking numbers and relative rotation rates can be determined simply by determining the zero crossings of the difference $x_A(t) - x_B(t + p \times T)$ ($p = 1, 2, \ldots, p_A \times p_B$). Similarly, the self-linking numbers and self-relative rotation rates are computed by locating the zero crossings of the difference $x_A(t) - x_A(t + p \times T)$ ($p = 1, 2, \ldots, p_A$).

6.4.2 Differential Embeddings

Our background as physicists encourages us to look for cause–effect relations among the components of an embedded vector. Our exposure to dynamical systems encourages us to think in terms of equations involving first derivatives. Our experience with driven dynamical systems suggests a natural method for creating three-dimensional embeddings from scalar data from autonomous dynamical systems. That is, we create an embedding in which each component is the derivative of the previous component:

$$\begin{aligned} x(t) \rightarrow y(t) &= (y_1(t), y_2(t), y_3(t)) \\ &= (x(t), dx/dt, d^2x/dt^2) \end{aligned} \tag{6.13}$$

This embedding leads directly to a set of equations of motion and a canonical way to model the data

$$\begin{array}{rcll} y_1 & = & x(t) & \text{definition of } y_1 \\ dy_1/dt & = & y_2 & \text{definition of } y_2 \\ dy_2/dt & = & y_3 & \text{definition of } y_3 \\ dy_3/dt & = & f(y_1, y_2, y_3) & \text{physics lives here} \end{array} \tag{6.14}$$

The entire modeling process reduces to attempts to construct one unknown scalar function: $f(y_1, y_2, y_3)$.

This differential embedding procedure has strengths and weaknesses. One strength is that it is Newtonian in spirit and leads directly to a simple modeling procedure. Furthermore, there is only one unknown function to estimate, rather than three.

A second strength is that the linking numbers can be computed by inspection in this embedding. In Fig. 6.8 we show two orbits projected onto the x–$\dot{x}$ plane, with $\ddot{x}$ out of the plane of the page. We compute the slopes at the intersection:

$$\text{slope} = \frac{d\dot{x}}{dx} = \frac{d\dot{x}/dt}{dx/dt} = \frac{y_3}{y_2} \tag{6.15}$$

As a result,

$$y_3 = \text{slope} \times y_2 \tag{6.16}$$

This means that the larger the slope at the crossing in the upper ($y_2 > 0$) half plane, the nearer the observer. Conversely, the larger the slope in the lower half plane, the farther from the observer. Thus all crossings in the upper half plane are left-handed and therefore negative by convention. All crossings in the lower half plane are positive.

Another nice feature of the differential embedding is the existence of a Poincaré section. In Fig. 6.9 we show the possible flow directions projected into the y_2–y_3 plane. The flow always crosses from left to right in the half plane $y_2 = 0$, $y_3 > 0$,

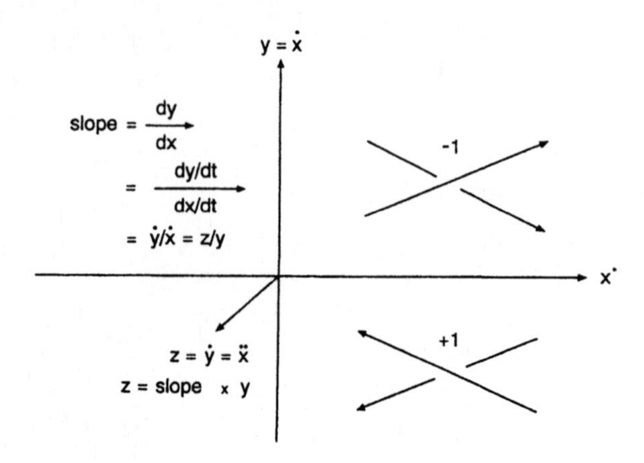

Fig. 6.8 The differential embedding involves three coordinates x, $y = \dot{x}$, and $z = \dot{y} = \ddot{x}$. In the projection onto the x–y plane all crossings in the upper half plane $y > 0$ are left handed and therefore negative. All crossing in the lower half plane are positive.

and in the opposite direction in the other half plane $y_2 = 0$, $y_3 < 0$. We may take either of these half planes as one component of the Poincaré section. We choose the positive half plane $y_3 > 0$, and call this H_1.

The flow may pass through the half plane $y_3 = 0$, $y_2 > 0$ from either direction. Passing from top to bottom provides no information not already provided by H_1. However, passing in the opposite direction, from bottom to top, does provide new information. Thus, as a second component of the Poincaré section we can choose the half plane $y_3 = 0$, $y_2 > 0$ on which $\dot{y}_3 > 0$. We call this half plane H_+. We define H_- similarly. Then

$$\begin{array}{llll} H_1: & y_2 = 0 & y_3 > 0 & \dot{y}_2 > 0 \\ H_+: & y_3 = 0 & y_2 > 0 & \dot{y}_3 > 0 \\ H_-: & y_3 = 0 & y_2 < 0 & \dot{y}_3 < 0 \end{array}$$

$$\mathcal{PS} = H_1 \cup H_+ \cup H_- \tag{6.17}$$

The differential embedding has two serious weaknesses. The first is theoretical, the second is practical.

At the theoretical level, all fixed points for a flow in this embedding must lie on the y_1 axis, $y_2 = y_3 = 0$. They can only occur at the zeros of $f(y_1, 0, 0) = 0$. Thus, this embedding can be used to describe dynamical systems whose fixed-point distribution is one-dimensional (i.e., of rank 1). That is, the singularities are restricted to cuspoid type A_k ($k = 2$: fold; $k = 3$: cusp, etc.). This embedding method will not allow a 1:1 embedding of a three-dimensional dynamical system whose singularity distribution is of type D_k or E_k [37].

At the practical level, this embedding requires construction of the second derivative. As a general rule of thumb, an order of magnitude is lost in the signal-to-noise

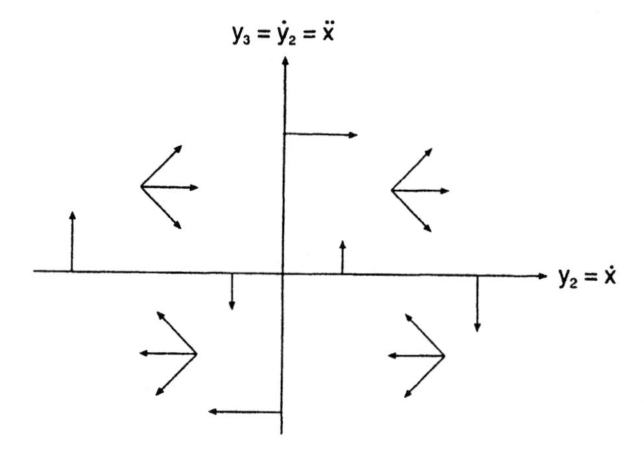

Fig. 6.9 The differential embedding is projected onto the y_2–y_3 or $\dot{x}$–$\ddot{x}$ plane. This projection clearly shows the flow directions. A Poincaré section consists of the union of three half planes: H_1, the half plane $y_3 > 0$, $y_2 = 0$; H_+, the half plane $y_3 = 0$ and $y_2 > 0$, where the flow enters with $\dot{y}_3 > 0$, and H_-, the half plane $y_3 = 0$ and $y_2 < 0$, where the flow enters with $\dot{y}_3 < 0$.

ratio each time a derivative (or integral) is taken. Loss of two orders of magnitude degrades the embedding to the extent that the topological organization of unstable periodic orbits may be difficult to compute, even with very clean data sets.

6.4.3 Differential–Integral Embeddings

Rather than abandon the differential phase-space embedding because of the signal-to-noise problem, we will define another embedding with the same virtues. This is the integral–differential embedding. In this case we define y_1 to be the integral of $x(t)$ and y_3 to be the differential of $x(t)$. To avoid secular trends, we perform the embedding on the zero-mean data set $x(t) - \langle x \rangle$:

$$
\begin{aligned}
y_1(t) &= \int_{-\infty}^{t} [x(t') - \langle x \rangle]\, dt' \\
y_2(t) &= \frac{dy_1}{dt} = x(t) - \langle x \rangle \\
y_3(t) &= \frac{dy_2}{dt} = \frac{dx}{dt}
\end{aligned}
\tag{6.18}
$$

The three variables y_1, y_2, y_3 are related to each other differentially in exactly the same way as in the differential embedding. As a result, this embedding possesses the same desirable features. One of the two undesirable features has now been mitigated. The signal-to-noise ratio of the two variables y_1 and y_3 is now only one order of magnitude smaller than the S/N ratio of the observational data set $y_2(t) = x(t)$.

However, a new headache appears with this embedding. Integration can accentuate secular trends in the data, as described above. This problem can be treated by filtering in either the time domain or the frequency domain. In the time domain we can replace the straightforward integral by an integral with memory loss, by defining

$$y_1(t) = \int_0^\infty x(t - t')e^{-t'/\tau}\, dt' \tag{6.19}$$

The derivative of y_1 must then be modified slightly:

$$\frac{dy_1}{dt} = x(t) - \frac{1}{\tau}y_1(t) \tag{6.20}$$

It is a simple matter to construct this embedding from a discretely sampled data set:

$$\begin{aligned} y_1(i) &= x(i) + e^{-1/\tau}y_1(i-1) \\ y_2(i) &= x(i) \\ y_3(i) &= x(i) - x(i-1) \end{aligned} \tag{6.21}$$

For τ sufficiently large, the three variables remain related to each other differentially. The beauty of this embedding is that it can be done online with simple circuit elements.

6.4.4 Embeddings with Symmetry

The presence of symmetry in an attractor creates possibilities for embeddings that we have not yet encountered. We illustrate these possibilities for the Lorenz equations. These equations are invariant under rotations by π radians about the z axis:

$$(x, y, z) \xrightarrow{R_z(\pi)} (-x, -y, +z) \tag{6.22}$$

This symmetry tells us that the x and y variables behave differently from the z variable. A differential phase-space embedding based on the z variable is shown in Fig. 6.10(a). It is clear from this figure that there is no simple smooth transformation that will map this attractor in a 1-1 way onto the more familiar Lorenz attractor (cf. Fig. 5.8): one has two holes, the other has only one.

On the other hand, a differential phase-space embedding based on the x or y variable has more possibilities. The projection onto the x–$\dot{x}$ plane for Lorenz data is shown in Fig. 6.10(b). The embedding

$$\begin{aligned} y_1 &= x \\ y_2 &= \dot{y}_1 = \dot{x} \\ y_3 &= \dot{y}_2 = \ddot{x} \end{aligned} \tag{6.23}$$

has inversion symmetry:

$$(y_1, y_2, y_3) \xrightarrow{P} (-y_1, -y_2, -y_3) \tag{6.24}$$

As a result, it also cannot be deformed to the familiar Lorenz attractor, which has rotation symmetry. On the other hand, the embedding

$$\begin{array}{rcccl} y_1 & = & x & & \\ y_2 & = & \dot{y}_1 & = & \dot{x} \\ y_3 & = & dy_2^2/dt & = & 2y_2\dot{y}_2 \end{array} \tag{6.25}$$

possesses the desired rotation symmetry

$$(y_1, y_2, y_3) \xrightarrow{R_z(\pi)} (-y_1, -y_2, +y_3) \tag{6.26}$$

suggesting that this embedding can be deformed to the familiar Lorenz attractor.

The difficulty with this embedding is that the dynamical system equations derived using it are singular (non-Lipschitz) unless the driving function f has the form $f(y_1, y_2, y_3) = y_2 g(y_1, y_2, y_3)$, where g is Lipschitz and has the appropriate symmetry.

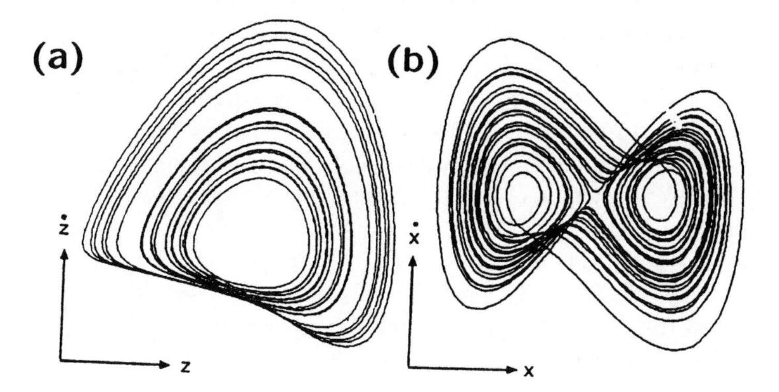

Fig. 6.10 The differential phase-space embeddings for the Lorenz attractor based on (a) the z variable and (b) the x variable are topologically inequivalent. One has one hole, the other two.

6.4.5 Time–Delay Embeddings

The time-delay embedding is the default embedding which has been used for basically *every* fractal dimension estimate that has ever been carried out. This embedding involves creating an n vector from a scalar time series using the map

$$\begin{array}{c} x(t) \rightarrow y(t) = (y_1(t), y_2(t), \ldots, y_n(t)) \\ y_k(t) = x(t - \tau_k) \qquad k = 1, 2, \ldots, n \end{array} \tag{6.27}$$

The parameters τ_k are called *time delays*. These are usually spaced equally (it is a matter of convenience) and expressed as multiples of a single parameter τ, called *the* time delay, so that

$$\tau_k = (k-1)\tau \tag{6.28}$$

but this is not necessary. For discretely sampled time series, the (equally spaced) embedding is

$$\begin{aligned} x(i) \to y(i) &= (y_1(i), y_2(i), \ldots, y_n(i)) \\ y_k(i) &= x(i-(k-1)d) \qquad k = 1, 2, \ldots, n \end{aligned} \tag{6.29}$$

where d is the delay.

The time-delay embedding has several advantages over other embedding methods. One virtue is that it can always be constructed, even if the embedding dimension is larger than necessary. Another very important virtue is that each coordinate in a delay embedding has the same signal-to-noise ratio. For other embeddings the signal-to-noise ratio usually decreases as additional coordinates are added.

This embedding method also has several disadvantages. One major disadvantage is that is that there is no obvious relationship between the coordinates as there is in both differential and integral–differential embeddings. As a result of this disadvantage, when it comes time to fit a quantitative model to the data, three functions must be constructed in place of the single function required for differential embeddings.

A second major disadvantage is that linking numbers and relative rotation rates are not as easy to compute. There are no rules that all crossings are left-handed (periodically driven systems), or that all crossings in the upper (lower) half space are negative (positive).

A third disadvantage is that results of a topological analysis are not independent of the time-delay parameter. In Fig. 6.11 we show time-delay embeddings of experimental data from the Belousov–Zhabotinskii reaction using three different time delays. Figure 6.11(a) shows the embedding using the smallest possible time delay: $d = 1$. This embedding uses triples of successive observations. Figure 6.11(c) shows the embedding using a time delay comparable to one cycle time T. Figure 6.11(b) shows the attractor obtained using an intermediate time delay. Roughly speaking, the attractor seems to be confined to a "fat" two-dimensional manifold that undergoes self-intersection as the time delay is increased from $d \simeq 0.01T$ in Fig. 6.11(a) to $d \simeq \frac{3}{4}T$ in Fig. 6.11(c). At the self-intersection, invariance of the topological indices is lost. So also is the uniqueness theorem.

Dependence of topological invariants on the time delay has been described in two related studies. Mindlin and Solari [111] studied a three-dimensional delay embedding in which $\tau_1 = 0$, τ_2 was fixed, and τ_3 was allowed to vary. They found values of τ_3 for which the attractor undergoes apparent self intersections and other regions without self intersections. The topological invariants were different on different sides of the self-intersection regions. Worse, periodic orbits that had zero topological entropy (cf. Section 4.5.2) for τ_3 small turned up with different symbolic names on the other side of the region of self-intersection. The new symbolic names suggested that these orbits had positive topological entropy. That there is no inconsistency is a consequence of a subtle theorem dealing with the topological properties of the map to the Poincaré section. That is, changing the time delay changes the boundary conditions on the mapping of the Poincaré section to itself. This in turn changes the properties of braids in the flow. In a related study, Mancho, Duarte, and Mindlin [112] studied a parametrically forced oscillator and found results comparable to those shown in Fig. 6.11 and those reported by Mindlin and Solari.

Delay embeddings with minimal delay are closely related to differential embeddings. The first k derivatives of a discretely sample time series are related to $k+1$ successive observations by simple difference rules. In particular, for a three-dimensional embedding,

$$\begin{aligned} x(i) &= x(i) \\ dx(i)/dt &\simeq x(i+1) - x(i-1) \\ d^2x(i)/dt^2 &\simeq x(i+1) - 2x(i) + x(i-1) \end{aligned} \tag{6.30}$$

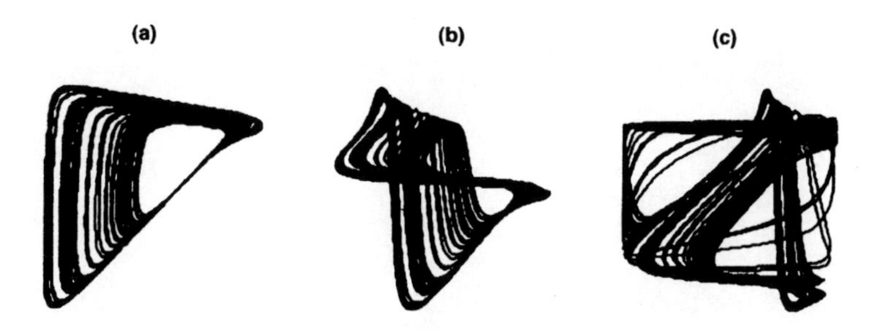

Fig. 6.11 These three embeddings of experimental data use different time delays: (a) minimal delay; (b) intermediate delay; (c) $\tau \sim \frac{3}{4}T$. The attractor undergoes self-intersection between (a) and (c). Topological invariants for the same orbits differ in the two embeddings (a) and (c).

6.4.6 Coupled–Oscillator Embeddings

At the present time the methods of topological analysis are restricted to three-dimensional phase spaces because knots "fall apart" in higher dimensions. There is the hope that topological methods will be developed so that stretching and squeezing mechanisms can be identified in higher-dimensional spaces. For this reason it is useful to search for embedding methods that can be extended to higher dimensions.

It is not likely that the differential embedding technique can be extended much beyond three dimensions because of the signal-to-noise problem. Therefore, we have to rely on other embedding methods. One natural fallback is the time-delay method. However, it is lacking in dynamics.

Another embedding method, with more dynamics, involves a combination of time delay and Hilbert transform pairs. In this procedure, 4, 6, ..., $2n$-dimensional embeddings are created from scalar data sets by the mapping

$$x(t) \to z(t) \to (z_1(t), z_2(t), \ldots, z_n(t)) \tag{6.31}$$

The first step is to create the complex time series $z(t)$ whose real part is $x(t)$ and imaginary part is its Hilbert transform $y(t)$. This is mapped into a complex n-vector by time-delay methods. The complex variables are related to each other by a standard delay [2]

$$z_{j\pm 1}(t) = z_j(t \pm \tau) \tag{6.32}$$

where τ is about $\frac{1}{4}$ of the characteristic cycle time of the time series.

The motivation behind this coupled oscillator representation is as follows: Each complex coordinate $z_j(t)$ behaves more or less like the coordinates $[x_j(t), y_j(t)]$ of a nonlinear oscillator. Adjacent oscillators $z_j(t)$ and $z_{j\pm 1}(t)$ which are $\pi/2$ radians out of phase might be expected to interact strongly with each other.

This embedding possesses the strength of the time-delay embedding (equal signal-to-noise ratio in all real coordinates) as well as the strength of the differential phase-space embedding (strong dynamic coupling between x_j and y_j; probably strong coupling between z_j and $z_{j\pm 1}$). Its weakness is that modeling the dynamics will involve n complex functions of n complex variables:

$$\frac{dz_j}{dt} = f_j(z_1, z_2, \ldots, z_n) \tag{6.33}$$

The functions $f_j(z)$ cannot be expected to be analytic in the n complex variables z.

6.4.7 SVD Projections

Some data sets include spatial and temporal signals. Such data sets have the form $d(i, r)$, where $i, j, \ldots$ are indices describing spatial position (in either one or more dimensions), and $r, s, \ldots$ are temporal indices. We are then blessed with the curse of too much information. It is necessary to reduce this large data set to three time series which can be embedded in a three-dimensional space.

The singular-value decomposition (SVD) provides a useful way to extract a small number of time series from a large amount of data. We treat the data set $d(i, r)$ as an $S \times T$ matrix, where $1 \leq i \leq S$ and $1 \leq r \leq T$. The method is based on an elegant theorem of linear algebra. This theorem states that it is always possible to write an $S \times T$ matrix in the form of a sum over outer products of pairs of dual vectors in the form

$$d(i, r) = \sum_{\alpha=1}^{\text{Min } S,T} \mu(\alpha) u(i, \alpha) v(r, \alpha) \tag{6.34}$$

The S-dimensional vector $u(i, \alpha)$ is dual to the T-dimensional vector $v(r, \alpha)$. The index α describes distinct modes. The vector $u(i, \alpha)$ describes the αth spatial mode, while its conjugate $v(r, \alpha)$ describes the temporal evolution of that mode. The eigenvalue $\mu(\alpha)$ is the amplitude with which the mode α occurs in the data set $d(i, r)$.

The spatial modes are mutually orthonormal, as are the dual temporal modes:

$$\sum_{i=1}^{S} u(i, \alpha) u(i, \beta) = \delta_{\alpha,\beta} \qquad \sum_{r=1}^{T} v(r, \alpha) v(r, \beta) = \delta_{\alpha,\beta} \tag{6.35}$$

Several stable computer algorithms exist for effecting a singular-value decomposition. A robust method involves constructing a real symmetric $S \times S$ matrix:

$$M(i,j) = \sum_{r=1}^{T} d(i,r)d(j,r) \tag{6.36}$$

This matrix can then be diagonalized using any of the methods suitable for symmetric matrices. The eigenvalues of this positive semidefinite matrix are nonnegative, and are

Eigenvalue	Eigenvector
$\lambda(\alpha)$	$u(i,\alpha)$

(6.37)

It is useful to order the eigenvectors by decreasing size of their eigenvalues: $\lambda(1) \geq \lambda(2) \geq \cdots$. The dual eigenvector $v(r,\alpha)$ is constructed simply by taking the inner product

$$\sqrt{\lambda(\alpha)}v(r,\alpha) = \sum_{i=1}^{S} u(i,\alpha)d(i,r) \tag{6.38}$$

A similar result is obtained by starting from the $T \times T$ matrix $M'(r,s) = \sum_i d(i,r)d(i,s)$. It is more efficient to diagonalized the smaller of the $S \times S$ matrix $M(i,j)$ or the $T \times T$ matrix $M'(r,s)$.

The nonzero eigenvalues $\lambda(\alpha)$ of the real symmetric matrices $M(i,j)$ and $M'(r,s)$ are equal. They are related to the amplitudes $\mu(\alpha)$ of the original rectangular matrix $d(i,r)$ by

$$\lambda(\alpha) = \mu(\alpha)^2 \tag{6.39}$$

This amplitude-intensity relation allows us to provide a useful probability interpretation to the results of the SVD. Define the norm, Z, of the data set by

$$Z = \sum_{i=1}^{S}\sum_{r=1}^{T} d(i,r)^2 = \text{tr } M(i,j) = \text{tr } M'(r,s) = \sum_{\alpha=1}^{\text{Min } S,T} \lambda(\alpha) = \sum_{\alpha=1}^{\text{Min } S,T} \mu(\alpha)^2 \tag{6.40}$$

Then $\Pr(\alpha) = \lambda(\alpha)/Z = \lambda(\alpha)/\sum_\beta \lambda(\beta)$ is the probability that the spatial-temporal mode $u(i,\alpha)v(r,\alpha)$ occurs in the original data set $d(i,r)$.

To summarize, the results of a singular-value decomposition of a data set $d(i,r)$ are:

Probability Amplitude	Relative Probability	Probability	Spatial Mode	Time Series
$\mu(\alpha)$	$\lambda(\alpha) = \mu(\alpha)^2$	$\lambda(\alpha)/Z$	$u(i,\alpha)$	$v(r,\alpha)$

(6.41)

Here $Z = \sum_\alpha \lambda(\alpha)^2$ is the "partition function," or norm $\sum_i \sum_r d(i,r)^2$, of the data set.

If the sum of the three largest probabilities Pr(1)+Pr(2)+Pr(3) is sufficiently large (e.g., greater than 0.90), most of the information in the original data set is preserved if only the three largest modes are retained. Then the time series that can be used for a three-dimensional embedding describing the dynamics are the vectors $v(r, \alpha)$, $\alpha = 1, 2, 3$, providing that no self-intersections occur.

Remark: The spatial eigenvectors $u(i, \alpha)$ are called *empirical orthogonal functions*. They are the normal modes of the system, determined directly from experimental data. If r is a one-dimensional index, they are easy to interpret visually. If r is a two- or three-dimensional index, the construction and interpretation of these modes can be subtle.

6.4.8 SVD Embeddings

The singular-value decomposition can also be used to construct three-dimensional embeddings from scalar data sets [110]. Starting from a scalar data time series $s(r)$, $1 \leq r \leq T$, a k-dimensional embedding is constructed by forming a set of k-vectors

$$s(r) \rightarrow \mathbf{v}(r) = (v(1,r), v(2,r), \ldots, v(k,r)) \qquad v(i,r) = s(r+i-1) \tag{6.42}$$

The $k \times (T - k + 1)$ matrix of data $v(i, r)$ can then be treated exactly as described above. Choosing the dimension, k, of the embedding and the delay [replacing $(i - 1)$ by $n(i - 1)$, where $n = 2$ or 3 or $\cdots$] can be elevated to an art form in the search for the ultimate three-dimensional embedding.

Remark: Embeddings created using SVD methods do not necessarily have nice topological properties, as do differential embeddings or x-$\dot{x}$-t embeddings for periodically driven systems.

6.4.9 Embedding Theorems

The embedding theorems of use for the analysis of dynamical systems have two parts [113–115]. One part involves the reconstruction of a faithful n-dimensional dynamics from a single, or else a small number of, time series. The second part involves mapping n-dimensional dynamics into manifolds of higher dimension.

The second part of this program can be understood in terms of the idea of transversality. If two manifolds of dimensions d_1 and d_2 are invertibly mapped into a manifold of dimension D, then typically (generically [37]) either:

1. They do not intersect, or

2. If they do intersect, their intersection is a manifold of dimension $d = d_1 + d_2 - D \geq 0$. Under perturbation, their intersection remains a manifold of dimension $d \geq 0$.

If $d_1 + d_2 - D < 0$, then typically the two manifolds do not intersect. If they do, it is an accident (*nongeneric*) and the intersection is removed by an arbitrary perturbation.

Example: Two one-dimensional curves in a plane either do not intersect or, if they do, intersect at a point $(1 + 1 - 2 = 0)$ or several isolated points. The intersection(s) cannot be removed by a small perturbation. Alternatively, a one-dimensional curve may have nonremovable zero-dimensional self-intersections when mapped into a plane. However, a one-dimensional curve typically will not have self-intersections when mapped into R^k, $k > 2$ since $1 + 1 - k < 0$. Any self-intersection in this larger space is accidental and can be removed by an arbitrarily small perturbation.

The part of this theory that is useful for dynamical systems is as follows. An n-dimensional dynamical system is described by a flow in an n-dimensional manifold. If this manifold is mapped into a k-dimensional manifold, for example R^k, then self-intersections are possible unless $k > 2n$. Self-intersections are bad for dynamical systems: At a self-intersection the uniqueness theorem is not valid. Therefore, embeddings of dynamical systems must avoid self-intersections. Transversality guarantees that an n-dimensional dynamical system generically avoids self-intersections when $k \geq 2n + 1$.

The first part of the embedding program addresses the question of how to reconstruct an n-dimensional dynamics from a single time series. Whitney suggests using the original time series and its first $n - 1$ derivatives. These derivatives can be approximated from the scalar time series by taking appropriate differences (first differences, second differences, ...). The time series and its first $n - 1$ derivatives contain information equivalent to that contained in an n-vector whose components are $s(i), s(i + 1), s(i + n - 1)$. This delay embedding was proposed by Packard et al. [114] and by Takens [115]. However, this n-dimensional reconstruction of dynamics might involve self-intersections. These can be avoided by embedding the dynamics in a space of dimension $2n + 1$, using the embedding $s(i) \to \mathbf{v}(i) = (s(i), s(i + 1), \ldots, s(i + 2n))$.

For an n-dimensional dynamical system, the actual dynamics occurs in a subspace of R^n, so that it might be anticipated that the bound $k \geq 2n + 1$ is too conservative. If motion occurs on a fractal attractor of dimension d_A, it might be anticipated that a better bound would be $k > 2d_A$ or $k = [2d_A] + 1$. This result was shown to be true for $d_A = D_0$ in [116].

For the Lorenz attractor, $n = 3$, so the earlier bounds guaranteed that an embedding always exists for $k = 2 \times 3 + 1 = 7$. At the usual parameter values, the Lorenz attractor has $D_0 \simeq 2.06$, so that a tighter bound is $k = 5$. However, since the Lorenz attractor is three-dimensional, we would prefer an embedding with $k = 3$. Although not guaranteed by embedding theorems, it is not forbidden, either.

Remark: Embedding theorems are useless for the topological analysis program. If the dynamics cannot be squeezed into a three-dimensional manifold, the topological organization of the unstable periodic orbits in the strange attractor cannot be computed. We prefer to believe that if a dynamical system is three-dimensional, one of the embedding procedures described above will provide a three-dimensional embedding without self-intersections. The fact that a seven- or five-dimensional embedding of Lorenz dynamics can be guaranteed by embedding theorems is of no use to us—only a three-dimensional embedding is useful. The embedding methods described above have never failed to provide us with a three-dimensional embedding for data generated by a three-dimensional dynamical system [i.e., data flows on a $(2+\epsilon)$-dimensional strange attractor].

6.5 PERIODIC ORBITS

Unstable periodic orbits that reside in a strange attractor can be located by the method of close returns. Segments can then be extracted from the data set and used as surrogates for unstable periodic orbits. These surrogates are used to compute linking numbers. If the strange attractor is embedded in a torus $D^2 \times S^1$, more refined topological invariants, the relative rotation rates, can also be computed from these surrogates. The topological invariants are used as fingerprints for strange attractors [117].

6.5.1 Close Returns Plots for Flows

Close returns plots are based on the observation that the difference $|\mathbf{x}(t) - \mathbf{x}(t+\tau)|$ remains small over some time interval $t_i \leq t \leq t_j$ when $\mathbf{x}(t)$ is near a periodic orbit of period τ. Searches for periodic orbits based on this observation were first proposed by Eckmann and Ruelle [118] in terms of recurrence plots, which are plots of $\Theta(\epsilon - |\mathbf{x}(t_i) - \mathbf{x}(t_j))|)$ as a function of times t_i and t_j. Here Θ is the Heaviside, or indicator, function: $\Theta(x) = 0$ if $x \leq 0$, $\Theta(x) = 1$ if $x > 0$. In recurrence plots, the (i, j) pixel is colored black if $|\mathbf{x}(t_i) - \mathbf{x}(t_j)| \leq \epsilon$ and white otherwise. The parameter ϵ is usually chosen to be about 1% of the diameter of the strange attractor. In such plots the diagonal $t_i = t_j$ is black. Close returns segments appear as line segments parallel to the diagonal.

We have modified recurrence plots and instead plot $\Theta(\epsilon - |\mathbf{x}(t) - \mathbf{x}(t+\tau))|)$ as a function of t, the location in the data set, and τ, the period. Close returns segments appear as horizontal line segments, which are more easily recognized visually than segments parallel to the diagonal. In addition, their location in the close returns plot easily identifies their location in the data set and the period of the surrogate orbit.

Close returns plots and recurrence plots are normally carried out on vectors $\mathbf{x}(t) \in R^n$, which are coordinates of points in a strange attractor. If only a scalar data set is available, it might seem that an embedding of the data is required before a close returns plot can be constructed. This is not so. An embedding of a scalar data set contains no information not already contained in the original data set. As a result, close returns plots on scalar data are often effective in locating surrogates for unstable periodic orbits. In Fig. 6.12 we show a close returns plot of a scalar time series obtained from the Belousov–Zhabotinskii reaction. The horizontal line segments indicating periodic orbits stand out clearly. The upward-curving segments occur as rising ($\dot{x} > 0$) parts of the data set cross descending ($\dot{x} < 0$) parts. Such artifacts are examples of false nearest neighbors [100, 101]. They are easily removed by embedding the data.

The method of close returns becomes increasingly effective in locating surrogate periodic orbits as the dimension, $2+\epsilon$, of the strange attractor approaches 2. The argument can be seen in the context of the proof of the Poincaré–Bendixon theorem. This theorem states that if a flow in R^2 is bounded and does not decay to a fixed point,

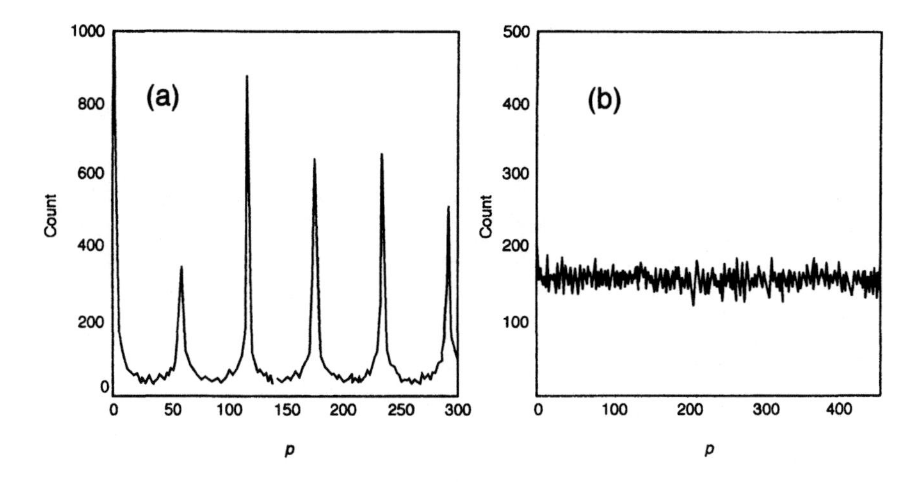

Fig. 6.12 Close returns plot $\Theta(\epsilon - |x(i) - x(i+p)|)$ of scalar time series from the Belousov–Zhabotinskii reaction. Horizontal line segments indicate regions of the data set that are near unstable periodic orbits.

it follows a periodic orbit. The idea of the proof is to construct a line segment in the plane that the flow returns to again and again. Then Cauchy-type arguments are used to show that the intersections converge to a point on this line. This limit point is the intersection of the periodic orbit with a Poincaré section.

For a strange attractor of dimension $2 + \epsilon$, the analogous Poincaré section has dimension $1 + \epsilon$. The smaller ϵ, the closer the return flow comes to intersecting a line, and the easier it is to locate the close returns. As ϵ approaches 1, the Poincaré section approaches a plane, and the more difficult it becomes for the flow to exhibit close returns.

Remark: In some way, it is poetic justice that it is only in low dimensional attractors $2 + \epsilon < 3$ that the method of close returns is useful for locating periodic orbits. In higher dimensions this method is useless—on the other hand, so are periodic orbits, since they fall apart.

6.5.1.1 Close Returns Histograms When there are more data than can conveniently be fitted on the screen of a computer, a histogram of the close returns plot is useful for determining the periods of surrogates for closed orbits. A close returns histogram is conveniently generated by "summing sideways" in the close returns plot. The close returns histogram is defined explicitly by

$$H(t_j) = \sum_i \Theta(\epsilon - |\mathbf{x}(t_i) - \mathbf{x}(t_i - t_j)|) \tag{6.43}$$

The close returns histogram for the close returns plot of Fig. 6.12 is shown in Fig. 6.13(a). It is obtained simply by counting the number of black pixels as a function of the time difference, p. This histogram clearly shows the presence of closed orbits whose period is an integer multiple of some smallest value p_1 of p, which can be taken as the duration of a period-1 orbit, counted in terms of the sample time.

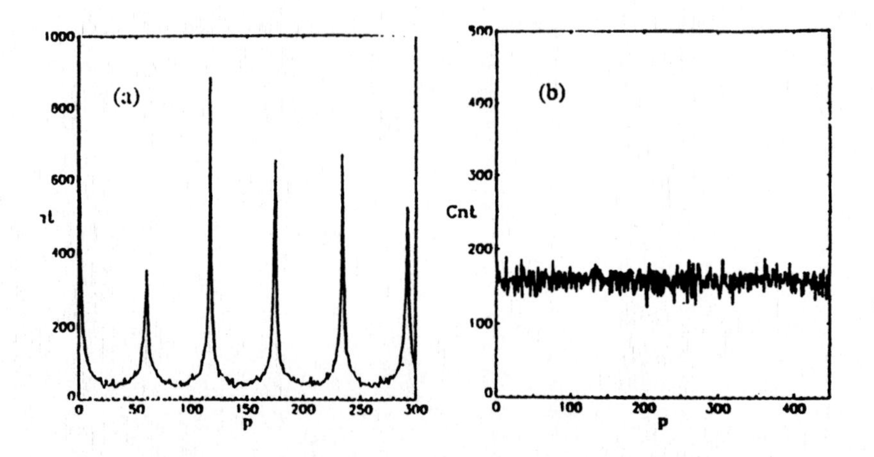

Fig. 6.13 (a) Close returns histogram for the close returns plot shown in Fig. 6.12. The peak at the smallest time-delay produces a good estimate of both the cycle time and the period of the lowest (period-1) orbit. (b) Close returns histogram of a stochastic data set. The distribution is compatible with a uniform distribution. Standard statistical tests can be used to test the null hypothesis H_0 that these distributions are uniform. In case (a) the null is resoundingly rejected; in case (b) the null cannot be rejected.

6.5.1.2 Tests for Chaos Close returns histograms have been used to distinguish chaotic from stochastic behavior [119,120]. The close returns plot for a chaotic flow, shown in Fig. 6.12, differs from a close returns plot for a stochastic data set. Such a plot appears to be a distribution of black pixels uniformly distributed over the area of the plot. Histograms of the two close returns plots are also markedly different. A close returns histogram for a stochastic data set (Gaussian random numbers $[N(0,1)]$) is shown in Fig. 6.13(b). This histogram is essentially a uniform distribution. Standard statistical tests [99] can then be used to test the null hypothesis H_0 that the histogram $H(t_j)$ is uniform. These tests were applied to the two histograms shown in Fig. 6.13. The null hypothesis H_0 was thoroughly rejected in the case of the chaotic data set. We failed to reject the null for the stochastic data set.

Statistical tests cannot be used to prove that a data set is chaotic, but they can be used to reject the alternative hypothesis that it is stochastic. At the present time, the most convincing way to prove that a data set is chaotic is to identify at least one unstable periodic orbit with positive topological entropy (Section 2.9.2).

6.5.2 Close Returns in Maps

Close returns searches can be applied to maps as well as flows. In fact, they are more easily implemented for maps than for flows. To create a map from flow data, it is necessary to construct a Poincaré section. For our preferred embedding of scalar data (the differential embedding), Poincaré sections always exist and can easily be found. One is simply defined by $\dot{x} = 0$ for $\ddot{x} < 0$. From a long file of flow data, we construct a file (M_i, t_i) containing the values of the ith local maximum, M_i, and the location in the data set, t_i, at which this maximum occurs. For a file of length 10^4 and a sampling rate of about 10^2 samples/cycle, the original flow file of length 10^4 is compressed to a file containing 10^2 local maxima, along with their locations in the initial data set.

6.5.2.1 First Return Map A first return map $M_i \rightarrow M_{i+1}$ can be constructed from the compressed file. The first return map can be used to locate period-1 orbits in the data set. The method is to plot M_{i+1} vs. M_i, and then search for points in this plot that lie closest to the diagonal. Two first return plots are shown in Fig. 7.3 for data from the Belousov–Zhabotinskii reaction. In Fig. 7.3(a) the Poincaré section has been taken at the minima; in Fig. 7.3(b) it has been taken at the maxima. The two figures are first return plots of successive minima (m) and maxima (M). The points closest to the diagonal identify the best surrogates for period-1 orbits. If the point (m_j, m_{j+1}) lies closest to the diagonal, the segment between t_j and t_{j+1} in the original flow data provides the best surrogate for a period-1 orbit starting and ending at a minimum.

In general, it is useful to plot several surrogate period-1 orbits. The surrogates may fall into classes, each class representing a different period-1 orbit. It is usually possible to distinguish among classes by eyeball. The best surrogate within each class should be retained. Several surrogates within a class can be used to determine the local torsion of the unstable periodic orbit for which they are all surrogates.

The first return plot can often be used to identify an alphabet of symbols to label the orbits. Critical points, or their approximations, separate parts of the return map that are labeled by different letters. The return maps for the Belousov–Zhabotinskii reaction, shown in Fig. 7.3, exhibit a single critical point (minimum or maximum). This suggests that a two-letter alphabet is sufficient to label the flow in this dynamical system.

In general, the more dissipative a system is (i.e., the smaller ϵ), the easier it is to make an error-free identification between data segments and symbols. Ultimately, orbit identification by symbol sequence must be consistent with template identification.

Remark: Having a symbolic encoding for the periodic orbits is useful in guessing the structure of a branched manifold, but ultimately it is not a necessary input. What is necessary is a set of periodic orbits with their associated linking numbers. The output of a search for a branched manifold is a consistent solution of the simultaneous problems of identifying a symbolic dynamics for the periodic orbits and the branches of a template that reproduces the spectrum of topological invariants. A general algorithm for accomplishing this task is described in Appendix A.

6.5.2.2 pth Return Map Surrogates for unstable period-p orbits can be obtained in the same way from pth return maps. All values of t_i for which $|M_i - M_{i+p}|$ is sufficiently small are candidates for initial conditions for unstable period-p orbits. When possible, the symbol sequence for these orbits should be determined, as many topologically inequivalent period-p orbits can exist in a flow.

The search for period-p orbits can turn up period-$p/2$ orbits that "go around twice," period-$p/3$ orbits if p is divisible by three, and so on. Finding such orbits is not useless, for they make computation of the local torsion of the period-$p/2$ orbit very simple.

By slowly increasing the rejection threshold for $|M_i - M_{i+p}|$, it is possible to find several representatives of a period-p orbit. Usually, we choose the best (smallest $|M_i - M_{i+p}|$) as the surrogate orbit. However, we can often use the other surrogates for the same orbit to determine the local torsion of the nearby unstable period-p orbit.

6.5.3 Metric Methods

For the close returns procedure to work, the unstable Lyapunov exponent cannot be too large. A useful rule of thumb for period-p orbits is $\lambda(t_{i+p} - t_i) < 1$, where λ is the positive. If the Lyapunov exponent is too large, it is not very likely that an initial condition near a period-p orbit will evolve in the neighborhood of the period-p orbit for the entire time interval $t_{i+p} - t_i$. In this case it is still possible to find surrogates for period-p orbits. We have used the following method successfully when necessary, but prefer to avoid it when possible.

The basic idea is to push the search for periodic orbits into the symbol-sequence space. The first step is to encode the entire time series by a symbol sequence $\sigma_1\sigma_2\sigma_3\cdots\sigma_N$, where the data between t_i and t_{i+1} is encoded by the symbol σ_i. As above, t_i is the time of the ith intersection of the flow with a Poincaré section. As mentioned above, the identification of a symbol with a data segment is often simple after a first return map has been created.

Now suppose that the symbols are drawn from a small alphabet A, B, C and that the branches are ordered alphabetically in a template: $A < B < C$. To find a period-2 orbit AB we proceed as follows. The future of A is $(A)BABAB\cdots$. We identify a symbol A in the symbol sequence representing the data and compare its future with the future of the symbol A in the period-2 orbit AB:

$$
\begin{array}{lcccccccccccc}
\text{Data} & \hat{A} & B & A & B & C & \tilde{A} & B & A & B & B & \cdots \\
\text{Period 2} & A & B & A & B & A & \cdots & & & & & \\
\text{Period 2} & & & & & & A & B & A & B & A & \cdots
\end{array}
\qquad (6.44)
$$

The segment of data starting from $\hat{A}$ agrees with the period-2 orbit for the next three periods, but is different (C vs. A) at the fourth period in the future. On the other hand, the segment of data starting at $\tilde{A}$ is also different at the fourth place (B vs. A), but is less different (B vs. C) than the future of $\hat{A}$. On this basis the data segment starting from $\tilde{A}$ would be more like A in the period-2 orbit AB than the data segment starting at $\hat{A}$.

Forward and backward objective functions measuring the degree of similarity can be defined. By optimizing these objective functions in some useful way, it is possible to choose from the data set the best approximation to each of the p symbols in a period-p orbit. This search can be carried out as follows [32, 121]: The distance between two symbol sequences $\mu_1\mu_2\cdots\mu_n$ and $\nu_1\nu_2\cdots\nu_n$ at symbol i can be defined by

$$d_i(\mu,\nu) = \sum_{k=0,\pm1,\pm2,\ldots} \frac{1}{2^{|k|}} f^{|k|}(\mu_{i+k},\nu_{i+k}) \tag{6.45}$$

where $f(\mu_j,\nu_j) = 0$ if $\mu_j = \nu_j$ and has some nonzero value otherwise, depending on how far apart μ_j and ν_j are. There are many closely related ways to impose a metric on the space of symbols.

If suitable thresholds are satisfied for all symbols in the symbol name for a period-p orbit, it is possible to string the associated p segments of data together to get a reasonable surrogate for the period-p orbit. One should plot the orbit $x_{\sigma_i}(t)$ vs. t to verify that it is a suitable surrogate. That is, one should verify that the discontinuities between contiguous segments are sufficiently small that the surrogate actually looks like a continuous trace. If one (or more) of the thresholds is not satisfied, no surrogate for the period-p orbit can be constructed by this method, and we can assume that this unstable periodic orbit is "not in the data."

We have used this method to construct surrogates for orbits of periods 6, 7, and 8 which are predicted to exist in the Belousov–Zhabotinskii data set but were not recovered by the method of close returns. These orbits are discussed more fully in Section 9.5 (cf. Fig. 9.10).

6.6 COMPUTATION OF TOPOLOGICAL INVARIANTS

The central step in the topological analysis program is the determination of the topological organization of the unstable periodic orbits in a low-dimensional strange attractor. This organization is determined by the linking numbers and relative rotation rates of these orbits.

6.6.1 Embed Orbits

These topological invariants of unstable periodic orbits can be computed only in three dimensional spaces. The first step in their computation involves embedding the strange atractor in R^3 or some other three-dimensional manifold. A number of different embedding procedures have been described in Section 6.4. Our preference for nondriven systems runs to the x-$\dot{x}$-$\ddot{x}$ embedding, or the integral–differential embedding, as described in Section 6.4.3. The reason is that crossings are easily determined simply by noting whether they occur in the upper or lower half space. For periodically driven systems our preference is for x-$\dot{x}$-t mod 2π embeddings, since all crossings are the same sign in this embedding.

6.6.2 Linking Numbers and Relative Rotation Rates

Linking numbers are computed by counting signed crossings, as described in Section 4.2. In practice, surrogates for two unstable periodic orbits are overlaid. It is useful to plot the surrogates in different colors. The sign of each crossing of one orbit by the other is determined. The linking number of the two orbits is then half the sum of the signed crossings.

When the strange attractor is embeddable in a torus $R^2 \times S^1$ or $D^2 \times S^1$, relative rotation rates can also be computed. Four methods for computing these topological invariants have been presented in Section 4.3.

6.6.3 Label Orbits

It is necessary to assign a name to each of the surrogates for unstable periodic orbits extracted from the data. This is because we eventually wish to identify each surrogate with a periodic orbit in a branched manifold. All orbits on a branched manifold are uniquely labeled by the (order of the) branches through which they travel.

A preliminary identification of surrogates can be made by locating the successive intersections in a Poincaré section. If the first return map on a Poincaré section has well defined critical points, then each segment of the return map between critical points can be labeled by a letter in a small alphabet. For example, the return maps shown in Fig. 7.3 strongly suggest that two letters are sufficient. The location of each of the p intersections of the surrogate for an unstable periodic orbit A in one or the other of the two segments of this return map provides a preliminary identification of the name of the orbit A. Identifications of orbit segments with intersections near the critical points may be problematic and subject to "renormalization" in subsequent work. This preliminary labeling of orbits by symbol sequences is increasingly reliable as the dissipation increases, that is, the closer the dimension of the strange attractor $2 + \epsilon$ approaches 2.

Once again, the orbit labeling and template identification problems constitute a single global problem. It is convenient, but not necessary, to have symbol names for periodic orbits in order to identify a branched manifold. See Appendix A for a global solution to this problem.

6.7 IDENTIFY TEMPLATE

Periodic orbits that have been extracted from the data can be used to identify the underlying branched manifold. The number of branches is equal to the number of letters in the alphabet needed to label the orbits extracted from the data.

6.7.1 Period-1 and Period-2 Orbits

If the dynamics occurs in a torus $D^2 \times S^1$, the orbits of period 1 and period 2 are sufficient to identify the template. Each branch contains one period-1 orbit. The linking numbers and local torsions determine the matrix elements of the topological matrix according to Section 5.3. The order in which the branches are connected at branch lines is determined from the linking numbers of the period-2 orbits. The grammar that the alphabet obeys defines the transition matrix.

6.7.2 Missing Orbits

It often occurs that the full spectrum of N period-1 orbits and $N(N-1)/2$ period-2 orbits predicted by a branched manifold with N branches cannot be extracted from the data. In these cases, the missing information can be supplied by using orbits of higher period. For example, in horseshoe dynamics with symbolic dynamics $0, 1$, it often happens that the period-1 orbit 0 is missing, while the other period-1 orbit 1 and the single period-2 orbit 01 are present. In such cases, one of the two period-3 orbits 011 or 001 can be used in place of the missing orbit 0. If neither of the period-3 orbits is available, some higher-period orbit can be used.

6.7.3 More Complicated Branched Manifolds

If the strange attractor cannot be embedded in a torus, it is no longer true that each branch must contain a period-1 orbit. An example is shown in Fig. 5.2. In such cases the Poincaré section consists of the union of two or more disjoint two-dimensional surfaces in the embedding space. Branches flow from one of these disjoint surfaces to another, possibly the same, surface.

Orbits of low period can be used to construct the topological matrix T. Branches that flow from one component of the Poincaré section back to the same component each contain one period-1 orbit. These can be analyzed as described above for strange attractors that can be embedded in a torus. Branches joining different components of the Poincaré section are analyzed as follows. Two or more segments in one such branch are used to compute a crossing number, as described in Section 4.2. This index is the local torsion of that branch. The crossing index for one segment in each of two different branches determines the crossing index of these branches. These observations are sufficient to determine the topological matrix of the branched manifold. The joining array determines the order in which branches are joined at each branch line or on each component of the Poincaré section. The joining array is determined by computing the linking numbers of some closed orbits of low period. The transition matrix is determined by inspection of the symbol sequence.

6.8 VALIDATE TEMPLATE

Once a template has tentatively been identified, the identification must be confirmed. This step is analogous to the χ^2, or goodness-of-fit, test which should conclude any data-fitting problem for a general linear model. However, the template identification step for nonlinear systems is more stringent than the χ^2 test for linear systems. This comes about because a template carries a lot of predictive capability. Unless *all* its predictions are consistent with the data, the original identification must be rejected.

6.8.1 Predict Additional Toplogical Invariants

Once a template has tentatively been identified from a small number of surrogate periodic orbits, it can be used to predict the topological organization for all the unstable periodic orbits described by the template. These predictions can be made algorithmically. FORTRAN and C++ codes for these predictions are available at the authors' web sites.

6.8.2 Compare

These predictions must be compared with the topological organization of all remaining surrogates extracted from the data. If all predictions based on the template are compatible with the topological organization determined from surrogate orbits, the template can be accepted (more formally, the template "cannot be rejected"). Otherwise, the original identification must be rejected.

6.8.3 Global Problem

The identification of a symbolic dynamics for periodic orbits and the naming of a template consistent with the spectrum of orbits observed in the data is a global, nonlocal problem. Our experience has been that most of the surrogate periodic orbits are easy to identify, based on the intersections with a Poincaré section. However, intersections near the critical points of a return map sometimes induce problems. We have chosen to label segments with such intersections by special symbols (e.g., ?) and then compute the topological organization of such orbits by replacing this symbol with the two nearby possibilities. For example, suppose that the critical point separates branches labeled 0 and 1, and we find an orbit with symbol sequence 0010?1. We compute linking numbers of the two possible orbits 001001 and 001011 with all other surrogates, and compare these two sets with the linking numbers of the mystery orbit 0010?1 with all other surrogates. This comparison provides a useful way to resolve any uncertainties that come up in identifying surrogates. It has always resolved the intertwined global problems of orbit labeling and template identification.

6.9 MODEL DYNAMICS

The topological analysis program ends with the steps described above. However, we usually prefer to proceed beyond the qualitative understanding of a dynamical system provided by topology, to an analytic understanding of the system as provided by a quantitative model [90].

The first step in this refined understanding involves the construction of a quantitative model. A dynamical system model is proposed. At the present time this is always a dynamical system model of the form (6.2):

$$\frac{dx_i}{dt} = F_i(x,c) = \sum_{\beta=1}^{n} A_{i\beta}\Phi_\beta(x;c) \tag{6.46}$$

It should be emphasized here that this is a *general linear model* (it is linear in the unknown coefficients $A_{i\beta}$ and generally nonlinear in the state variables x). The unknown coefficients $A_{i\beta}$ in this model can be determined by fitting the data to the model. The usual fitting procedure involves least-squares fitting. That is, for an embedding of length N, N values of dx_i/dt and the n values of $\Phi_\beta(x;c)$ are computed and used to determine each of the coefficients $A_{i\beta}$ for the three values of $i = 1, 2, 3$. The derivatives dx_i/dt can be estimated by first differences (this gives only $N - 1$ values) or computed using spline fits.

There is no guarantee that a general linear model will be a good model for a nonlinear dynamical system. However, until better modeling methods are proposed, such models will be adopted. Many different methods exist for choosing the basis functions $\Phi(x;c)$. These have been discussed extensively in classic reviews [100, 101].

Most procedures involve some choice of functions $\Phi_\beta(x;c)$ together with an invariant measure $\rho(x)$ on the strange attractor. The measure is estimated from the data and the embedding. With respect to this measure, there is a natural inner product

$$g_{\beta\beta'} = \int \Phi_\beta(x;c)\Phi_{\beta'}(x;c)\rho(x)\,dx \simeq \frac{1}{N}\sum_j^N \Phi_\beta(x_j;c)\Phi_{\beta'}(x_j;c) \tag{6.47}$$

The parameters $A_{i\beta}$ in the general linear model (6.46) are then estimated by standard statistical techniques: the maximum likelihood method or, more often, the least-squares method. These methods involve computing the inverse of the matrix $g_{\beta\beta'}$ of overlap integrals. Inverting this matrix can be delicate. If the functions $\Phi_\beta(x;c)$ are constructed to be orthogonal, say by the Gram–Schmidt procedure, then $g_{\beta\beta'} = \delta_{\beta\beta'}$ and the matrix inversion is simple. However, the problem has been pushed back to the construction of an orthonormal basis set, which is then difficult.

To indicate the flavor of this difficulty, we sketch two examples. The first is classical. The Legendre polynomials $P_n(x)$ can be constructed from the matrix of overlaps:

$$g_{ij} = \langle x^i x^j \rangle = \int_{-1}^{+1} x^i x^j\,dx = \begin{cases} \dfrac{2}{i+j+1} & i+j \text{ even} \\ 0 & i+j \text{ odd} \end{cases} \tag{6.48}$$

An LDU (lower triangular, diagonal, upper triangular) decomposition [99] is used to rewrite the overlap:

$$g_{ij} = \sum_{\tau \le \min(i,j)} L_{i\tau} D_\tau U_{\tau j} \tag{6.49}$$

where L and U are lower and upper triangular matrices with $+1$ on the diagonal. The diagonal matrix element D_τ provides the normalization factor. The matrix elements of D decrease very rapidly ($D_\tau \sim 10^{-\tau}$), making this numerical procedure for constructing these classical functions very sensitive to noise in the measure.

As a second example, we try to model data generated by the logistic map $x_{i+1} = \lambda x_i(1 - x_i)$ by a model with a basis set $x_i, x_i^2, x_{i-1}, x_{i-1}^2$:

$$x_{i+1} = Ax_i + Bx_i^2 + Cx_{i-1} + Dx_{i-1}^2 \tag{6.50}$$

Since $x_i = \lambda x_{i-1} - \lambda x_{i-1}^2$, the basis set has a degeneracy, and any attempt to invert the matrix of overlaps $g_{\beta\beta'}$ or to orthonormalize the basis set is subject to instabilities due to this degeneracy among the basis vectors.

Rather than fight these singularities, we have developed a model-fitting procedure that effectively avoids them. More precisely, it explicitly exhibits degenerate directions in the space of basis vectors, including one direction in which dx_i/dt is maximally degenerate with the expansion functions $\Phi_\beta(x;c)$. The idea is to create a model in which the time derivative dx/dt is also treated as a basis vector (Φ_0), on the same footing as the other basis vectors $\Phi_\beta(x;c)$. For each i, the dynamical system equation (6.46) can be written as

$$B_0\Phi_0 + \sum_\beta B_\beta\Phi_\beta(x;c) \sim 0 \tag{6.51}$$

with $B_0 \neq 0$. An $N \times (1+n)$ matrix is generated for the $1+n$ coefficients B_0 and B_β for each of the embedding coordinates i in the usual way [122]. A singular value decomposition is performed [99]. Eigenvectors corresponding to nonzero (large) eigenvalues are not model candidates. Eigenvectors corresponding to small eigenvalues express degeneracies among the basis vectors, one of which is now dx_i/dt. To determine the appropriate eigenvector for the model, we search for the minimum value of $[\lambda(\alpha)/B_0(\alpha)]^2$. For this eigenvector $(B_0(\alpha), B_1(\alpha), \ldots, B_\beta, \ldots, B_n(\alpha))$,

$$B_0(\alpha)\frac{dx_i}{dt} + \sum_\beta B_\beta(\alpha)\Phi_\beta(x;c) \simeq \epsilon \tag{6.52}$$

The model (6.46) is then given explicitly in terms of this SVD eigenvector by

$$\frac{dx_i}{dt} = F_i(x;c) = \sum_\beta A_{i\beta}\Phi_\beta(x;c) = -\frac{1}{B_0(\alpha)}\sum_\beta B_\beta(\alpha)\Phi_\beta(x;c) \tag{6.53}$$

where the stochastic term is $\epsilon' = \epsilon/B_0(\alpha)$ [2].

The same singular-value decomposition procedure can be used to construct rational fractional models using as a basis set the functions Φ_0, $\Phi_\beta(x;c)$, and $\Phi_0\Phi_\beta(x;c)$ and constructing vectors of the form

$$D_0\Phi_0 + \sum_\beta C_\beta\Phi_\beta(x;c) + \sum_\beta D_\beta\Phi_0\Phi_\beta(x;c) + \epsilon \tag{6.54}$$

Following the procedure just described, we find a best-fit model

$$\frac{dx_i}{dt} = \frac{-\sum_\beta C_\beta(\alpha)\Phi_\beta(x;c)}{D_0(\alpha) + \sum_{\beta\neq 0} D_\beta(\alpha)\Phi_\beta(x;c)} + \epsilon' \tag{6.55}$$

The least-squares residual is once again $\epsilon' = [\lambda(\alpha)/D_0(\alpha)]^2$.

If monomials of the form $x_1^{n_1} x_2^{n_2} \cdots x_k^{n_k}$, $n_1 + n_2 + \cdots + n_k \leq d$ are used as the basis functions $\Phi_\beta(x; c) = \Phi_{\mathbf{n}}(x; c)$, the simple model (6.53) involves ${}_{d+k}C_k$ coefficients B_β for each value of i. Here ${}_*C_*$ is a binomial coefficient. The more complicated rational function approximation (6.55) requires ${}_{p+k}C_k$ coefficients C_β and ${}_{q+k}C_k$ coefficients D_β to represent polynomials of degree p in the numerator and degree q in the denominator of the rational function approximation of each of the driving terms $F_i(x; c)$.

For a three-dimensional embedding of degree 5, $k = 3$, $d = 5$, each forcing function $F_i(x; c)$ is described by an eigenvector containing $(5 + 3)!/5!(3)! = 56$ coefficients. For a rational model in R^3 with polynomials in the numerator and denominator of degrees 5 and 4, a total of ${}_8C_3 + {}_7C_3 = 89$ degrees of freedom are required for each of the three driving terms.

In practice, many of the monomials are coupled to amplitudes that are small, compatible with zero. In such cases, these monomials can be dropped from the basis set and the fit tried again. If there is little difference in the residual ϵ' and the amplitudes of the optimal eigenvectors for the remaining basis functions are unchanged, the fit with fewer degrees of freedom can be used. Alternatively, each lower-degree monomial dropped from the basis set can be replaced by one of higher degree, keeping the number of degrees of freedom constant at a not too unreasonable value.

In general, eigenvectors consisting of an array with a long string of small values are very difficult to interpret physically. Eigenvectors with only a small number of large amplitudes are much more conducive to providing physical information for the quantitative dynamics of a system.

6.10 VALIDATE MODEL

As usual, it is not sufficient to make a model, find the parameter values that provide the best fit among all the models in that class, and then walk away from the problem. It is necessary to decide whether the best-fit model is any good. Testing models of nonlinear dynamical systems can be carried out at both a qualitative and a quantitative level.

6.10.1 Qualitative Validation

At the qualitative level, a good model must reproduce the stretching and squeezing mechanisms that generate the strange attractor observed. Therefore, a good test for qualitative agreement involves running the model to generate a chaotic data set. This itself is a highly nontrivial exercise. Experience has shown that models derived from chaotic data more often than not do not behave chaotically: They typically generate stable limit cycles. However, if the model is a reasonable representation of the dynamics, it will generate chaotic data. The model output that is assumed to be the surrogate for observed data is then analyzed exactly as the experimental data was analyzed. Both sets, the experimental data and the model simulation, should give the same template. In principle, we should get more: the same spectrum of unstable periodic orbits (as measured, for example, by a basis set of orbits; Section 9.3.4), the same spectrum of Lyapunov exponents, and the same spectrum of metric invariants.

As usual, topological comparisons cannot be used to "accept" a model. The positive outcome of a test is simply to "fail to reject" a model. A negative outcome is a rejection criterion. We have used the topological analysis method to reject models in the past.

Remark: Dynamical and metric tests for chaos do not have model rejection criteria.

6.10.2 Quantitative Validation

The most important statistical problem facing nonlinear theory in the next century involves developing a quantative rejection criterion for models of nonlinear dynamical systems. Such a test should have the ease of use of the methods currently available for general linear models as well as comparable quantitative rejection criteria. In short, we need to develop the nonlinear analog of the χ^2 test.

A promising test has been described in Section 6.2.7. It is based on entrainment. The entrainment is implemented as shown in Eq. (6.4). The test that we have in mind involves developing a universal plot of the type illustrated in Fig. 6.14. This figure shows the value of some entrainment parameter, λ, plotted as a function of c, a multidimensional control parameter $c \in R^k$. The solid curve, anchored at the correct value $c = c'$, probably has the shape of a Hopf–Arnol'd tongue. It describes the minimum value of the entrainment parameter λ at which the data entrain the model behavior as a function of "$|c - c'|$." Dashed curves within the tongue show how the presence of noise at various levels (1.0%, 2.5%, 5.0%) degrade entrainment by requiring larger values of the coupling parameter λ to entrain the model. What is missing is:

1. A standard, optimally useful method for coupling the data with the model output (i.e., a standard way to introduce a single parameter λ).

2. A universal way for measuring distance d between the proposed model c and the "actual" model c' (i.e., a standard way to introduce a single parameter $d = |c - c'|$).

3. A method for measuring noise levels.

4. The solid curve for noise-free data and the dashed curves for noise at various levels.

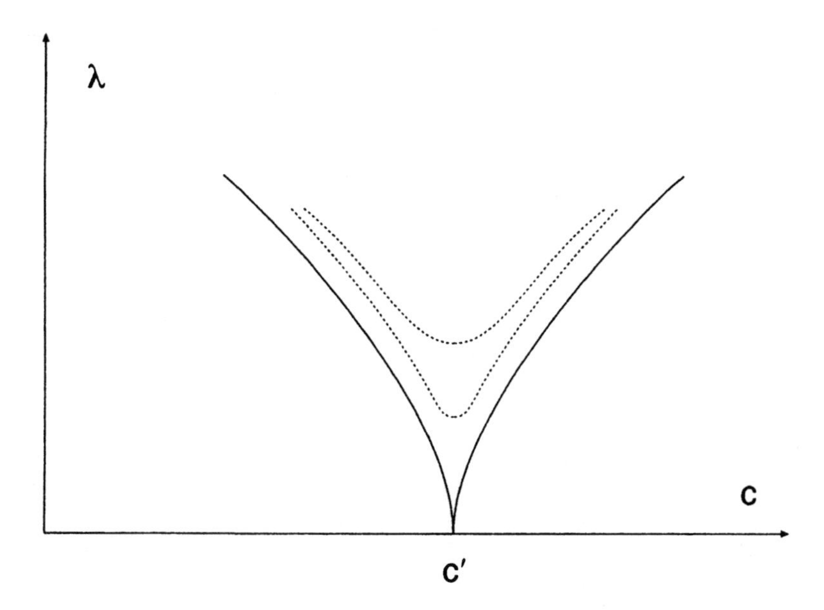

Fig. 6.14 Some time in the twenty-first century, a plot of this type will be the basis for quantitative tests of goodness of fit for nonlinear models. Vertical axis λ: strength of coupling between model output and data input which is required for model entrainment by data. Horizontal axis $d = |c - c'|$: universal way to measure distance between the real model with control parameter c' which generates data and the model that is proposed to describe physical data generation mechanism with control parameter c. The solid line is the entrainment envelope in the absence of noise. Dashed curves describe degradation of entrainment by noise at various levels.

6.11 SUMMARY

The topological analysis program is a prescription for extracting information about the stretching and squeezing mechanisms responsible for creating chaos from chaotic data. The procedure has been broken down into five relatively simple steps. These are: Locate periodic orbits in the data; find a suitable embedding; compute the topological invariants of these periodic orbits; identify a suitable branched manifold on the basis of a subset of these topological invariants; and confirm this identification by predicting the remaining invariants and comparing them to the invariants determined from the remaining orbit pairs. The identification of a branched manifold which summarizes the stretching and squeezing mechanisms that take place in the phase space also serves to classify the strange attractor. The classification is discrete and given in terms of integers.

7

Folding Mechanisms: A_2

The first topological analyses that were carried out on experimental data showed that a single mechanism was responsible for creating chaos. This is the Smale horseshoe mechanism, generated by a simple stretch and fold deformation of phase space. Ultimately, more complicated mechanisms were unearthed. However, these all belonged to the same class: the stretch and roll mechanism. The reason that this class of mechanisms is so extensive is explored in Chapter 11.

The first analysis presented in this chapter is devoted to the first data set subjected to a topological analysis. The data set was generated by the Belousov–Zhabotinskii chemical reaction. We describe the analysis of this data set in detail. The next three sections treat data generated by a laser with saturable absorber (LSA), a stringed instrument, and the problem of signals from class B lasers with intervals of very low intensity output. In each case, chaos is generated by a simple stretch and fold.

This mechanism seemed so widespread that for a time, many wondered whether it was even useful to carry out topological analyses on experimental data. This worry was dissipated when the YAG laser was studied. The initial study showed that under reasonable operating conditions, the strange attractor generated by the laser model was described by a branched manifold with three branches. Subsequent experimental work showed that the strange attractor was much more interesting. It was described by two or three branches, depending on the control parameters. Specifically, as the driving frequency of the external modulating force changed, the branches describing the strange attractor changed in a systematic and well-determined way. The mechanism involved is essentially the same as the mechanism responsible for generating a jellyrole (in French, a *gâteau roulé*). This mechanism had previously been observed in the Duffing oscillator. It was also later discovered to be responsible for chaotic behavior observed in a model of a neuron exhibiting subthreshold oscillations. In this chapter we describe, in more or less detail, the six systems mentioned above.

7.1 BELOUSOV–ZHABOTINSKII CHEMICAL REACTION

Oscillating chemical reactions were apparently first discovered by Belousov in 1951 [123, 124]. The oscillations were described for closed, homogeneous chemical systems. In these systems the oscillations eventually die out as the system approaches thermodynamic equilibrium. Inhomogeneities in chemical reactions were first reported by Zhabotinskii [125].

Oscillations in chemical concentrations could be sustained in open systems. These are systems that are fed continuously. That is, chemical species are allowed to flow into a reaction vessel [a tank (industrial scale) or beaker (laboratory scale)] at a constant rate, and the reactants are allowed to flow out at the same rate. To maintain a homogeneous state, the reactants are stirred. Such a system is called a continuously stirred tank reactor (CSTR). A schematic diagram for such a reactor is shown in Fig. 7.1. The topological analysis described below was carried out on a segment of data measured by the Texas group [126, 127]. The data consist of 64K digitized observations of the bromine ion concentration taken at a single value of the control parameters. This data segment represents a day-long experiment. It contains about 543 oscillations with an average interpeak distance of 124 measurements/cycle. The noise level was determined by digitization of the signal.

7.1.1 Location of Periodic Orbits

Periodic orbits were located in two ways. We first used a close returns search in the flow data [1]. As described in Section 6.5, this search involves plotting $\Theta(\epsilon - |x(i) - x(i+p)|)$ as a function of i and p, where $1 \leq i \leq\sim$ 64K indexes observations, $1 \leq p \leq 1000$ indexes period in measurement units (period 1 $\sim$ 124 observations), and ϵ is of the order of 1% of the dynamic range of the signal. The dynamic range of the signal is the maximum difference $|x(i) - x(j)|$. A portion of this close returns plot is shown in Fig. 7.2. This portion of the close returns plot suggests a strong signal for a period-3 orbit.

Close returns plots of the type shown in Fig. 7.2 can display only a small portion of the entire data set at a time. This makes locating periodic orbits inefficient. To facilitate the location of periodic orbits, we searched the data set for successive

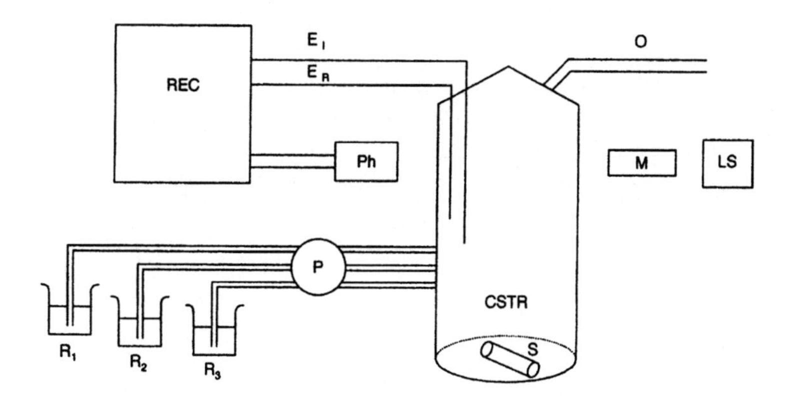

Fig. 7.1 Schematic diagram shows a continuously stirred chemical reaction. Reactants (R_1,R_2,R_3) are fed by a pump P into the reaction vessel (CSTR) at a constant flow rate, are continuously mixed by a stirrer S, remain in the vessel for a characteristic residence time, and the product removed at O to keep the volume in the vessel constant. Oscillating chemical concentrations are monitored by either spectroscopic means (laser source, LS; monochromator, M; photodetector, Ph) or by measuring specific ion potentials (ion potential, E_I; reference potential E_R). All signals are recorded (REC).

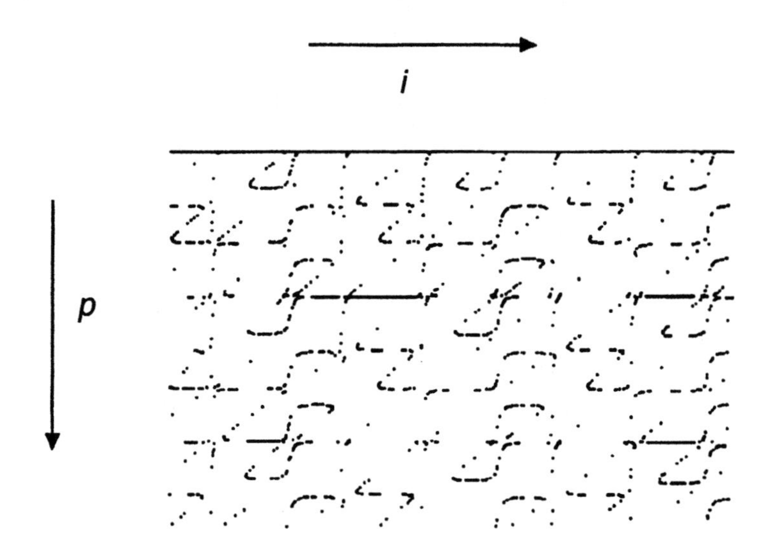

Fig. 7.2 In a typical close returns plot, points (i, p) are plotted black when the measure $|x(i) - x(i+p)|$ is less than some value. This close returns plot is constructed for a segment of data from the Belousov–Zhabotinskii reaction.

minima ($x_m(j)$) and maxima ($x_M(j)$). These smaller data sets each contained 543 pairs of measurements: the location, j, in the original data set, and the data value: $(i(j), x_m(j))$ and $(i(j), x_M(j))$, $1 \leq j \leq 543$. The first return plots for successive minima and successive maxima are shown in Fig. 7.3. The point(s) in these close returns plots closest to the diagonal provide candidates for period-1 surrogate orbits. The point nearest the diagonal in the first return plot for minima and the point in the first return plot for maxima both identify the same period-1 surrogate orbit.

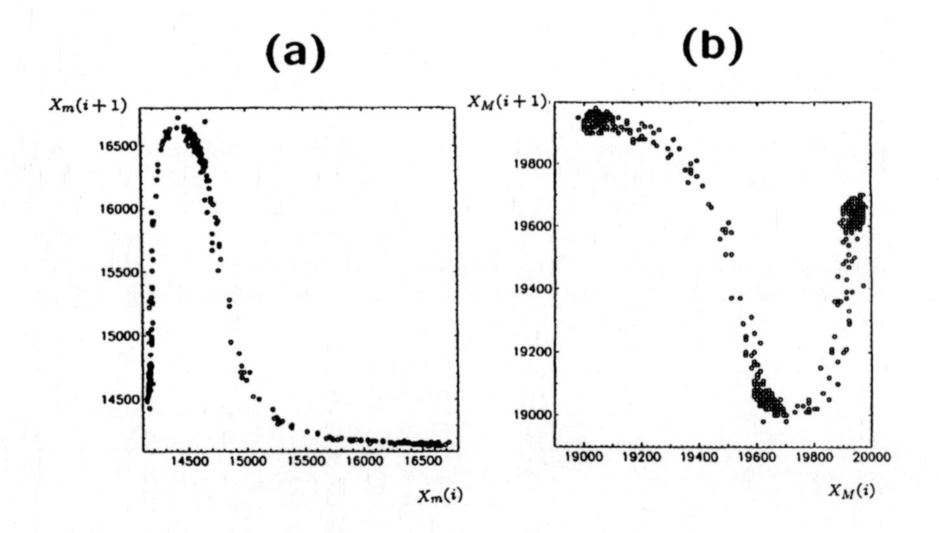

Fig. 7.3 First return maps are constructed from successive minima (a) and successive maxima (b) in the BZ data.

Period-p ($p = 1, 2, 3, 4, \ldots$) orbits were located by searching for small values of the differences $|x_m(j) - x_m(j+p)|$ and $|x_M(j) - x_M(j+p)|$. For $p = 1$ only one data segment was a suitable surrogate for a closed orbit. However, segments with larger values of $|x_m(j) - x_m(j+1)|$ and $|x_M(j) - x_M(j+1)|$ were extracted from the data and used subsequently to determine the local torsion of the period-1 orbit after an embedding was constructed. This was also the case for $p = 2$. For $p = 3$ there was an abundance of suitable surrogates. Orbits of period up to 17 were identified in this manner. These orbits were identified by their symbolic dynamics on the $x_m(i+1)$ vs. $x_m(i)$ return map and the $x_M(i+1)$ vs. $x_M(i)$ return map [Fig. 7.3(a) and (b)]. The orbits extracted from the data are listed in Table 7.1.

Table 7.1 Periodic orbits located in data from the Belousov–Zhabotinskii reaction[a]

Orbit	Name	Symbolics	Local Torsion	Self-Linking
1	1_1	1	1	0
2	2_1	01	1	1
3	3_1	011	2	2
4	4_1	0111	3	5
5	5_1	01 011	3	8
6	6_2	011 0M1	3	9
7	7_2	$(01)^2011$	4	16
8a	8_1	$(01)^20111$	5	23
8b	8_3	$01(011)^2$	5	21
9	9_3	$(01)^3011$	5	28
10a	10_6	$(011)^20101$	6	33
10b	10_6	$(011)^20111$	7	33
11	11_9	$01(011)^3$	7	40
13a		$(01)^2011\ 01\ 0111$	8	62
13b		$(01)^3011\ 0111$	8	60
13c		$(011)^30101$	8	56
13d		$(011)^30111$	9	56
13e		$(01)^2011\ 011111$	9	62
14		$01(011)^4$	9	65
15		$01(011)^20111\ 011$	10	78
16a		$(01)^3(011)^20111$	10	89
16b		$(011)^40101$	10	85
16c		$(011)^40111$	11	85
16d		$(01)^2(011)^2011111$	11	91
17a		$(01)^3011\ 01(011)^2$	10	102
17b		$01(011)^5$	11	96
17c		$(01)^2011\ 01(0111)^2$	11	108

[a]The orbits are identified by the symbol sequence according to the first return maps $x_m(i+1)$ vs. $x_m(i)$ and $x_M(i+1)$ vs. $x_M(i)$. The symbol M in the orbit 6_2 identifies the cut point between the orientation-preserving (0) and orientation-reversing (1) branches of the first return map. The universal-sequence name is provided for the lower-period orbits. All topological indices are negative for the embedding adopted later.

7.1.2 Embedding Attempts

The Belousov–Zhabotinskii chemical reaction involves a large number of chemical species as well as a large number of chemical reactions—more than three in both cases. It is therefore not obvious that a three-dimensional embedding is possible.

The existence of a thin first return map (cf. Fig. 7.3) suggests that this reaction is highly dissipative, and as a result the flow in phase space rapidly relaxes into a low dimensional inertial manifold. The default embedding, using time delays, has been used to construct a strange attractor from the time series [1, 128]. Three dimensions were sufficient to construct an induced strange attractor without self-intersections. As a result, we know that the strange attractor can be embedded in a three-dimensional inertial manifold. The tools of the Birman–Williams theorem are applicable, linking numbers can be computed for periodic orbits, and these topological invariants determine the organization of all the unstable periodic orbits in the strange attractor.

To facilitate computation of linking numbers, we prefer a differential embedding to a time-delay embedding. The first attempt to construct a differential embedding involved constructing the three embedding variables (y_1, y_2, y_3) by taking first and second differences:

$$\begin{aligned} y_1(i) &= x(i) \\ y_2(i) &= x(i+1) - x(i-1) \\ y_3(i) &= x(i+1) - 2x(i) + x(i-1) \end{aligned} \tag{7.1}$$

A projection of this embedding on the y_1–y_2 plane is shown in Fig. 7.4. In this embedding, about two orders of magnitude in the signal-to-noise ratio were lost in taking the first difference, and another two were lost in taking the second difference. The original signal-to-noise ratio was four orders of magnitude. As a result, the second difference coordinate y_3 was in the noise level. In particular, orbit crossings, which all occur in the horizontal region at the top of the embedding, could not be resolved. This differential embedding was unsuitable, since it was impossible to compute topological invariants of unstable periodic orbits.

In an attempt to circumvent the signal-to-noise problem while retaining the virtues of the differential embedding, we tried to construct an integral–differential embedding in the form

$$\begin{aligned} y_1(i) &= \textstyle\sum_{j=1}^{i}(x(j) - \bar{x}) \\ y_2(i) &= x(i) \\ y_3(i) &= x(i+1) - x(i-1) \end{aligned} \tag{7.2}$$

In this embedding the average value, $\bar{x} = \frac{1}{N}\sum_{j=1}^{N} x(j)$, is removed from each observation to create a stationary time series. Figure 7.5 shows a projection of this embedding onto the y_1–y_2 plane. The plot was carried out on a slow, reliable old computer (Mac II running at 16 MHz). It was possible to watch the attractor drift horizontally (in the y_1 direction) from side to side during the construction of the embedding. This suggested that the data and their first differences were (visually) stationary but that the integral of the data was not stationary. Nonstationarity is due

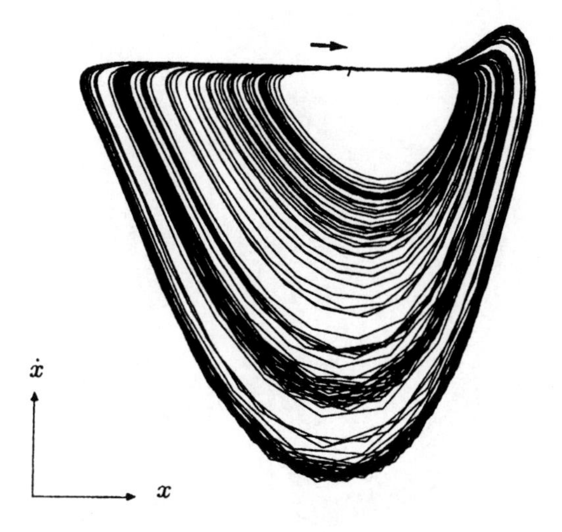

Fig. 7.4 Projection of the differential embedding (7.1) onto the y_1–y_2 = x–$\dot{x}$ plane shows that all crossings occur in the thin horizontal region at the top of the attractor. Crossings cannot be resolved in this region.

to long-term variations in the electrical grid. These variations were invisible in the data itself, and were only picked up by integrating the data.

The third attempt at constructing an embedding that allowed resolution of the crossings while retaining the nice topological properties of the differential embedding was successful. This embedding was a slight modification of the integral–differential embedding, in which the integral was replaced by an integral with memory: $y_1(t) = \int_{\text{``}-\infty\text{''}}^{t} x(t')e^{-(t-t')/\tau}\, dt'$. The variable $y_1(t)$ is related to $x(t)$ by $dy_1(t)/dt = x(t) - y_1(t)/\tau$. This embedding is easily implemented online as follows:

$$\begin{aligned} y_1(i) &= x(i) + \lambda y(i-1) \\ y_2(i) &= x(i) \\ y_3(i) &= x(i) - x(i-1) \end{aligned} \tag{7.3}$$

It is useful to choose a decay time on the order of several cycle times. The embedding obtained by choosing $\lambda = 1 - 10^{-3}$ is shown in Fig. 7.6. In this embedding, folding is seen to occur at the bottom center of the projection.

7.1.3 Topological Invariants

Once a suitable embedding was constructed, it was possible to compute the linking numbers of surrogate periodic orbits extracted from the data. Since it is possible to embed the strange attractor in a solid torus $D^2 \times S^1$, it has a hole in the middle. Therefore, it was also possible to compute relative rotation rates for these orbits as well.

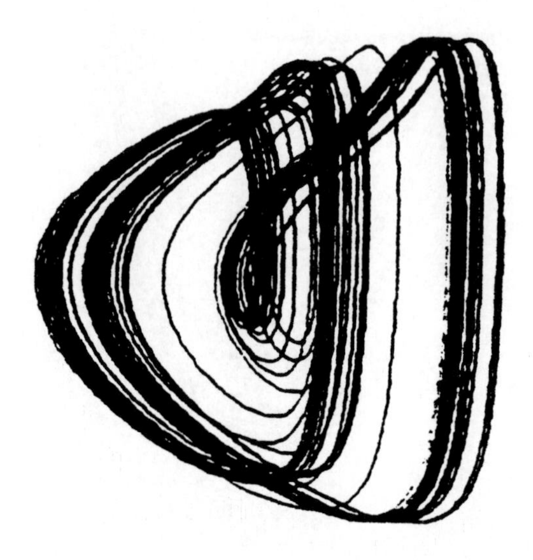

Fig. 7.5 The first attempt at creating an integral–differential embedding (7.2) is projected onto the y_1–y_2 plane. This projection drifted horizontally, from side to side, indicating that the integral of the time-averaged data $y_1 = x(i) - \bar{x}$ was not stationary.

In Fig. 7.7 we show surrogates for orbits of periods 2 through 5 in the embedding adopted. Each orbit is identified by its symbol name. We also show a superposition of all orbits up to period 12 extracted from the data. It should not be surprising that this superposition provides an excellent surrogate for the strange attractor itself.

Linking numbers were computed for all orbits up to period 8 by counting crossings. Self-linking numbers were also computed in the same way. These topological invariants for the orbits extracted from the data are reported in Table 7.2. Local torsions were computed by computing linking numbers of two or more surrogates for the same periodic orbit.

Relative rotation rates for pairs of periodic orbits were computed as described in Section 4.3. The orbits were superposed and a crossing matrix was constructed for the two orbits. This crossing matrix was then similarity transformed by a cyclic matrix representing time evolution of the signal, and the sum of the similarity transformed matrices computed and divided by the appropriate integer $(2 \times p_A \times p_B)$ (cf. Fig. 4.17). The procedure produced the relative rotation rates of the two orbits A and B as well as the self-relative rotation rates of each of the orbits A and B. These fractions are reported in Table 7.3 for all surrogate orbits up to period 8 which were extracted from the experimental data.

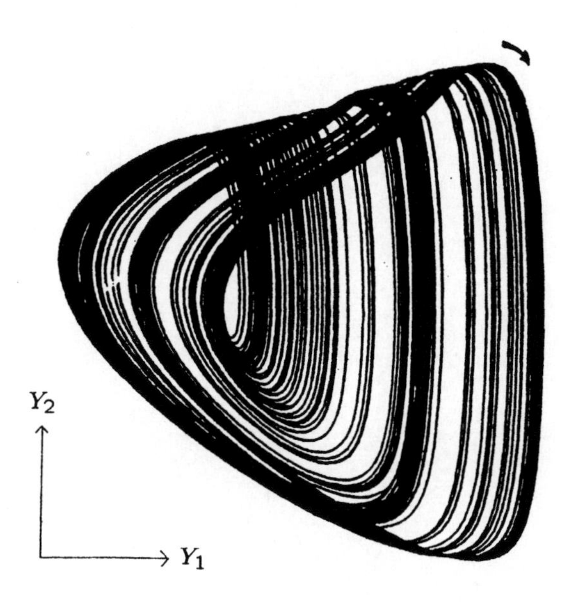

Fig. 7.6 The strange attractor obtained from the near integral–differential embedding (7.3) is projected onto the y_1–y_2 plane. In this embedding, all crossings are resolved and the crossing properties that facilitate computation of the linking numbers are preserved.

Table 7.2 Linking numbers for all the surrogate periodic orbits, to period 8, extracted from Belousov–Zhabotinskii data[a]

Orbit	Symbolics	1	2	3	4	5	6	7	8a	8b
1	1	0	1	1	2	2	2	3	4	3
2	01	1	1	2	3	4	4	5	6	6
3	011	1	2	2	4	5	6	7	8	8
4	0111	2	3	4	5	8	8	11	13	12
5	01 011	2	4	5	8	8	10	13	16	15
6	011 0M1	2	4	6	8	10	9	14	16	16
7	01 01 011	3	5	7	11	13	14	16	21	21
8a	01 01 0111	4	6	8	13	16	16	21	23	24
8b	01 011 011	3	6	8	12	15	16	21	24	21

[a] All indices are negative.

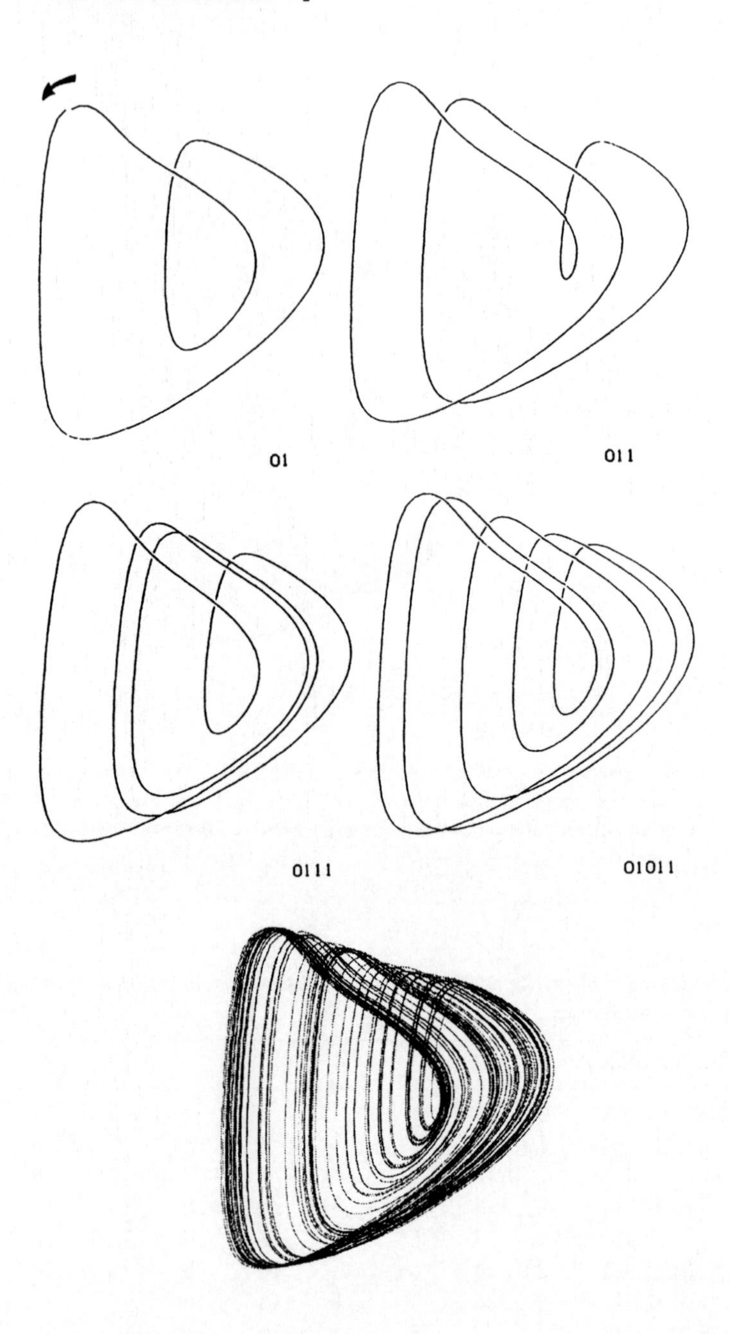

Fig. 7.7 Surrogates for four of the lowest-period orbits in the BZ data are shown in a differential–integral embedding. Each orbit is identified by its symbolic name. Below we show a superposition of all surrogate orbits up to period 12 extracted from the BZ data. This superposition bears a striking resemblance to the original strange attractor.

Table 7.3 Relative rotation rates for all the surrogate periodic orbits, to period 8, which were extracted from Belousov-Zhabotinskii data[a]

	1	2	3	4	5	6	7	$8a$	$8b$
	1	01	011	0111	01 011	011 0M1	01 01 011	01 01 0111	01 011 011
1	0	$\frac{1}{2}$	$\frac{1}{3}$	$\frac{1}{2}$	$\frac{2}{5}$	$\frac{1}{3}$	$\frac{3}{7}$	$\frac{1}{2}$	$\frac{3}{8}$
2	$\frac{1}{2}$	$\frac{1}{2}, 0$	$\frac{1}{3}$	$\frac{1}{2}, \frac{1}{4}$	$\frac{2}{5}$	$\frac{1}{3}$	$\frac{5}{14}$	$\frac{1}{2}, \frac{1}{4}$	$\frac{3}{8}$
3	$\frac{1}{3}$	$\frac{1}{3}$	$(\frac{1}{3})^2, 0$	$\frac{1}{3}$	$\frac{1}{3}$	$\frac{1}{3}$	$\frac{1}{3}$	$\frac{1}{3}$	$\frac{1}{3}$
4	$\frac{1}{2}$	$\frac{1}{2}, \frac{1}{4}$	$\frac{1}{3}$	$(\frac{1}{2})^2, \frac{1}{4}, 0$	$\frac{2}{5}$	$\frac{1}{3}$	$\frac{11}{28}$	$(\frac{1}{2})^2, \frac{3}{8}, \frac{1}{4}$	$\frac{3}{8}$
5	$\frac{2}{5}$	$\frac{2}{5}$	$\frac{1}{3}$	$\frac{2}{5}$	$(\frac{2}{5})^4, 0$	$\frac{1}{3}$	$\frac{13}{25}$	$\frac{2}{5}$	$\frac{3}{8}$
6	$\frac{1}{3}$	$\frac{1}{3}$	$\frac{1}{3}$	$\frac{1}{3}$	$\frac{1}{3}$	$(\frac{1}{3})^4, \frac{1}{6}, 0$	$\frac{1}{3}$	$\frac{1}{3}$	$\frac{1}{3}$
7	$\frac{3}{7}$	$\frac{5}{14}$	$\frac{1}{3}$	$\frac{11}{28}$	$\frac{13}{25}$	$\frac{1}{3}$	$(\frac{3}{7})^4, (\frac{2}{7})^2, 0$	$\frac{3}{8}$	$\frac{3}{8}$
$8a$	$\frac{1}{2}$	$\frac{1}{2}, \frac{1}{4}$	$\frac{1}{3}$	$(\frac{1}{2})^2, \frac{3}{8}, \frac{1}{4}$	$\frac{2}{5}$	$\frac{1}{3}$	$\frac{3}{8}$	$(\frac{1}{2})^4, \frac{3}{8}, (\frac{1}{4})^2, 0$	$\frac{3}{8}$
$8b$	$\frac{3}{8}$	$\frac{3}{8}$	$\frac{1}{3}$	$\frac{3}{8}$	$\frac{3}{8}$	$\frac{1}{3}$	$\frac{3}{8}$	$\frac{3}{8}$	$(\frac{3}{8})^7, 0$

[a] All indices are negative.

7.1.4 Template

Since the return map (cf. Fig. 7.3) has two branches, it appeared that the template that describes the strange attractor had only two branches. A two-branch template can be identified by investigating the organization of three orbits. We used the three lowest-period orbits to make the initial template identification. A simple left-handed Smale horseshoe template describes the organization of the period-1 orbit 1, the period-2 orbit 01, and the period-3 orbit 011.

This template was then used to compute the topological organization of all other surrogate orbits extracted from the data. In particular, we identified each surrogate orbit with an orbit on the horseshoe template by using the symbolic name of the surrogate. The symbolic name was determined from the return maps. We then computed the self-relative rotation rates of both the surrogate and its presumed counterpart on the horseshoe template. In every case there was agreement. In one case this comparison resolved an uncertainty in the symbol assignment of a surrogate. This was the orbit $0110M1$, where the symbol M occurred at the critical point of the first return map. This orbit appeared compatible with a period-3 orbit 011011, which goes around twice, or the nearby period-6 orbit 001011, which is the period-doubled version of the period-3 node 001. The self-relative rotation rates showed that the surrogate orbit is the latter: 001011. All surrogate orbits extracted from the data set are compatible with periodic orbits on the left-handed Smale horseshoe template with zero global torsion.

7.1.5 Dynamical Properties

This same data set had previously been analyzed by Lathrop and Kostelich [128] in order to determine some of the dynamical properties of this attractor. In particular,

Table 7.4 Eigenvalues of the linear fit to neighborhoods of the low-period surrogate orbits in the Belousov–Zhabotinskii attractor

Period	Eigenvalue	Period	Eigenvalue
1	3.66	5	7.10
2	5.34	6	3.40
3	3.18	7	9.56
4	4.17	8	5.56

their study was motivated by the attempt to support a conjecture that most, if not all, of the properties of a strange attractor could be determined from a study of the unstable periodic orbits contained in the attractor.

They first constructed a three-dimensional time-delay embedding, with a delay equal to the cycle time of the data. Then they extracted periodic orbits with periods from 1 to 8. In order to determine the positive Lyapunov exponent for an orbit of period p, they chose a reference point $\mathbf{x}_{ref}$ on that orbit. Then they chose a large number ($\sim$ 50) of points $\mathbf{x}_i$ in the neighborhood of the reference point. This neighborhood was followed p periods into the future, and the image points $f^{(p)}(\mathbf{x}_i)$ identified. Since the image points were near the original points, it was assumed that the forward mapping was well approximated by a linear map: $f^{(p)}(\mathbf{x}_i) = \mathbf{A}\mathbf{x}_i + \mathbf{b}$, where $\mathbf{A}$ is a 3×3 matrix and $\mathbf{b}$ a column vector. The 12 unknowns were fitted by standard least-squares methods, and the eigenvalues of the matrix $\mathbf{A}$ were computed. These eigenvalues are reproduced in Table 7.4. These eigenvalues were weighted by an approximation to the invariant density on the attractor to provide an estimate of the stretching per period: $1.474 = 2^{0.56} = e^{0.39}$. The Lyapunov exponent was estimated as 0.56 bit/period or 0.39/period.

Lathrop and Kostelich also estimated the topological entropy. Their first bound was based on the transition matrix for period-1 orbit segments:

$$T = \begin{array}{c} \\ 0 \\ 1 \end{array} \begin{array}{c} \begin{array}{cc} 0 & 1 \end{array} \\ \left[\begin{array}{cc} 0 & 1 \\ 1 & 1 \end{array} \right] \end{array}$$

The largest eigenvalue of this matrix is the golden mean $\lambda = (1+\sqrt{5})/2 = 1.618034$. The logarithm of this number provides an upper bound on the topological entropy, $h_T = 0.481212$. They made a more refined estimate using a grammar containing five three-letter words:

$$T = \begin{array}{c} \\ 010 \\ 011 \\ 101 \\ 110 \\ 111 \end{array} \begin{array}{c} \begin{array}{ccccc} 010 & 011 & 101 & 110 & 111 \end{array} \\ \left[\begin{array}{ccccc} 0 & 0 & 1 & 1 & 1 \\ 1 & 1 & 1 & 0 & 0 \\ 1 & 1 & 1 & 1 & 1 \\ 0 & 0 & 1 & 1 & 1 \\ 1 & 1 & 1 & 0 & 1 \end{array} \right] \end{array}$$

This leads to a largest eigenvalue $\lambda = 1.48$ and an estimate $h_T \simeq 0.392 = 0.563$ bit.

If we assume that the grammar is made up of symbol sequences 01, 011, 0111, and 01111, and these symbol sequences can occur in any order, the topological entropy is estimated from

$$1 = \frac{1}{x^2} + \frac{1}{x^3} + \frac{1}{x^4} + \frac{1}{x^5}$$

so that $X_0 = 1.534158$ and $h_T = 0.427982$.

A slightly better estimate can be obtained by constructing a grammar based on the sequences, and their order of occurrence, observed in the periodic orbits listed in Table 7.1. This results in the following capacity matrix:

$$\begin{bmatrix} \frac{1}{x^2} - 1 & \frac{1}{x^2} & \frac{1}{x^2} & 0 & 0 \\ \frac{1}{x^3} & \frac{1}{x^3} - 1 & \frac{1}{x^3} & 0 & \frac{1}{x^3} \\ \frac{1}{x^4} & \frac{1}{x^4} & -1 & 0 & 0 \\ 0 & 0 & 0 & -1 & 0 \\ \frac{1}{x^6} & 0 & 0 & 0 & -1 \end{bmatrix}$$

so that $X_0 = 1.452626$ and $h_T = 0.373373$.

7.1.6 Models

The flow on the strange attractor in the embedding phase space was modeled by the methods described in Section 6.9. The models assumed the following form [2]:

$$\begin{aligned} \dot{y}_1 &= y_2 - \tfrac{1}{\tau} y_1 \\ \dot{y}_2 &= y_3 \\ \dot{y}_3 &= f(y_1, y_2, y_3) \end{aligned} \tag{7.4}$$

Only the function f in the third equation had to be fitted to the data. The three variables were scaled to the interval $[-1, +1]$, and then monomials of the form $y_1^{k_1} y_2^{k_2} y_3^{k_3}$ were used as basis functions to determine an appropriate representation of the function f. Fitting was carried out by using singular-value decomposition methods as described in Section 6.9. Both analytic representations and rational fractional function fits were carried out. The analytic fits involved polynomials of degree $k_1 + k_2 + k_3 \leq K$, with $K = 2, 3, 4$, and 5. These models involved 10, 20, 35, and 56 coefficients, respectively. Rational fractional fits were also carried out, with polynomial functions of degree K_n in the numerator and K_d in the denominator. The number of coefficients required for such fits is $\begin{pmatrix} K_n + 3 \\ 3 \end{pmatrix} + \begin{pmatrix} K_d + 3 \\ 3 \end{pmatrix}$. Typically, we chose $K_d = K_n - 1$.

In essence, these models were column vectors, eigenvectors of a singular value decomposition. No immediate physical interpretation for any of the components of these eigenvectors could be discerned.

7.1.7 Model Verification

The models developed according to the description above were tested in various ways.

Each was run using as initial conditions points in the attractor. None of the analytic models ($K_d = 0$) reproduced the attractor. The lower-degree models relaxed to a fixed point, while the higher-degree models relaxed to a limit cycle. The rational fraction models with $0 < K_d = K_n - 1$ eventually "blew up." This occurred because of the occurrence of zeros in the denominator.

The models were also subject to entrainment tests [2]. The polynomial model with $K_n = 2$ could not be entrained. Those with $K_n = 3, 4, 5$ could be entrained, with the strength of the entrainment parameter λ [cf. Eq. (6.4)] decreasing with increasing degree. In Fig. 7.8 we show plots of the model output $y_1^m(t)$ and the data input $y_1^d(t)$ for the polynomial model with $K_n = 5$. The two are offset for comparison. We also show plots of $y_3^m(t)$ and $y_3^d(t)$ for the same model. Finally, we show plots of $y_1^m(t)$ vs. $y_1^d(t)$ and $y_3^m(t)$ vs. $y_3^d(t)$. The entrainment occurred for $\lambda \sim 0.01$.

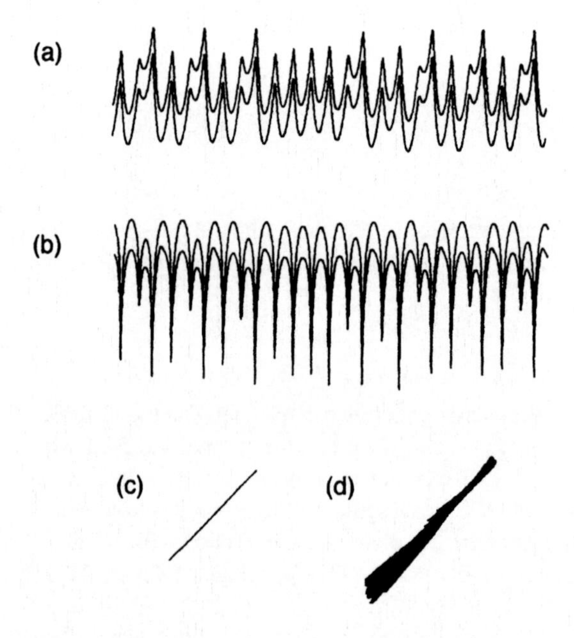

Fig. 7.8 Time traces of (a) $y_1^m(t)$ and $y_1^d(t)$ and (b) $y_3^m(t)$ and $y_3^d(t)$ for a model based on Belousov–Zhabotinskii data. The traces are slightly offset for purposes of comparison. Plots of (c) $y_1^m(t)$ vs. $y_1^d(t)$ and (d) $y_3^m(t)$ vs. $y_3^d(t)$ show that the data signals entrain the model output. Calculations were done for a 5 degree polynomial model of the data.

Similar entrainment tests were carried out for the rational fractional models. In general, rational models with $K_n = K_d + 1 = K$ tracked data more closely than polynomial models with $K_n = K$ ($K = 3, 4, 5$), until somewhere in the data set the fit blew up.

We conclude that polynomial fits can provide reasonable approximations to flows on this strange attractor, but when run autonomously (without entrainment by the

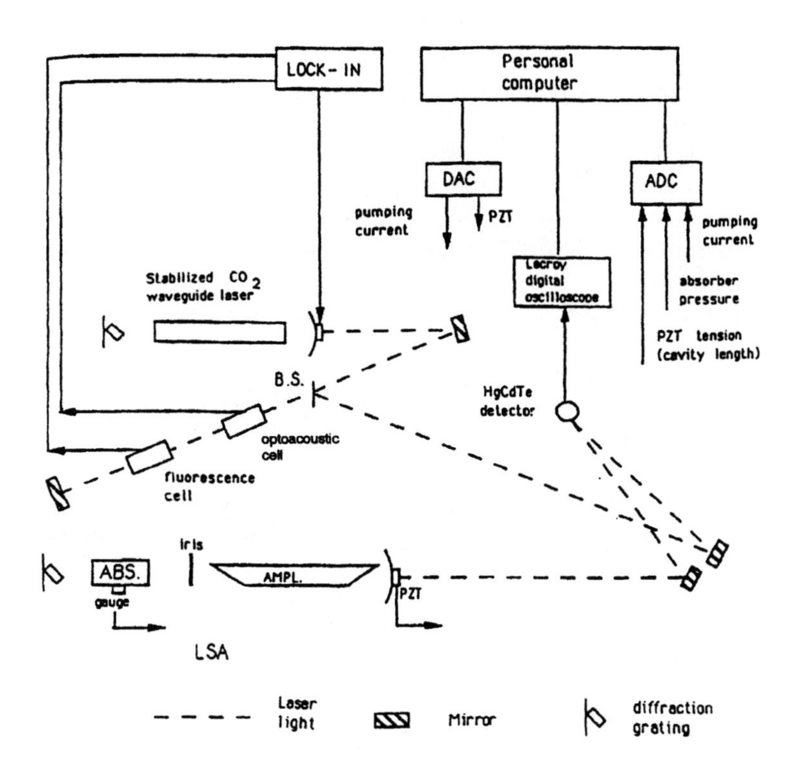

Fig. 7.9 Schematic shows the experimental arrangement for the laser with saturable absorber. Reprinted from the thesis of F. Papoff with permission [129].

data), low-degree models were not able to reproduce the flow on the induced strange attractor.

7.2 LASER WITH SATURABLE ABSORBER

7.2.1 Experimental Setup

The laser with saturable absorber (LSA) consists of a CO_2 discharge tube with Brewster angle windows immersed in an infrared cavity. The cavity is defined by a pair of highly reflective infrared mirrors. An absorbing cell is placed within the cavity to drive the laser into the chaotic regime. The absorbing cell contained CH_3I:He and OsO_4:He in the ratio 1:20 as absorbers. The current discharge and the pressure in the absorber cell could be modified. A schematic of the experimental setup is shown in Fig. 7.9 [88, 129, 130].

7.2.2 Data

Once the laser had been coaxed into behaving chaotically, output intensity data, $I(t)$, were recorded. Data were recorded at a rate of about 80 samples/cycle. Time series of up to 32K data points were taken per run. Discharge currents, the absorber, and the absorber pressure were varied from run to run.

Several segments of data from the laser with saturable absorber are shown in Fig. 7.10. In each case, a portion of the time series near an unstable periodic orbit is shown. The orbit is shown between the arrows. Each "goes around" the unstable periodic orbit twice. The data segments shown shadow orbits of period 5 (10010), 7 (1011010), and 10 (1010010010).

A typical intensity time series contained a set of large and small peaks, followed by a deep intensity minimum. The peaks were identified as large (L) or small (S). Each deep minimum was followed by a large peak. Between the deep minima, each minimum was followed by a local maximum: The deeper the preceding minimum, the larger the succeeding maximum. Sequences of the form L, LS^n, and $LS^{n-1}L$, with $n = 1, 2, \ldots, 15$ were observed. Other sequences which involved combinations of the foregoing pulses with $n = 1, 2$ were also seen in several data sets. About 25 data sets taken in this way were subjected to a topological analysis.

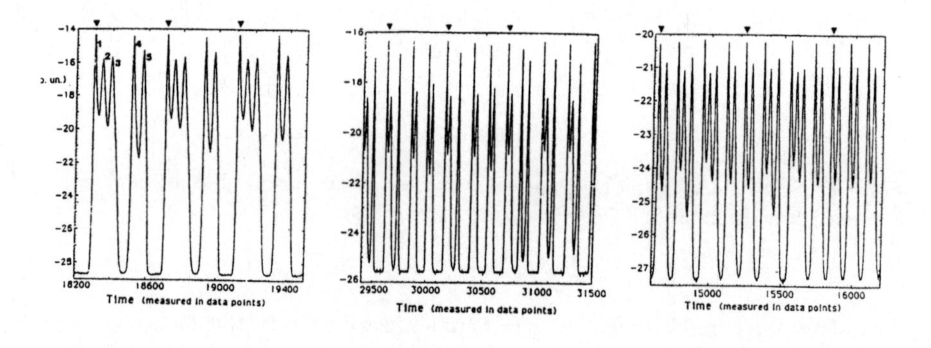

Fig. 7.10 Segments of time series observed in the LSA exhibit near periodic orbits of periods 5, 7, and 10. Each orbit goes around twice and is shown between arrows. Reprinted from the thesis of F. Papoff with permission [129].

7.2.3 Topological Analysis

In each of the data sets, only peaks of the type L and S (and their combinations) were observed. Two symbols sufficed to describe the behavior exhibited in all data.

As a first step, periodic orbits were extracted from each of the data sets. The number of orbits extracted varied from a minimum of four orbits for a few of the data sets to about a dozen orbits for several of the others. Since two symbols sufficed to describe all time series, the underlying template had two branches and three orbits were required to identify the template. There was always at least one additional orbit available to confirm the initial template identification for each data set [131].

A single embedding method was used for all time series. This was the integral–differential embedding, described in Eq. (7.3). The decay time for the integral was about two periods.

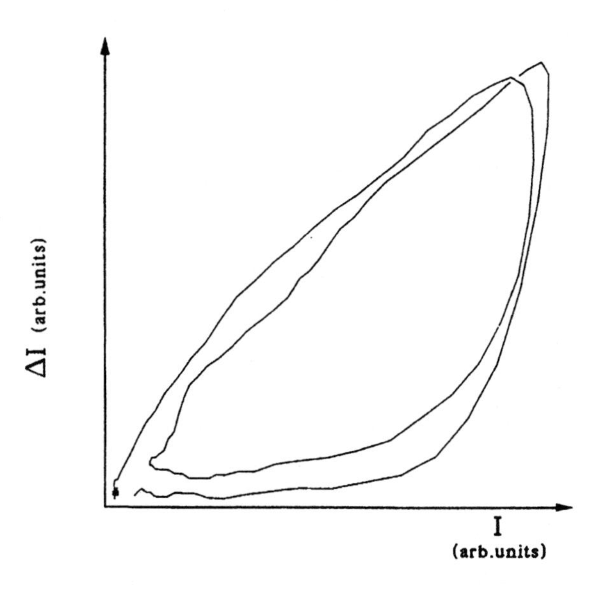

Fig. 7.11 The embedded orbit segment LL exhibits one crossing and forms the boundary of a Möbius strip. L belongs on an orientation-reversing branch of any branched manifold used to describe this system.

To determine whether the orbit segments L and S correspond to orientation-preserving or orientation-reversing flows, we extracted segments of type LL and SS and plotted their embeddings. The embedded segment LL is shown in Fig. 7.11. This embedded segment has one crossing and clearly forms the boundary of a Möbius strip. The embedded segments SS had no crossings. The orbit segment S is therefore orientation-preserving with zero global torsion, and was labeled 0; the orbit segment L is orientation-reversing, and labeled 1.

For each of the data sets a table of linking numbers was established. For example, in Fig. 7.12(a) we show a segment of data exhibiting a period-1 orbit L or 1 and back-to-back period-3 orbits LSL or 101. These two orbits are shown in the integral–differential embedding in Fig. 7.12(b). The linking number of these two orbits was computed counting crossings. All other crossings were computed in the same way.

For all data sets studied, an initial identification using three orbits pointed to a simple Smale horseshoe template, with $S = 0$ and $L = 1$. Subsequent tests using the remaining orbits extracted from the data set always confirmed the initial template identification.

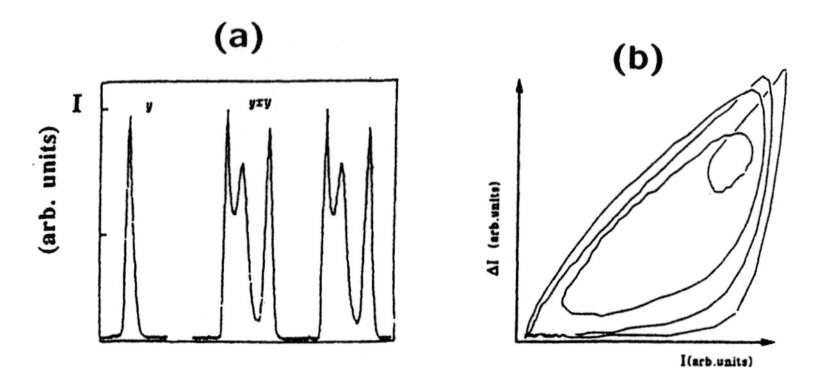

Fig. 7.12 (a) A single peak of type L and a back-to-back pair of peaks of type LSL are shown from a typical data set. (b) These two orbits are embedded in a three-dimensional phase space using an integral–differential embedding. The linking number for this pair of orbits is -1.

7.2.4 Useful Observation

During the course of this analysis, it was observed that certain sets of orbits seemed to appear together. It seemed almost as if the presence of one orbit "encouraged" (or *forced*) the presence of a coterie of specific other orbits. This observation, based on orbits extracted from experimental data, encouraged a study of forcing in horseshoe dynamics. The results of this analysis are presented in Chapter 9. In brief, the results are as follows. Some orbits (A_s, A_n) (saddle, node) that exist in Smale horseshoe dynamics cannot be created until other (B_s, B_n) orbits already exist. We say that the orbits A *force* the orbits B. The forcing is topological in nature. The theory of topological forcing in the horseshoe has not been completed, but forcing of orbits up to period 8 is known. Orbit forcing for other mechanisms that generate chaos (other templates) is in the rudimentary stages of understanding. It would really be nice if there existed an algorithm whose inputs were (1) a branched manifold and (2) one or more orbits in this branched manifold, and whose output consisted of all periodic orbits forced by the initial data.

7.2.5 Important Conclusion

Before this study was carried out, it was felt that template type was invariant under variation of control parameters. The results of this study showed with clarity that the templates underlying all the LSA data sets were the same. Changing control parameters (discharge current, operating pressure in the absorber, even the absorber itself) served only to "push the flow around" on the underlying branched manifold, but it did not modify the branched manifold itself. The branched manifold describes a stretch and fold mechanism for generating chaos. Changing control parameter values did not modify the mechanism: It served only to modify the spectrum of unstable periodic orbits which the invariant mechanism generated. Although this makes sense from a theoretical perspective, it is gratifying that this conclusion was drawn from the analysis of experimental data.

7.3 STRINGED INSTRUMENT

We normally think of a vibrating string as the most pristine example of a linear phenomenon: a harmonic solution of a linear wave equation subject to fixed boundary conditions and producing a sound of crystalline clarity and harmony. However, strings are "human," too, in that they have a nonlinear dimension to them. This nonlinearity can generate chaotic behavior if handled delicately enough. In the sections below we describe a beautiful experiment designed to explore and describe the chaotic behavior which can be exhibited by a vibrating string [132].

7.3.1 Experimental Arrangement

The partial differential equation that properly describes an oscillating string is complicated and nonlinear. The second-order PDE which is normally used to describe the oscillations of a string is a lowest-order truncation of a more complicated nonlinear equation. This truncation is linear. We often lose sight of the fact that the appropriate equation is nonlinear. The nonlinearities come about because of the deformation (stretching) of the string and because of the coupling of two transverse modes of oscillation.

A nonlinear equation describing string oscillations was first derived by Kirchhoff in 1883. Later versions of this equation were studied, and some chaotic orbits were identified in solutions of these equations. The chaotic behavior is manifest in slow variations of the amplitude of higher-frequency linear normal modes.

An experiment was designed to detect these amplitude oscillations. A schematic diagram of the experimental arrangement is shown in Fig. 7.13. In this experiment a tungsten wire is rigidly mounted. The wire is 0.15 mm in diameter, 7 cm in length, has density 3.39×10^{-3} g/cm, and its Young's modulus is 1.98×10^{14} dyne/cm. The wire is nonmagnetic and immersed in a static transverse magnetic field of 0.2 T. It has a natural vibration frequency of about 1350 Hz. The wire is forced by passing an alternating current through it with a frequency of about 1300 Hz. The current flow in the magnetic field forces the wire to vibrate near its resonant frequency. The amplitude varies with a lower frequency in the range of about 10 Hz. This amplitude variation is the interesting part of the experiment. The amplitude variation is detected by sampling the displacement of the wire once per forcing cycle at the driving frequency. This method of sampling effectively filters out the high-frequency (linear) response, sampling only the lower-frequency nonlinear response. Given the frequency ratios, this sampling method provides about 130 samples/cycle. Data sets of length 64K and 128K were taken; each observation was 16 bits. A typical data set recorded about 500 or 1000 cycles in the amplitude oscillations. The noise level was about 1% of the rms amplitude variation. Successful attempts were made to clean the data further.

Standard tests (false nearest neighbors [100, 101]) were used to determine a suitable embedding dimension. These tests showed that the data could be embedded successfully in three dimensions. A time delay for the embedding was determined by locating the first minimum in the mutual information. This suggested that a delay corresponding to a quarter of a cycle was optimum. This is not surprising in view of our previous comments about coupled oscillator embeddings in Section 6.4.6. Strange attractors obtained by this embedding procedure were constructed and rotated to determine the most suitable projection onto the plane (computer screen). Two strange attractors obtained from runs under different operating conditions are shown in Fig. 7.14.

7.3.2 Flow Models

Before analysis of the data was undertaken, a model for the flow was created. The motivation for this step was the following. If a model could be developed that was an accurate representation of the dynamics, the model might be able to provide additional subtle bits of information about the dynamics which were either not available from the data, or else not so reliable because of corruption by noise or the sparseness of the data.

In broad principles, the modeling effort follows the outline presented in Section 6.2.6. The details are different. A model of the form

$$\frac{d\mathbf{y}}{dt} = \mathbf{F}(\mathbf{y}) \tag{7.5}$$

was constructed. Since a three-dimensional delay embedding was adopted, all three driving functions $F_i(\mathbf{y})$ had to be modeled. Each function was modeled in the form

$$\mathbf{F}(\mathbf{y}) = \sum_{I=0}^{N_P} \mathbf{p}^{(I)} \pi^{(I)}(\mathbf{y}) \tag{7.6}$$

where the $\pi^{(I)}$ form a basis set of polynomials orthonormal over the attractor [with respect to the experimentally estimated invariant measure; cf. Eq. (6.49)]. The expansion coefficients $\mathbf{p}^{(I)}$ were determined by adopting an implicit Adams integration scheme [104]:

$$\mathbf{y}(t+\tau) = \mathbf{y}(t) + \tau \sum_{j=0}^{M} a_j^{(M)} \mathbf{F}(\mathbf{y}(t-(j-1)\tau)) \tag{7.7}$$

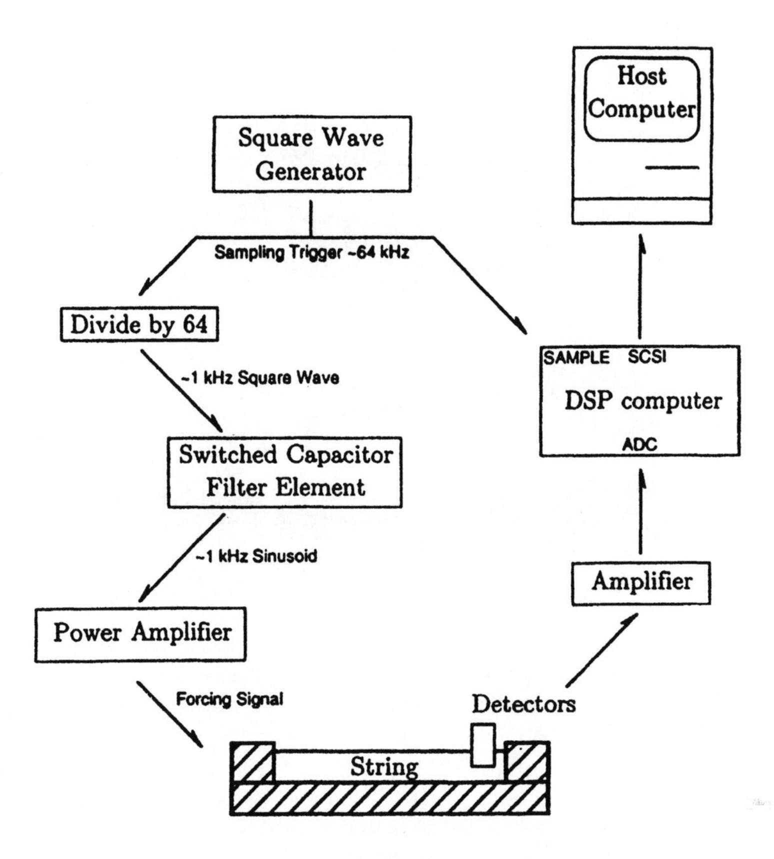

Fig. 7.13 Experimental arrangement for the driven string experiment showing string, detector, signal generator, power amplifier, signal amplifier, computer, scope, and recorder. Reprinted with permission from Tufillaro et al. [132].

Here the coefficients $a_j^{(M)}$ are implicit Adams coefficients of order M.

Models of this type were developed for some of the data sets taken. The usefulness of these models was tested using the entrainment test. The data did entrain the model outputs with sufficiently small coupling that several of the models were considered to be accurate representations of the dynamics [104, 132].

7.3.3 Dynamical Tests

Lyapunov exponents were computed both for the strange attractors constructed from the time-delay embedded data and for the models of these strange attractors. The analysis was extensive but was essentially a modified form of the analysis described

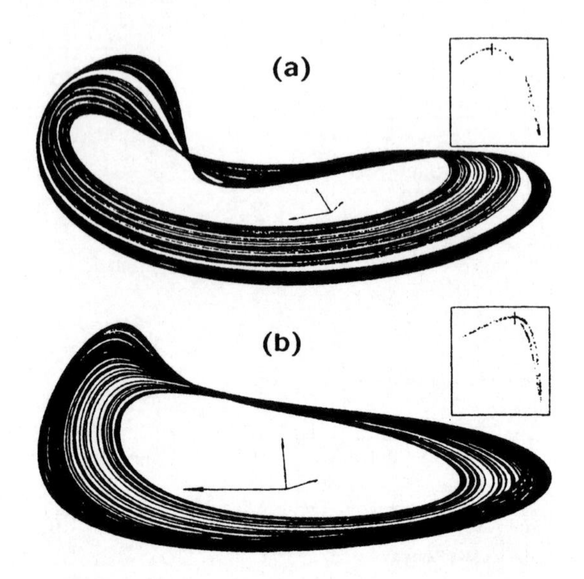

Fig. 7.14 Two strange attractors, for two experimental runs under different operating conditions, are projected from the embedding space R^3 onto a plane, which is the screen of the recording computer. Insets show the first return maps on a Poincaré section. Reprinted with permission from Tufillaro et al. [132].

in Section 7.1.5. For the embedded strange attractors, $\lambda_1 = 4.97 \times 10^{-2}$ and $\lambda_3 = -0.702$. The corresponding Lyapunov dimension is $d_L \sim 2.071$.

For the strange attractors generated by the model, the corresponding statistics are $\lambda_1 = 4.31 \times 10^{-2}$, $\lambda_3 = -0.576$, and $d_L = 2.075$. The agreement is not too bad considering the difficulty in computing these statistics. The salient feature is that the Lyapunov dimension is much less than 3, reinforcing the conviction that the dynamics is three -dimensional. In particular, these results justify the topological analysis that follows.

7.3.4 Topological Analysis

The experimental attractors shown in Fig. 7.14 are very thin. With imagination, it is possible to see the stretching and folding that occur in phase space on the far side of these attractors. The two first return maps also strongly suggest that the dynamics is generated by the development of a simple stretch and fold mechanism. This conclusion was confirmed by topological analysis of three experimental data sets and one data set generated by a model that was validated by an entrainment test.

The topological analyses proceeded as usual. The method of close returns was used to identify segments of data that followed unstable periodic orbits. These data segments were extracted and used as surrogates for the unstable periodic orbits "in" the strange attractor. Each surrogate orbit was tentatively named by symbol sequence. The name was determined by position on the first return map, taking the cut at the maximum. Six orbits extracted from an experimental data set are shown in Fig. 7.15.

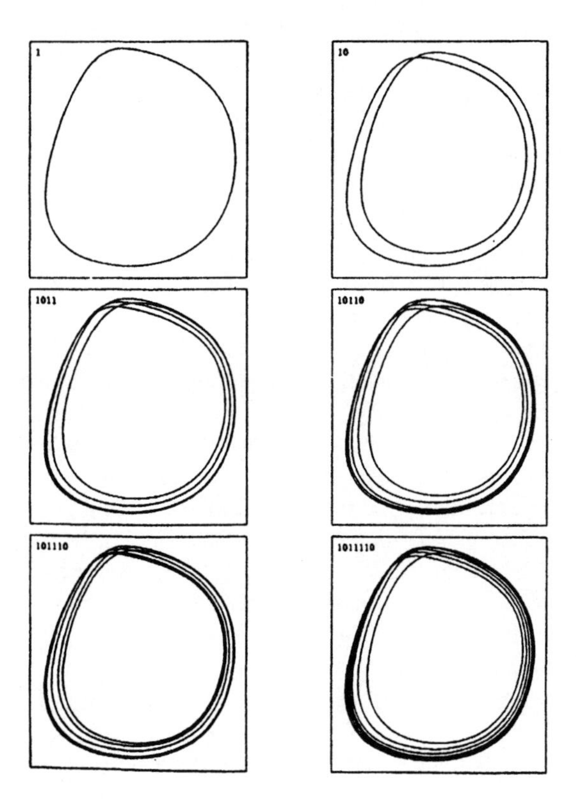

Fig. 7.15 Six surrogates for unstable periodic orbits. In this projection, the flow is clockwise, with stretching evident at 7 o'clock and folding at high noon. The name of each orbit is given. Reprinted with permission from Tufillaro et al. [132].

For each set of orbits extracted from the data (or simulations), the linking numbers were computed and compared with the linking numbers of the corresponding orbits on the zero-torsion Smale horseshoe template, which folds over from outside to inside. Disagreements only occurred for orbits where (noise-limited) resolution brought segments of separate orbits sufficiently close that crossing information was problematic. Neglecting these cases, there were no disagreements between the linking numbers observed from experimental data sets and simulated data sets, and those predicted for the corresponding orbits on the horseshoe template. The spectrum of unstable periodic orbits extracted from three different experimental data sets and one simulation are summarized in Table 7.5.

The Lyapunov dimensions of all attractors are very close to 2, and the return maps are very thin. It was therefore expected that this dynamical system exhibits characteristics very similar to those shown by unimodal maps of the interval. The highest entropy orbit extracted from the simulations was 5_1. Up to period 11 this orbit forces: 1_1, 2_1, 4_1, 8_1, 10_1, 6_1, 10_2, 8_2, 10_3, 11_1, 9_1, 11_2, 7_1, 11_3, 9_2, 11_4 [this is the universal-sequence (U-sequence) order]. All these orbits, and no others (to period 11), were observed in the simulation.

In the case of the first experimental data set, the highest entropy orbit observed is 10_2. This forces 1_1, 2_1, 4_1, 8_1, 10_1, 6_1, all of which are also observed, except for the fundamental saddle 1_1. In the second experimental data set, the highest entropy orbit 9_2 forces 1_1, 2_1, 4_1, 8_1, 10_1, 6_1, 10_2, 8_2, 10_3, 11_1, 9_1, 11_2, 7_1, 11_3. Of these, at least one of the two saddle-node partners was observed for all but the two high-period orbits 11_2 and 11_3. In the third experimental set, only the orbit 11_2, which is forced by the highest-entropy orbit 11_4, was not observed in the data. In view of the difficulty in finding close returns for high-period orbits, these results are not surprising and do not contradict the hypothesis that the U-sequence order is preserved in these experiments.

Topological analysis of the dynamics of the envelope of amplitude variations of the normal-mode oscillations of a stringed instrument has shown that a simple stretch and fold (Smale horseshoe mechanism) is responsible for generating the observed chaotic behavior. The small Lyapunov dimension of ~2.07 suggest that the dynamics is highly dissipative. The attractor has a hole in the middle, so it is possible to construct a global Poincaré section. The return map on the Poincaré section strongly recalls a unimodal map of the interval. The thinness of this return map suggests that periodic orbits are created in the U-sequence order. Analysis of the periodic orbits extracted from three data sets, and from a simulation of one of them, strongly supports this conjecture.

As the control parameters are varied, the underlying Smale horseshoe template remains unchanged: it is only the spectrum of unstable periodic orbits that changes.

7.4 LASERS WITH LOW-INTENSITY SIGNALS

The laser with modulated losses was described in Section 1.1. Its bifurcation diagram is shown in Fig. 1.3, and some low-period orbits are exhibited in Fig. 1.4. Time series from this laser were the first data on which a topological analysis was attempted. The attempt failed.

The reason is the following. The laser intensity output consists of several peaks, with each group of peaks separated by a very low intensity output. So low, in fact, that all intensities measured during these dead time intervals wound up in the same last channel of the multichannel analyzer used to record the data. As a result, there is no honest way to determine orbit crossing information in those regions.

This type of behavior is characteristic of many class B lasers. For example, it is also exhibited by the laser with saturated losses (LSA) [cf. Figs. 7.10 and 7.12(a)]. To carry out a successful topological analysis, this *deep minimum problem* must be

Table 7.5 Spectrum of low-period orbits extracted from three experimental data sets and one simulation[a]

P_j	Symbol	ES	h_T	Exp. 1	Exp. 2	Exp. 3	Model
1_1	1	0	0.000	m	x	x	x
2_1	01	1	0.000	x	x	x	x
4_1	0111	5	0.000	x	x	x	x
5_1	01111	8	0.414				x
5_1	01101	8	0.414				x
6_1	011111	13	0.241	x	x	x	x
6_1	011101	13	0.241	x	x	x	x
7_1	0111111	18	0.382		x	x	x
7_1	0111101	18	0.382		x	x	x
8_1	01110101	23	0.000	x	x	x	x
8_2	01111111	25	0.305		m	m	x
8_2	01111101	25	0.305		x	x	x
9_1	011111111	32	0.366		m	m	x
9_1	011111101	32	0.366		x	x	x
9_2	011110111	30	0.397		m	m	x
9_2	011110101	30	0.397		x	x	x
10_1	011111111	37	0.207	x	x	x	x
10_1	0111111101	37	0.207	x	x	x	x
10_2	0111110111	39	0.272	x	x	x	x
10_2	0111110101	39	0.272	m	x	x	x
10_3	0111010111	41	0.328		m	m	x
10_3	0111010101	41	0.328		x	x	x
11_1	01111111111	50	0.357		m	m	x
11_1	01111111101	50	0.357		x	x	x
11_2	01111110111	48	0.374		m	m	x
11_2	01111110101	48	0.374		m	m	x
11_3	01111010111	46	0.390		m	m	x
11_3	01111010101	46	0.390		m	x	x
11_4	01111011111	46	0.403			m	x
11_4	01111011101	46	0.403			x	x

[a]For each orbit we provide: the U-sequence name P_j (period, order of occurrence); symbol name; exponent sum (ES); and topological entropy (h_T). An x indicates that the orbit has been located in the data set. Saddle-node pairs have the same U-sequence name, exponent sum, and topological entropy. Orbits indicated m are predicted by the U-sequence order but not observed in the data.

resolved. Two ways now exist to resolve this problem. The first involves sophisticated theoretical methods (this section), the second involves sophisticated experimental methods (Section 7.5.2 and Fig. 7.32). The first method has been carried out offline on data that had previously been taken. It provides information about orbit organization that is unique up to global torsion. The second method must be carried out online and provides information about global torsion.

7.4.1 SVD Embedding

A *semitopological analysis* was carried out successfully by Solari, Natiello, and Vazquez [133] on LSA data taken in Pisa by Arimondo's group [129]. They first created an SVD embedding of the data. Then they constructed a first return map for the peaks in the embedding space. This return map was used to construct a branched manifold that described the dynamics. The branched manifold was not unique: Its global torsion was not determined. In this subsection we describe the SVD embedding.

Each peak in the data was located. The value of the intensity at the peak and at k ($\sim$10) equally spaced points on both sides of the peak were determined. Since the signal usually wasn't recorded exactly at a peak, these measurements usually had to be interpolated. To these $2k + 1$ pieces of information one other important piece of information was added: the time interval to the next peak. Thus each of the N peaks was identified by a point in a $(2k+2)$-dimensional space. This reduced the long time series to an $N \times (2k + 2)$ matrix $Z(\mu, i)$, with $1 \leq \mu \leq N$ and $1 \leq i \leq 2k + 2$.

This rectangular matrix was subjected to a singular-value decomposition. Normally, only a small number of eigenvalues was sufficient to recreate the input data matrix $Z(\mu, i)$ from the left and right eigenvectors. To be explicit, the real symmetric matrix $Z^t Z = \sum_{\mu=1}^{N} Z(\mu, i) Z(\mu, j)$ was created and diagonalized. The eigenvalues λ_α were ordered, starting from the largest: $\lambda_1 > \lambda_2 > \cdots$. Each eigenvector $v_\alpha(i)$ of the $(2k + 2) \times (2k + 2)$ matrix was used to construct a dual eigenvector $z_\alpha(\mu)$ in the usual fashion:

$$z_\alpha(\mu) = \sum_{j=1}^{2k+2} Z(\mu, j) v_\alpha(j) \tag{7.8}$$

If d eigenvalues described a sufficient part of the original matrix Z (e.g., $\sum_{\alpha=1}^{d} \lambda_\alpha /$ Tr $Z^t Z > 0.95$), only those vectors with $\alpha = 1, 2, \ldots, d$ were retained in the analysis. This allowed each peak to be described by a d-component vector $\mathbf{z}(\mu) = (z_1(\mu), z_2(\mu), \ldots, z_d(\mu))$.

7.4.2 Template Identification

This SVD embedding allowed for the construction of a first return map according to

$$\mathbf{z}(\mu) \xrightarrow{R} \mathbf{z}(\mu + 1) \tag{7.9}$$

Once a return map was available, it was possible to locate periodic orbits. These orbits were used to determine a template up to global torsion.

The starting point is the construction of a braid for the action of the first return map on the periodic orbits. The process is illustrated in Fig. 7.16. In Fig. 7.16(a) we show the five intersections of a period-5 orbit with the Poincaré section. The Poincaré section is the truncated embedding space R^d. The five intersections are labeled by integers $1 \to 5$. The action of the first return map is to generate a permutation, for example, the permutation $(12435) = 1 \to 2 \to 4 \to 3 \to 5$ is shown in Fig. 7.16(b). This mapping deforms the Poincaré section. The deformation is outlined by first projecting R^d into R^3. Then each point i is connected to its image $R(i)$ under the first return map. The curves that make these connections outline a flow in R^3. This flow deforms R^3. The deformation can be determined as follows. Links that connect adjacent points i and $i+1$ are drawn. The evolution of these links under the flow outline the deformation that the flow induces on the phase space. Finally, the evolution is projected from R^3 to R^2, preserving crossing information.

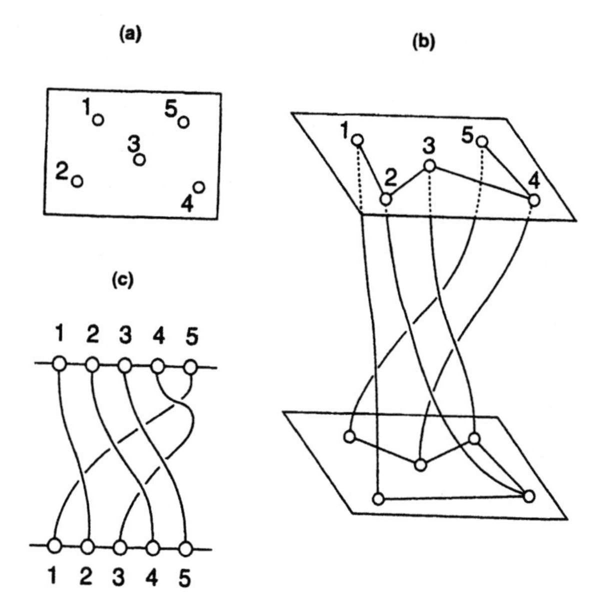

Fig. 7.16 (a) Five intersections of a period-5 orbit with a Poincaré section (R^d). (b) A braid is constructed by projecting the return map into R^3 and then connecting point i with its image $R(i)$. The deformation induced by the flow is revealed by connecting points i and $i+1$ and then computing the image of these segments under the flow. (c) The braids are projected to a plane, $R^3 \to R^2$, preserving crossing information. This projection outlines the branches of a template. Additional periodic orbits, or braid segments, fill in details.

The result for a period-5 orbit is shown in Fig. 7.16(c). In this projection, the front three segments, $1 \to R(1) = 2$, $2 \to R(2) = 4$, and $3 \to R(3) = 5$, outline an orientation-preserving branch of a template with two branches. The remaining two components of this braid, $4 \rightarrow R(4) = 3$ and $5 \rightarrow R(5) = 1$, outline an orientation-reversing branch of the template, which folds behind the first in this projection. The branch $3 \rightarrow R(3) = 5$ plays a pivotal role in this projection.

The period-5 orbit provides a skeleton that outlines a two-branch Smale horseshoe template. Additional periodic orbits can be added to this projection to fill out the details. Flow segments not belonging to periodic orbits can be introduced to determine the local torsion in some parts of the flow.

7.4.3 Results of the Analysis

The projections of the segments $R^d \rightarrow R^3$, and then $R^3 \rightarrow R^2$, play the same role here as the identifications (5.3) and (5.1) in the Birman–Williams theorem. The first is equivalent to projecting a strongly contracting system in R^d into R^3. The second deflates flows in R^3 down to two-dimensional branched manifolds. This procedure is not completely equivalent to a topological analysis because global torsion information is lost. It is not always safe to assume that global torsion is zero if it is not otherwise available, as the analyses in the following two sections will make clear.

As with so many other simple physical systems, this analysis showed that chaotic behavior was generated by a simple stretch and fold mechanism characteristic of the Smale horseshoe. The deformation of the Poincaré section, as outlined by the braid (the more periodic orbits, or braid segments, the better), was then used to construct an estimate for the topological entropy of this return map. The topological entropy is independent of the global torsion.

7.5 THE LASERS IN LILLE

A series of experiments has been carried out at the Laboratoire de Spectroscopie Hertzienne in Lille, France to apply the topological analysis procedure to experimental data generated by class B lasers operating under a variety of conditions. A theoretical effort to model the behavior of the lasers was undertaken in parallel with the development of the experiments. This was done to provide guidance for some aspects of the data analysis program.

In the first subsection we describe the laser model. We also describe the data sets which this model generates, the topological analysis of these data sets, and the template that governs the chaotic behavior of this model in the range of control parameters under which it was run.

In the following subsections we describe three different lasers that were operated in the laboratory. The first is a CO_2 laser with modulated losses similar to the one described in Chapter 1. The second one is a neodynium-doped yttrium–aluminum–garnet laser (Nd:YAG). The third is a neodynium-doped fiber optic laser (Nd:FL). The three of them are class B lasers, so their behavior is described more or less accurately by the model developed in the first subsection. The chaotic behaviors observed in the model, the CO_2 laser, the Nd:YAG laser, and the Nd:FL operating in three different regimes have been found to be described by different templates, depending on the system and on the control parameters investigated. Nevertheless, there is a systematic relation among these templates. These differences are put into satisfying context in the final subsection. Specifically, they are all subtemplates of one simple, organizing template, called variously a *gâteau roulé* or jellyrole template.

7.5.1 Class B Laser Model

A laser operates when at least one mode of the electromagnetic field increases its amplitude as it propagates through an excited medium. In the simplest laser models the excited medium is modeled by a set of two-level atoms. The appropriate variables are the radiation field, the population inversion of the two-level medium, and the atomic polarization that couples these two variables. If the time scale of the polarization is much faster than the time scale of either of the other two variables, the polarization can be adiabatically eliminated. Lasers that exhibit this property are called class B lasers.

If a single mode is excited, the amplitude $\mathbf{E}$ can be replaced by the intensity $I = \mathbf{E} \cdot \mathbf{E}$, and the laser equations expressed in the simple form

$$\begin{aligned} \frac{dI}{dt} &= I(D-1) \\ \frac{dD}{dt} &= \gamma[A - D(1+I)] \end{aligned} \tag{7.10}$$

The population inversion is represented by $D(t)$, t (reduced time) is measured in units of the photon lifetime in the cavity, γ is the decay rate of the population inversion, and A describes the pump parameter, normalized so that $A = 1$ at the onset of lasing. Typically, γ is on the order of 10^{-3} for a CO_2 laser and 10^{-4} for a Nd:YAG laser. Written in a different form, these equations are equivalent to Eqs. (1.2).

Above threshold, the only dynamical state is the stationary regime $(I, D) = (A - 1, 1)$. When perturbed, the laser behaves as a damped oscillator, relaxing to the fixed point with oscillations at the characteristic angular frequency $\omega_r \sim \sqrt{\gamma(A-1)}$.

The phase space is two-dimensional. There is not enough room in this phase space for chaotic motion to occur. The dimension of this phase space is enlarged by one when one of the control parameters is modulated, such as the pump parameter A or the cavity losses. Periodic modulation of the cavity losses is modeled simply by replacing the first of equations (7.10) by $dI/dt = I[D - 1 - m\cos(\omega t)]$, where m ($|m| < 1$) and ω are, respectively, the amplitude and angular frequency of the periodic loss modulation.

These equations are "stiff": They are difficult to integrate because of abrupt jumps in the output variables, in particular, the intensity. They are softened by a logarithmic transformation $I = e^L$, or $L = \log(I)$. Since this is a smooth scale transformation, it has no effect on the topology of the attractor. The resulting equations are

$$\begin{aligned} \frac{dL}{dt} &= D - 1 - m\cos(\omega t) \\ \frac{dD}{dt} &= \gamma[A - D(1+e^L)] \end{aligned} \tag{7.11}$$

These equations were integrated for $\gamma = 1.1 \times 10^{-3}$, $A = 1.1$, $m = 2.5 \times 10^{-2}$, and $\omega = 10^{-2}$ using standard techniques.

The phase space for this useful model is three-dimensional, with variables $\{L(t), D(t), \phi(t)\}$, where $\phi(t) = (t/T) \bmod 1$ and $\omega T = 2\pi$. Topologically, the phase space is a donut: $D^2 \times S^1$. There is a hole in the middle so that (1) a global Poincaré section always exists, and (2) the relative rotation rates can be computed for all periodic orbits. The phase space is shown in Fig. 7.17. One possible Poincaré section is shown, together with a pair of period-1 orbits with a linking number of -1.

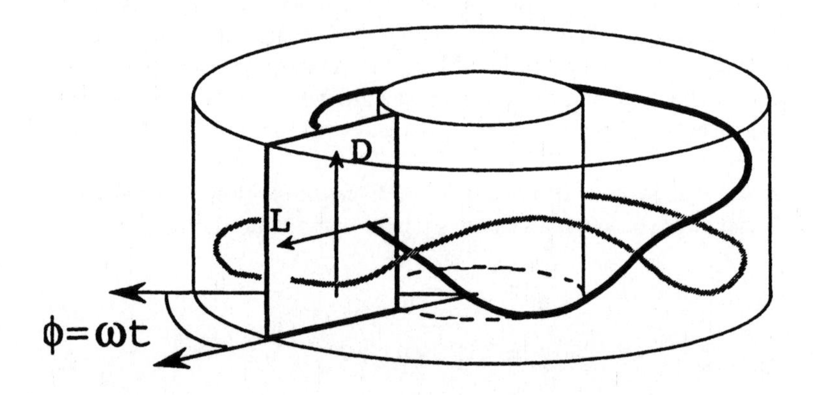

Fig. 7.17 The phase space of the modulated class B laser (7.11) is a donut $R^2 \times S^1$. Two period-1 orbits are shown. They have linking number -1. The plane shown at constant ϕ can be used as a Poincaré section. Reprinted from Boulant *et al.* [134] with permision of World Scientific.

In numerical simulations, different templates are obtained depending on the values of control parameters such as modulation frequency and amplitude. These templates are consistent with those observed experimentally and described in the next subsections. In this section we restrict ourselves to the particular regime described in [134]. The template observed in this work includes as subtemplates both the horseshoe template observed in the first experiment discussed below, the reverse horseshoe described in the second experiment, and a scroll template, also observed in experiments.

The usual steps were followed [134]. The equations were integrated until the transients died out. Then time-series data were used to construct a strange attractor in $D^2 \times S^1$. Intersections of the strange attractor on a number of Poincaré sections ($\phi =$ const.) were recorded. Figure 7.18 shows the intersection of the strange attractor with one such Poincaré section ($\phi = 0$). Also shown in this figure are the intersections of one period-1 orbit and three different period-2 orbits with this section.

A large number of periodic orbits were identified. The method of close returns was not used. Since the system is not a real experiment, there existed the luxury of searching for periodic orbits using the dynamical equations themselves. The Poincaré

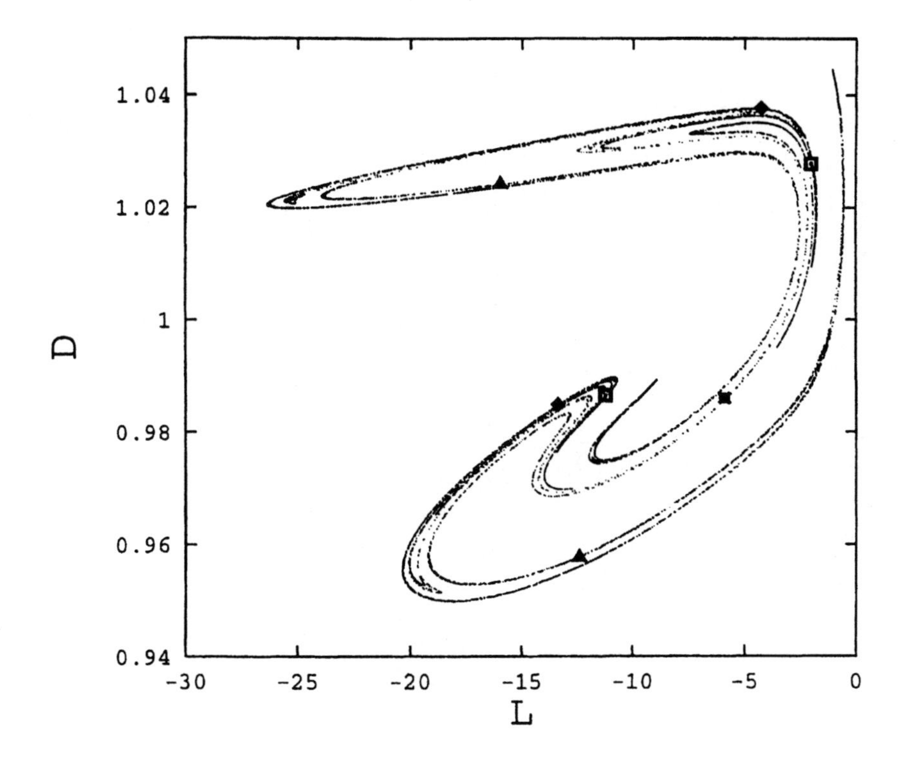

Fig. 7.18 The strange attractor intersects a Poincaré section at $\phi = 0$ in 10^4 points. The intersection of a period-1 orbit α (star) and the two intersections of the three period-2 orbits β (diamond), γ (square), and δ (triangle) with this plane are shown. Reprinted from Boulant *et al.* [134] with permision of World Scientific.

section was divided into a 64×64 grid of small squares. In each of the squares where no lower-period orbit had been found, a point on the attractor was used as an initial condition. The equations were integrated until the evolution brought the phase-space point back to the original square or one of its nearest neighbors after P periods. At that point a Newton–Raphson procedure was invoked to walk into the periodic orbit of period P. In this way a total of 27 orbits of period up to 10 were located in the strange attractor. The linking numbers and local torsions of the orbits up to period 7 are collected in Table 7.6.

There are three period-2 orbits. The template that describes the dynamics must have at least three branches. Because the system is weakly dissipative, no simple symbolic coding was available, and the general procedure detailed in Appendix A was used to determine the simplest template describing the organization of periodic orbits. Two three-branch templates were consistent with the invariants of the detected

Table 7.6 Seventeen orbits up to period 7 found in the attractor generated by equations (7.11)[a]

	1_1	2_1	2_2	2_3	4_1	4_2	6_1	6_2	6_3	6_4	6_5	6_6	7_1	7_2	7_3	7_4	7_5
1_1	1																
2_1	1	2															
2_2	1	2	3														
2_3	1	2	2	1													
4_1	2	4	4	3	3												
4_2	2	4	5	4	8	5											
6_1	3	6	6	5	10	12	4										
6_2	3	6	6	5	10	12	15	6									
6_3	3	6	6	5	10	12	15	16	5								
6_4	3	6	7	6	12	14	18	18	18	7							
6_5	3	6	6	5	10	12	15	16	16	18	5						
6_6	3	6	7	6	12	14	18	18	18	21	18	8					
7_1	3	7	7	6	12	14	18	18	18	21	18	21	6				
7_2	3	7	7	6	12	14	18	18	18	21	18	21	21	6			
7_3	3	7	7	6	12	14	18	19	18	21	19	21	21	21	7		
7_4	3	7	7	6	12	14	18	18	18	21	18	21	21	21	21	5	
7_5	3	7	7	6	12	14	18	18	18	21	18	21	21	21	21	21	7

[a] Linking numbers for pairs of orbits up to period 7 are shown off the diagonal. The diagonal contains the local torsion of each orbit.

orbits. One of these templates was a scroll, as shown in Fig. 7.19. The template matrix and squeezing array for this template are

$$T = \begin{bmatrix} 0 & 0 & 0 \\ 0 & 1 & 2 \\ 0 & 2 & 2 \end{bmatrix}$$

$$A = \begin{bmatrix} 0 & 2 & 1 \end{bmatrix} \tag{7.12}$$

Another possible solution for the template matrix and array was the set

$$T = \begin{bmatrix} 0 & 0 & 0 \\ 0 & 2 & 2 \\ 0 & 2 & 1 \end{bmatrix}$$

$$A = \begin{bmatrix} 0 & 2 & 1 \end{bmatrix} \tag{7.13}$$

This solution cannot describe a continuous stretching and folding deformation in phase space, since the local torsions of adjacent branches 0 and 1 differ by more than 1. This solution was therefore rejected.

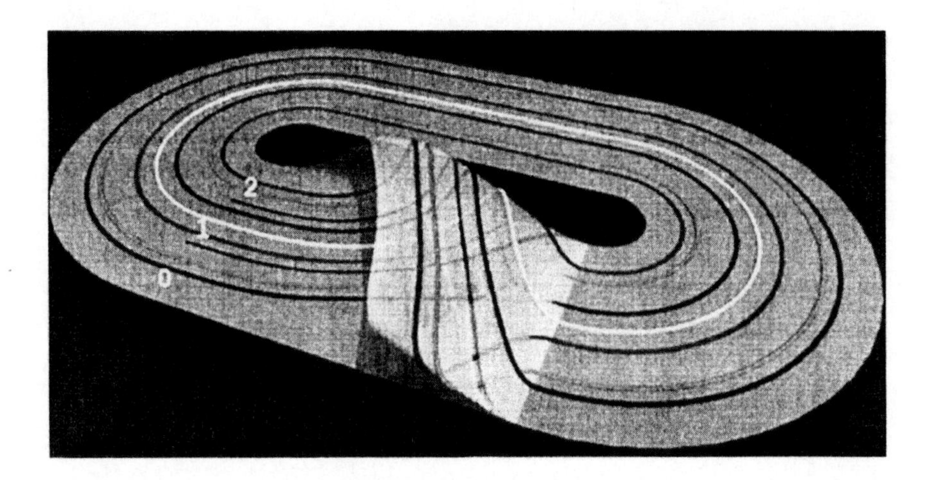

Fig. 7.19 This template describes the topological organization of the chaotic attractor generated by the dynamical system (7.11). It displays three branches that are rolled up by the stretching and folding (rolling) mechanism. Branches 0, 1, and 2 are labeled. The four low-period orbits used to identify this template were originally shown in color [$1_1 = 0$ (yellow), $2_1 = 02$ (green), $2_2 = 12$ (red), $2_3 = 01$ (blue)]. Reprinted from Boulant *et al.* [134] with permision of World Scientific.

Additional solutions with matrices T' and A' related to the original matrices T and A by $T'_{ij} = T_{2-i,2-j}$ and $A'_i = A_{2-i}$ simply describe the same template from the opposite side.

The scroll template defined by the matrices (7.12) was used to predict the set of linking numbers for all other orbits that were located. This effort helped to identify possible symbolic names of the orbits. The symbolic coding could not be obtained in the usual way for lack of a suitable return map. After appropriate orbit labeling, there were no discrepancies between the linking numbers computed for orbits identified in the dynamics and the corresponding orbits on the three-branch scroll template.

It is possible to become more comfortable with the scroll template mechanism by following the Poincaré section around through one cycle. Figure 7.20 presents the intersection of the strange attractor with 12 equally spaced Poincaré sections at phases $\phi = \frac{i}{12}$, with $i = 0, 1, \ldots, 11$. With some imagination, it is possible to visualize the scrolling as the transverse section winds around the torus $D^2 \times S^1$. As an aid to the imagination, in Fig. 7.21 we have duplicated this figure with a dark one-dimensional curve running through the attractor. As the phase advances, this curve winds up and is mapped back onto itself, with many points having three preimages.

The stretching and squeezing mechanism is extracted from this figure as follows. The dark curve in Fig. 7.21(a) is topologically an interval. The location of points along this curve can be parameterized by a real number s, $0 \leq s \leq 1$. We choose $s = 0$ at the left-hand end of this interval [near (a)]. Under evolution through one period ($a \to a$) the curve is elongated and folded over, so that $s \to s'$. The return map under forward evolution is shown in Fig. 7.22(a). This return map has three branches. The outer two are orientation preserving, the middle branch is orientation reversing.

The interval shown in Fig. 7.21(a) can be straightened out and embedded in a long thin rectangle. The mapping of this rectangle into itself generated by the flow is shown in Fig. 7.22(b). The scroll nature of this deformation is clear from this figure.

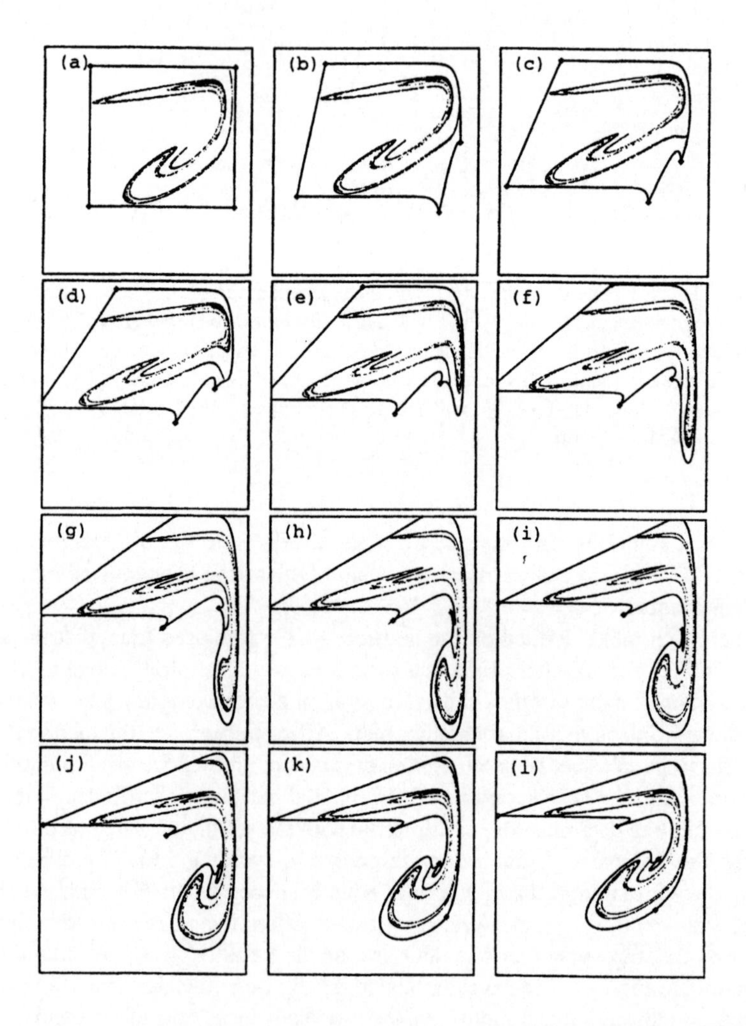

Fig. 7.20 As the Poincaré section advances from $\phi = 0$ at (a) to $\phi = \frac{11}{12}$ at (l), the shape of the intersection of the strange attractor changes by rolling up. Two folding mechanisms can be seen. The first becomes visible in the right-hand section of (d), and the second is initiated in the bottom section of (g). Reprinted from Boulant *et al.* [134] with permision of World Scientific.

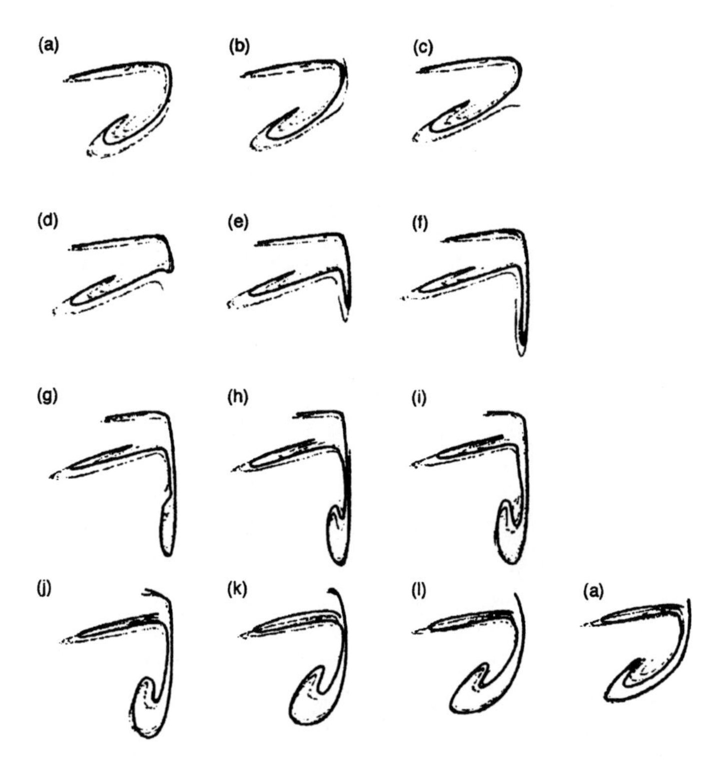

Fig. 7.21 A thick one-dimensional curve has been drawn through the attractor shown in Fig. 7.20. As the phase advances, this curve is deformed and finally mapped back into itself. The deformation exhibits the two nonlocal folds that create the three-branch scroll template.

7.5.2 CO_2 Laser with Modulated Losses

The CO_2 laser used in the first experiment is based on a 235-mm-long sealed-off waveguide containing a gaseous mixture of He (70%), CO (9%), Xe (4%), and CO_2 (17%), the latter being responsible for laser action. The total gas pressure is about 100 mm Hg. The active medium is pumped by a radio-frequency discharge. The waveguide is enclosed in Fabry–Perot cavity delimited by a plane and a spherical mirror. A distinctive feature of this laser originally designed for optical communications is a signal-to-noise ratio on the order of a few hundreds.

The modulating device consists of a Brewster plate and of an electro-optical modulator made of a CdTe crystal. As with the laser of Section 1.1, the polarization state of the laser radiation is modified when a voltage is applied to the crystal because of the induced birefringence. This modifies the intracavity losses because the Brewster plate transmits only one polarization direction (Fig. 1.1). A small loss modulation can induce large variations in the ouput intensity of the laser if the modulation frequency

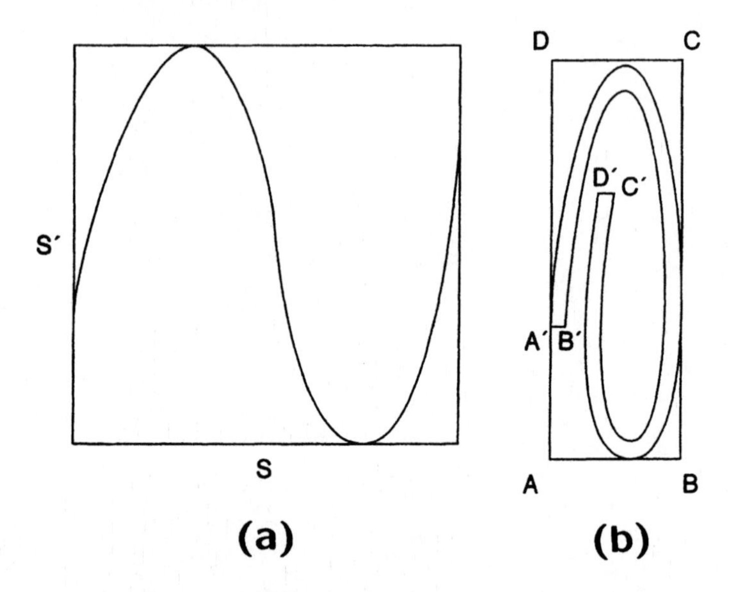

Fig. 7.22 (a) The interval shown in Fig. 7.21(a) is parameterized by a real number s, $0 \leq s \leq 1$. The flow generates a return map $s \rightarrow s'$. This can be seen by comparing the deformed interval in Fig. 7.21(l) and (a) to the original interval in Fig. 7.21(a). (b) The interval in Fig. 7.21(a) can be embedded in a long, thin rectangle. The return map of the rectangle $ABCD$ to its image $A'B'C'D'$ reveals the beginning of a scroll.

(which was fixed at 382.5 kHz in this experiment) is chosen close to the relaxation frequency of the laser. Under these conditions, the laser follows a classical period-doubling cascade leading to chaos as the modulation amplitude is increased. The signal-to-noise ratio is sufficiently high that the period-16 orbit of this cascade can clearly be observed.

Both the very low noise and a high dissipation rate (attractors have fractal dimension close to 2) make it relatively easy to extract unstable periodic orbits from the chaotic signals delivered by this laser. Figure 7.23 shows beautiful signatures of unstable periodic orbits embedded in a chaotic regime. They have been detected using the close-return technique described in the Chapter 6. In particular, the period-1 orbit in the first row is shadowed closely by the trajectory of the system during more than seven modulation periods. Similar sequences lasting up to 14 periods have been observed. Note also the peculiar case of the second row, where after a few cycles near a period-3 orbit, the systems enters directly into the neighborhood of a period-2 orbit.

At first glance, these bursts of almost-periodic behavior look like exceptional events. However, they carry most of the information needed to characterize the chaotic regime. This might seem counterintuitive if one looks at the time-series

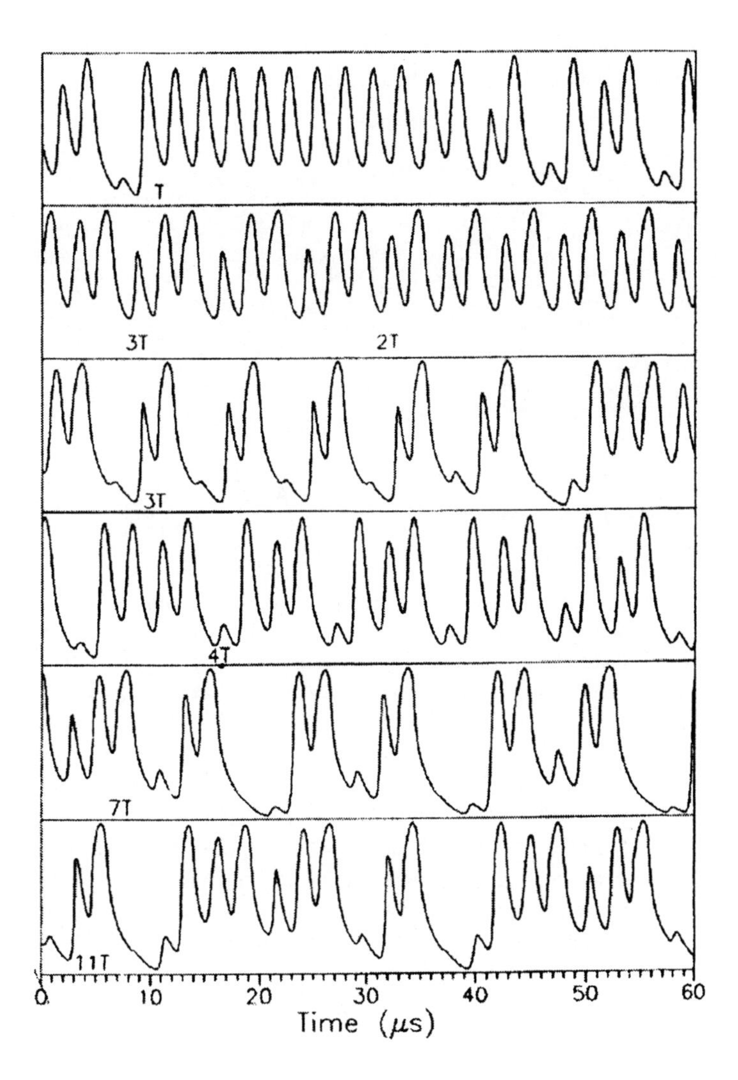

Fig. 7.23 Time-series segments closely shadowing unstable periodic orbits. The periods of these orbits are indicated at the beginning of segments. Reprinted from Lefranc and Glorieux [89] with permision of World Scientific.

segments of Fig. 7.23, but it becomes obvious when a geometric representation of the dynamics in a phase space is used, as can be seen in Fig. 7.24.

Figure 7.24(a) shows a strange attractor reconstructed from a long chaotic trajectory. The phase space is similar to that of Fig. 7.17, with planes of constant phase modulation ϕ being parameterized by time-delay coordinates $X(t)$ and $X(t+\tau)$. Figure 7.24(b) shows the same phase space but this time filled with the (almost) closed curves associated to periodic time series extracted from the time series (Fig. 7.23). The two representations are almost indistinguishable to the eye.

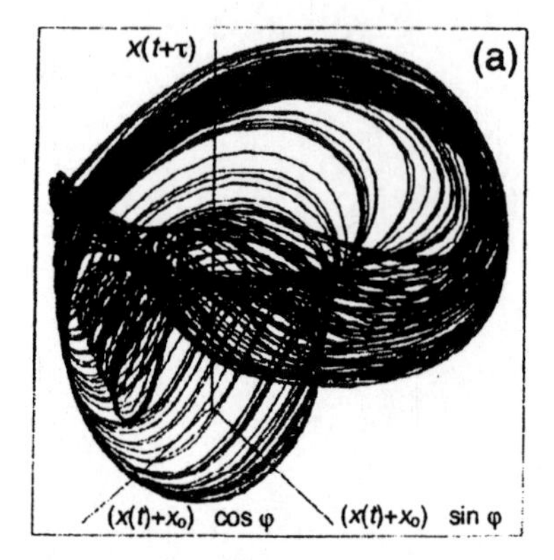

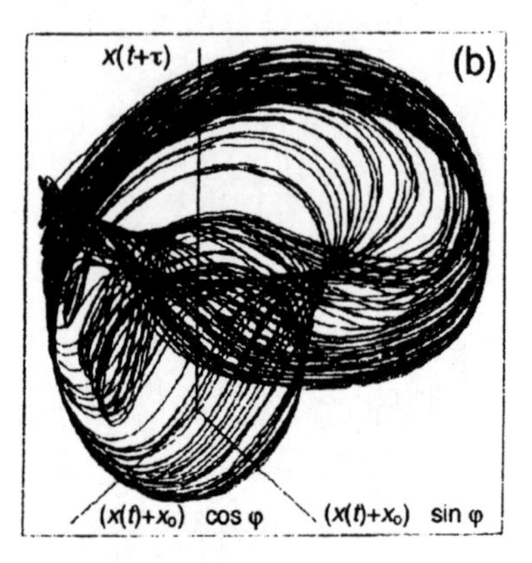

Fig. 7.24 (a) Experimental strange attractor of the CO_2 laser reconstructed using cylindrical coordinates $\{X(t), X(t+\tau), \phi\}$ for the phase space ($\tau = T/4$); (b) Superposition of unstable periodic orbits embedded in this attractor and extracted from the time series provide an excellent representation of the attractor. Reprinted from Lefranc and Glorieux [89] with permision of World Scientific.

Since unstable periodic orbits are densely embedded in a chaotic attractor, it is not surprising that it can be well approximated by a relatively small number of orbits. Yet, Fig. 7.24 shows that it is not unreasonable to rely on this for characterizing an experimental system: Every point in the attractor is close to a detected periodic orbit, except in a few places. It has recently been observed that low-period orbits are exceedingy rare at the border between two parts of the attractor that are folded over each other [49, 135]. When one plots periodic orbits up to a given period, one finds gaps precisely at these regions. It is plausible that the small regions not visited by an unstable periodic orbit in Fig. 7.24(b) correspond to such gaps. Gaps can also correspond to very unstable periodic orbits whose neighborhood is more rarely visited by the system.

Once a sufficient number of periodic orbits have been reconstructed in phase space, their topological invariants can be computed. Table 7.7 displays the relative rotation rates of 8 periodic orbits up to period 6 (the analysis of a more complete set can be found in [89]). It was found that these invariants are compatible with the standard Smale's horseshoe template [Figs. 5.7(d), 5.20, and 5.22].

Since the attractor has fractal dimension close to 2, this result can be verified by inspecting the evolution of Poincaré sections as the modulation phase ϕ is swept from 0 to 1, as was done for the laser model in Fig. 7.20. This is shown in Fig. 7.25. Folding occurs from Fig. 7.25(f) to (l), where the Poincaré section has the characteristic form of a horseshoe. It can be checked that the left branch of the section of Fig. 7.25(l) is

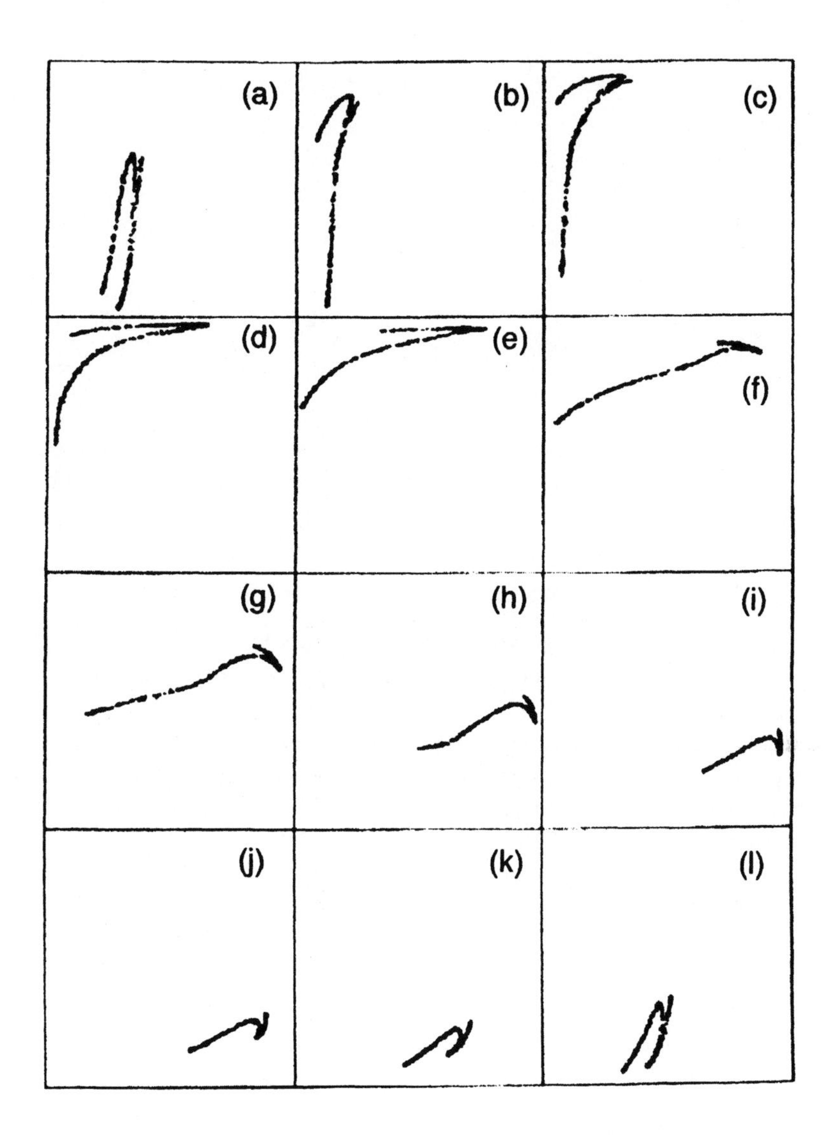

Fig. 7.25 Poincaré sections of constant modulation phase of the chaotic attractor of Fig. 7.24. The phase increases from 0 at (a) to $\frac{11}{12}$ at (l). The stretching and squeezing mechanisms organizing the strange attractor are easily discerned. They correspond to the ones described by the standard Smale's horseshoe. Reprinted from Lefranc and Glorieux [89] with permision of World Scientific.

Table 7.7 Relative rotation rates of extracted orbits up to period 6 for a regime of the CO_2 laser

Orbit	1	2	3	4	5a	5b	6a	6b
	y	xy	xy^2	xy^3	xy^4	$xyxy^2$	$xyxy^3$	xy^5
1	0							
2	$\frac{1}{2}$	$0, \frac{1}{2}$						
3	$\frac{1}{3}$	$\frac{1}{3}$	$0, \frac{1}{3}$					
4	$\frac{1}{2}$	$\frac{1}{2}, \frac{1}{4}$	$\frac{1}{3}$	$0, \frac{1}{2}, (\frac{1}{2})^2$				
5a	$\frac{2}{5}$	$\frac{2}{5}$	$\frac{1}{3}$	$\frac{2}{5}$	$0, (\frac{2}{5})^4$			
5b	$\frac{2}{5}$	$\frac{2}{5}$	$\frac{1}{3}$	$\frac{2}{5}$	$\frac{2}{5}$	$0, (\frac{2}{5})^4$		
6a	$\frac{1}{2}$	$\frac{1}{2}, \frac{1}{3}$	$\frac{1}{3}$	$\frac{1}{2}, \frac{1}{3}$	$\frac{2}{5}$	$\frac{2}{5}$	$0, (\frac{1}{3})^2, (\frac{1}{2})^3$	
6b	$\frac{1}{2}$	$\frac{1}{2}, \frac{1}{3}$	$\frac{1}{3}$	$\frac{1}{2}, \frac{1}{3}$	$\frac{2}{5}$	$\frac{2}{5}$	$\frac{1}{2}, \frac{1}{3}$	$0, (\frac{1}{3})^2, (\frac{1}{2})^3$

not rotated over one modulation period, which indicates that the global torsion of the flow is zero.

7.5.3 Nd-Doped YAG Laser

The Nd:YAG laser experiment is shown schematically in Fig. 7.26 [136]. It consists of a Nd-doped YAG rod (the laser) which is 10 mm long, 7 mm in diameter, and contains a Nd concentration of 1.1%. It is in a Fabry–Perot laser cavity and pumped by a continuous-wave laser diode operating at 812 nm. The ends of the rod are plane. One is highly reflective at the laser wavelength (1064 nm), and the other is antireflection coated. The pump power is sinusoidally modulated, $A(t) = A_0(1 + m \cos \omega t)$.

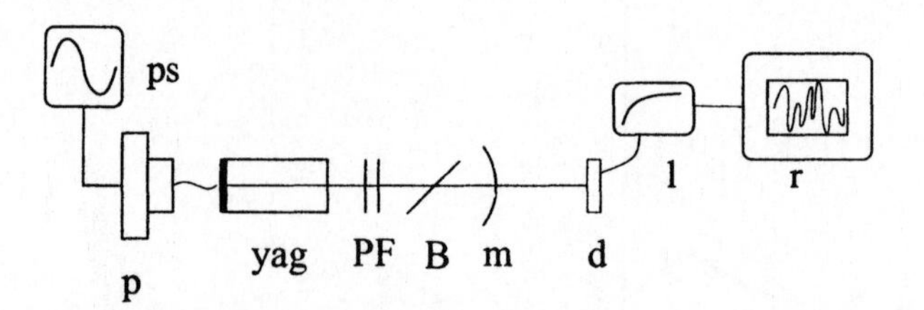

Fig. 7.26 Schematic diagram of the Nd:YAG laser shows (p) diode laser (pump); (ps) diode laser power supply; (yag) Nd:YAG rod; (PF) Fabry–Perot etalon; (B) Brewster plate; (m) output mirror; (d) detector; (l) logarithmic amplifier; (r) data recorder. The laser cavity is located between the left (reflective) end of the rod and the output mirror.

When the modulation frequency ω is near the relaxation frequency ω_r of the laser (~100 kHz), there is a standard period-doubling cascade into chaos as the pumping amplitude is increased. The chaos is described by a standard Smale horseshoe template.

The experiment was run at a pump amplitude $A_0 = 1.2$, with a modulation $m = 0.5$, and at a frequency almost half the relaxation frequency, $\omega = 54$ kHz. The laser output was monitored with an $In_xGa_{1-x}As$ detector, processed with a logarithmic amplifier (described in the next section) and then recorded. A time series of length 5×10^5 was recorded at about 100 samples/cycle, providing data for about 5000 modulation periods. A segment of the time series is shown in Fig. 7.27. A projection of a phase-space embedding is shown in Fig. 7.28.

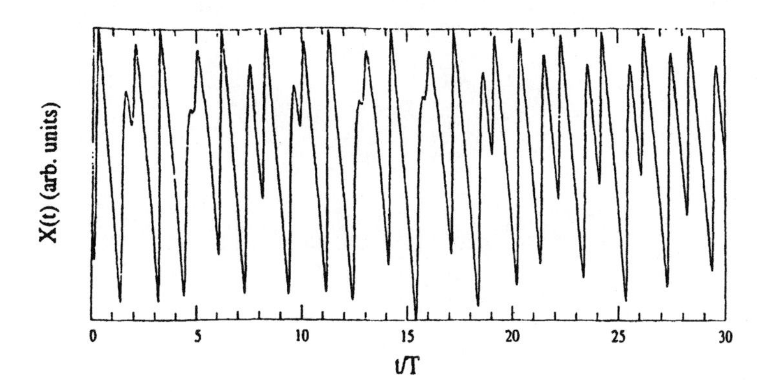

Fig. 7.27 Section of the time series $X(t) = \log(I(t) + I_0)$. T is the period of the pump modulation.

Orbits of low period (up to 9) were located in the data set by the method of close returns. The linking numbers of these orbits are shown in Table 7.8. Not all of the linking numbers could be determined unambiguously. In the cases left blank in this table [e.g., $Lk(4, 5)$], different surrogates for the same orbit gave different linking numbers when used. Only those linking numbers that could be determined without ambiguity are reported. It is apparent just by comparing Table 7.8 with Table 4.1 that the Smale horseshoe template does not describe the chaotic dynamics for the laser under the operating conditions that have been specified. That the Smale horseshoe template cannot describe the dynamics is made clear by inspection of the period-3 orbit. This orbit is shown in Fig. 7.29. This orbit exhibits four crossings. The period-3 orbit on a Smale horseshoe template with global torsion n has $4n + 2$ crossings. Therefore, the underlying template cannot be a Smale horseshoe even with nonzero global torsion.

A topological analysis showed that the template matrix and array for this data set are

$$T = \begin{bmatrix} 1 & 2 \\ 2 & 2 \end{bmatrix}$$

$$A = \begin{bmatrix} 2 & 1 \end{bmatrix} \tag{7.14}$$

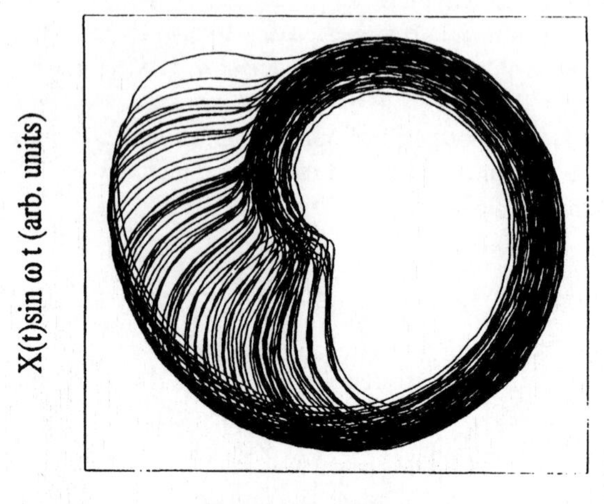

Fig. 7.28 Projection of the attractor for the Nd:YAG laser in the $y_1 - y_2$ plane. The coordinates of a point on the attractor are $(X(t) \cos \phi, X(t) \sin \phi)$, where $\phi = \omega t$.

Table 7.8 Linking and self-linking numbers for orbits extracted from the strange attractor generated by the Nd:YAG laser[a]

Orbit	1	2	3_1	3_2	4	5	7	9
1	0							
2	1	1						
3_1	2	4	4					
3_2	2	4	6	4				
4	2	5	8		7			
5	3	6	10	10		12		
7		8	14	14			24	
9	5		18					

[a]The entries left blank correspond to invariants whose values could not be determined unambiguously, since different surrogates for the same orbit gave different values.

This template is a *reverse* horseshoe. It contains only branches 1 and 2 of the scroll template shown in Fig. 7.19. This is the very first instance in which topological analysis of experimental data showed that the underlying stretch and fold mechanism was not of simple Smale horseshoe (zero-torsion) type.

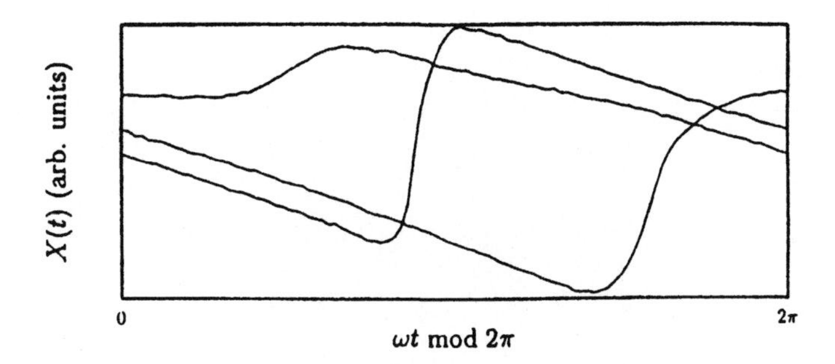

Fig. 7.29 The projection of the period-3 orbit onto the phase plane has four crossings. The period-3 orbit on a Smale horseshoe template has two crossings. This alone suffices to show that the underlying template is not a Smale horseshoe.

7.5.4 Nd-Doped Fiber Laser

The central element of the Nd-doped fiber laser is a 4-m-long silica fiber doped with 300 ppm Nd^{3+} [137]. It is pumped by a diode laser that emits a single polarized mode at 810 nm. The pumped fiber operates at a wavelength of 1.08 μm. The output consists of a single transverse mode in each of the polarization states. As a consequence of the broad, inhomogeneous gain profile, about 10^4 longitudinal modes operate simultaneously. At the output end, one of the two polarization states is selected for analysis by a polarization beam splitter. The laser was operated at a pump power up to $A = 5$ and and driving frequencies near the subharmonics $\frac{1}{2}\omega_r$, $\frac{1}{3}\omega_r$, and $\frac{1}{4}\omega_r$, where ω_r is the relaxation frequency, about 36 kHz. A schematic of the experiment is shown in Fig. 7.30.

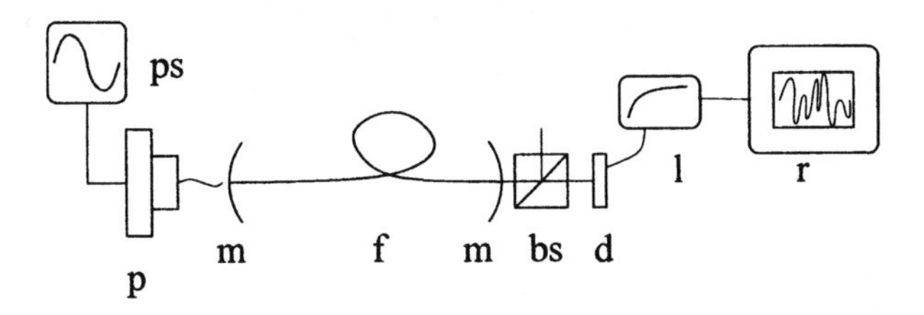

Fig. 7.30 Schematic of the fiber optic laser shows (p) diode laser (pump); (ps) diode laser power supply; (f) Nd-doped silica fiber; (m) cavity mirror; (bs) beamsplitter; (d) detector; (l) logarithmic amplifier; (r) data recorder.

Chaos was induced in the behavior of the laser by modulating the pump power. The experiment was run with $A_0 \simeq 2.7, m \simeq 0.6$. Chaotic behavior was not observed at the relaxation frequency $\omega_r \simeq 36$ kHz. However, chaos was observed in the first three subharmonic windows $C_{1/n}$ ($n = 2, 3, 4$), located at $\omega_{1/2} \simeq 18$ kHz, $\omega_{1/3} \simeq 12$ kHz, and $\omega_{1/4} \simeq 9$ kHz. The experimentally observed bifurcation diagram in each of these three windows is shown in Fig. 7.31.

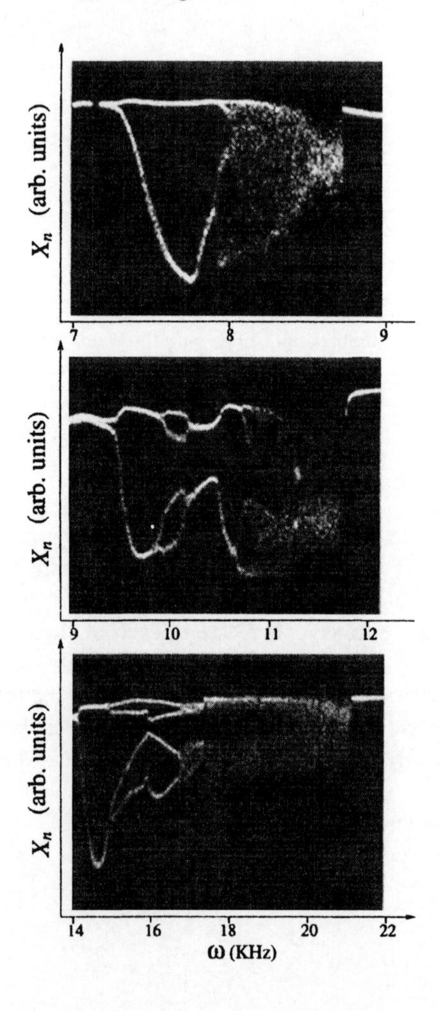

Fig. 7.31 Experimentally observed bifurcation diagram for the fiber optic laser in the subharmonic windows near $\omega_{1/4}$, $\omega_{1/3}$, and $\omega_{1/2}$.

Data were recorded in the chaotic regime within each of these three windows. For each of the three chaotic attractors, the Lyapunov dimensions were estimated. The results are summarized in Table 7.9. In each of the three cases, the first Lyapunov exponent was positive, the second close to zero, the third negative, and the sum of all three was negative. The Lyapunov dimensions were estimated using the Kaplan–Yorke conjecture [138] ($d_L = 2 + \lambda_1/|\lambda_3|$) and ranged around 2.5. The value of the second Lyapunov exponent provides some measure of the level of accuracy of these estimates. It should be zero; in the three cases studied, it is small but not always zero. The Lyapunov dimension is larger than expected from the thinness of the first return maps created for the three attractors. However, the salient feature is that the Lyapunov dimension is less than 3 in all cases. This justifies the topological analyses that were carried out.

Table 7.9 Three Lyapunov exponents were estimated for the chaotic attractors in the three windows $C_{1/4}$, $C_{1/3}$, and $C_{1/2}$[a]

Window	λ_1	λ_2	λ_3	d_L
$C_{1/4}$	0.37	−0.06	−0.57	2.6
$C_{1/3}$	0.34	−0.00	−0.92	2.3
$C_{1/2}$	0.52	−0.03	−0.66	2.7

[a]The Lyapunov dimension estimate was made from the Kaplan–Yorke conjecture that $d_L = 2 + \lambda_1/|\lambda_3|$.

Data as measured directly suffered from the problem of low intensities. That is, between bursts of peaks the intensity settled into a very low value. Essentially, all measurements in such regions wound up in the same channel of a multichannel analyzer. This makes it impossible to carry out a topological analysis, since crossings of different orbits cannot be determined in these low-intensity regions.

The problem was resolved as shown in Fig. 7.32. The problem of low intensities appears in the neighborhood of $\phi = 0$ in Fig. 7.32(a). Resolution in the low-intensity region is increased by measuring the logarithm of the signal, as shown in Fig. 7.32(b). Still, resolution is not optimum, since there are a lot of bare regions in this data record. These can be reduced by removing some sort of time average from the signal, and expanding the difference signal. The signal that was removed is shown in Fig. 7.32(c). It was obtained as the mean of the envelopes of the logarithms of the intensities. More specifically, if $L_T^+(\phi)$ is the maximum observed logarithm at phase angle $\phi = t/T \bmod 1$, and $L_T^-(\phi)$ is the minimum, the reference signal that was removed was constructed as their average $L_T^{\text{ref}}(\phi) = \frac{1}{2}(L_T^+(\phi) + L_T^-(\phi))$. The signal actually recorded, and then used for the topological analyses to be described, was $\alpha \times (L(I(t)) - L_T^{\text{ref}}(\phi))$. This signal is shown in Fig. 7.32(d) for the data segment shown in Fig. 7.32(a). The processing was carried out with analog devices so that a well-formed signal could be sent to the digitizer.

The strange attractors in the three windows $C_{1/4}$, $C_{1/3}$, and $C_{1/2}$ are shown projected into the y_1–y_2 plane in Fig. 7.33. Here $y_1 = X(t)\cos(\omega t)$ and $y_2 = X(t)\sin(\omega t)$, where $X(t)$ is the recorded signal: the scaled difference between the log of the intensity and a reference signal. Also shown in this figure are the return maps for these three attractors on a suitable Poincaré section. The thinness of these return maps indicates that the flow is more dissipative than suggested by the Lyapunov dimension computed. The location of one periodic orbit in each of the three return maps is shown. Each return map is typical of a horseshoe, consisting of an orientation-preserving branch (x) and an orientation-reversing branch (y).

With some imagination, it is possible to see two *nodal regions* or *zones of convergence* in the strange attractor in window $C_{1/2}$. These are the two regions in Fig. 7.33(c) where the flow crosses over itself. By the same rules of imagination, it is possible to see four and six regions in the attractors in windows $C_{1/3}$ [Fig. 7.33(b)] and $C_{1/4}$ [Fig. 7.33(a)], respectively. These regions are closely connected with the global torsion exhibited by the templates for these attractors. Specifically, the number of convergence regions is $2(N-1)$, where N is the global torsion of the template describing the attractor. In fact, $N = n$ for $n = 2, 3, 4$.

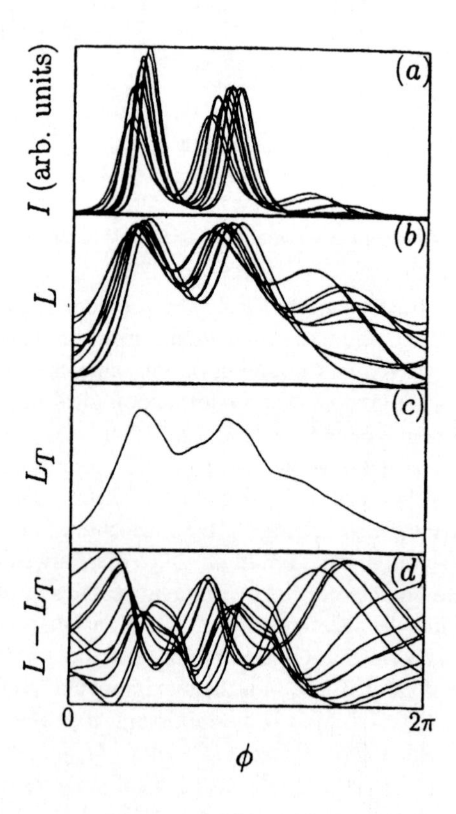

Fig. 7.32 Transformation steps from the raw to recorded data are shown from the top to the bottom. (a) Intensity over several drive periods shows low values over part of the cycle. (b) The logarithm of the intensity reduces the degeneracy. (c) The signal $L_T^{\mathrm{ref}}(\phi)$ is subtracted from the signals shown in (b). (d) These are the processed signals that are recorded.

In general, the larger the value of n, the more dissipative the attractor, the easier to extract unstable periodic orbits, and the more orbits it was possible to extract. Twelve unstable periodic orbits up to period 8 were located for the attractor in the $C_{1/4}$ window; 12 to period 9 for the attractor in the $C_{1/3}$ window; and five to period 10 in the $C_{1/2}$ window. In all cases, the relative rotation rates and the linking numbers

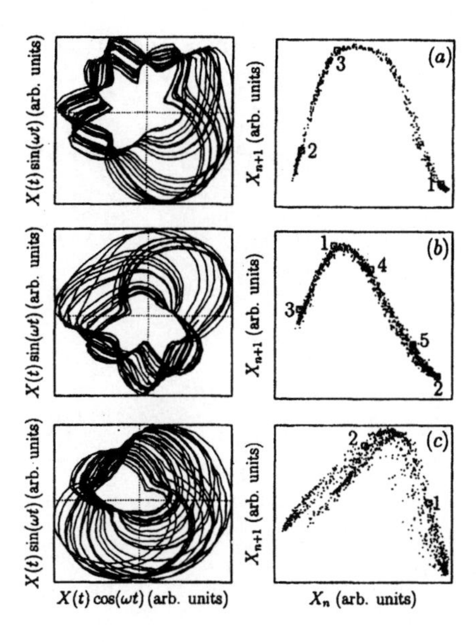

Fig. 7.33 The strange attractor in each of the three windows is projected onto a suitable plane. The return map on a Poincaré section is shown, along with the intersection points of one low-period orbit: (a) $C_{1/4}$, showing orbit yxx; (b) $C_{1/3}$, showing orbit $xyxyy$; (c) $C_{1/2}$, showing orbit yx. The branch x has even parity, branch y is orientation reversing.

were computed. The orbits were also tentatively identified by their symbolic name, for subsequent comparison with corresponding orbits on an appropriate branched manifold. The relative rotation rates for the orbits in the $C_{1/4}$ attractor are shown in Table 7.10. When these invariants could not be determined without ambiguity, the entry has been left blank.

The linking numbers for the orbits in the $C_{1/2}$, $C_{1/3}$, and $C_{1/4}$ attractors are summarized in Tables 7.11 to 7.13. A template for each of these attractors was determined. In each instance, the template matrix and array are

$$T = \begin{bmatrix} 0 & 0 \\ 0 & 1 \end{bmatrix} + (n-1) \times \begin{bmatrix} 2 & 2 \\ 2 & 2 \end{bmatrix}$$
$$A = \begin{bmatrix} 0 & 1 \end{bmatrix} \tag{7.15}$$

These are precisely Smale horseshoe templates with global torsion n for the strange attractor in the window $C_{1/n}$. In Fig. 7.34 we show a series of Smale horseshoe templates with global torsion increasing from zero (which is not observed in this experiment) to three. The dynamics changes in a systematic way as we proceed down through the subharmonic windows $C_{1/2} \to C_{1/3} \to C_{1/4}$ by decreasing the driving frequency to $\omega_{1/n} = \omega_r/n$. This is equivalent to increasing the drive period to $T_n = nT_r$. In short, the global torsion is proportional to the ratio T/T_r.

Table 7.10 Symbol name and relative rotation rates for orbits up to period 8 extracted from the chaotic attractor of the fiber optic laser in the $C_{1/4}$ window

Orbit	1	2	3_a	3_b	4	5_a
	y	xy	xy^2	x^2y	xy^3	x^2y^3
1	0					
2	$\frac{7}{2}$	$\frac{7}{2},0$				
3_a	$\frac{10}{3}$	$\frac{10}{3}$	$(\frac{10}{3})^2,0$			
3_b	$\frac{10}{3}$	$\frac{10}{3}$	$\frac{10}{3}$	$(\frac{10}{3})^2,0$		
4	$\frac{7}{2}$	$\frac{7}{2},\frac{13}{4}$	$\frac{10}{3}$	$\frac{10}{3}$	$(\frac{7}{2})^2,\frac{13}{4},0$	
5_a	$\frac{17}{5}$	$\frac{33}{10}$	$\frac{49}{15}$	$\frac{10}{3}$	$\frac{67}{20}$	$(\frac{17}{5})^2,(\frac{16}{5})^2,0$
5_b	$\frac{17}{5}$			$\frac{10}{3}$	$\frac{67}{20}$	
5_c	$\frac{17}{5}$	$\frac{17}{5}$	$\frac{10}{3}$			$(\frac{17}{5})^3,(\frac{16}{5})^2$
6_a	$\frac{10}{3}$	$\frac{10}{3}$	$(\frac{10}{3})^2,\frac{19}{6}$	$\frac{10}{3}$	$\frac{10}{3}$	$\frac{33}{10}$
6_b	$\frac{10}{3}$	$\frac{10}{3}$	$(\frac{10}{3})^2,\frac{19}{6}$	$\frac{10}{3}$	$\frac{10}{3}$	$\frac{33}{10}$
6_c	$\frac{10}{3}$	$\frac{10}{3}$		$\frac{10}{3}$	$\frac{10}{3}$	$\frac{33}{10}$
8	$\frac{27}{8}$	$\frac{27}{8},\frac{13}{4}$				

	5_b	5_c	6_a	6_b	6_c	8
	x^2yxy	$xyxy^2$	x^2y^2xy	x^2y^4	x^2yxy^2	$x^2y^3x^2y$
5_b	$(\frac{17}{5})^2,(\frac{16}{5})^2,0$					
5_c	$(\frac{17}{5})^2,(\frac{16}{5})^3$	$(\frac{17}{5})^4,0$				
6_a	$\frac{33}{10}$	$\frac{10}{3}$	$(\frac{10}{3})^4,\frac{19}{6},0$			
6_b	$\frac{33}{10}$	$\frac{10}{3}$		$(\frac{10}{3})^4,\frac{19}{6},0$		
6_c			$(\frac{10}{3})^5,\frac{19}{6}$		$(\frac{10}{3})^4,\frac{19}{6},0$	
8						$(\frac{27}{8})^3(\frac{13}{4})^4,0$

7.5.5 Synthesis of Results

Regularities exist in the sequence of Smale horseshoe templates that describe the experimental strange attractors generated by the fiber optic laser operating at $\omega_{1/n} = \omega_r/n$. The regularities can be seen in the number, $2(n-1)$, of rotation-type crossings in the phase-space projections of the strange attractors shown in Fig. 7.33, in the systematic increase in the global torsion shown in Fig. 7.34 and in Eq. (7.15), and in the results presented in Tables 7.11 to 7.13. These regularities are summarized in Table 7.14. This table lists all the orbits found in the three experiments by their

Table 7.11 Symbol name and linking numbers for orbits up to period 8 extracted from the chaotic attractor of the fiber optic laser in the $C_{1/4}$ window

Orbit	1	2	3_a	3_b	4	5_a	5_b	5_c	6_a	6_b	6_c	8
$1 = y$	0											
$2 = xy$	7	7										
$3_a = xy^2$	10	20	20									
$3_b = x^2y$	10	20	30	20								
$4 = xy^3$	14	27	40	40	41							
$5_a = x^2y^3$	17	33	49	50	67	66						
$5_b = x^2yxy$	17			50	67		66					
$5_c = xyxy^2$	17	34	50			83	83	68				
$6_a = x^2y^2xy$	20	40	59	60	80	99	99	100	99			
$6_b = x^2y^4$	20	40	59	60	80	99	99	100		99		
$6_c = x^2yxy^2$	20	40		60	80	99			119		99	
$8 = x^2y^3x^2y$	27	53										185

symbol sequence, x, y, where x and y carry even and odd parity, and by their symbol sequence in the appropriate template. The corresponding orbit in the zero-torsion lift of the horseshoe is also given. For each orbit in the zero-torsion template and in each of the three observed templates, we also present the relative rotation rates.

These results make it clear that something systematic is taking place. We illustrate the systematic evolution of the dynamics in Fig. 7.35 [139]. In the upper part of the figure we present three experimentally observed strange attractors. These are presented in order of increasing driving frequency, from (a) to (c). Below each attractor is the branched manifold that summarizes the dynamics. On the right [Fig. 7.35(c)] the dynamics is described by a simple Smale horseshoe template, $\mathcal{T}(0,1)$. The two branches are identified by their local torsion. As the driving frequency decreases, or equivalently, the period of the drive increases, a third branch is formed. The attractor and its three-branched template $\mathcal{T}(0,1,2)$ are shown in Fig. 7.35(b). The new branch is rolled inside the two previously existing branches. As the driving frequency is decreased further, the flow abandons branch 0, leaving behind the template $\mathcal{T}(1,2)$, shown in Fig. 7.35(a). This is a reverse horseshoe. Continuing on in this way, we encounter a series of transformations (perestroikas) of the form

$$\mathcal{T}(0,1) \to \mathcal{T}(0,1,2) \to \mathcal{T}(1,2) \to \mathcal{T}(1,2,3) \to$$

$$\mathcal{T}(2,3) \to \mathcal{T}(2,3,4) \to \mathcal{T}(3,4) \to \mathcal{T}(3,4,5) \to \cdots$$

The two-branch templates of the form $\mathcal{T}(2n, 2n+1)$ are Smale horseshoe templates with global torsion n. These appear in Fig. 7.34: (a) $n = 0$; (b) $n = 1$; (c) $n = 2$; (d)

Table 7.12 Symbol name and linking numbers for orbits up to period 9 extracted from the chaotic attractor of the fiber optic laser in the $C_{1/3}$ window

Orbit	1	2	4	5_a	5_b	6_a	6_b	7_a	7_b	7_c	9_a	9_b
$1 = y$	0											
$2 = xy$	5	5										
$4 = xy^3$	10	19	19									
$5_a = xy^4$	12	24	48	48								
$5_b = xyxy^2$	12	24	48	60	48							
$6_a = xy^5$	15	29		72	72	73						
$6_b = xyxy^3$	15	29	58	72	72		73					
$7_a = xyxyxy^2$	17	33	67	84	83	101	100	100				
$7_b = xy^6$	17	34	68	84	84		102	118	102			
$7_c = xyxy^4$	17	34	68	84	84	102				102		
$9_a = xy^2xyxy^3$	22	43	86	108		130				151	172	
$9_b = xyxyxy^4$	22	43	87									174

Table 7.13 Symbol name and linking numbers for orbits up to period 10 extracted from the chaotic attractor of the fiber optic laser in the $C_{1/2}$ window

Orbit	1	2	3	4	10
$1 = y$	0				
$2 = xy$	3	3			
$3 = x^2y$	4	8	8		
$4 = x^2y^2$	5	10	15	15	
$10 = xy^2x^2y^2x^2y$	13				

$n = 3$. Grosso modo, the attractor at $T/T_r + 1$ is the same as the attractor at T/T_r, except that the global torsion is larger by one.

All of these templates can be treated as subtemplates of a larger template $\mathcal{T}(0, 1, 2, 3, 4, \ldots)$ which rolls up from outside to inside. We call this a *scroll template* (Fig. 7.36). Other suggestive names for this object are *gâteau roulé* and *jellyrole template*. Such a template is illustrated in Fig. 7.37. The branches of this template are numbered successively 0, 1, 2, ... The integer is the local torsion of the period-1 orbit in that branch. An initial condition at the left of this template travels toward the right, winding around a number of times equal to the integer of the branch in which it resides. At the end of this stage, all the branches are squeezed together and the

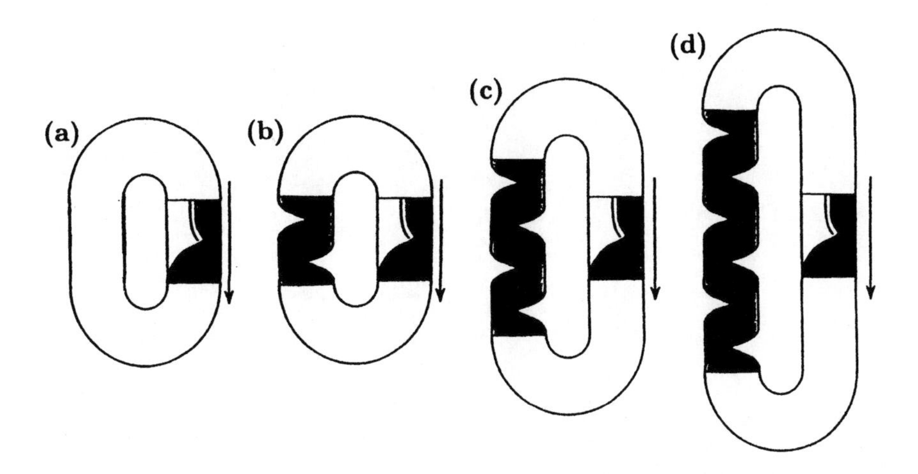

Fig. 7.34 Sequence of Smale horseshoe templates with global torsions: (a) $n = 0$; (b) $n = 1$; (c) $n = 2$; (d) $n = 3$. These describe the strange attractors observed in the fiber laser when driven at subharmonics $T_n = nT_r$ of the principal relaxation period.

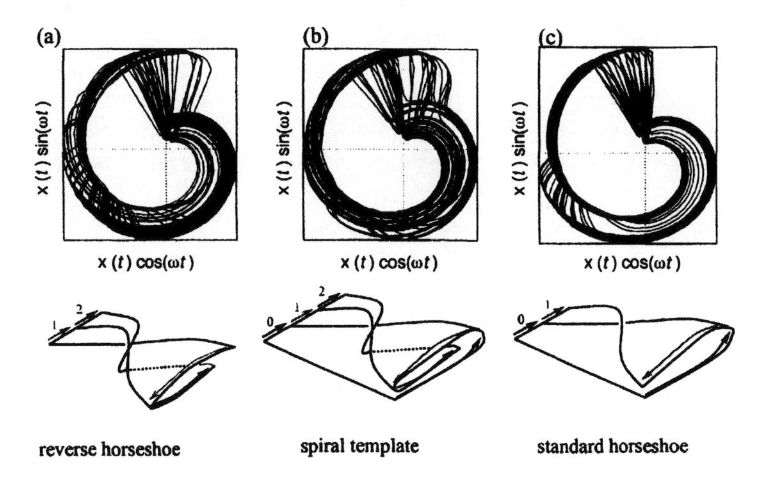

Fig. 7.35 Top row: Three strange attractors observed in the Nd:YAG laser are shown as a function of increasing frequency ω of the driving term, from (a) to (c). Bottom row: Below each of the strange attractors is the branched manifold that describes it. As the period T of the driving term increases [from (c) to (a)] a new branch is introduced and an old branch is removed. Reprinted from Boulant *et al.* [139] with permision of World Scientific.

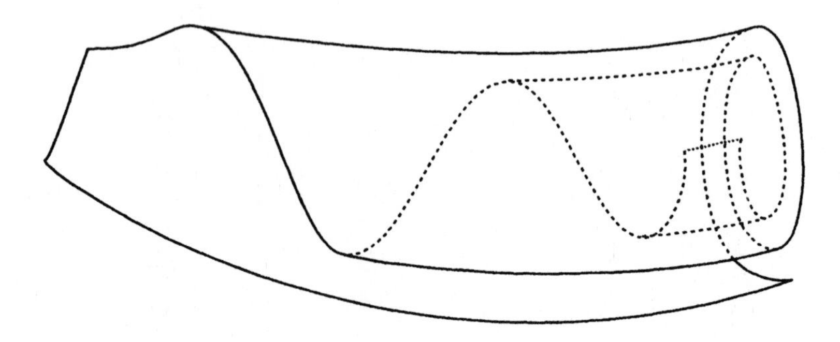

Fig. 7.36 Representation of a *gâteau roulé* branched manifold. The flow begins at the left of this figure. An initial condition flows to the right in one of the branches, winding around as it does. At the right-hand edge, the branches are squeezed together and the flow is returned to the left of the template to continue anew.

flow returned to the left of the figure. This template encompasses the results of all the experiments and simulations described in this section. Specifically, the zero-torsion lift of the horseshoe, the original three-branched *gâteau roulé* described in Section 7.5.1 and shown in Fig. 7.19, the reverse horseshoe template described in Section 7.5.2, and the three horseshoes with global torsion $n = 2, 3, 4$ described in Section 7.5.3 and shown in Fig. 7.34 are all subtemplates of the large *gâteau roulé* shown in Fig. 7.36.

The systematic behavior exhibited above can be described qualitatively by a simple two-parameter model. We obtain this model by unwinding the large *gâteau roulé*. The unrolled template is shown in Fig. 7.37. In this representation, the flow is from left to right. An initial condition begins in the vertical line (it is a branch line) on the left. The trajectory moves to the right (time evolution) and drifts upward (increasing torsion). The right-hand edge of this model is determined by the period of the drive. More accurately, it is determined by the ratio T/T_{ref}. The larger this ratio, the longer the horizontal axis and the farther upward the drift of the trajectory. As initial conditions along the branch line on the left evolve in time, they stretch out (sensitivity to initial conditions). The ratio of the length of the image on the right-hand branch line to the original length on the left-hand branch line is the stretch factor s, related to the Lyapunov exponent of the attractor by $s = e^{\lambda_1}$.

The two parameters that account qualitatively for the diversity of results presented in this section are the drift time T/T_{ref} and the stretch factor s. The longer the drift time, the greater the global torsion. The larger the stretch factor, the more branches are traversed in the flow, and the more branches occur in the subtemplate that describes the dynamics [140]. The zero-global-torsion template (0, 1) will occur for $T/T_{\text{ref}} < 1$ and $s < 2$. The original *gâteau roulé* (0, 1, 2) occurs for $T/T_{\text{ref}} < 1$ and $2 < s < 3$. The three templates with global torsion $n = 2, 3, 4$ occur for $T/T_{\text{ref}} \sim n$ and $s < 2$.

As the drift time T/T_{ref} is increased systematically, template branches formerly traversed (with lower n) are abandoned, and those with higher torsion, not formerly traveled, are visited. We could say that old branches are destroyed in saddle-node-type inverse bifurcations, and new branches with higher global torsion are created systematically in direct-saddle-node bifurcations. Or we could say that all branches are present but the flow is restricted to a small number (depending on s) of them, and redirected from one set to another by changing the value of the drift time T/T_{ref}. The two interpretations are equivalent, they are simple, and both provide a lot of predictive capability.

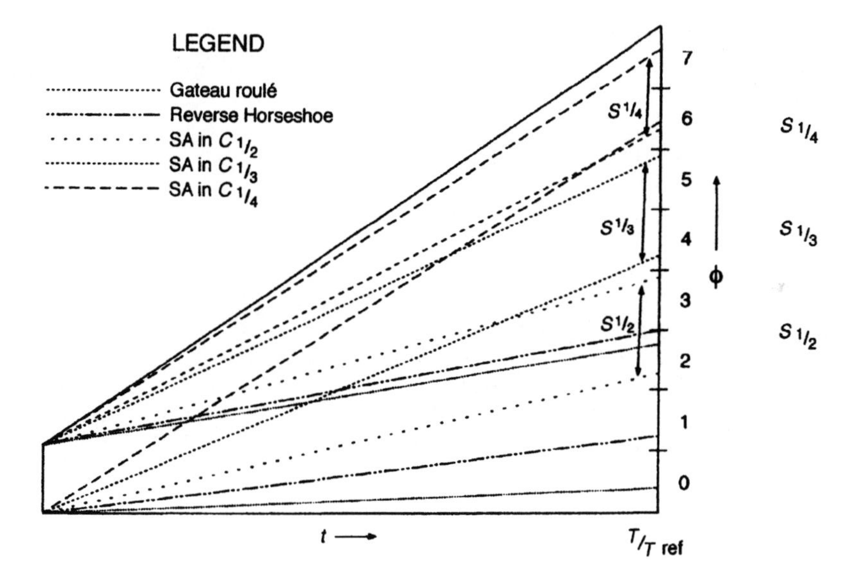

Fig. 7.37 The large *gâteau roulé* of Fig. 7.36 is unwound to the quadrilateral shown. The branch lines on the left and right are identified. Flow begins on the branch line at the left. As the flow evolves, it moves to the right and spreads out. The flow also drifts upward, increasing its global torsion as it does. The upward drift is proportional to T/T_{ref}, the spread is equal to $s = e^{\lambda_1}$. These two parameters describe all the qualitative properties of the laser models and experiments treated in Section 7.5. For the three experiments reported on the fiber laser, the stretch factors are $s_{1/2} = 1.68$, $s_{1/3} = 1.40$, and $s_{1/4} = 1.45$.

Table 7.14 Periodic orbits extracted from the three strange attractors of the fiber laser compared with orbits in the U-sequence and the zero-global-torsion Smale horseshoe template[a]

P_j	$U-Sequence$ Name	$x-y$ Name	$\mathcal{T}$	(0,1)	$\mathcal{T}$	(2,3)	$\mathcal{T}$	(4,5)	$\mathcal{T}$	(6,7))
1_1		x	0							
1_1		y	1							
2_1	R	xy	01	$\frac{1}{2},0$	23	$\frac{3}{2},0$	45	$\frac{5}{2},0$	67	$\frac{7}{2},0$
3_1	RL	x^2y	001	$(\frac{1}{3})^2,0$	223	$(\frac{4}{3})^2,0$			667	$(\frac{10}{3})^2,0$
3_1	RL	xy^2	011	$(\frac{1}{3})^2,0$					667	$(\frac{10}{3})^2,0$
4_1	RLR	xy^3	0111	$(\frac{1}{2})^2,\frac{1}{4},0$	2333	$(\frac{3}{2})^2,\frac{5}{4},0$	4555	$(\frac{5}{2})^2,\frac{9}{4},0$	6777	$(\frac{7}{2})^2,\frac{13}{4},0$
4_2	RLL	x^2y^2	0011	$(\frac{1}{4})^3,0$					6677	$(\frac{13}{4})^3,0$
5_1	RLR^2	xy^4	01111	$(\frac{2}{5})^4,0$			45555	$(\frac{12}{5})^4,0$		
5_1	RLR^2	xy^2xy	01101	$(\frac{2}{5})^4,0$			45545	$(\frac{12}{5})^4,0$	67767	$(\frac{17}{5})^4,0$
6_1	RLR^3	xy^5	011111	$(\frac{1}{2})^3,(\frac{1}{3})^2,0$			455555	$(\frac{5}{2})^3,(\frac{7}{3})^2,0$		
6_1	RLR^3	xy^3xy	011101	$(\frac{1}{2})^3,(\frac{1}{3})^2,0$			455545	$(\frac{5}{2})^3,(\frac{7}{3})^2,0$		
6_2	RL^2RL	x^2yxy^2	001011	$(\frac{1}{3})^4,\frac{1}{6},0$					667677	$(\frac{10}{3})^4,\frac{19}{6},0$
6_3	RL^2R^2	x^2y^4	001111	$(\frac{1}{3})^4,\frac{1}{6},0$					667777	$(\frac{10}{3})^4,\frac{19}{6},0$
6_3	RL^2R^2	x^2y^2xy	001101	$(\frac{1}{3})^4,\frac{1}{6},0$					667777	$(\frac{10}{3})^4,\frac{19}{6},0$
7_1	RLR^4	xy^6	0111111							
7_1	RLR^4	xy^4xy	0111101							
8_6	RL^2R^3L	$x^2y^3x^2y$	00111001	$(\frac{3}{8})^3,(\frac{1}{4})^4,0$					66777667	$(\frac{27}{8})^3,(\frac{13}{4})^4,0$
9_2	RLR^4LR	$xyxyxy^4$	011110101							
9_3	RLR^2LRLR	xy^2xyxy^3	011010111							
10_{21}	$RL^2R^2L^2RL$	$xy^2x^2y^2x^2y$	0011001011 ?						6677667677	

[a]The symbol name and relative rotation rates are provided for all orbits observed in the experiments.

7.6 NEURON WITH SUBTHRESHOLD OSCILLATIONS

The electrical properties of neurons were studied extensively by Hodgkin, Huxley, and Katz in an important and elegant series of papers beginning in 1949 [141–148]. These research studies eventually lead to the proposal of a model for the electrical signals produced by a neuron, since known as the *Hodgkin–Huxley equations*. These equations are the starting point for all subsequent studies of the electrical properties of neurons.

These equations describe "platonic" neurons: no input, no output. More specifically, if the input falls below a certain threshold, the neuron does not fire. Some real neurons do not behave this way. In particular, sensory neurons are known to produce output even in the absence of electrical input. These neurons exhibit subthreshold oscillations. It is almost as if these neurons talk to each other even in the absence of signals just to keep informed of the status of their surroundings.

Braun and co-workers have studied sensory neurons from dogfish, catfish, and rats that exhibit subthreshold oscillations [149–151]. These studies have resulted in a mathematical model describing the output of a neuron that exhibits subthreshold oscillations [152]. This model is based on the transfer of charged ions through the membrane. The transfer rate depends on the potential difference across the membrane (membrane polarization) V. It also depends on two mechanisms.

There is a pump mechanism. Pumping is carried out by proteins that span the wall of the membrane and slowly transmit molecules through the membrane against the concentration gradient.

There is a gate mechanism. This is essentially a hole in the membrane wall that can be opened and closed. The hole is also a protein. When closed, no ions flow out. When open, ions of a particular type (Na^+, K^+) flow out quickly, down the concentration gradient. The gate state (open, closed) is controlled by the membrane polarization V.

A relatively simple version of this model is based on the difference in time scales involved in the transmission of ions through neural membranes. The model consists of five coupled ordinary differential equations. One describes the time evolution of the membrane polarization, the other two pairs describe the fast- and slow-time evolution of two ion types, Na^+ and K^+. The model was tested by comparing its output with the output of several different types of sensory neurons that exhibit subthreshold oscillations. No model output–data entrainment tests have been carried out so far.

The output of the model suggests that the model generates chaotic behavior in some regions of control parameter space. In Fig. 7.38 we present a bifurcation diagram for this model in a form standard for the discipline [152, 153]. The bifurcation parameter is the ambient temperature, T. Plotted on the state-variable axis is the interspike time interval. The membrane potential output consists of a series of bursts (inset). Each burst consists of one or more spikes. After the last spike in a burst, the membrane potential returns to a low, resting value, and then the next burst begins. The bursts shown in the inset consist of a spike train consisting of either three or four spikes. The bifurcation diagram is a plot of the time interval between successive spikes. The upper envelope roughly measures the time interval between successive bursts.

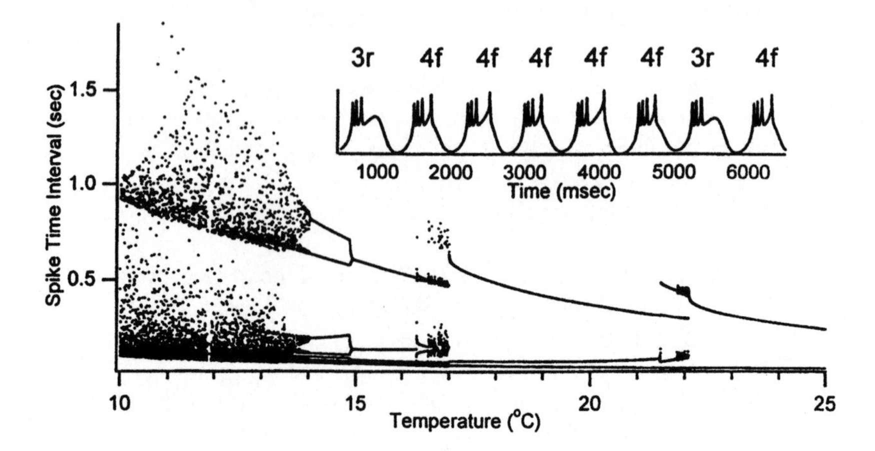

Fig. 7.38 Bifurcation diagram for a model of a sensory neuron exhibiting subthreshold oscillations shows regions of periodic behavior and regimes with chaotic behavior. Inset shows the morphology of a membrane potential signal.

The bifurcation diagram shows that at high temperatures ($22 < T < 25$) there are two spikes per burst and a time interval of about a half second between bursts. There are three spikes per burst in the range $17 < T < 21$, then four for $15 < T < 16.5$. Below 15°C there seems to be the beginning of a period-doubling cascade. The model output seems to be chaotic in short temperature intervals between regimes with n spikes per burst and $n + 1$ spikes per burst ($n = 2, 3$) and below 15°C, with the possibility of a window near 12°C. Does this model exhibit chaotic behavior? If so, what is the mechanism that generates chaos?

So far, it is only possible to determine mechanisms for strange attractors that can be embedded in R^3. The dynamical model for neurons with subthreshold oscillations is five-dimensional. Before an attempt to identify mechanisms can be initiated, we must first determine that the motion in the five-dimensional phase space relaxes into a three-dimensional inertial manifold. This was done by locating a chaotic regime and plotting the first three coordinates of a point in this strange attractor against each other. The results are shown in Fig. 7.39 [153]. These three plots show that only one of the three variables y_1, y_2, y_3 is independent. This reduces the dimensionality from five to three, so that topological methods can be used to determine mechanism.

Strange attractors in the chaotic regimes were plotted in a large number of projections. Two of these are shown in Fig. 7.40. The attractor on the left is a projection into the y_4–y_5 plane. The attractor on the right is a differential embedding based on the variable y_4. The attractor is projected into the $\dot{y}_4$–y_4 plane. Integral–differential embeddings based on y_4 and other variables were also constructed. The template that describes the chaotic dynamics is independent of variables and the embedding procedure used.

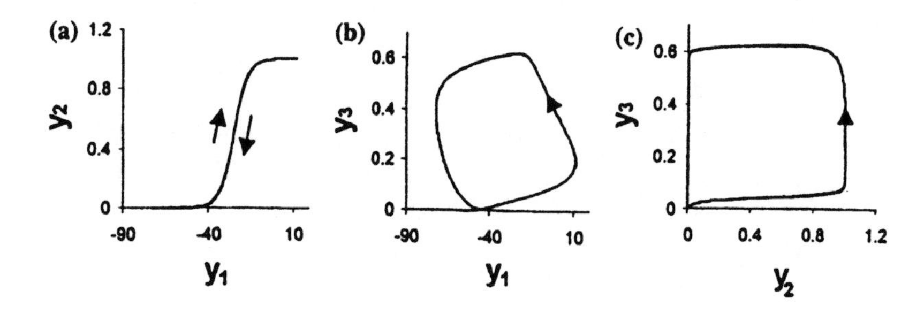

Fig. 7.39 Pairwise plots of the three coordinates y_1, y_2, and y_3 of a point in a strange attractor show that only three of the five coordinates can be independent. Reprinted from Gilmore and Pei [154] with permission of Elsevier Science.

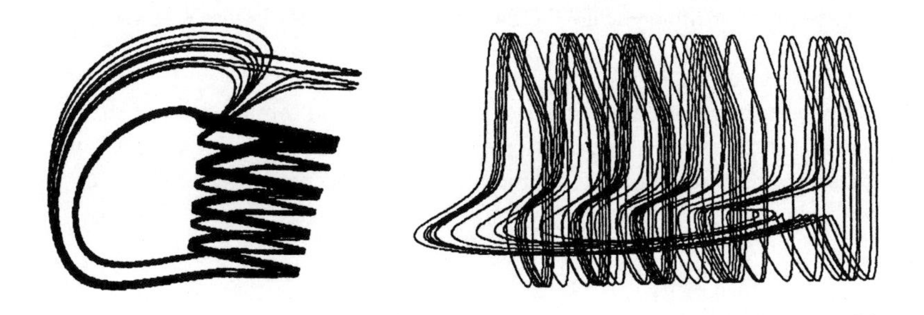

Fig. 7.40 Two projections of chaotic strange attractors are shown. On the left is an attractor projected into the y_4–y_5 plane. On the right is a differential embedding projected into the $\dot{y}_4$–y_4 plane. Both embeddings point to a *gâteau roulé* as the determining mechanism. Reprinted from Gilmore and Pei [154] with permission of Elsevier Science.

Preliminary information about the dynamics was obtained from the first return map, based on successive minima of the variable y_3. Three return maps are shown for three different chaotic regimes in Fig. 7.41. These were constructed for ambient temperatures $T = 12, 13.5$, and 16.5°C. The branches of this return map are labeled by the morphology of the burst. The first label is the integer identifying the number of spikes per burst. The second, f or r (later identified as flip saddle or regular saddle), distinguishes between two different types of bursts with the same number of spikes. For example, burst type $3r$ consists of a burst with three spikes, followed by a failed attempt to form a fourth spike before a return to the deep polarization minimum. Burst type $4f$ consists of four spikes followed by a fast return to the deep polarization minimum. These two distinct types can be distinguished in the inset to Fig. 7.38.

Period-1 orbits were extracted from each of the branches identified in the return maps. In each case, two time series near the period-1 orbit were embedded and their

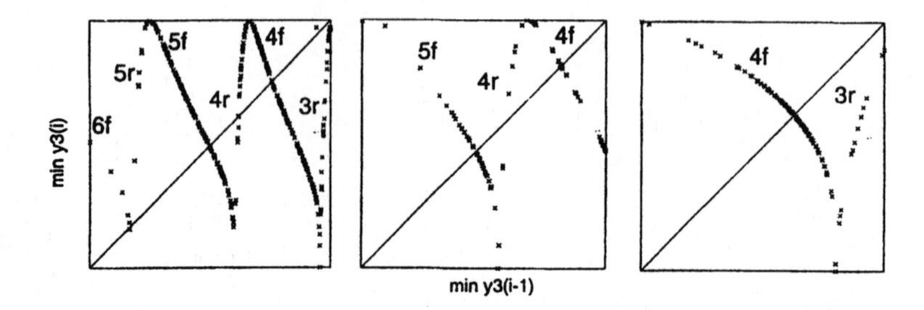

Fig. 7.41 First return maps on a Poincaré section in three chaotic regimes. The return map is on the minimum value of the variable y_3. Each branch is labeled by the morphology of the burst. Temperature increases from left to right: (a) $T = 12°\text{C}$; (b) $T = 13.5°\text{C}$; (c) $T = 16.5°\text{C}$. As the temperature increases, the return map slides to the left and changes shape slightly. Reprinted from Gilmore and Pei [154] with permission of Elsevier Science.

local torsion determined by counting crossings. The local torsion of the orbit nr is $2n$ and that of nf is $2n - 1$. The orbits nr have even parity (regular saddles), while orbits nf have odd parity. The local torsion changes systematically as follows:

$$\begin{array}{cccccccccc} 6f & 5r & 5f & 4r & 4f & 3r & 3f & 2r & 2f & 1r \\ 11 & 10 & 9 & 8 & 7 & 6 & 5 & 4 & 3 & 2 \end{array} \tag{7.16}$$

The linking numbers of adjacent period-1 orbits with local torsion n and $n+1$ is $n/2$ or $(n+1)/2$, whichever is integer.

The regularity in the behavior of the local torsion and the behavior of linking numbers of adjacent period-1 orbits is a clear signature that the underlying template has a rolled structure. It is a *gâteau roulé*. The only question is whether the template rolls up from outside to inside (as in the fiber laser) or the other way around, from inside to outside. The template matrices and arrays for the two scrolled templates are shown in Fig. 7.42.

To distinguish between these two templates, three period-1 orbits and one period-2 orbit were located in the strange attractor at 12°C. These periodic orbits, $4f, 4r, 5f$, and $(4f, 5f)$ are shown in Fig. 7.43. The linking numbers of these orbits were computed and compared with those predicted by the outside-to-inside and inside-to-outside scrolls, with the following results:

Linking Number	Neuron Model	Outside to Inside	Inside to Outside
$Lk(4f,(4f,5f))$	7	7	8
$Lk(4r,(4f,5f))$	8	8	8
$Lk(5f,(4f,5f))$	8	8	9

(7.17)

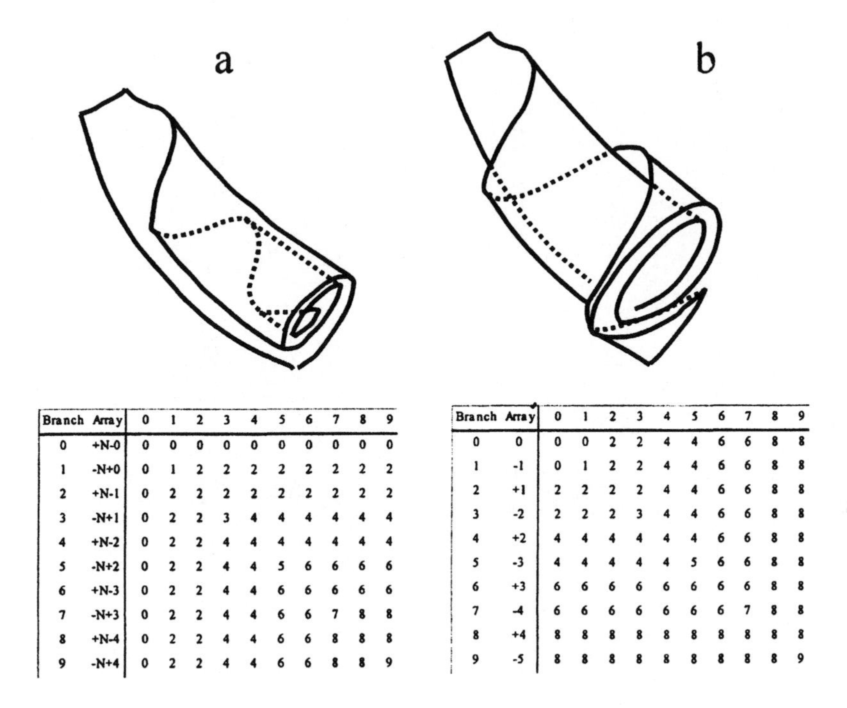

Branch	Array	0	1	2	3	4	5	6	7	8	9
0	+N-0	0	0	0	0	0	0	0	0	0	0
1	-N+0	0	1	2	2	2	2	2	2	2	2
2	+N-1	0	2	2	2	2	2	2	2	2	2
3	-N+1	0	2	2	3	4	4	4	4	4	4
4	+N-2	0	2	2	4	4	4	4	4	4	4
5	-N+2	0	2	2	4	4	5	6	6	6	6
6	+N-3	0	2	2	4	4	6	6	6	6	6
7	-N+3	0	2	2	4	4	6	6	7	8	8
8	+N-4	0	2	2	4	4	6	6	8	8	8
9	-N+4	0	2	2	4	4	6	6	8	8	9

Branch	Array	0	1	2	3	4	5	6	7	8	9
0	0	0	0	2	2	4	4	6	6	8	8
1	-1	0	1	2	2	4	4	6	6	8	8
2	+1	2	2	2	2	4	4	6	6	8	8
3	-2	2	2	2	3	4	4	6	6	8	8
4	+2	4	4	4	4	4	4	6	6	8	8
5	-3	4	4	4	4	4	5	6	6	8	8
6	+3	6	6	6	6	6	6	6	6	8	8
7	-4	6	6	6	6	6	6	6	7	8	8
8	+4	8	8	8	8	8	8	8	8	8	8
9	-5	8	8	8	8	8	8	8	8	8	9

Fig. 7.42 The templates roll up from outside to inside (a) and inside to outside (b). Bottom: The template matrices and arrays show systematic behavior. Reprinted from Gilmore and Pei [154] with permission of Elsevier Science.

This comparison was able to distinguish between the two scroll templates [153].

The relation between the number of spikes per burst and the torsion is illustrated in Fig. 7.44. The underlying idea is that there is a lot of twisting going on in phase space. The twisting is measured by the linking number of nearby period-1 orbits. There is an elegant relation between two geometric quantities, twist and writhe, and a topological quantity, the linking number (see also Section 4.2.9 and Fig. 4.14):

$$\text{twist} + \text{writhe} = \text{linking number} \tag{7.18}$$

Twist and writhe can be exchanged, as the following thought exercise illustrates. Take a rubber band, stretch it, and then twist it four full twists about its long axis. Then relax the tension. As the tension relaxes, the rubber band deforms to a state with no twist but lots (= 4) of writhe. Writhe is represented as big loops in space. A similar phenomenon takes place in the phase space of the neural model. We determine linking numbers of the low-period orbits as topological invariants. In fact, although the link of adjacent period-1 orbits is large, their twist (a geometric quantity) is small,

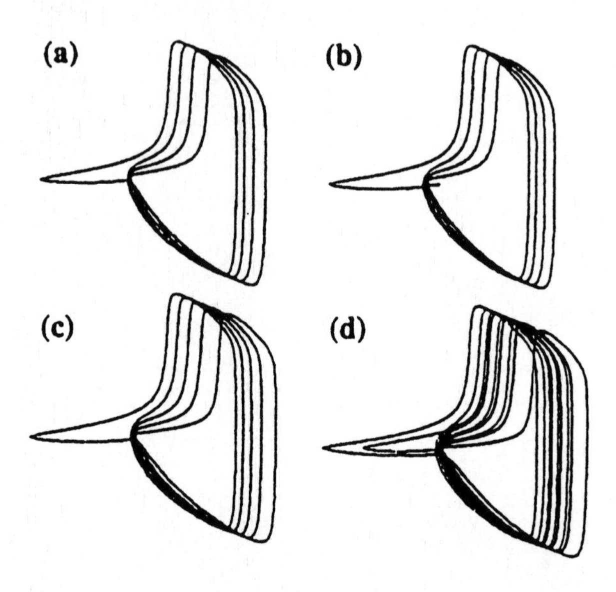

Fig. 7.43 Period-1 orbits (a) $4f$, (b) $4r$, (c) $5f$; and period-2 orbit (d) $(4f, 5f)$ in the strange attractor shown in Fig. 7.39(b).

near zero. The small twist is compensated by a large writhe. The writhe appears in projections (cf. the spikes in y_5 in the projection shown in Figs. 7.40) as large spikes.

This is very similar to a result found in the fiber laser [137]. The number of convergences in Fig. 7.33 is related to the global torsion by $2(n-1)$ and to the number of large-amplitude peaks in the data by $n-1$. Spiking occurs in the fiber optic laser, but it is "softer," not as extreme as in the neural model.

There is another similarity between the fiber laser and the neural model. When the *gâteau roulé* for the neural model is unwound, it appears as shown in Fig. 7.45. This figure compares the unrolled templates for the fiber laser and the neural model. In the fiber laser the drift time is determined by the external forcing period T. In the neural model it is determined by the length of the period-1 orbits. The bursts nf are fast, the bursts nr are slowed down by the very slow time scale of the last, unrealized peak ("critical slowing down"). Once again, two parameters provide a good qualitative understanding of this complicated nonlinear system under variation of the control parameters. In the present case, these are the phase ϕ and the stretch R, both of which are illustrated in Fig. 7.45 and which increase with decreasing temperature [154].

We close this chapter with a slight bit of speculation. When two or more neurons interact, the possibility of synchronization of their output arises. Phase synchronization may easily be possible because the different period-1 orbits have slightly different

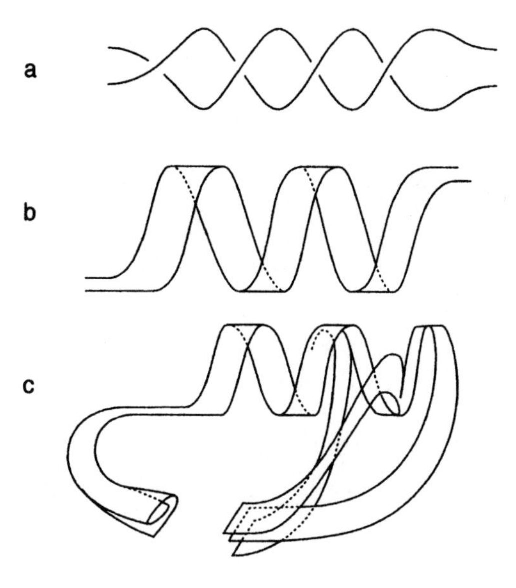

Fig. 7.44 (a) A stretched rubber band shows a lot of twist about its axis. (b) When the tension is relaxed, twist is converted to writhe. (c) Schematic of the template for a strange attractor generated by the neural model shows little twist and lots of writhe. The writhe is expressed in measurements as sharp spikes in the output potential. Reprinted from Gilmore and Pei [154] with permission of Elsevier Science.

temporal periods. This may be nature's way to design nervous communication networks that can respond quickly to differing environmental conditions.

7.7 SUMMARY

Topological analyses have been carried out on a large number of experimental data sets. In more than half the cases studied to date a single mechanism has been implicated in the creation of chaos: the simple stretch and fold mechanism. This leads to the Smale horseshoe in the simplest cases. In more complicated cases reverse horseshoes may be seen. In yet more complicated cases, horseshoes with nonzero global torsion have been identified.

These are all special cases of a very general mechanism: the stretch and roll mechanism. This mechanism is illustrated for branched manifolds in Fig. 7.36 [93, 153, 154]. When this mechanism is operative, the gross features of the dynamics are described by only two parameters. These are the *stretch* per period $s = e^{\lambda_1}$ and the *drift* per period. The first determines the number of branches in the branched manifold, and the second determines the torsion of these branches. These parameters

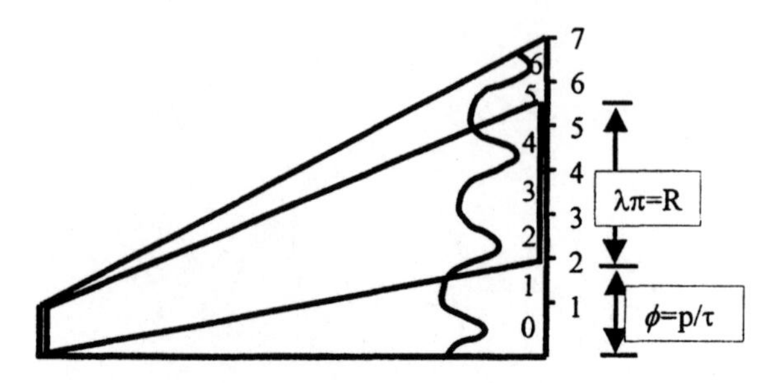

Fig. 7.45 The unscrolled *gâteau roulé* template has the form shown. For the fiber laser the period is fixed: All period-1 orbits have the same temporal length. For the neuron, all period-1 orbits have a slightly different period. The right-hand edge (branch line) in this case is not vertical, as for the fiber laser. As the temperature T decreases, the drift (ϕ) increases and the stretch also increases. Reprinted from Gilmore and Pei [154] with permission of Elsevier Science.

change in a predictable way when experimental control parameters are varied. This allows us to predict perestroikas in experiments. When the torsion is large, it may be converted to *writhe*. This appears in experimental data as spikes. These phenomena have been seen in data from widely different fields. For all of the experimental systems exhibiting chaos through the stretch and roll mechanism (*gâteau roulé* or jellyroll mechanism), the inertial manifold, or phase space, is a torus $R^2 \times S^1$.

8

Tearing Mechanisms: A_3

If the Rössler system, with its folding mechanism (A_2), is the hydrogen atom problem of nonlinear dynamics, then the Lorenz system, with its tearing mechanism (A_3), is its hydrogen molecule problem. In this chapter we treat two physical systems that are governed by this mechanism for creating chaotic behavior. These are the optically pumped molecular laser and a fluid heated in a toroidal geometry. Both systems exhibit a twofold symmetry. It is the symmetry, in fact, which is responsible for the tearing mechanism. In addition to the tearing mechanism, both systems can exhibit the folding mechanism if the control parameters are pushed too far.

We begin this chapter with a review of the Lorenz equations, placing special emphasis on properties that are useful for a discussion of the two experimental data sets.

8.1 LORENZ EQUATIONS

The Lorenz equations were discussed extensively in Chapter 3, particularly Section 3.3. These equations are

$$\begin{aligned} \frac{dx}{dt} &= -\sigma x + \sigma y \\ \frac{dy}{dt} &= rx - y - xz \\ \frac{dz}{dt} &= -bz + xy \end{aligned} \tag{8.1}$$

These equations are usually studied for fixed values of the control parameters $\sigma = 10$, $b = 8/3$ as a function of the single control parameter r.

These equations are a consequence of a severe truncation of the Navier–Stokes equations (3.16) or (3.17). The standard procedure for truncating these partial differential equations is to introduce a normal-mode representation for the velocity field $\mathbf{u}(\mathbf{x}, t)$ or the velocity potential $\psi(\mathbf{x}, t)$: for example, $\psi(\mathbf{x}, t) = \sum_i a_i(t) M_i(\mathbf{x})$. The normal modes $M_i(\mathbf{x})$ can be taken from theory as Fourier modes, or from experiment as empirical orthogonal modes, following some sort of singular-value decomposition. The normal-mode ansatz is plugged into the Navier–Stokes evolution equations [it is an "evolution equation" since the time dependence is manifest in a first-order time derivative, $\partial \mathbf{u}(\mathbf{x}, t)/\partial t$]. The result involves first-order derivatives of the normal-mode amplitudes (da_i/dt) coupled to the normal modes and their derivatives. The spatial dependence is integrated away using the orthogonality of the modes and the fact that products over more than two modes, or modes and their derivatives, are essentially Clebsch–Gordan coefficients. There results a system of coupled first-order ordinary differential equations of the form $da_i/dt = f_i(a)$. The forcing functions on the right-hand side are polynomial functions. In fact, they are of degree no greater than 2, since the Navier–Stokes equations have only a quadratic nonlinearity in the velocity field.

Several many-mode truncations (52, 7) of the Navier–Stokes equations for a restricted geometry were initially studied by Saltzman [65]. He found that all but three of the mode amplitudes eventually died out. The volume of control parameter space studied was small because the number of modes involved was large. Lorenz eliminated the mordant modes and extended the volume of the control parameter space that could practically be studied. In these studies, Lorenz stumbled on the phenomenon of "sensitive dependence on initial conditions," a property that is at the heart of the concept of chaos.

Before plunging into a brief review of the Lorenz equations and their properties, we emphasize two properties that *any* truncation of the Navier–Stokes equations will *always* possess:

1. All forcing terms $f_i(a)$ are of degree 2 (or lower) in the mode amplitudes, since the original PDEs have only a quadratic nonlinearity.

2. The equations will exhibit a symmetry (at least twofold) that is induced from the symmetry possessed by the Navier–Stokes equations.

8.1.1 Fixed Points

A broad understanding of the structure of a flow is often determined by locating the fixed points (if any) and determining their stability. The fixed points for the Lorenz equations were determined in Fig. 3.5. They are $(x, y, z) = (0, 0, 0)$ and

$$(x, y, z) = (\pm\sqrt{bz}, \pm\sqrt{bz}, r - 1)$$

The fixed points exhibit a cusp (A_3) bifurcation, as opposed to the fold (A_2) bifurcation that the fixed points of the Rössler system exhibit. The cusp bifurcation is ultimately responsible for tearing in Lorenz dynamics.

For all r there is a real fixed point at $(x, y, z) = (0, 0, 0)$. For $r < 1$ it is the only real fixed point. There is an imaginary pair of fixed points at $(i\sqrt{b|z|}, i\sqrt{b|z|}, z = r - 1)$ and $(-i\sqrt{b|z|}, -i\sqrt{b|z|}, r - 1)$. As r approaches 1, z approaches 0, the complex conjugate pair approach each other and are responsible for all the phenomena that ghosts typically generate. For $r > 1$ there are three real fixed points. The symmetric pair separate from each other and the fixed point at $(0, 0, 0)$ on a canonical parabolic trajectory.

8.1.2 Stability of Fixed Points

The stability of the fixed points has been determined in Section 3.5.4. In particular, the stability around a fixed point is determined by the eigenvalues of the Jacobian (3.48). The fixed point at the origin is a stable node for $r < 1$ and a saddle for $r > 1$. The saddle has one unstable and two stable directions. The unstable eigenvalue is in the direction of the symmetric pair of fixed points.

The eigenvalue equation for the symmetric pair of fixed points is

$$\lambda^3 + (\sigma + b + 1)\lambda^2 + (r + \sigma)b\lambda + 2b\sigma(r - 1) = 0$$

As $r \to 1^+$ these eigenvalues approach $0^-, -(\sigma+1), -b$. As r increases above $r = 1$, the three eigenvalues remain negative, real, and unequal until a value $r_f \sim 1.345$ is reached, at which value two of the negative eigenvalues become equal, after which they become complex with negative real part (near r_f). At r_f the symmetry-related stable nodes become stable foci. At $r_4 \simeq 24.74$ the real part of these complex eigenvalues crosses through zero, and the stable foci become unstable foci. For larger values of r the three fixed points are all unstable, the motion is bounded, recurrent, nonperiodic, and sensitive to initial conditions.

8.1.3 Bifurcation Diagram

A schematic bifurcation diagram for the Lorenz equations is shown in Fig. 8.1. This diagram shows the following sequence of events:

r_1 A cusp bifurcation (A_3) occurs at $r_1 = 1.0$. The original stable fixed point at the origin becomes unstable as r increases through $+1$, and two symmetry-related stable nodes are created. The fixed point at the origin remains unstable for all $r > 1$.

r_f Two negative eigenvalues of the stable nodes become equal and scatter into the complex plane at $r_f = 1.345$. The stable nodes become stable foci.

r_2 A nonlocal bifurcation takes place at $r = 13.926$. This bifurcation cannot be studied by local (Taylor series) techniques. The first extended orbits occur. These are orbits that have initial conditions on one side of the plane that is the perpendicular bisector of the line joining the symmetry-related fixed points but which decay asymptotically to the fixed point on the other side of this plane.

r_3 A pair of symmetry-related fold bifurcations A_2 occur at $r = 24.06$. These create pairs of unstable periodic orbits.

r_4 An inverse Hopf bifurcation takes place at $r_4 = 24.74$ between the stable foci and one of the unstable periodic orbits created at r_3 in a saddle-node bifurcation. The two fixed points become unstable foci.

r_0 For very large values of r there is a stable symmetric periodic orbit.

r_a At r_a an A_3 bifurcation creates a symmetry-related pair of asymmetric periodic orbits. The symmetric orbit from which they bifurcate becomes unstable.

r_b Each of the asymmetric periodic orbits initiates a period-doubling cascade of period-doubled orbits.

r_c, r_d, r_∞ The period-doubling cascade continues to accumulation. Between r_4 and r_∞ there is chaos.

8.1.4 Templates

The Lorenz equations have been studied most extensively in the parameter range $25 < r < 50$ and $b = 8/3$, $\sigma = 10$. The behavior has been sampled in other parameter ranges. It seems likely that the full panoply of behavior which this simple equation, with only three fixed points, can assume has not been fully explored or completely understood.

The Lorenz attractor is shown in Fig. 8.2(a). It is invariant under $(x, y, z) \to (-x, -y, +z)$. The attractor has been projected onto the $(x = y)$–z plane. The two off-axis fixed points are counterrotating unstable foci. The fixed point at the origin is stable in the z direction and the direction perpendicular to the plane $x = y$. It is unstable in the direction of the two unstable foci.

The two unstable foci and the unstable fixed point organize the flow. In the neighborhood of a focus, the flow is around the focus, expanding away from it since the focus is unstable. Eventually, the flow extends sufficiently far from the original focus

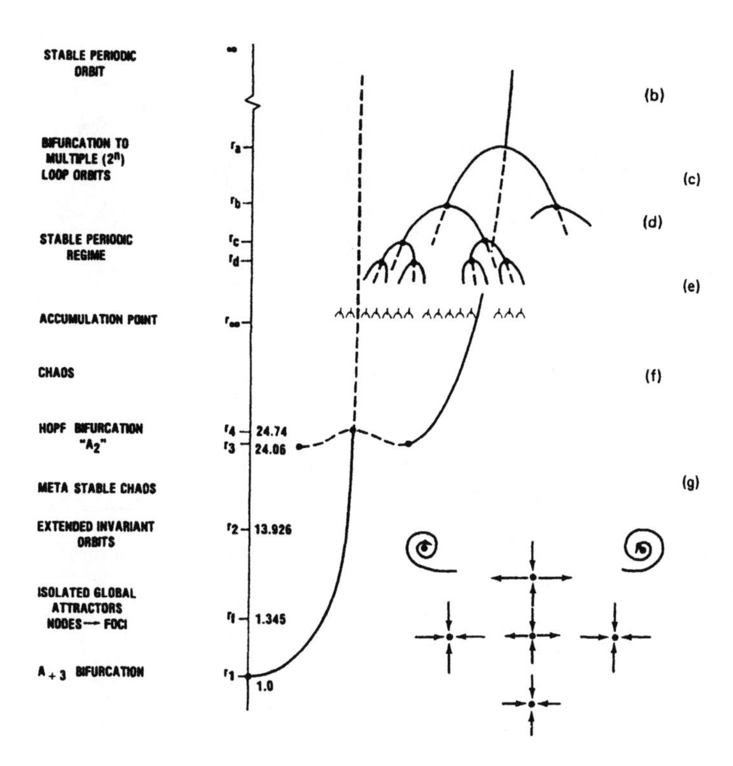

Fig. 8.1 Bifurcation diagram for the Lorenz equations as a function of r for fixed values of $b = \frac{8}{3}$ and $\sigma = 10$. A period-doubling cascade is initiated at r_a and accumulates at r_∞.

that it enters the neighborhood of the *splitting point.* This is the fixed point at the origin that directs the flow back to one of the two foci, depending on where the flow enters the neighborhood of this fixed point. This point is responsible for tearing the flow.

Figure 8.2(b) provides another perspective of the Lorenz attractor. Here the coordinates of the attractor are $x, \dot{x}, \ddot{x}$, and the attractor is projected into the $\dot{x}$–x plane. In this projection the flow around each focus is in the clockwise direction. The insets and outsets of the saddle in the center guide the flow into and away from this saddle.

The projection of the Lorenz attractor shown in Fig. 8.2(a) has rotation symmetry about the z–axis. The projection shown in Fig. 8.2(b) has inversion symmetry with respect to the fixed point in the middle. These two attractors are not topologically equivalent.

Three branched manifolds for the Lorenz attractor are shown in Fig. 8.3. At the top we show a branched manifold that describes the flow shown in the standard representation of the Lorenz attractor, presented in Fig. 8.2(a). Alongside this branched

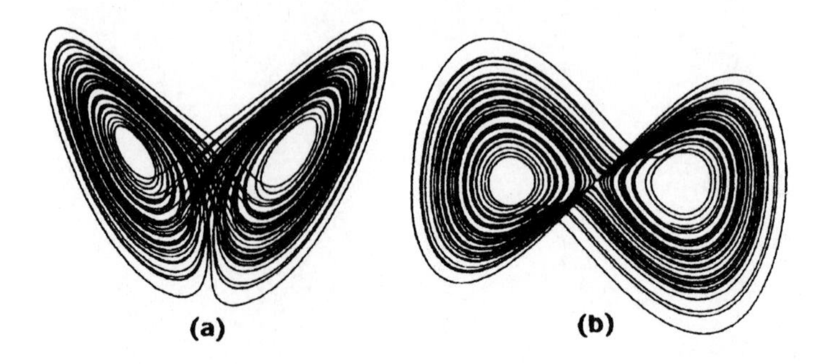

Fig. 8.2 (a) The Lorenz attractor is projected onto the $(x = y)$–z plane. The attractor is invariant under rotation about the z axis. (b) The Lorenz attractor is projected onto the $\dot{x}$–x plane. The attractor is invariant under inversions. Parameter values: $(\sigma, b, r) = (10, \frac{8}{3}, 30)$.

manifold, we present the template matrix and array for this representation. Below the Lorenz template in Fig. 8.3 we show two alternative representations of this flow. The one on the left [Fig. 8.3(b)] has rotation symmetry $R(\pi)$ about an axis through the symmetric fixed point and perpendicular to the plane of projection. This branched manifold can be obtained from Fig. 8.3(a) by rotating the right-hand lobe by π radians about a horizontal axis, with the upper part of the right-hand lobe coming out of the page (cf. Fig. 8.4). The third branched manifold, shown in Fig. 8.3(c), is similar in many respects to Fig. 8.3(b), except that it has inversion symmetry through the symmetric fixed point. This last branched manifold is not topologically equivalent to the other two.

When Lorenz dynamics are studied using simulations, the branched manifold which is obtained is that shown in Fig. 8.3(a). When physical systems exhibiting Lorenz-type dynamics are studied, usually only scalar time series are available. In this case the embedded attractor *must* have inversion symmetry (unless the time series corresponds to the z direction), and thus must appear as a variant of Fig. 8.3(c). We emphasize that the difference between the two representations is one of symmetry (rotation vs. inversion).

8.1.5 Shimizu–Morioka Equations

A modified form of the Lorenz dynamical system has been studied by Shimizu and Morioka [2,98]:

$$\begin{aligned}\frac{dx}{dt} &= y \\ \frac{dy}{dt} &= x - \lambda y - xz \\ \frac{dz}{dt} &= -\alpha z + x^2\end{aligned} \tag{8.2}$$

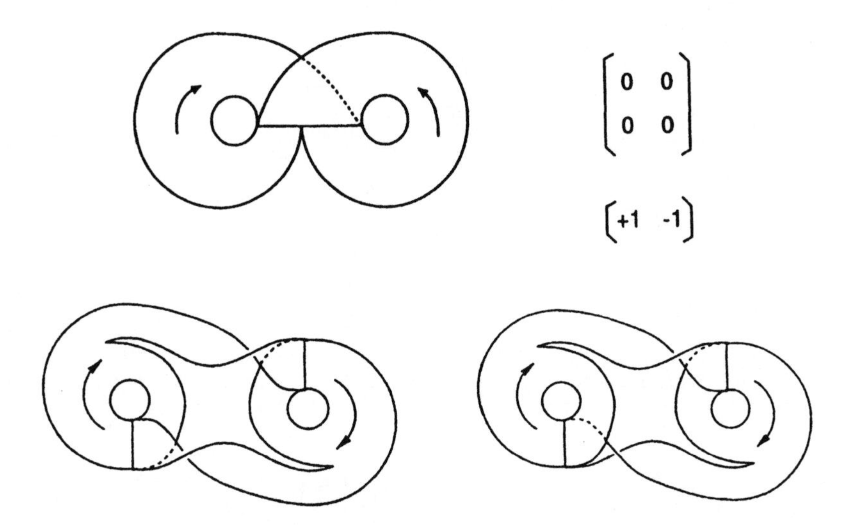

Fig. 8.3 Three branched manifolds describing the Lorenz attractor. (a) This is a clear representation of the flow in R^3. The template matrix and array are shown for this branched manifold. (b) This branched manifold is obtained from the more familiar one in (a) by rotating the lobe on the right about a horizontal axis by π radians, with the top of the lobe coming out of the plane. (c) This branched manifold describes a representation of the Lorenz flow obtained from a differential embedding $(x, \dot{x}, \ddot{x})$ of time series from one of the nonsymmetric variables. It has inversion symmetry and is not topologically equivalent to either (a) or (b).

An attractor generated by these equations is shown in Fig. 8.5. These equations, and the attractor they generate exhibit the same symmetry as the Lorenz equations. This attractor has the same variety of branched manifolds as the Lorenz attractor.

8.2 OPTICALLY PUMPED MOLECULAR LASER

In 1975, Haken shocked the laser physics community by showing that one of the standard laser models was equivalent to the Lorenz equations in a certain limit [155]. This stimulated a great deal of experimental and theoretical work that has not yet subsided.

Haken showed that the equations describing a coherently pumped, homogeneously broadened ring laser reduce to the Lorenz equations under the following conditions:

1. The atomic resonance is equal to a cavity resonance (zero detuning).

2. The rotating-wave approximation is made.

3. The field is uniform.

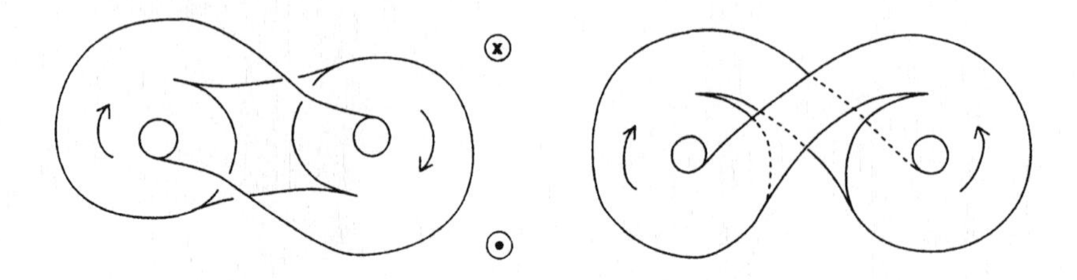

Fig. 8.4 The two branched manifolds representing the Lorenz flow in Fig. 8.3(a) and (b) can be deformed into each other. A rotation of the right-hand lobe by π radians about an axis through the three critical points transforms the branched manifold on the left to the standard representation of the flow in the Lorenz system on the right.

Fig. 8.5 Shimizu–Morioka strange attractor in two projections: (a) Lorenz projection; (b) differential embedding. Control parameter values: $(\alpha, \lambda) = (0.55, 0.85)$.

4. The slowly varying envelope approximation is valid.

This observation about the properties of a laser model introduced the ideas of nonlinear dynamics into the field of laser physics. It stimulated the use of lasers as tools to study nonlinear phenomena. Since the Haken–Lorenz model is so elementary, questions were raised as to whether:

1. Lasers actually behaved this way
2. Realistic models of lasers could be constructed that actually behaved this way

Early experiments to address these questions were carried out by Lawandy and Koepf [156], Lefebvre, Dangoisse, and Glorieux [157], Weiss and Klisch [158], Abraham, Dangoisse, and Glorieux [159], Harrison, Al-Saidi, and Biswas [160], Harrison and Biswas [161], Weiss [162], and Weiss and Brock [163]. These experiments, and the theoretical descriptions of these experiments, were carried out on optically pumped molecular lasers. This class of lasers seemed most likely to operate in a regime compatible with dynamics of Lorenz type. The first review of this field was by Biswas et al. [164]. Thereafter, work on this problem accelerated. The first review was succeeded by many others: Abraham, Narducci, and Mandel [165], Narducci and Abraham [166], Weiss and Vilaseca [167], Arecchi and Harrison [168], Khanin [169], and Roldan et al. [170].

8.2.1 Models

An optically pumped molecular laser consists of a molecular gas (NH_3, CH_2F_2, CH_3F, HF, HCOOH, ...) confined to a laser cavity. The gas is coherently excited by another laser, typically a CO_2 or NO_2 laser. The molecule is excited to a vibrational state; lasing involves deexcitation through a rotational state. This produces an output signal in the range 30 to 100 μm. Since both vibrational and rotational transitions occur, a minimum of three molecular levels is involved.

The dynamics is expressed in terms of an equation of motion for the electric field $E(t)$ and the molecular density operator ρ. This is a 3×3 Hermitian matrix ($\rho^\dagger = \rho, \operatorname{Tr} \rho = 1$) that describes the occupation probabilities and correlations among three molecular states: $|0\rangle$, $|1\rangle$, and $|2\rangle$. The dynamical equations for the molecular subsystem are

$$\begin{aligned}
\dot{\rho}_{00} &= \gamma_0(n_0 - \rho_{00}) - 2\alpha \operatorname{Im} \rho_{10} - 2\beta \operatorname{Im} \rho_{20} \\
\dot{\rho}_{11} &= \gamma_1(n_1 - \rho_{11}) + 2\alpha \operatorname{Im} \rho_{10} \\
\dot{\rho}_{22} &= \gamma_2(n_2 - \rho_{22}) + 2\beta \operatorname{Im} \rho_{20} \\
\dot{\rho}_{10} &= -[\gamma_{10} + i(\Delta_c + \dot{\phi})]\rho_{10} + i\alpha(\rho_{00} - \rho_{11}) - i\beta\rho_{12} \\
\dot{\rho}_{20} &= -[\gamma_{20} + i\,\Delta_2]\rho_{20} - i\alpha\rho_{12} + i\beta(\rho_{00} - \rho_{22}) \\
\dot{\rho}_{12} &= -[\gamma_{12} + i(\Delta_c + \dot{\phi} - \Delta_2)]\rho_{12} + i(\alpha\rho_{02} - \beta\rho_{10}) + i\beta(\rho_{00} - \rho_{22})
\end{aligned} \tag{8.3}$$

In these equations, the ground state $|0\rangle$ is shared by the vibrational ($2 \leftrightarrow 0$) and the rotational ($1 \leftrightarrow 0$) transitions. The field evolution is coupled to the molecular evolution through

$$\begin{aligned}
E(t) &= \alpha(t)\, e^{-i\phi(t)} \\
\dot{\alpha} &= -\kappa\alpha + g \operatorname{Im} \rho_{10} \\
\dot{\phi} &= -g \operatorname{Re} \rho_{10}/\alpha
\end{aligned} \tag{8.4}$$

The parameters in these equations have meanings shown in Table 8.1.

The dynamical system (8.3)–(8.4) comprises a set of 10 real ordinary differential equations. Dupertuis, Salomaa, and Siegrist [171] showed that these equations reduce to the Lorenz equations when the following conditions are satisfied:

- The fields are resonant ($\Delta_c = \Delta_2 = 0$).
- The relaxation rates are all equal: $\gamma_0 = \gamma_1 = \gamma_2$.
- $n_0 = n_1$ at equilibrium.
- The pump is weak: $\beta \ll \gamma_i, \gamma_{ij}$.

Table 8.1 Significance of the parameters in the molecular laser model

Symbol	Significance
γ_i	Decay rate, level i
n_i	Population of level i in thermal equilibrium
γ_{ij}	Decay rate for transition $i \leftrightarrow j$
Ω_2	Pump frequency
ω_c	Cavity resonance
Δ_2	$\Omega_2 - \omega_{20}$ cavity detuning with respect to transition $2 \leftrightarrow 0$
Δ_c	$\omega_c - \omega_{20}$ cavity detuning
g	Unsaturated gain
κ	Cavity loss rate
2α	Rabi frequency of output field
2β	Rabi frequency of pump field

- The density matrix elements ρ_{02} and ρ_{12} can be eliminated adiabatically ($\gamma_{01}, \gamma_{12} \gg \gamma_{01}$).

- The equation of motion for ρ_{22} can be decoupled ($\gamma_{02}\gamma_{12} > |\alpha|^2$).

In an optically pumped laser operating under typical conditions, the pump line is much narrower than either the width of the vibrational transition or the Doppler line width of the thermal molecules. As a result, laser gain is velocity dependent. To make the model (8.3) more realistic, molecules were divided into 81 velocity groups. The gains and losses for each group were made velocity dependent. Under these Doppler broadening conditions, the laser model consisted of 730 coupled ordinary differential equations. The question then is: Do the dynamics of this high-dimensional model remain Lorenz-like?

This set of equations was numerically integrated under resonance pump assumptions ($\Delta_2 = 0$) under several conditions. The laser output amplitude $E(t)$ and intensity $I(t) = |E(t)|^2$ were recorded and analyzed (Roldan et al. [170], Gilmore et al. [172]).

8.2.2 Amplitudes

Under the resonance condition ($\Delta_2 = 0$) the phase ϕ changes by π whenever the amplitude approaches zero. The phase change was incorporated into the field amplitude to give a real field $E(t)$. The amplitude time series obtained under two different operating conditions are shown in Fig. 8.6. In Fig. 8.6(a) the time trace behaves in the same way as the x and y variables for the Lorenz equations operating in the range $r \sim 30$. However, the behavior shown in Fig. 8.6(b) is slightly different. When the amplitude changes sign, the envelope of the oscillations first decreases before spiraling out again. These two behaviors are topologically equivalent.

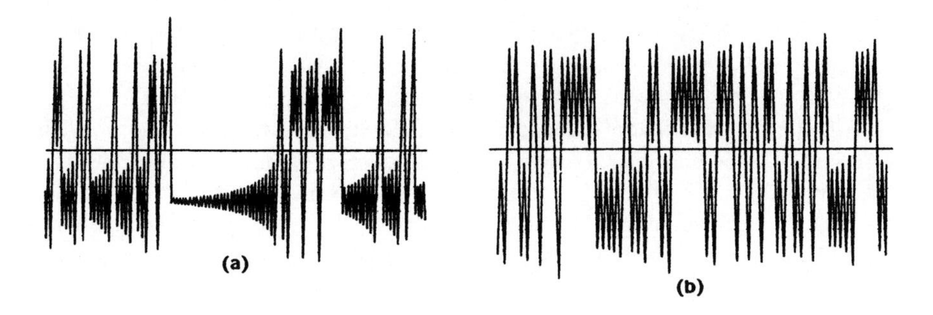

Fig. 8.6 Two time series for the amplitude $E(t)$ obtained from the model equations (8.3) are plotted for two different operating conditions.

8.2.3 Template

We first study the properties of the time series shown in Fig. 8.6(a). An $\dot{x}$ vs. x ($x = E$) projection of the differential embedding of this time series is presented in Fig. 8.7. This embedding has inversion symmetry. The linking numbers of periodic orbits in this embedding are closely related to the linking numbers of orbits with the same symbol names in the embedding with $R_z(\pi)$ symmetry.

8.2.4 Orbits

A number of unstable periodic orbits in the attractor were located by the method of close returns. A symbol name was applied to each surrogate periodic orbit. The

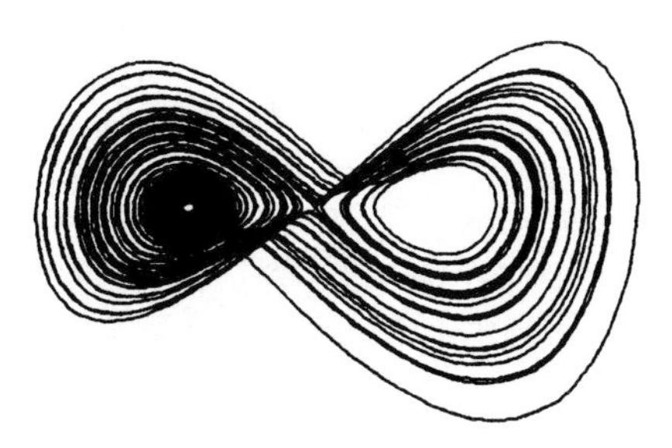

Fig. 8.7 The time series shown in Fig. 8.6(a) is projected onto the $\dot{x}$–x plane. All crossings occur in the first and third quadrants.

symbol identification was easy: An orbit segment around the unstable focus on the left (right) was given the symbol name L (R). Up to period 6, the surrogate orbits that were identified were LR; L^2R, LR^2; L^3R, LR^3, L^2R^2; L^4R, LR^4, L^3R^2, $LRLR^2$; and L^5R, LR^5, L^4R^2, L^2R^4, $LRLR^3$. The period-1 orbits are the two fixed points at $(\pm x_0, 0, 0)$. The linking numbers for these periodic orbits were computed and compared with those expected from the inversion symmetric branched manifold shown in Fig. 8.3(c). There were no discrepancies.

Although the differential embedding used to induce the strange attractor from the time series displays an inversion symmetry, it is not obvious that the attractor itself has this symmetry (cf. Fig. 8.7). The attractor could lack symmetry for two reasons: (1) the underlying dynamics might not be symmetric under $E \rightarrow -E$; (2) the time series might be too short to build up an an invariant density that is symmetric to suitable statistical precision.

Two approaches can be taken to test for the presence of symmetry. One involves statistical procedures. This is less convincing than the one that we have adopted. Our test depends on the properties of a strange attractor: that embedded in it are unstable periodic orbits. If the attractor possesses a symmetry, the surrogate orbits extracted from it must possess this symmetry. In this case, the orbits must be symmetric under left–right conjugation: $L \leftrightarrow R$. We therefore compared conjugate pairs of orbits and self-conjugate orbits that were extracted from the data. These orbits include:

Conjugate Pairs	Self – Conjugate Orbits
$L^2R \leftrightarrow R^2L$	$LR \leftrightarrow RL$
$L^3R \leftrightarrow R^3L$	$L^2R^2 \leftrightarrow R^2L^2$
$L^4R \leftrightarrow R^4L$	
$L^5R \leftrightarrow R^5L$	
$L^4R^2 \leftrightarrow R^4L^2$	

In each of these cases, the rotated image of one orbit was indistinguishable from its conjugate. In Fig. 8.8 we show the self-conjugate orbit L^2R^2 and the conjugate pair L^2R and R^2L. The x's are fiducial marks, which allow unbiased comparisons to be made. This test produced convincing evidence that the attractor induced from the amplitude time series by the integral–differential (inversion symmetric) embedding did in fact possess the required symmetry. The lack of apparent symmetry of the strange attractor is a consequence only of the relatively short time series that was recorded.

Remark: This is an instance where a data file containing more than about 100 cycles might have been useful. However, the problem of symmetry was resolved by relying on the properties of unstable periodic orbits in the strange attractor rather than on its statistical properties.

We turn our attention now to the second time series shown in Fig. 8.6. When the amplitude changes phase, it first spirals in toward an unstable fixed point before it begins to spiral away. This time series was treated in the same way as the time series shown on the left in Fig. 8.6.

Fig. 8.8 The self-conjugate orbit L^2R^2 and the conjugate pair L^2R and R^2L were extracted from a chaotic time series. They show the expected invariance under inversion symmetry. The x's are fiducial marks to facilitate comparisons.

To verify that this identification is correct, the following unstable periodic orbits were extracted from the chaotic time series: L, R; L^2R, R^2L; L^3R, R^3L, L^2R^2; L^4R, R^4L, L^2RLR, R^2LRL, L^3R^2, R^3L^2; and L^5R, R^5L, L^3RLR, R^3LRL, L^2R^2LR. We were unable to locate the orbits L^2R^4 and R^2L^4 to the precision used to find the other unstable periodic orbits. Many of these orbits had the same name as corresponding orbits found in the first data set. The presence of the fold merely served to push some crossings from the first or third quadrant (in the projection of the embedding) into the fourth or second quadrant without changing the sign of the crossing or, therefore, the linking numbers of the corresponding orbits. As a result, the orbits in the second data set had the same spectrum of linking numbers as those in the first set. Thus both flows are topologically equivalent. A schematic representation of the flow in the $\dot{x}$–x embedding with rotation symmetry is shown in Fig. 8.9. The twists in the second and fourth quadrants can be pushed back to the third and first quadrants without changing any topological indices. The branched manifold can also be deformed into a Lorenz-like flow [Fig. 8.9(b)] by rotating the lobe on the right through π radians about a horizontal axis, as shown in Fig. 8.9(a).

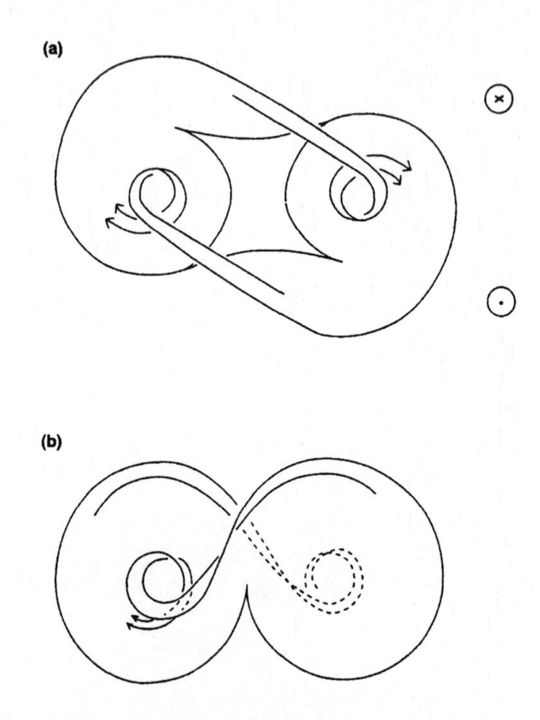

Fig. 8.9 (a) The branched manifold for the data set in Fig. 8.6(b) is shown for a differential embedding. The inversion symmetry has been transformed to rotation symmetry. (b) This branched manifold can be deformed into a standard Lorenz-like form by rotating the right-hand lobe, as indicated.

We also tested for the symmetry of the induced attractor. The conjugate pairs and self-conjugate orbits which were compared were

Conjugate Pairs	Self – Conjugate Orbits
$L \leftrightarrow R$	$LR \leftrightarrow RL$
$L^2R \leftrightarrow R^2L$	$L^2R^2 \leftrightarrow R^2L^2$
$L^3R \leftrightarrow R^3L$	
$L^4R \leftrightarrow R^4L$	
$L^2RLR \leftrightarrow R^2LRL$	
$L^3R^2 \leftrightarrow R^3L^2$	
$L^5R \leftrightarrow R^5L$	
$L^3RLR \leftrightarrow R^3LRL$	

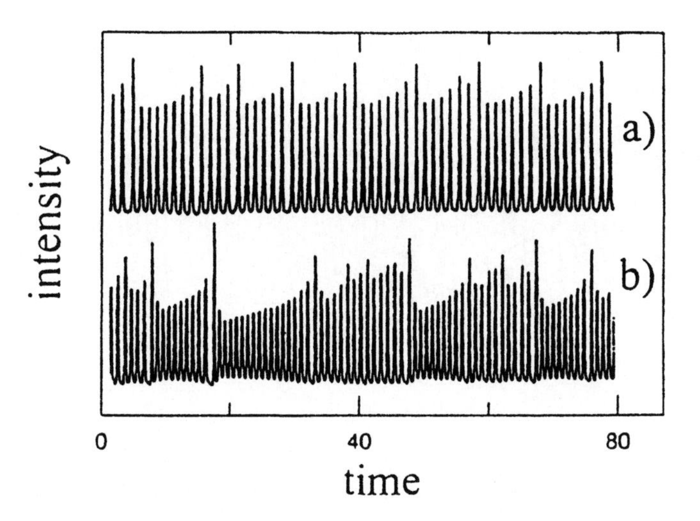

Fig. 8.10 Intensity output from an optically pumped molecular laser under two different operating conditions.

Orbit comparisons were very similar to those shown in Fig. 8.8. Even though the strange attractor did not exhibit the required symmetry, the conjugate pairs and self-conjugate unstable periodic orbits extracted from the (short) chaotic time series did exhibit the required symmetry. We therefore concluded that the attractor would have appeared symmetric if additional data had been available.

8.2.5 Intensities

Two time series of intensities were also analyzed. These are shown in Fig. 8.10(a) and (b). A differential plot of dI/dt vs. $I(t)$ shows a simple fold, suggesting horseshoe dynamics for the intensity. However, the dynamical equations of motion involve the amplitude $E(t)$, whose absolute square is the intensity. Therefore, it was necessary to extract square-root information from intensity data before a topological analysis could be carried out.

To locate the points at which the amplitude changes sign, we plotted $\sqrt{I(t)}$ vs. t. This plot is shown in Fig. 8.11. When the minimum occurs at I_0, the shape of the square root in the neighborhood of the minimum is

$$|E(t)| = \sqrt{I_0 + \Delta I(t)} \begin{array}{ll} \stackrel{I_0 \neq 0}{\longrightarrow} & \sqrt{I_0} + \dfrac{1}{2\sqrt{I_0}} \Delta I(t) \quad \text{parabolic} \\ \stackrel{I_0 = 0}{\longrightarrow} & \sqrt{\Delta I(t)} \qquad \text{linear} \end{array}$$

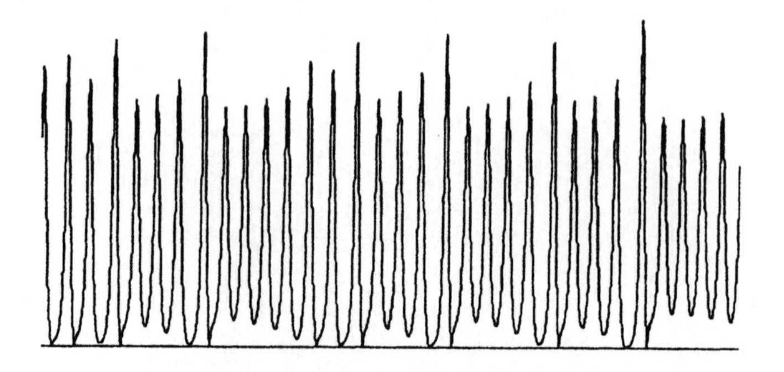

Fig. 8.11 The square root is taken of the intensity output in Fig. 8.10(a).

It is a simple matter to distinguish minima of $\sqrt{I(t)}$ where the local behavior is parabolic from those where it is linear.

This test for assignment was implemented numerically. We chose the sign assignment that minimized the estimate of $dE/dt \sim E_{i+1} - E_i$. Specifically, at the ith element in the data field, we defined

$$a_3 = \text{ISIGN} \times \sqrt{I_i}$$

Then we tested the two differences

$$\begin{aligned} S_1 &= |(+a_3 - a_2) - (a_2 - a_1)| \\ S_2 &= |(-a_3 - a_2) - (a_2 - a_1)| \end{aligned}$$

After each test, we updated the variables according to $a_2 \to a_1$ and $a_3 \to a_2$. Normally, $S_1 < S_2$ and there is no sign change. Whenever $S_1 > S_2$, we change the sign: ISIGN $\to$ $-$ISIGN. Implementation of this algorithm is illustrated in Fig. 8.12. At the bottom of this figure we show intensity data. At the top we show amplitude data extracted with this algorithm. The differential embedding of this data set is shown projected onto the $\dot{x}$–x plane in Fig. 8.13. The following unstable periodic orbits were extracted: L, R; LR; L^2R, R^2L; L^3R, R^3L, L^2R^2; L^4R, R^4L, L^2RLR, R^2LRL, L^3R^2; and L^5R, R^3LRL, R^2L^2RL. All conjugate pairs and self-conjugate orbits were invariant under the expected symmetry. A similar analysis was carried out on the intensity data shown in Fig. 8.10(b), with identical results.

8.3 FLUID EXPERIMENTS

A simple experiment has been designed and carried out that exhibits Lorenz dynamics. In this experiment a torus with large radius R and small radius r is filled with water (Creveling et al. [173]; Gorman, Widman, and Robins [174, 175]; Singer, Wang, and Bau [176]). The bottom half was wrapped with a thermal heating ribbon, which

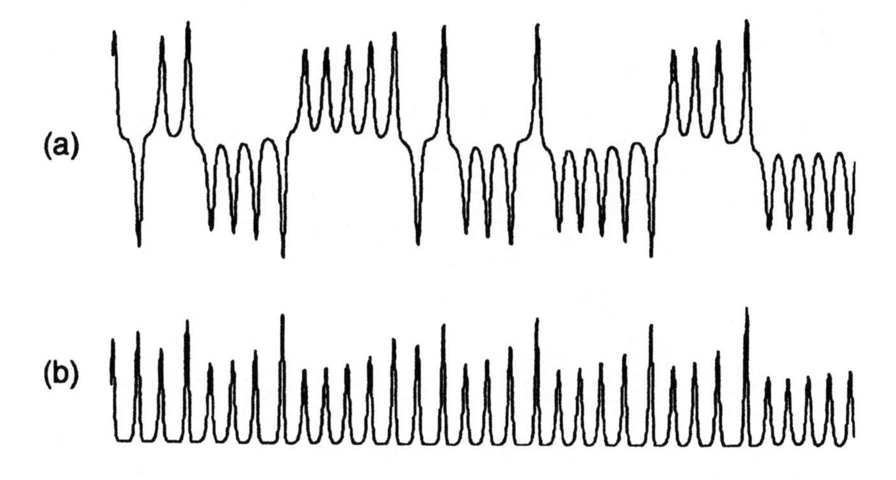

Fig. 8.12 The E-field amplitude (a) is recovered from intensity data (b) using the algorithm described in the text.

Fig. 8.13 The amplitude data constructed from intensity data are shown in a differential embedding.

generated a constant, uniform heat flux. The top half of the torus was surrounded by a water jacket, which kept that part of the surface at approximately the coolant temperature. Heat sensors were embedded within the torus at the 3 o'clock and 9 o'clock positions. The temperature difference was monitored as a function of time for different heating rates. The experimental arrangement, and the data it generated, are shown in Fig. 8.14.

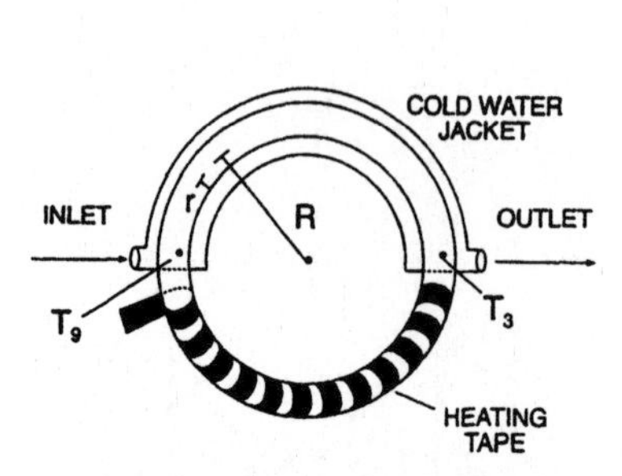

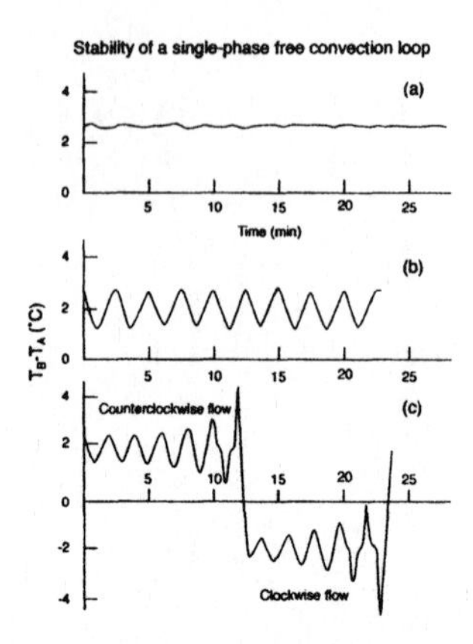

Fig. 8.14 Left: Fluid is heated in the lower half of the torus and cooled in the upper half. This creates thermal instability. The temperature is monitored at the 9 o'clock and 3 o'clock positions. Right: The temperature difference between the two probes is plotted as a function of time. The imposed temperature difference ΔT between the temperature maintained by the heating tape and the cooling jacket increases from (a) to (b) to (c). The flow in (c) varies unpredictably between the clockwise and counterclockwise directions. The flow morphology is described by the Lorenz equations. Reprinted with permission from Creveling et al. [173]. Copyright 1975 by American Institute of Physics.

8.3.1 Data

The heated fluid in the bottom half of the torus was less dense than the fluid in the top half. This density difference set up thermal instability. For low heating, the fluid circulated in the clockwise or counterclockwise direction. The direction of circulation can be inferred from the temperature difference: If $T_9 - T_3 > 0$, the flow is clockwise [Fig. 8.14(a)].

As heating was increased, the rotation maintained its direction (either clockwise or counterclockwise) but became oscillatory [Fig. 8.14(b)]. As heating was increased above some threshold, the direction of rotation changed unpredictably from clockwise to counterclockwise and back again [Fig. 8.14(c)].

8.3.2 Template

The behavior shown in Fig. 8.14(c) is chaotic. An $\dot{x}$ vs. x ($x = T_9 - T_3$) embedding will have the structure shown in Fig. 8.2(b). It is therefore possible to construct an embedding in which the attractor is equivalent to the strange attractor generated by the Lorenz system.

It is worthwhile to make a few remarks at this stage:

1. The experimental apparatus has reflection symmetry in a vertical plane. It is this symmetry that forces the Lorenz mechanism (A_3) rather than the Rössler mechanism (A_2).

2. The identification of the thermal instability behavior observed in Fig. 8.14(c) with the Lorenz mechanism was made possible by observing the phase-space plot $\dot{x}$ vs. x. In fact, we did not even have to perform this plot: The identification was already possible by inspecting the morphology (shape) of the time series alone. This is often possible for very dissipative dynamical systems.

8.4 WHY A_3?

The phase space plots in this chapter all show three critical points very clearly. Two symmetric saddle foci at $(x, \dot{x}) = (\pm x_0, 0)$ are separated by a regular saddle at the origin.

The Lorenz model has three critical points. They may all be real, or two may be hidden in the complex plane. They are "unmasked" by a cusp catastrophe (A_3). In one parameter unfoldings, it is typical to encounter only fold catastrophes. However, in the presence of symmetry, the cusp catastrophe is also generic [37]. It is for this reason that physical systems whose dynamical equations of motion exhibit symmetry may exhibit mechanisms leading to chaos that are more complicated than the A_2 fold mechanism: that is, heteroclinic as opposed to homoclinic connections (Ott [177]; Solari, Natiello, and Mindlin [178]).

It is not surprising that the differential phase-space embedding of the intensity $I = E^2$ shows a fold, while that of the amplitude shows a heteroclinic connection. There is a 2:1 relation between the Lorenz branched manifold and the Rössler branched manifold. Dynamics on a Rössler branched manifold are locally isomorphic to dynamics on a Lorenz branched manifold, but globally, two regions on a Lorenz attractor map to a single region of a Rössler attractor. Another way to state this is that the Lorenz system is a 2:1 covering of the Rössler system. We describe cover and image dynamical systems in Chapter 10.

8.5 SUMMARY

The stretch and fold (*gâteau roulé*) mechanism is the mechanism for creating chaos which is most frequently encountered in the analysis of experimental data (Chapter 7). It occurs typically when the phase space for the dynamics is a torus $R^2 \times S^1$. The rotation axis of the torus can be considered as containing a pair of fixed points (of focus type) which generate rotation. Mechanisms of this type are described by singularities of type A_2.

The mechanism treated in this chapter depends on stretching and tearing. In this case a saddle point splits the flow into two parts. Each part is directed to the influence of a different focus. It is convenient to describe such flow types by singularities of type A_3 involving three fixed points. Such singularities arise naturally in physical systems that exhibit some type of symmetry. In the two cases treated in this chapter, the physics of the laser is invariant under a sign reversal of the electric field amplitude. In the fluid example the physics is invariant under reversal of the direction of the velocity field. These symmetries are responsible for the tearing observed in the dynamics. In Chapter 10 we describe transformations (local 2:1 diffeomorphisms) that relate these two distinct classes of chaos producing mechanisms.

9

Unfoldings

Unfolding is a technical term for a beautiful idea. An unfolding summarizes all possible consequences of the most general possible perturbation. The concept is fundamental to the study of singularities [13, 36, 37, 179–184]. Unfoldings occur in the study of dynamical systems in two ways:

1. Branched manifolds describe the topology of the flow. Under perturbation of the flow they often remain unchanged. However, sometimes they change by adding one or more branches or by having some branches annihilated. Branched manifolds act in some sense like the *germ* of a flow.

2. The spectrum of unstable periodic orbits in a flow refines the rough description in terms of a branched manifold. Under perturbation, even if the branched manifold remains unchanged, the spectrum of unstable periodic orbits in the flow does change. In this sense, the spectrum of unstable periodic orbits behaves like the *unfolding* of the flow.

9.1 CATASTROPHE THEORY AS A MODEL

The important concepts *germ* and *unfolding* first appear in the study of *catastrophes*. We therefore begin our discussion of these ideas by reviewing the main features of catastrophe theory.

9.1.1 Overview

Catastrophe theory is a study of how the critical points (equilibria) of a family of functions are created, move about, collide, and annihilate, as the control parameters in the family are changed. Catastrophe theory is important in the physical sciences because it facilitates reduction of complicated families of functions to simple canonical forms. For example, the Ginzburg–Landau canonical form for a potential $[V(x;\lambda) = -\frac{1}{2}\lambda x^2 + \frac{1}{4}x^4]$ can be intuited on the basis of simple general arguments such as symmetry, instability, and bimodality. Catastrophe theory, its concepts, and its development will serve as a model for many of the features of dynamical systems that we introduce in the future.

9.1.2 Example

The general ideas of catastrophe theory are most simply introduced through a simple example. We consider a two (control)-parameter family of functions depending on 100 state variables: $V(x_1, x_2, \ldots, x_{100}; c_1, c_2)$. We ask the simple question: What is the worst kind of behavior that this family can typically exhibit?

At a typical point x_0 in phase space, $\nabla V \neq 0$. At such a point it is possible to find a smooth change of variables, $y = y(x)$, so that in the neighborhood of x_0, $V \doteq y_1$, where $y(x_0) = 0$. The implicit function theorem guarantees this. There isn't much more that can be said about such points.

It is useful to look for the equilibria of V. The location of these equilibria, and their stability, largely determine all the qualitative properties of V. At equilibria, $\nabla V = 0$, but typically the stability matrix (Hessian) $\partial^2 V/\partial x_i \partial x_j$ is nonsingular. This real symmetric matrix typically has nonzero eigenvalues $\lambda_1 \leq \lambda_2 \leq \cdots \leq \lambda_{100}$ with $\lambda_k < 0$ and $0 < \lambda_{k+1}$. At such an equilibrium another theorem (the Morse lemma) guarantees that there be a smooth change of variables $y = y(x)$ so that $V \doteq -y_1^2 - \cdots - y_k^2 + y_{k+1}^2 + \cdots + y_{100}^2$. This canonical quadratic form is called a *Morse normal form*: $M_k^{100}(y)$. The quadratic form has k unstable directions and $100 - k$ stable directions. The coordinate y_i is in the direction of the eigenvector with eigenvalue λ_i.

For most control parameter values, the family V has only isolated or nondegenerate equilibria with nonzero eigenvalues, and the function has a Morse quadratic form in the neighborhood of each of these equilibria. It may be possible to locate degenerate critical points by exploiting the degrees of freedom available from the control parameter dependence that is built into this family of functions. Here is how. The eigenvalues are functions of the control parameters c. It may be possible to "tune" the values of the c so that one or more eigenvalues become zero. If

l eigenvalues become zero at a critical point, it is degenerate, and there is yet another theorem (the Thom lemma) which guarantees that there is a smooth change of variables $y = y(x)$, so that in the neighborhood of the degenerate critical point, $V \doteq f_{NM}(y_1, \ldots, y_l; c) + M_k^{100-l}(y_{l+1}, \ldots, y_{100})$. The function f_{NM} is a non-Morse function. Its Taylor series has no linear terms (it is an equilibrium) and no bilinear terms either. The lowest terms in its Taylor series expansion must be of degree 3 or higher.

The nature of the singularity of V can now be understood by studying the singularity of the simpler function $f_{NM}(y_1, \ldots, y_l; c)$. When there are only two control parameters, l can only be 1. The function of the single variable y_1 is expanded in a power series. It has the form

$$f_{NM}(y_1; c_1, c_2) = t_0 + t_1 y_1^1 + t_2 y_1^2 + t_3 y_1^3 + t_4 y_1^4 + t_5 y_1^5 \cdots \tag{9.1}$$

The constant term can be removed by changing the origin of the ordinate. The first -degree term is absent because we have searched through the state-variable space for particular points that are in equilibria. The leading term in the Taylor series expansion at an equilibrium is $t_2 y_1^2$. Now the two degrees of freedom available through the control parameters can be used to annihilate no more than two of the Taylor series coefficients. If they are used to annihilate t_2 and t_3, the leading term in $f_{NM}(y_1)$ is $t_4 y_1^4$. It is also possible to find another smooth change of variables, $\tilde{y}_1 = \tilde{y}_1(y_1)$, which eliminates all terms of degree greater than 4 in this Taylor series expansion and gives the fourth-degree term a canonical value $\pm\tilde{y}_1^4$. The final result of this series of transformations is that at the most degenerate critical point

$$V(x_1, x_2, \ldots, x_{100}; c_1, c_2) \doteq \tilde{y}_1^4 + M_k^{100-1}(y_2, \ldots, y_{100}) \tag{9.2}$$

The function $\tilde{y}_1^4$ is the *germ* of the degeneracy.

The next part of the question is: What happens to this degenerate critical point under the most general perturbation? This question also has a very nice answer.

The Taylor series of an arbitrary perturbation will have powers of all terms $y_1, y_2, \ldots, y_{100}$. We can block these terms into three groups: one group involving powers of the terms $y_2, \ldots, y_{100}$, another group involving powers of the single variable y_1, and a third group involving cross-product terms.

The first group, involving only terms in the Morse canonical form, can be transformed away by a smooth change of variables. The same is true of the cross terms. This leaves only perturbations involving powers of the single variable y_1 (we replace $\tilde{y}_1$ by y_1 for simplicity). This perturbation has the form

$$\pm y^4 + \text{pert}(y) = p_0 + p_1 y^1 + p_2 y^2 + p_3 y^3 + (\pm 1 + p_4) y^4 + p_5 y^5 + \cdots \tag{9.3}$$

We proceed as before. A smooth transformation can be used to eliminate all terms in the Taylor series of degree greater than 4 and to return the fourth-degree term to the canonical form $\pm 1 \times y_1^4$. The constant term can be removed by readjusting the origin in the image space. Finally, the origin in the y_1 space can be displaced to eliminate one of the three remaining Taylor series coefficients. The coefficient of the cubic term can always be so removed. In this way, the most general perturbation (= universal unfolding) of the cusp germ leads to the surprisingly simple answer to the two questions posed above:

$$V(x_1, x_2, \ldots, x_{100}; c_1, c_2) \doteq \pm\, y_1^4 + ay_1^2 + by_1 + M_k^{100-1}(y_2, \ldots, y_{100}) \quad (9.4)$$

Here the unfolding coefficients a and b are functions of the two control parameters c_1 and c_2. The function $ay_1^2 + by_1$ is the *unfolding* of the germ $\pm y_1^4$.

9.1.3 Reduction to a Germ

There is a general procedure for reducing a family of functions depending on control parameters to some organizing singularity (the germ), and then determining the universal unfolding of that germ. This procedure is illustrated in Fig. 9.1. In the top half of Fig. 9.1, we show how to compute the germ of a singularity. A critical point of a family of functions is located, and the function is expanded about that critical point. At the critical point the first-degree terms in the Taylor expansion are all zero. The constant term can be removed by readjusting the origin. The control parameter degrees of freedom are used to annihilate the lower-degree terms in the expansion, and a general coordinate transformation is used to eliminate the "Taylor tail" of the expansion. To the extent possible, the terms remaining in the middle are given canonical values such as ± 1 or 0. These terms constitute the germ of the singularity. This germ is the functional form of the potential in the neighborhood of the degenerate critical point.

There is only one type of germ depending on one state variable. It is called a *cuspoid*:

$$A_{\pm n} \qquad f_{NM}(x) = \pm x^{n+1}$$

The cuspoid $A_{\pm 1}$ is not degenerate: it describes a Morse canonical form. All higher cuspoids describe n-fold-degenerate critical points in R^1.

In two dimensions there is one infinite family of germs and three exceptional germs:

$$\begin{array}{lll} D_{\pm n} & x^2 y \pm y^{n-1} & n \geq 4 \\ E_{\pm 6} & x^3 \pm y^4 & \\ E_7 & x^3 + xy^3 & \\ E_8 & x^3 + y^5 & \end{array}$$

The infinite family of two state-variable germs $D_{\pm n}$ is called the *umbilic series*. In three dimensions there is one infinite family:

$$T_{p,q,r} \qquad x^p + y^q + z^r + axyz$$

where the integers p, q, r satisfy $1/p + 1/q + 1/r > 1$. The real parameter a cannot be given a canonical value, as for the case of the simple germs described above. This singular germ is therefore called a *unimodal germ*. In addition to this one infinite family of unimodal germs, there are three boundary members of the family ($T_{3,3,3}, T_{4,4,2}, T_{6,3,2}$) and 14 exceptional unimodal singularities.

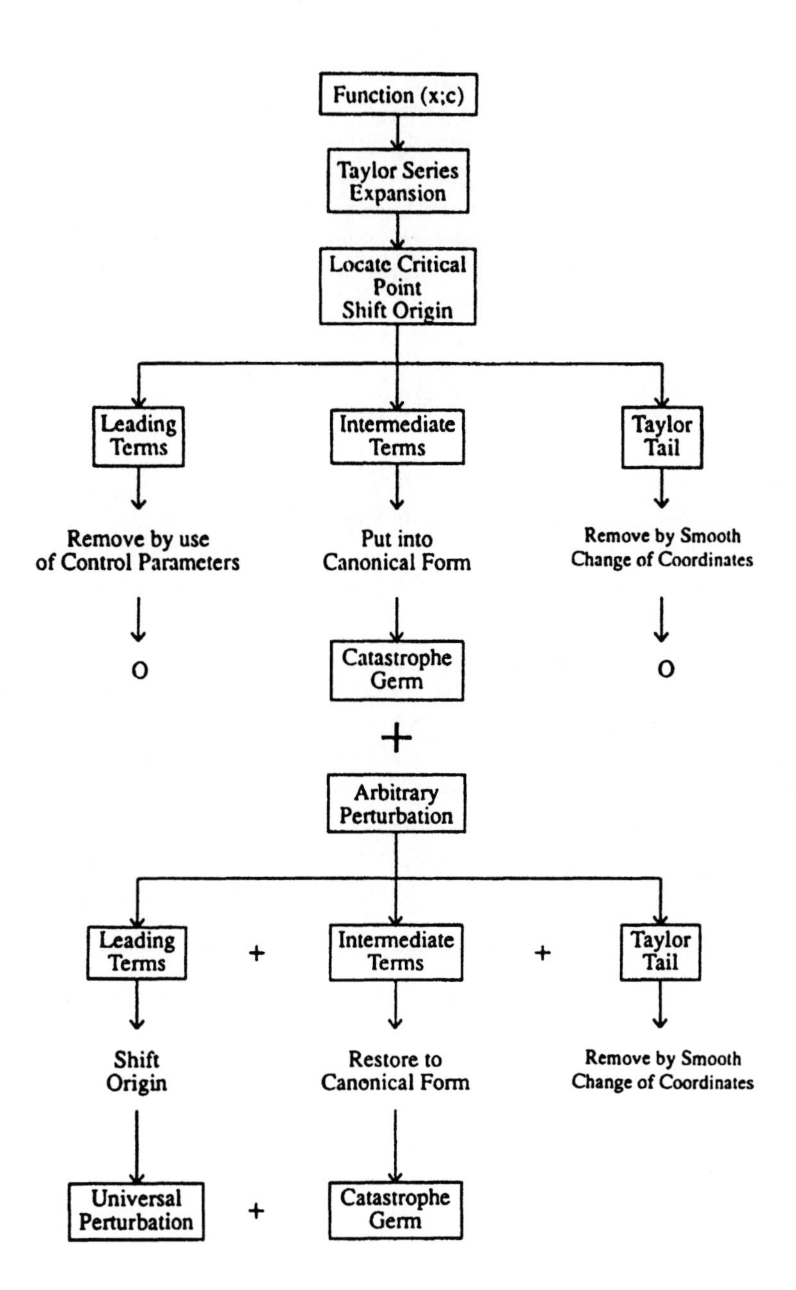

Fig. 9.1 The general procedures for determining the germ of a singularity, and the universal unfolding of that germ, consist of the steps shown.

9.1.4 Unfolding the Germ

The steps involved in computing the universal unfolding of a germ are shown in the bottom part of Fig. 9.1. An arbitrary perturbation is added to the germ. Whatever transformation was used to transform away all higher-degree terms and to put the germ into canonical form in the first half of the algorithm is used again for the same purpose. This leaves as the only remaining terms in the perturbation the ones that were annihilated by using the control parameters. Some of these can also be eliminated by shifting the origin of ordinate and coordinates. The terms that remain constitute the universal unfolding of the singularity represented by the germ. The universal perturbations of the cuspoids and the umbilics are

$$\begin{array}{lll} A_{\pm n} & \pm x^{n+1} & +\sum_{j=1}^{n-1} a_j x^j \\ D_{\pm n} & x^2 y \pm y^{n-1} & +a_1 x + a_2 x^2 + \sum_{j=3}^{n-1} a_j y^{j-2} \end{array}$$

The coefficients $a_1, \ldots, a_{n-1}$ are the unfolding parameters.

9.1.5 Summary of Concepts

Catastrophe theory has introduced two important new concepts: germ and unfolding. A *germ* is an algebraic representation of a degenerate critical point. A germ tightly constrains the organization of *all* critical points in its neighborhood. The *unfolding* of a germ includes the results of *every* possible perturbation of that singularity. The concepts of germ and unfolding will shortly be extended to enlarge our understanding of dynamical systems and their behavior under perestroika (change in control parameter values).

9.2 UNFOLDING OF BRANCHED MANIFOLDS: BRANCHED MANIFOLDS AS GERMS

To a great extent, branched manifolds play the same role for dynamical systems (in R^3) as germs play in catastrophe theory. They organize the topological properties of the flow, just as germs organize the structure of all the lower singularities in the neighborhood of the germ.

When a dynamical system is perturbed, generally the branched manifold is unchanged. This is the lesson learned from analysis of about 25 data sets from the LSA (Section 7.2): Under change of operating conditions, the underlying branched manifold remains unchanged—only the spectrum of unstable periodic orbits in the strange attractor is altered. However, under sufficiently large perturbation, the underlying branched manifold can change. This is the message drawn from analysis of the YAG laser, the fiber optic laser, and the nerve cell. However, the changes in the structure of the underlying branched manifold are circumscribed in ways first made apparent by the previous analysis of Duffing oscillator [93].

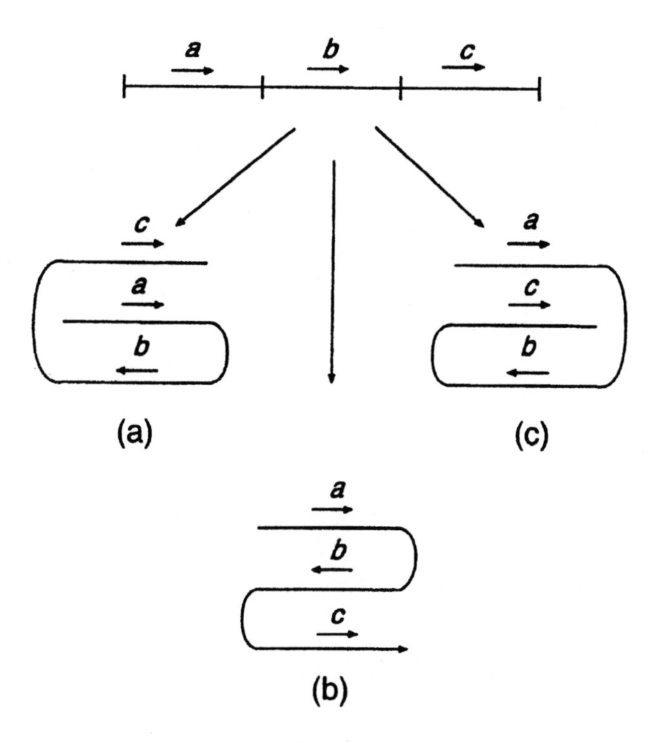

Fig. 9.2 The two branches a and b describe folding in a Smale horseshoe template. A third branch c can be added in only one of three ways.

9.2.1 Unfolding of Folds

We begin by treating a concrete example, relevant for the interpretation of all the data discussed in Chapter 7. Assume that a dynamical system generates a strange attractor in a torus $R^2 \times S^1$ and that the underlying branched manifold is a Smale horseshoe template with branches a and b in Fig. 9.2. If control parameters are varied so that the expansion per period increases above 2, one new branch (at least) is required to describe the strange attractor. This new branch c must be connected continuously to the branches a and b and can appear with respect to a and b in the branched manifold in only the three ways shown. The organization represented in Fig. 9.2(a) is the beginning of an inside-to-outside spiral template, while that shown in Fig. 9.2(c) is the beginning of an outside-to-inside spiral template. The organization shown in Fig. 9.2(b) is the third possibility. There are only three possible ways to add a third branch to a folded two-branch manifold. The number of ways of adding additional branches is constrained by continuity considerations but grows rapidly with n, the number of branches.

In the analysis of experimental data from strange attractors in $R^2 \times S^1$, we have only seen Smale horseshoe and scroll templates. There is no theorem which says that a dynamical system that initially is organized by a horseshoe must evolve into a scroll template as the control parameters are changed. In exactly the same spirit, there is no theorem which says that a second period-doubling bifurcation must follow a first. However, these two cases are very similar: The more period-doubling bifurcations that occur, the more likely that they will be followed by yet another. In the same way, the more branches added to a scroll template, the more likely that the next branch added will continue the scroll. This phenomenon has been observed in lasers, nerve cells, and the Duffing oscillator.

9.2.2 Unfolding of Tears

It is also possible to discuss unfoldings of branched manifolds underlying flows with symmetry, such as the Lorenz flow. The results are richer if only because they are less obvious.

In Fig. 9.3 we present a representation of an attractor with rotation symmetry. As some control parameter is increased, the outer edge of the flow in one lobe is reinjected closer and closer to the center of the opposite focus. As the control is pushed harder, the reinjection "passes through" the focus. In fact, if it passes through in front of the focus, it can be rotated around underneath the original translobe branch, as shown in Fig. 9.3(a). On the other hand, if it passes behind the focus, it can be rotated back to the original branch line above the original translobe branch, as shown in Fig. 9.3(b). These are the only two possibilities, by continuity.

These figures also show how the new pair of branches join the two pairs of branches already present. In addition, a return map of the two branch lines to themselves is provided. The return maps for these two different unfoldings of the Lorenz branched manifold appear identical until the branches in these maps are "dressed" with the local torsion in each of the branches. In the one case, the third branch continues twisting in the same direction as the second branch (first translobe branch), so the local torsion increases from +1 to +2. In the other case, the third branch twists in the opposite direction, so the local torsion decreases from +1 to 0.

Remark 1: The branched manifold for the Lorenz attractor unfolds according to the scenario shown on the right in Fig. 9.3.

Remark 2: In principle it is possible to find (an unfolding of) a dynamical system for which one set of control parameter values generates a flow of the type shown on the left in Fig. 9.3, while another set of control parameter values pushes the junction past the focus to generate a flow of the type shown on the right. The bifurcation that takes place as the branch line passes through the focus is a type not studied previously. At the bifurcation, an entire pair of branches (with local torsion +2) is annihilated and another pair of branches (with local torsion 0) is created. These annihilations and creations are accompanied by the destruction of all periodic orbits passing through the annihilated branches and simultaneous creation of all periodic orbits with segments passing through the newly enfranchised branches.

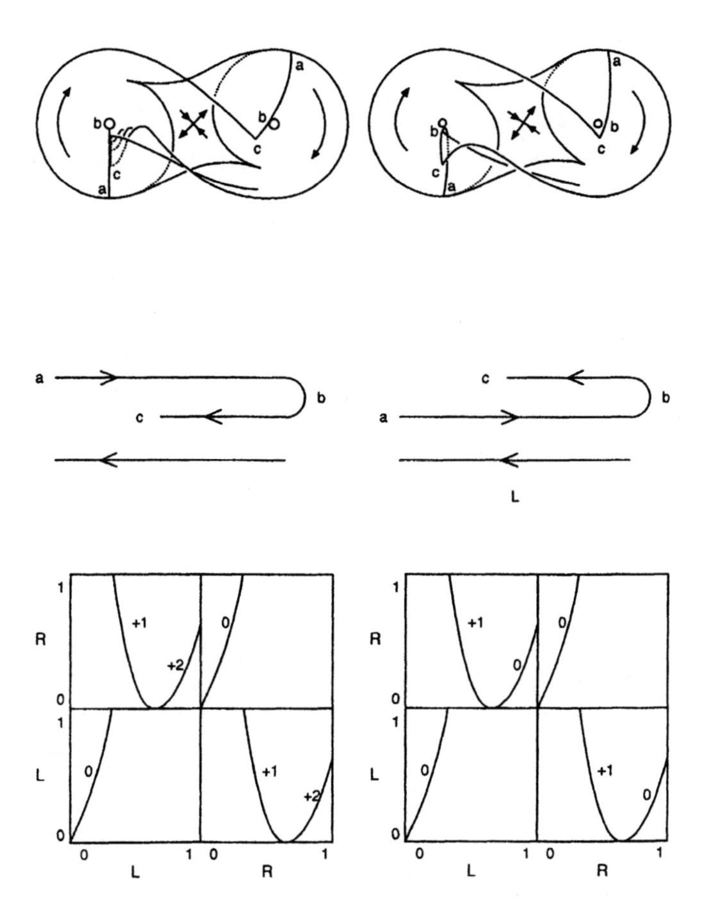

Fig. 9.3 (a) Top: The flow on the branched manifold extends past the focus, joining the lobe on the right in front of the focus. This can be deformed, as shown on the left, where the new translobe branch folds underneath the old. Middle: The manner in which the branch on the left and the two translobe branches from the right join on the left is shown as seen from the fixed point at the origin, looking toward the lobe on the left. Bottom: The return map for the branch lines on the left and right is shown. Integers indicate the torsion of each of the six branches. (b) The same information is provided for the case where the flow extends past the focus, joining the lobe on the right on the far side of the focus.

9.3 UNFOLDING WITHIN BRANCHED MANIFOLDS: UNFOLDING OF THE HORSESHOE

Our objective is to understand how periodic orbits are created as a Smale horseshoe is formed. More specifically, we assume that at one extreme a flow exhibits only a stable period-1 orbit, and as some control parameter(s) are increased, additional periodic orbits are created by period-doubling and saddle-node bifurcations, until the full spectrum of periodic orbits exhibited in a flow described by a horseshoe is present. What are the constraints on the creation of this ensemble of orbits?

This problem is treated more simply by progressing in the reverse direction. Begin with the full spectrum of orbits in a horseshoe flow. Then run the control parameters in the reverse direction. The horseshoe orbits are annihilated but in an order governed by topological constraints. In this section we discuss the topology of the constraints and use this to determine possible orders in which orbits can be created and annihilated in the formation or destruction of a horseshoe. The results are summarized in a forcing diagram (Fig. 9.8).

9.3.1 Topology of Forcing: Maps

We introduce the subject by describing forcing in one dimensional maps of the interval, such as the logistic map $x' = \lambda x(1-x)$ (or $x' = a - x^2$). We begin in the hyperbolic limit $\lambda = 4$ or $(a = 2)$. In this limit all allowed)orbits are present. In Fig. 9.4(a) we show the location of the points in the saddle-node pairs 3_1 and 4_2. When the saddle-node pair is created or annihilated, pairs of points, one from each orbit, become degenerate and then separate (orbit creation) or disappear (orbit annihilation). The pairs of points that are born together are joined in this figure.

As the control parameter λ is decreased, one of these orbit pairs is annihilated before the other. The topological organization of this pair of saddle-node pairs [Fig. 9.4(a)] requires that 4_2 be annihilated before 3_1. The reason is as follows. If the orbit pair 3_1 were to be annihilated first, the two pairs of points on 3_1 (the two pairs on the left) separated by two pairs of points on 4_2 (the two pairs in the middle) would have to travel through the points in the period-4 orbits. By continuity, at some stage in this process a point on the interval would be on both a period-3 and a period-4 orbit. The *determinism* of the map and *uniqueness* of orbits forbids this. As a result, the orbits with the "interior points" (e.g., 4_2) must be annihilated before the orbits with the "exterior points" can be annihilated. This argument runs in reverse. The orbit pair 3_1 must be created before the orbit pair 4_2 can be created.

A similar construction is made for the orbit pair 4_2 and 5_3 in Fig. 9.4(b). This diagram shows that the second, third, and fourth pair of points on the saddle-node pair 5_3 come between the first, second, and third pair of points of 4_2. As a result, 5_3 must be annihilated before 4_2 can be annihilated. In reverse, 4_2 must be created by saddle-node bifurcation before 5_3 can be created.

These topological arguments can be used to determine the order of creation for the entire universal sequence. Fortunately, simpler means are available.

9.3.2 Topology of Forcing: Flows

Analogous arguments can be used with flows. These arguments also exploit determinism and uniqueness. However, the arguments are not quite as straightforward.

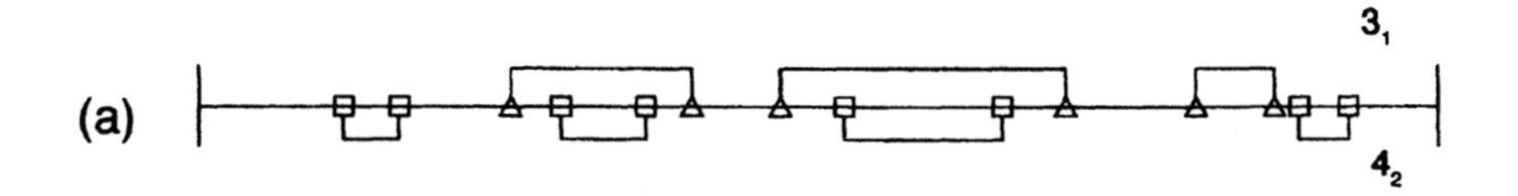

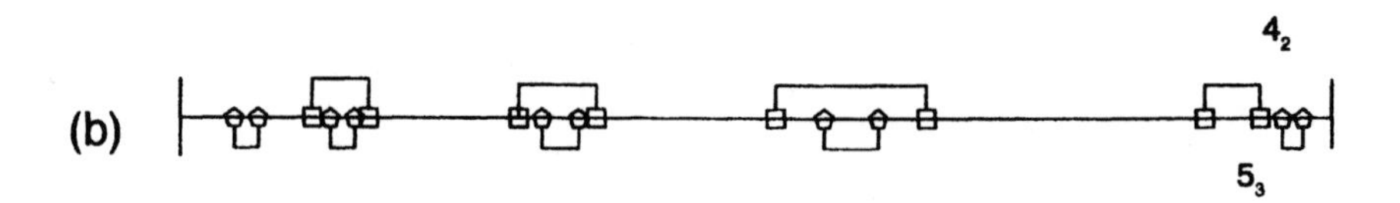

Fig. 9.4 Pairs of points in the logistic map that are created together at saddle-node bifurcations are linked. As the control parameter is decreased, inner pairs must be annihilated before outer pairs can be annihilated. These topological considerations determine the order of orbit creation and annihilation in one-dimensional maps. (a) The orbits 4_2 must be annihilated before the orbit pair 3_1 can be. (b) The orbits 5_3 must be annihilated before the orbit pair 4_2 can be.

As control parameters for a dynamical system are varied, periodic orbits are created or annihilated in either saddle-node or period-doubling bifurcations. Suppose that we have two saddle-node pairs of orbits $\{A_R, A_F\}$ and $\{B_R, B_F\}$. Here R refers to the regular saddle (even parity) and F to the flip saddle.

The topological organization of periodic orbits in flows is determined by linking numbers, as opposed to topological organization of periodic orbits in maps, where it is determined by *betweenness*. Linking numbers cannot change as long as the orbits exist by the uniqueness theorem (essentially the same arguments as used above for maps). In computing the linking numbers of the orbit pair $\{A_R, A_F\}$ with the orbit pair $\{B_R, B_F\}$, three possibilities occur. These are summarized in Fig. 9.5:

1. The pair $\{A_R, A_F\}$ cannot undergo an inverse saddle node bifurcation until the orbit pair $\{B_R, B_F\}$ does. Conversely, $\{B_R, B_F\}$ cannot be created until the orbit pair $\{A_R, A_F\}$ already exists. We summarize this situation by saying that B forces A and writing $B \Longrightarrow A$.

2. Neither orbit pair forces the other.

3. $A \Longrightarrow B$.

Period-doubling bifurcations can be treated similarly. In this case a saddle-node pair is replaced by a mother–daughter pair (M–D) of period p and $2p$. Both have odd parity. The orbits $MM = 2M$ (the p symbols of M are repeated twice) and D can then be treated like A_R and A_F in the procedure described above.

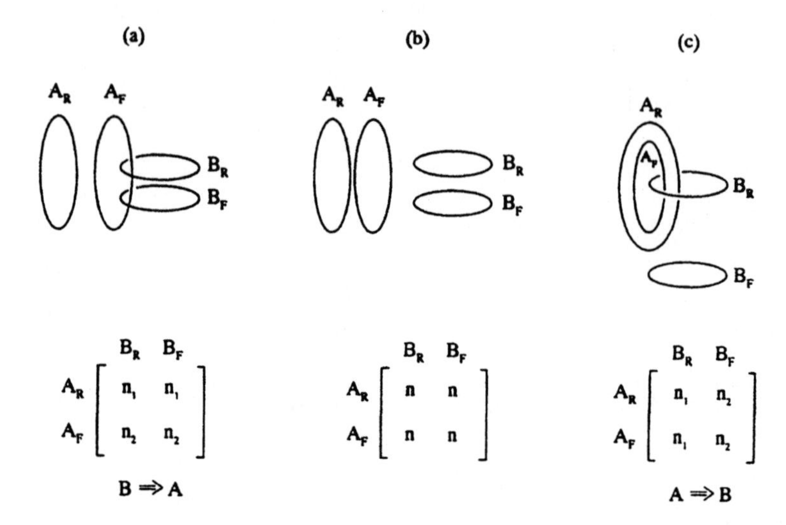

Fig. 9.5 Pair of saddle-node pairs in three configurations. (a) If the linking numbers between A_R and the pair $\{B_R, B_F\}$ are equal but different from the linking numbers of A_F with the pair $\{B_R, B_F\}$, the pair $\{A_R, A_F\}$ must be created before the pair $\{B_R, B_F\}$ can be created. (b) When all four linking numbers are equal, neither pair forces the other. (c) If the linking numbers between B_R and $\{A_R, A_F\}$ are equal but different from the linking numbers of B_F with $\{A_R, A_F\}$, the pair $\{A_R, A_F\}$ must be created after the pair $\{B_R, B_F\}$. A similar analysis holds with mother–daughter pairs of period-doubled orbits.

The topology of forcing is subject to two additional considerations: transitivity and exchange. The first is similar to the situation for maps, the second is not.

Transitivity is illustrated in Fig. 9.6. If $A \Longrightarrow B$ and $B \Longrightarrow C$, then $A \Longrightarrow C$. This can be shown by constructing the square matrix of linking numbers for the orbit pairs A and C.

Exchange is illustrated in Fig. 9.7. A linking number analysis may indicate that $A \Longrightarrow B$ and $A \Longrightarrow C$. However, if B and C have the same braid type [133, 185–187], the orbits B_R and C_R can exchange saddle-node partners so that $\{B_R, C_F\}$ forms one saddle-node pair while $\{C_R, B_F\}$ forms another. In the case shown, orbit pair A cannot force any of the four orbits in the quartet $\{B, C\}$. This means that the linking numbers must be computed between orbit multiplets of the same braid type rather than just between saddle-node pairings which occur in the U-sequence. Period is a braid type invariant, so only orbits of the same period can be of the same braid type and participate in exchange.

Remark: In the U-sequence, orbits are created with unique partners. This pairing is relaxed in flows described by horseshoes. It is this relaxation of topological

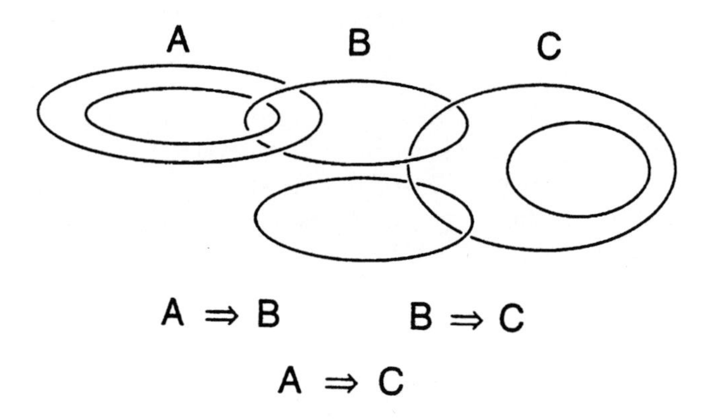

Fig. 9.6 Forcing is transitive. If $A \Longrightarrow B$ and $B \Longrightarrow C$, then $A \Longrightarrow C$.

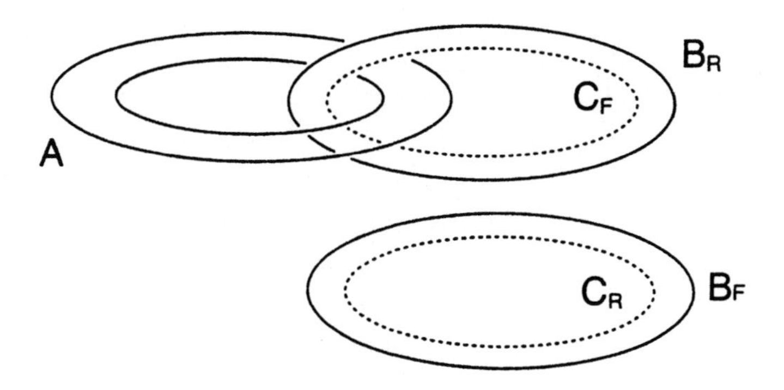

Fig. 9.7 Forcing is complicated by Exchange. If orbits B and C have the same braid type, the orbit pairs may exchange partners. Then A forces neither B nor C if both exist, while it forces one saddle node pair if the other does not exist.

constraints that opens up a rich but constrained spectrum of routes to chaos in the formation of strange attractors governed by a horseshoe template.

9.3.3 Forcing Diagrams

The topological arguments outlined above can be applied to periodic orbits on any branched manifold. They have been carried out in detail only for the Smale horseshoe template. This is in part because the calculations are very difficult, but also in part because this particular branched manifold occurs so often in physical systems [185–188]. The forcing diagram for orbits up to period 8 on the horseshoe template is shown in Fig. 9.8. All orbits up to period 8 have been summarized in Table 9.1, along with their properties. The orbits are identified as P_j, where P is the period and j indicates the order of occurrence of this orbit in the universal sequence. The orbits in the Smale horseshoe template exist in 1:1 correspondence with orbits in the logistic map in the hyperbolic limit $\lambda = 4$.

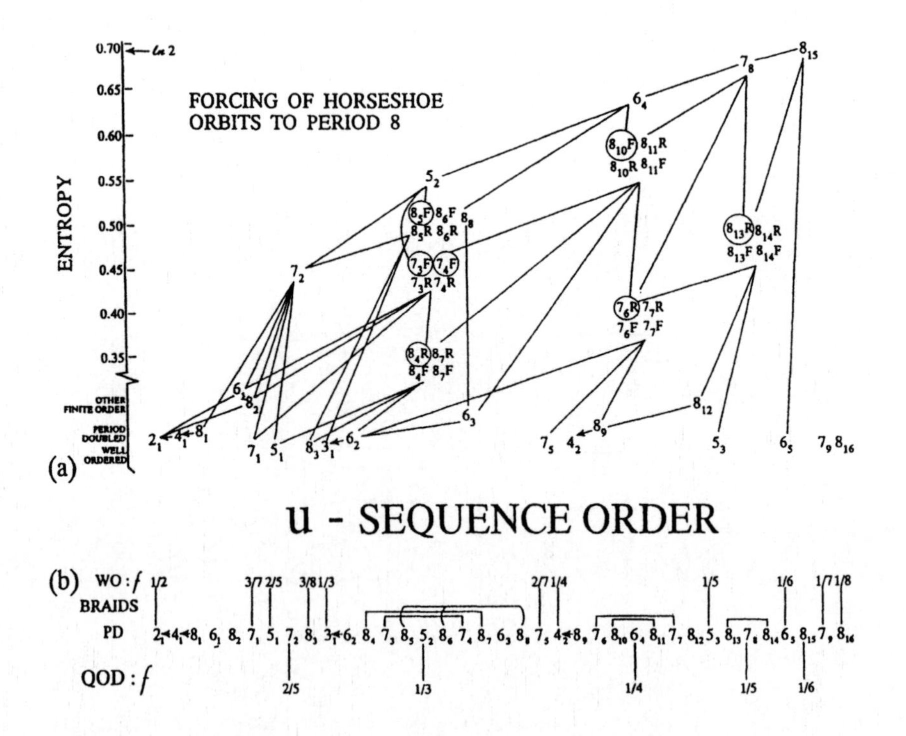

Fig. 9.8 (a) Forcing diagram shows the order in which periodic orbits in the Smale horseshoe template are organized. All orbits are identified by their U-sequence names. Orbits are arranged by their one-dimensional entropy (horizontal axis) and their topological entropy (vertical axis). Small distortions have been introduced to make the diagram more readable. Well-ordered orbits, finite-order orbits, and members of the initial period-doubling cascade all have zero entropy and occur along the horizontal axis. The backbone of the diagram consists of the well-ordered orbits (horizontal axis) and the quasi-one-dimensional orbits, whose topological entropy is the same as their one-dimensional entropy. Both types of orbits exist in 1:1 correspondence with rational fractions f in the interval $0 < f < \frac{1}{2}$. Transitivity of forcing requires only first-order forcing to be shown explicitly. (b) Orbits are shown in the order of their appearance in the U-sequence. The rational fractional values f are also shown for well-ordered orbits (above) and QOD orbits (below) the U-sequence.

Table 9.1 Most useful properties of horseshoe orbits up to period 8[a]

Orbit P_j	Symbol Sequence	Permutation	Remarks	Fraction	1D Entropy	Topological Entropy
2_1	01	12	PD of 1F		0	0
3_1	011	123	WO	$\frac{1}{3}$	0.481 212	0
4_1	01 11	1324	PD of 2_1		0	0
4_2	0011	1234	WO	$\frac{1}{4}$	0.609 378	0
5_1	011 11	134 25	WO	$\frac{2}{5}$	0.414 013	0
5_2	001 11	124 35	QOD	$\frac{1}{3}$	0.543 535	0.543 535
5_3	000 11	123 45	WO	$\frac{1}{5}$	0.656 256	0
6_1	011 111	143 526	FO		0.240 606	0
6_2	001 011	135 246	PD of 3_1		0.481 212	0
6_3	001 111	124 536	FO		0.583 557	0
6_4	000 111	123 546	QOD	$\frac{1}{4}$	0.632 974	0.632 974
6_5	000 011	123 456	WO	$\frac{1}{6}$	0.675 975	0
7_1	011 1111	145 3627	WO	$\frac{3}{7}$	0.382 245	0
7_2	011 0111	146 2537	QOD	$\frac{2}{5}$	0.442 138	0.442 138
7_3	001 0111	136 2547	PE		0.522 315	0.476 818
7_4	001 1111	125 4637	PE		0.562 400	0.476 818
7_5	001 1011	135 6247	WO	$\frac{2}{7}$	0.601 001	0
7_6	000 1011	124 6357	PE		0.618 362	0.382 245
7_7	000 1111	123 5647	PE		0.645 710	0.382 245
7_8	000 0111	123 4657	QOD	$\frac{1}{5}$	0.666 215	0.666 215
7_9	000 0011	123 4567	WO	$\frac{1}{7}$	0.684 905	0
8_1	0111 0101	1547 2638	PD of 4_1		0	0
8_2	011 111 11	1546 3728	FO		0.304 688	0
8_3	011 011 11	1472 5638	WO	$\frac{3}{8}$	0.468 258	0
8_4	001 011 11	1372 5648	PE		0.499 747	0.346 034
8_5	001 010 11	1364 7258	PE		0.539 792	0.498 093
8_6	001 110 11	1365 7248	PE		0.547 612	0.498 093
8_7	001 111 11	1256 4738	PE		0.574 865	0.346 034
8_8	001 101 11	1257 3648	PE		0.591 718	0.498 093
8_9	0001 0011	1357 2468	PD of 4_2		0.609 378	0
8_{10}	000 101 11	1247 3658	PE		0.626 443	0.568 666
8_{11}	000 111 11	1236 5748	PE		0.639 190	0.568 666
8_{12}	000 110 11	1246 7358	FO		0.651 766	0
8_{13}	000 010 11	1235 7468	PE		0.660 791	0.458 911
8_{14}	000 011 11	1234 6758	PE		0.671 317	0.458 911
8_{15}	000 001 11	1234 5768	QOD	$\frac{1}{6}$	0.680 477	0.680 477
8_{16}	000 000 11	1234 5678	WO	$\frac{1}{8}$	0.689 121	0

[a] Orbits are listed by period P and order of occurrence of the orbit in unimodal maps of the interval P_j. For each orbit the symbol sequence is given. If a saddle-node partner exists, it is obtained by changing the parity of the next-to-last symbol. The permutation of the orbit, either as an orbit in a unimodal map of the interval or on a Smale horseshoe template, is given. Orbits are identified according to type: WO, well ordered; PD, period doubled; FO, finite order; PE, positive entropy but not QOD; QOD, quasi-one-dimensional. Rational fractional values of WO and QOD orbits are given. The one- and two-dimensional entropy (= topological entropy) values are given. The two are equal only for QOD orbits.

Each orbit in the U-sequence has a one-dimensional entropy. The later the orbit is created, the higher its entropy. The same orbit has a two-dimensional entropy also, called its *topological entropy*. This entropy measures the number of orbits of period p whose existence is forced by the orbit when it appears in a two-dimensional map. Many two-dimensional maps contain this particular orbit: The mapping with *minimal* entropy is chosen, and the entropy of this map is identified as the two-dimensional or topological entropy of the orbit. The one- and two-dimensional entropies of a horseshoe orbit are generally not the same. The one-dimensional entropy is not less than the (two-dimensional) topological entropy.

9.3.3.1 Orbits with Zero Entropy Zero-entropy orbits either force no additional orbits (well-ordered orbits) or else force only a finite number (period doubled) or an algebraically growing (finite order) number of orbits. All zero-entropy orbits are isotopic to torus knots or iterated torus knots.

Well-Ordered Orbits: Well-ordered orbits do not force the existence of any other orbits, except for the period-1 orbits $\{1_F, 1_R\}$, which are forced by all horseshoe orbits and which are therefore not shown in the forcing diagram. The symbolic dynamics of a well-ordered orbit of period $q+2p$ consist of q symbols 0 and p symbols 11 "as equally spaced as possible." The irreducible rational fraction $f = p/(q+2p)$ identifies each well-ordered orbit uniquely. The symbol sequence identifying the well-ordered orbit with irreducible rational fraction $p/(q+2p)$ is $W(1)W(2)\cdots W(q+p)$, where

$$W(i) = \begin{cases} 0 & \text{if} \quad [ig] - [(i-1)g] = 0 \\ 11 & \text{if} \quad [ig] - [(i-1)g] = 1 \end{cases} \tag{9.5}$$

Here $g = p/(q+p)$ and $1 \leq i \leq q+p$. As usual, $[x]$ is the integer part of x. For example, the well-ordered orbits of period 7 are determined by the irreducible rational fractions $\frac{3}{7}$: $7_1 = (011\ 1111)$, $\frac{2}{7}$: $7_5 = (001\ 1011)$, and $\frac{1}{7}$: $7_9 = (000\ 0011)$. The node partner of each saddle is determined as usual by changing the next-to-last symbol from 1 to 0. The well-ordered horseshoe orbits to period 8 are $(7_1, 5_1, 8_3, 3_1, 7_5, 4_2, 5_3, 6_5, 7_9, 8_{16})$. They are identified by their fractional values, which decrease in the order of their creation in the U-sequence. Well-ordered orbits are easily identified through their spectrum of self-relative rotation rates: They are all $RRR_{ij} = f = p/(q+2p)$.

Period-Doubled Orbits: These orbits force only their mother, grandmother, ... orbits. The period-doubled orbits, to period 8, are $8_1 \Longrightarrow 4_1 \Longrightarrow 2_1, 6_2 \Longrightarrow 3_1$, and $8_9 \Longrightarrow 4_2$. The godmother orbit (e.g., 5_2) may have positive topological entropy, but her grandchildren force no more orbits than the head of the family and her immediate granddaughters, so have the same topological entropy.

Finite-Order Orbits: These orbits have zero topological entropy but do force some orbits of higher period. The number of such forced orbits grows algebraically rather than exponentially, so these orbits technically have zero entropy. The finite-order orbits, to period 8, are $(6_1, 8_2, 6_3, 8_{12})$.

9.3.3.2 Orbits with Positive Entropy Positive-entropy orbits force an exponentially growing number of orbits of higher period. Positive-entropy orbits fall into two classes.

Quasi-One-Dimensional Orbits: These orbits force all the orbits in two-dimensional maps which they force in one-dimensional maps of the interval. In this sense they are dual to the well-ordered orbits, which force only the period-1 orbits. Like the well-ordered orbits, there is a 1:1 correspondence between the QOD orbits and the irreducible rational fractions f in the interval $0 < f < \frac{1}{2}$. To each irreducible rational fraction $0 < f = m/n < \frac{1}{2}$ there corresponds a QOD orbit of period $n + 2$ with symbolic dynamics $0^{\kappa_1}110^{\kappa_2}11 \cdots 0^{\kappa_m}111$, where

$$\begin{aligned} \kappa_1 &= [1/f] - 1 = [n/m] - 1 \\ \kappa_i &= [i/f] - [(i-1)/f] - 2 \\ &= [in/m] - [(i-1)n/m] - 2 \quad 2 \leq i \leq m \end{aligned} \tag{9.6}$$

For example, the QOD orbits of period 7 are determined by the irreducible rational fractions $\frac{2}{5}$: $7_2 = (011\,0111)$ and $\frac{1}{5}$: $7_8 = (000\,0111)$. The saddle partner of each node is obtained as usual by changing the next-to-last 1 to 0.

The QOD orbits to period 8 are $(7_2, 5_2, 6_4, 7_8, 8_{15})$. Each QOD saddle-node pair belongs to a braid type containing no other horseshoe orbits. The entropy for QOD orbits is the same for three-dimensional flows and two-dimensional maps as for one-dimensional maps of the interval.

Positive-Entropy Orbits: These comprise all orbits not already discussed. For these orbits, two or more saddle-node pairs always belong to the same braid type. Orbits of the same braid type have been identified by their spectrum of relative rotation rates. This topological index is sufficient to identify braid type through orbits of period 10 but not period 11. Orbits of the same braid type have the same two-dimensional entropy but different one-dimensional entropies.

The positive entropy braids up to period 8, in order of increasing entropy, are $[8_4, 8_7]$, $[7_6, 7_7]$, $[8_{13}, 8_{14}]$, $[7_3, 7_4]$, $[8_5, 8_6, 8_8]$, and $[8_{10}, 8_{11}]$. In the sextet, linking number calculations show that orbits 8_5 and 8_6 are more similar to each other than to 8_8: in particular, 8_5 and 8_6 can participate in exchange, whereas 8_8 does not.

9.3.3.3 Additional Comments

The linking numbers of all 18 pairs of orbits with positive entropy were computed. We reproduce some of the submatrices in Fig. 9.9.

The table of linking numbers for the quartets $[8_4, 8_7]$ and $[7_3, 7_4]$ is shown in Fig. 9.9(a). These results show that $7_3 \Longrightarrow 8_4R$ and $7_4 \Longrightarrow 8_4R$. In fact, any orbit in the quartet $[7_3, 7_4]$ forces 8_4R, which may be paired with either 8_4F or 8_7F.

The table of linking numbers for the quartet $[7_6, 7_7]$ and the higher-entropy sextet $[8_5, 8_6, 8_8]$ is shown in Fig. 9.9(b). This appears to show that $7_6 \Longrightarrow 8_5$ and $7_6 \Longrightarrow 8_6$, and similarly for 7_7. However, once both 8_5 and 8_6 exist, 8_5R and 8_6R can exchange flip saddles, so that in fact $[7_6, 7_7]$ do not force $[8_5, 8_6]$, by the exchange phenomenon.

The linking numbers between the quartet $[8_4, 8_7]$ and the quartet $[8_{10}, 8_{11}]$ are shown in Fig. 9.9(c). This table clearly shows that $[8_{10}, 8_{11}] \Longrightarrow [8_4, 8_7]$. Each positive-entropy orbit forces a contiguous sequence of well-ordered orbits. That is, if a positive-entropy orbit forces well-ordered orbits corresponding to the irreducible

(a)

		7_3		7_4	
		F	R	R	F
8_4	R	18	18	18	18
	F	19	19	19	19
8_7	F	19	19	19	19
	R	19	19	19	19

(b)

		8_5		8_6		8_8	
		F	R	R	F	R	F
7_6	R	14	15	14	15	15	15
	F	14	15	14	15	15	15
7_7	F	14	15	14	15	15	15
	R	14	15	14	15	15	15

(c)

		8_{10}		8_{11}	
		F	R	R	F
8_4	R	18	18	18	18
	F	19	19	19	19
8_7	F	19	19	19	19
	R	20	20	20	20

(d)

		8_5		8_6		8_8	
		F	R	R	F	R	F
6_3	R	14	15	14	15	14	14
	F	14	15	14	15	15	15

Fig. 9.9 These tables of linking numbers were used to compute forcing in four parts of the forcing diagram. (a) This table of linking numbers between the quartets $[8_4, 8_7]$ and $[7_3, 7_4]$ shows that any of the orbits in $[7_3, 7_4]$ forces 8_4R, which may be paired with either 8_4F or 8_7F. (b) This table is for the lower-entropy quartet $[7_6, 7_7]$ and the higher-entropy sextet $[8_5, 8_6, 8_8]$. Once both 8_5 and 8_6 exist, 8_5R and 8_6R can exchange flip saddles, so that $[7_6, 7_7]$ do not force $[8_5, 8_6]$. (c) Linking numbers between the quartet $[8_4, 8_7]$ and $[8_{10}, 8_{11}]$ show that $[8_{10}, 8_{11}] \Longrightarrow [8_4, 8_7]$. (d) Linking numbers for the zero entropy finite-order orbit 6_3 and the sextet $[8_5, 8_6, 8_8]$ show that the zero-entropy orbit 6_3 can force either 8_5 or 8_6 when just one of the two pairs is present, cannot force either saddle-node pair $8_5R, 8_6F$ or $8_5F, 8_6R$, and cannot force either 8_5 or 8_6 when both are present.

rational fractions P/Q and P'/Q', it forces all well-ordered orbits with intermediate rational fractional values.

9.3.4 Basis Sets of Orbits

Embedded in a strange attractor is an infinite set of unstable periodic orbits. The spectrum of orbits changes as control parameters change. It is possible to identify the spectrum of orbits present by identifying a much smaller subset of orbits. This subset is a set of orbits that forces all orbits present. We call such a subset a *basis set of orbits*.

A basis set of orbits can be computed for strange attractors generated by a horseshoe mechanism, at least to period 8. A simple algorithm suffices to construct such a basis set. The algorithm consists of three steps:

1. List all the orbits present in order of increasing entropy. For orbits with the same entropy, order according to the U-sequence order (one-dimensional entropy).

2. Remove the last orbit in this list and all the orbits that this last orbit forces.

3. Repeat step 2.

The set of last elements removed in step 2 is the basis set of orbits, ordered by decreasing topological entropy, and for the same entropy, by decreasing one-dimensional entropy. We illustrate this algorithm with an example.

Example: Assume that all orbits that occur before 7_6 in the U-sequence, including 7_6, are present in the strange attractor for a flow. Then, to period 8, the list of orbits present, ordered by entropy and one-dimensional entropy, is

$$\underline{2_1}, \underline{4_1}, \underline{8_1}, \underline{6_1}, \underline{8_2}, \underline{7_1}, \underline{5_1}, \underline{8_3}, \underline{3_1}, \underline{6_2}, 6_3, 7_5, 4_2, 8_9, \underline{8_4}, 8_7, 7_6, \underline{7_2}, \underline{7_3}, 7_4, \underline{8_5}, 8_6, 8_8, 5_2$$

The last orbit in this list (5_2) is identified as a basis orbit. This orbit, and all orbits that it forces (they are undelined), are removed and the list is rewritten:

$$\underline{6_3}, 7_5, 4_2, 8_9, 8_7R, 7_6, 7_4F, 8_6F, 8_8$$

The second basis orbit is 8_8. Continuing in this way, the basis set of orbits that force all those initially present is (reading right to left)

$$8_7R, 7_6, 7_4F, 8_6F, 8_8, 5_2 \tag{9.7}$$

A lower bound on the topological entropy of the flow can be obtained by computing the topological entropy of the braid containing the basis set of orbits.

A forcing diagram exists for any branched manifold. It can be computed explicitly up to any period p. Then any flow on the branched manifold can be described by a basis set (up to period p). This allows a discrete topological classification of strange attractors by branched manifold and basis set.

9.3.5 Coexisting Basins

Whenever a new orbit is created by a period-doubling bifurcation, it is initially stable. When an orbit pair is created in a saddle-node bifurcation, the node of the pair is initially stable. The basin of attraction of the stable orbit "eats a hole" in the strange attractor. This means that motion in the strange attractor is bounded away from the newly created stable periodic orbit. As control parameters vary, the stable periodic orbit may undergo a series of period-doubling bifurcations that ultimately destroy the basin.

It is sometimes useful to know how many basins of attraction can coexist within a strange attractor. Suppose that an attractor is identified by a basis set consisting of t unstable periodic orbits up to period p. Then there is a perturbation that neither creates nor annihilates orbits by saddle-node bifurcations but which moves each of the basis orbits to the verge of saddle-node annihilation. At this point, each periodic node is stable and surrounded by its own basin of attraction. Therefore, under a perturbation, a strange attractor described by a basis set with t basis orbits can have t coexisting basins of attraction, each surrounding a stable node from the basis set.

9.4 MISSING ORBITS

The forcing theorem was confronted with experimental data in yet another constructive way. Up to period 8, the unstable periodic orbits extracted from the Belousov–Zhabotinskii data discussed in Chapter 7 are $(1_1,\ 2_1,\ 3_1,\ 4_1,\ 5_1,\ 6_2,\ 7_2,\ 8_1,\ 8_3)$ (cf. Table 7.1). Following the algorithm presented in Section 9.3.4, a basis set of orbits for the chaotic dynamics is $(8_3,\ 6_2,\ 7_2)$. This basis set forces 12 orbits up to period 8: the nine orbits listed above as well as three additional orbits $(6_1,\ 7_1,\ 8_2)$ not found by the method of close returns.

These three orbits might be missing for a variety of reasons: (1) the theorem could be wrong; (2) the close returns rejection threshold could be too stringent; or (3) the data set was too short. We discarded the first alternative and proceeded under the following assumption. These three orbits are, in fact, present in the strange attractor. However, during the experiment the system state never got close enough to any of these orbits to follow it around for the full six (or seven or eight) periods to return sufficiently close to the starting point that the method of close returns required. We therefore attempted to locate these three predicted orbits by alternative means.

Metric methods, as described in Section 6.5.3, were used to search for segments of the chaotic time series to use as surrogates for segments of these unstable periodic orbits. Specifically, the search proceeded as follows. The chaotic time series was reduced to a long symbol string (543 symbols) consisting of 0s and 1s. Each symbol described a segment of the time series ($\sim$120 samples) between a Poincaré section based on minima. A surrogate for the period-6 orbit 011101 was constructed as follows. The data segment most similar to $\underline{0}11101$ was identified by computing the distance between the segment $11101\underline{0}11101$ and each of the $543 - 2 \times 5$ symbols in the long symbol string for the entire chaotic time series. The distance was computed as indicated in (6.45). The computation was not done for the five symbols at the beginning and end of the long symbol string. The data segment giving minimal metric distance was chosen as the surrogate for the single symbol $\underline{0}$. This calculation was repeated for each of the remaining five symbols using strings $11010\underline{1}11010$, $10101\underline{1}10101$, and so on. In the case of the period-6 orbit 011101, the third, fourth, and fifth surrogate symbols (110) came from adjacent symbols in the long symbol string. The sixth, first, and second symbols (101) also came from three adjacent periods. As a result, the best surrogate for the period-6 orbit was constructed from two segments of the time series from widely separated parts of the time series. Each segment was three periods long. The reconstructed period-6 orbit is shown in Fig. 9.10. At the top is the symbolic name of the time series (below), which has purposely been broken in the "middle" of one of the continuous segments. This guarantees that there is no discontinuity at the endpoints of this time series. The two discontinuities occur at the first and fourth internal minima. The discontinuities are not visible at the resolution shown in the figure. Below the time series is an embedded version of this surrogate orbit. Again, neither discontinuity is visible at the resolution shown.

This search was repeated for the "missing" period-7 orbit $7_1 = 0111101$ and the period-8 orbit $8_2 = 01111101$, with similar results. In all three cases, the surrogate was constructed from just two continuous segments of lengths $3+3=6, 3+4=7$, and $3+5=8$ periods. The time series for these surrogates are all presented in Fig. 9.10. Below each time series is the embedded version of the surrogate orbit.

We conclude from these results that the three orbits are "missing" simply because the original time series was not sufficiently long.

9.5 ROUTES TO CHAOS

As control parameters for a dynamical system are varied, the strange attractors can change in two ways. Big changes involve the creation of additional branches through which the flow can travel or the annihilation of older branches through which the flow has traveled previously. Branched manifolds are *robust*, which means that most small control parameter changes do not involve change in the structure of the underlying branched manifold that describes the flow. However, when new branches are created or old branches are destroyed, these bifurcations occur in a systematic way. There are only a limited number of ways that new branches can be added, by continuity considerations (see Section 9.2).

One useful way to consider the bifurcations creating or annihilating branches is the following. These branches are always present. In fact, they are part of the unstable invariant manifold for the flow. However, the flow is guided onto a subset of these branches. As the control parameters change, the flow may be redirected onto "new" branches (i.e., branches not formerly visited) or redirected away from "old" branches. Or the flow may remain on branches already visited, but be redirected on these branches. We have observed these phenomena on scroll templates.

When the flow remains on a branched manifold, a change of control parameters changes the spectrum of unstable periodic orbits present in the strange attractor.

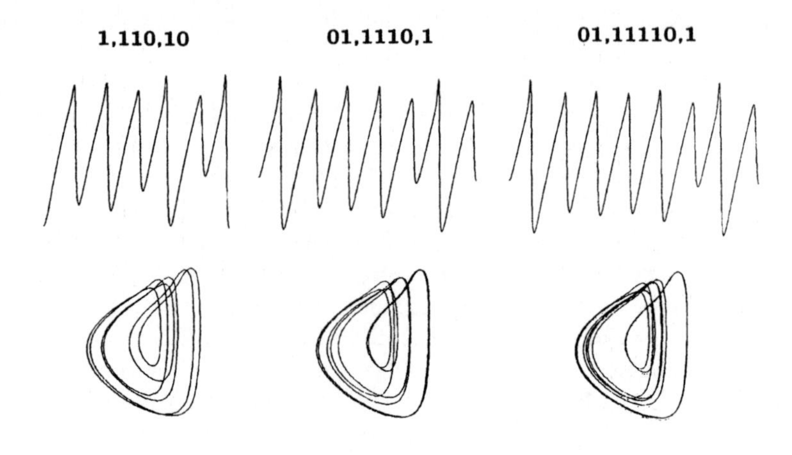

Fig. 9.10 Surrogates for the missing orbits $6_1, 7_1, 8_2$ were constructed by a metric search. Each surrogate was best represented by two segments of length at least three periods. To present the time series for each orbit, one of the segments was broken to avoid hiding discontinuities at the endpoints. Top, symbol sequence; middle, time series; bottom, embedded orbit (integral–differential embedding).

Therefore, the effect of the control parameters is represented by the change in the basis sets of orbits for the flow.

Different routes to chaos within a branched manifold are different unfoldings of the branched manifold. Although we cannot unfold a branched manifold analytically (as opposed to the simpler catastrophe theory case), we can describe the possible different routes to chaos by following different sequences of basis sets of orbits from a low-entropy to a high-entropy limit. At the present time, this can be done explicitly for the Smale horseshoe template. We present an example. The sequence

$$2_1 \to 4_1 \to 8_1 \to (8_1, 7_1) \to (8_1, 7_1, 8_3) \to (7_2, 8_3) \to 5_2 \to (5_2, 8_8)$$

$$\to 6_4 \to (6_4, 8_{12}) \to (6_4, 8_{12}, 5_3) \to (6_4, 8_{14}) \to 7_8 \to (7_8, 6_5)$$

$$\to 8_{15} \to (8_{15}, 7_9) \to (8_{15}, 7_9, 8_{16})$$

is one possible route from a laminar (low-entropy) flow to a chaotic flow on a horseshoe branched manifold. Other routes, some involving larger basis sets [cf. Eq. (9.7)], are also possible.

By this procedure we have a discrete classification (to finite period) of the routes to chaos in horseshoe dynamics. To any finite period the number of distinct routes to chaos is finite.

9.6 SUMMARY

The description of chaos, including classification of strange attractors, routes to chaos, and transitions from one chaotic state to another, involves two levels of structure. In other words, the unfoldings of dynamical systems and their strange attractors involve two levels of structure, both of which are discretely classified.

At the macroscopic level, strange attractors are described by their underlying branched manifold. This is essentially an unstable invariant manifold for the flow. Bifurcations can occur that add branches to, or remove branches from, the underlying branched manifold. Alternatively, these bifurcations simply involve redirecting the flow to different parts of this unstable invariant manifold.

At the microscopic level, strange attractors are classified in a more refined way by specifying the spectrum of unstable periodic orbits which they contain. The spectrum is conveniently summarized by a basis set of orbits. A simple algorithm exists for constructing this basis set. The microscopic unfolding of a strange attractor involves providing a discrete list of basis sets that describe the orbit spectrum as the control parameter values change. Each different sequence of basis sets leading from the laminar (zero entropy) to the chaotic ($h_T = \ln 2$) regime describes a different route to chaos in the Smale horseshoe.

10

Symmetry

Nature abounds with symmetry. These symmetries are often reflected in differential equations for dynamical systems. There is a systematic and very beautiful way to analyze dynamical systems with a discrete symmetry group. There are also very good physical reasons for carrying out such studies. We first describe the physical reasons. We then introduce the cover and image problem. The cover dynamical system is equivariant (unchanged) under a discrete symmetry group; its image retains no symmetry. The two are related by local diffeomorphisms. Were the two related by a global diffeomorphism, they would be globally equivalent. As it is, cover and image are *locally* but not globally equivalent. For the cases studied in this chapter, the symmetry group has two elements, so one dynamical system is a $2 \rightarrow 1$ image of the other. Locally, it is impossible to distinguish one from the other. It is only at a global level that the two can be distinguished. Three of the four standard dynamical systems introduced in Chapter 3 possess twofold symmetry. These are the Lorenz dynamical system and the two periodically driven nonlinear oscillators. The fourth dynamical system, the Rössler attractor, can be understood as an image of both the Lorenz attractor and the Duffing attractor.

10.1 INFORMATION LOSS AND GAIN

There are several important reasons for studying dynamical systems with symmetry and their counterparts without.

10.1.1 Information Loss

Many dynamical systems are properly described by amplitudes (in optics, classical electrodynamics, quantum mechanics, laser physics), but only intensities are measured. If the behavior is chaotic, the measured intensities are chaotic. So are the amplitudes, which have not been measured. Some information is lost in the transition from amplitudes to intensities: the information has been "squared away" (cf. Fig. 8.12). It is important to determine the spectrum of possible chaotic behavior of the unobserved amplitudes, which is compatible with the chaotic behavior of the observed intensities.

10.1.2 Exchange of Symmetry

It is possible that a physical system and the strange attractor that it generates possess one type of symmetry [e.g., the Lorenz system with $R_z(\pi)$ symmetry about the z axis], but that the strange attractor induced from a single scalar time series possesses a different symmetry [e.g., the Lorenz attractor induced from $x(t)$ data has inversion symmetry in the origin]. Proper dynamics must be obtained by group continuation: exchanging inversion for rotation symmetry. *Group continuation* is a natural extension of the idea of *analytic continuation* to dynamical systems with symmetry.

10.1.3 Information Gain

Constructing a branched manifold for a dynamical system with symmetry can be difficult. Systematically eliminating the symmetry greatly facilitates construction of an underlying branched manifold. Although this does not help too much for the standard Lorenz branched manifold, elimination of symmetry greatly facilitates the enumeration of the unfoldings of the Lorenz branched manifold under control parameter variation. Reduction of symmetry greatly simplifies construction of the branched manifolds for the Duffing and van der Pol oscillators.

10.1.4 Symmetries of the Standard Systems

Of the four canonical dynamical systems introduced in Chapter 3, three possess a twofold symmetry. The Lorenz equations exhibit an invariance ("equivariance") under rotations around the z axis: $(x, y, z) \to (-x, -y, +z)$. In fact, every Galerkin projection of the Navier–Stokes equations to a smaller set of ODEs exhibits at least a twofold symmetry when subject to appropriate boundary conditions. Both the Duffing and the van der Pol oscillators also obey a twofold symmetry under the group operation $(x, y, t) \to (-x, -y, t + \frac{1}{2}T)$. Here T is the period of the sinusoidal driving term. In fact, any nonlinear oscillator satisfying the undriven equations $\dot{x} = f(x, y)$, $\dot{y} = g(x, y)$ satisfies this symmetry when an additive driving term of the form $a \cos \omega t$ is added to either source term, provided that both source terms are odd: $f(-x, -y) = -f(x, y)$, $g(-x, -y) = -g(x, y)$.

Of the four standard dynamical systems introduced in Chapter 3, only the Rössler equations do not exhibit symmetry. We will see that when the twofold rotation symmetry of the Lorenz equations is eliminated, the resulting dynamical system is topologically equivalent to the Rössler dynamical system. We will also see a close relation between the reduced Duffing dynamical system and scroll dynamics, as exhibited by the Rössler equations under control parameter variation.

10.2 COVER AND IMAGE RELATIONS

The general problem that we introduce is the cover and image problem for dynamical systems. A covering dynamical system is a dynamical system with symmetry: The equations of motion are left unchanged by (equivariant under) a discrete symmetry group. It is mapped to a locally equivalent dynamical system without symmetry by a local diffeomorphism. The local diffeomorphism maps the cover to an image dynamical system. With suitable care, this also works in the reverse direction. An image dynamical system can be "lifted" to a covering dynamical system with some prespecified symmetry by a local diffeomorphism. Both dynamical systems are locally indistinguishable. However, they are globally distinct.

10.2.1 General Setup

We begin with a dynamical system $D = D(X)$ in R^N which is equivariant under a discrete symmetry group G. The equations of motion are $\dot{X}_i = F_i(X)$. Next, a local diffeomorphism $u = u(X)$ is introduced that removes the symmetry. In the new coordinate system, the reduced dynamical system is $D/G = \underline{D} = \underline{D}(u) \subset R^N$, with equations $\dot{u}_r = h_r(u)$.

The Jacobian of the local diffeomorphism provides a simple relation between the two dynamical systems:

$$\frac{du_r}{dt} = \frac{\partial u_r}{\partial X_i}\frac{dX_i}{dt} = \frac{\partial u_r}{\partial X_i}F_i(X) = h_r(u) \tag{10.1}$$

$$\frac{dX_i}{dt} = \frac{\partial X_i}{\partial u_r}\frac{du_r}{dt} = \left(\frac{\partial u_r}{\partial X_i}\right)^{-1} h_r(u) = F_i(X) \tag{10.2}$$

In general, computing the matrix elements of the forward transformation $\partial u_r/\partial X_i$ is simple, whereas computing the matrix elements $\partial X_i/\partial u_r$ of the inverse transformation is difficult. However, inverting the forward transformation is usually not difficult.

The strategy we follow is straightforward but not familiar. First, we use the symmetry group to construct an *integrity basis* [69,189]. This is a set of $K \geq N$ polynomials that are invariant under the group action. Then we use the relations among these polynomials (*syzygies*) to construct an N-dimensional subspace $M^N \subset R^K$ on which to construct the image dynamical system. The final step is to map the equivariant dynamical system down to this invariant subspace. The resulting equations are locally identical to the original set of equations, but the symmetry group has been "modded out."

We illustrate these ideas by carrying out the reduction of symmetry for the Lorenz equations in the following section and for the Duffing and van der Pol equations in Sections 10.7 and 10.8.

10.3 ROTATION SYMMETRY 1: IMAGES

We consider the problem of mapping an equivariant dynamical system to its image at three levels: mapping of the equations and the flows they generate, mappings of the branched manifolds that characterize the strange attractors generated by the equations, and relations between the unstable periodic orbits in the strange attractors and their branched manifolds.

10.3.1 Image Equations and Flows

Three-dimensional dynamical systems

$$\begin{aligned} \dot{X} &= F_1(X,Y,Z) \\ \dot{Y} &= F_2(X,Y,Z) \\ \dot{Z} &= F_3(X,Y,Z) \end{aligned} \tag{10.3}$$

that are equivariant under rotations about the Z axis $R_Z(\pi)$ satisfy the condition

$$\begin{aligned} F_1(-X,-Y,+Z) &= -F_1(X,Y,Z) \\ F_2(-X,-Y,+Z) &= -F_2(X,Y,Z) \\ F_3(-X,-Y,+Z) &= +F_3(X,Y,Z) \end{aligned} \tag{10.4}$$

An integrity basis for this two-element symmetry group consists of monomials up to degree 2 (there is a theorem by Noether [69]) which are left invariant under the actions of this group. Each function that is invariant under the group is a function of the K integrity basis functions. The four monomials in the integrity basis are

$$u_1 = X^2 \qquad u_2 = XY \qquad u_3 = Y^2 \qquad u_4 = Z$$

Since $K = 4$ and $N = 3$, there is only one syzygy. It is

$$u_1 u_3 - u_2^2 = 0$$

It is useful to choose the following three linear combinations of the four integrity basis functions as coordinates in the reduced symmetry space $R^3(u,v,w) \subset R^4(u_i)$:

$$u = u_1 - u_3 = X^2 - Y^2 \qquad v = 2u_2 = 2XY \qquad w = u_4 = Z \tag{10.5}$$

The Jacobian of this transformation is

$$\frac{\partial(u,v,w)}{\partial(X,Y,Z)} = \begin{bmatrix} 2X & -2Y & 0 \\ 2Y & 2X & 0 \\ 0 & 0 & 1 \end{bmatrix} \tag{10.6}$$

The transformation is singular on the Z axis $(X,Y) = (0,0)$. Off the Z axis the transformation is everywhere 2:1 and locally 1:1. The Jacobian (10.6) will play an important role in projecting an equivariant set of equations down to the image equations without symmetry. Its inverse, as well as the singular axis, will play an important role in lifting an invariant set of equations to a twofold cover equivariant under the rotation group $R_Z(\pi)$.

Equivariant equations are projected to invariant equations according to (10.1). For the Lorenz equations, the invariant image equations are [189, 190]

$$\frac{d}{dt}\begin{bmatrix} u \\ v \\ w \end{bmatrix} = \begin{bmatrix} 2X & -2Y & 0 \\ 2Y & 2X & 0 \\ 0 & 0 & 1 \end{bmatrix}\begin{bmatrix} -\sigma X + \sigma Y \\ rX - Y - XZ \\ -bZ + XY \end{bmatrix} \tag{10.7}$$

These equations should be expressed entirely in terms of the integrity basis functions. The result is

$$\begin{aligned} \dot{u} &= (-\sigma - 1)u + (\sigma - r)v + vw + (1 - \sigma)\rho \\ \dot{v} &= (r - \sigma)u - (\sigma + 1)v - uw + (r + \sigma)\rho - w\rho \\ \dot{w} &= -bw + \tfrac{1}{2}v \end{aligned} \tag{10.8}$$

The additional function ρ is needed because the original coordinates X, Y, Z span R^3, while the integrity basis spans R^4. This function is $\rho = X^2 + Y^2 = \sqrt{u^2 + v^2}$.

The flows generated by the Lorenz equations and the image equations are shown in Fig. 10.1. This figure makes it clear that modding out ("mod out" means "divide by") the twofold rotation symmetry of the Lorenz equations results in a dynamical system that is topologically equivalent to the Rössler system [191]. Eliminating the symmetry involves identifying the two unstable foci. It also replaces the tearing mechanism, operative in Lorenz dynamics, with a folding mechanism, operative in horseshoe dynamics ($A_3 \to A_2$).

Another set of equations that exhibits twofold rotation symmetry under $R_Z(\pi)$ was studied by Burke and Shaw. The equivariant equations are

$$\begin{aligned} \dot{X} &= -S(X + Y) \\ \dot{Y} &= -Y - SXZ \\ \dot{Z} &= SXY + V \end{aligned} \tag{10.9}$$

The invariant equations obtained from this equivariant system are

$$\begin{aligned} \dot{u} &= -(S+1)u - S(1-w)v + (1-S)\rho \\ \dot{v} &= S(1-w)u - (S+1)v - S(1+w)\rho \\ \dot{w} &= \tfrac{1}{2}Sv + V \end{aligned}$$

The flows generated by the equivariant Burke and Shaw equations and the invariant image equations are shown in Fig. 10.2. In this case, eliminating the symmetry identifies two unstable foci. However, in this case an iterated double fold is replaced by the single (reverse) fold characteristic of a Smale horseshoe mechanism.

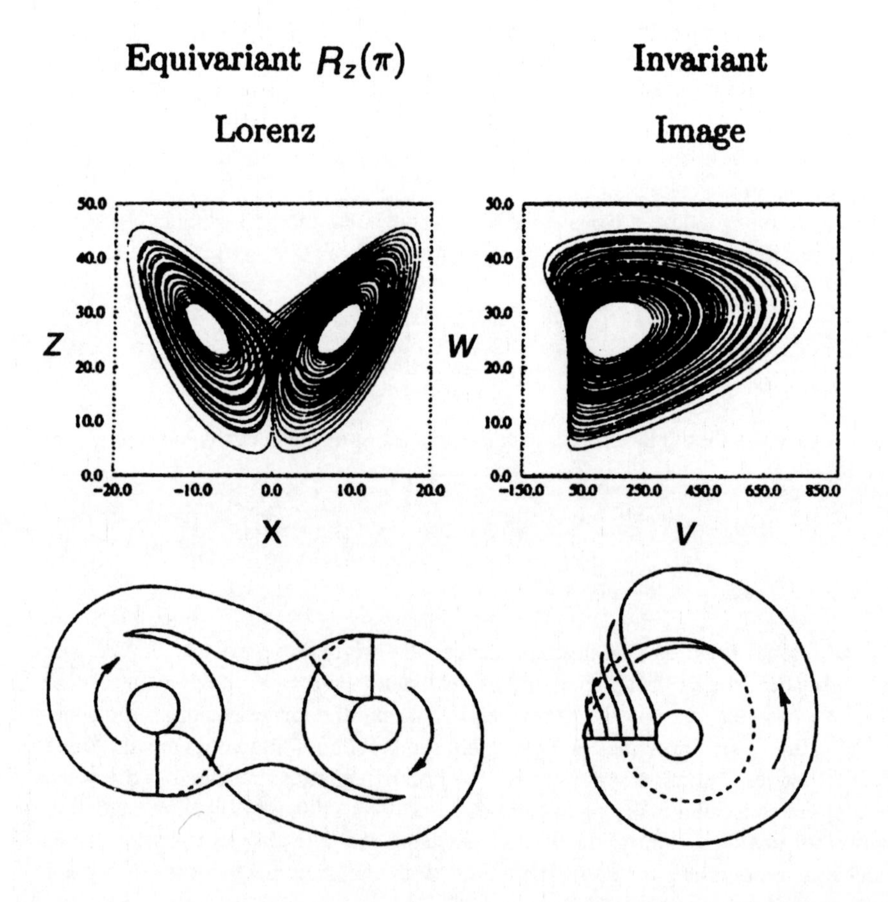

Fig. 10.1 Top: The strange attractors generated by the equivariant Lorenz equations (left) and the invariant image equations (right) are shown projected onto the X–Z plane and the v–w plane, respectively. Parameter values: $(\sigma, b, r) = (10.0, \frac{8}{3}, 28.0)$. Bottom: Branched manifolds with four and two branches, respectively, characterize the cover and image strange attractors.

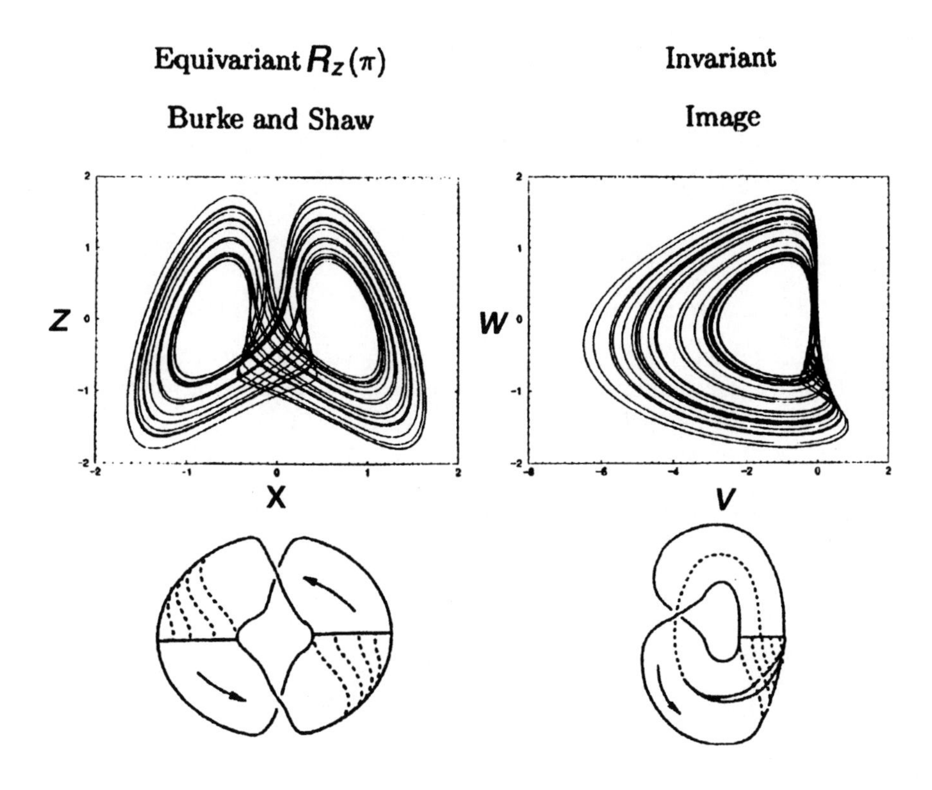

Fig. 10.2 Top: The strange attractors generated by the equivariant Burke and Shaw equations (left) and the invariant image equations (right) are shown projected onto the X–Z plane and the v–w plane, respectively. Parameter values: $(S, V) = (10.0, 4.271)$. Bottom: Branched manifolds with four and two branches, respectively, characterize the cover and image strange attractors. The Smale horseshoe template has a half twist: It describes a reverse horseshoe.

10.3.2 Image of Branched Manifolds

The effect of modding out the rotational symmetry can be visualized in terms of the branched manifolds that characterize the chaotic dynamics. In Fig. 10.1 we show a branched manifold with $R_Z(\pi)$ symmetry that characterizes Lorenz dynamics, and the image of that branched manifold under the $2 \rightarrow 1$ local diffeomorphism. We label the zero-torsion branches on the left and right as $0_l, 0_r$. We label the translobe branch from the left to the right lobe as 1_l and its image under $R_Z(\pi)$ as 1_r. The local $2 \rightarrow 1$ diffeomorphism produces the identification

$$\begin{matrix} 0_l \\ 0_r \end{matrix} \longrightarrow 0 \qquad\qquad \begin{matrix} 1_l \\ 1_r \end{matrix} \longrightarrow 1$$

The transition matrices for the four-branch manifold of the Lorenz system and its two branch image under the $2 \rightarrow 1$ local diffeomorphism are

	0_l	1_l	0_r	1_r
0_l	1	1	0	0
1_l	0	0	1	1
0_r	0	0	1	1
1_r	1	1	0	0

$\longrightarrow$

	0	1
0	1	1
1	1	1

The four-branch attractor for the equivariant Burke and Shaw equations and its two-branch image after modding out the rotation symmetry are shown at the bottom of Fig. 10.2. The four branches are conveniently labeled $1_l, 1_r, 2_l, 2_r$ and the image branches are $1, 2$. The local diffeomorphism produces the identification

$$\begin{matrix} 1_l \\ 1_r \end{matrix} \longrightarrow 1 \qquad \begin{matrix} 2_l \\ 2_r \end{matrix} \longrightarrow 2$$

The transition matrices are

	1_l	2_l	1_r	2_r
1_l	0	0	1	1
2_l	0	0	1	1
1_r	1	1	0	0
2_r	1	1	0	0

$\longrightarrow$

	1	2
1	1	1
2	1	1

The Lorenz system and the Burke and Shaw system are clearly topologically inequivalent. This is clear by comparing the flows shown in Figs. 10.1 and 10.2. It is also clear by comparing their branched manifolds and their transition matrices. However, their images are very similar: equivalent if one disregards the extra half twist that converts the horseshoe to a reverse horseshoe for the image of the Burke and Shaw dynamical system. This raises a question to which we return in Section 10.4: How many inequivalent dynamical systems with a specific symmetry possess equivalent images? Or, to put this in reverse perspective: How many topologically distinct covers with prespecified symmetry can an image dynamical system have?

10.3.3 Image of Periodic Orbits

The local $2 \rightarrow 1$ diffeomorphism maps periodic orbits in the covering system down to periodic orbits in the image system. However, the mapping is different for different covering systems. We illustrate the differences by discussing the images of periodic orbits for the Lorenz equations and the Burke and Shaw equations.

In the Lorenz system an orbit is either invariant under twofold symmetry, or else is mapped into a symmetry-related orbit under the symmetry. As an example, the period-4 orbit $0_l0_l1_l1_r$ is mapped into its symmetric partner $0_r0_r1_r1_l$ by the twofold symmetry, while the period-4 orbit $0_l1_l0_r1_r$ is mapped to itself under this symmetry. Under the $2 \longrightarrow 1$ local diffeomorphism, these three orbits are mapped onto periodic orbits in the image system as follows:

$$\begin{matrix} 0_l0_l1_l1_r \\ 0_r0_r1_r1_l \end{matrix} \longrightarrow 0011 \qquad\qquad 0_l1_l0_r1_r \longrightarrow 0101 = (01)^2$$

Two symmetry-related orbits of period-4 are mapped to a period-4 saddle in the image manifold, and one symmetric period-4 orbit is mapped to a period-2 orbit which goes around twice in the image manifold. For the Lorenz system these results are general. Odd-period orbits occur in symmetric pairs. Even-period orbits are either self-symmetric or occur in symmetric pairs. A symmetric pair of orbits of period p (even or odd) in the cover map to a single orbit of period p in the image. A self-symmetric orbit of period p in the cover (p must be even) maps to an orbit of period $p/2$ in the image.

The relation between periodic orbits in the symmetric Burke and Shaw dynamical system and its image is different. In this case all orbits in the covering sytem have even period. The two self-symmetric period two orbits 1_l1_r and 2_l2_r are mapped to the two period-1 orbits 1 and 2 in the image. The period-2 orbit 1_l2_r has a symmetric image 1_r2_l. The image of these two orbits is the single period-2 orbit 12.

The difference between the two dynamical systems with $R_Z(\pi)$ symmetry can be exhibited in yet another way. The four-branch manifold for the Lorenz system has topological entropy $\log(2)$. This is also true for the four-branch manifold describing the Burke and Shaw dynamical system. However, in the latter case it is possible to slide one of the two branch lines around the branched manifold until it coincides with the other branch line. The result is a branched manifold with four branches and one branch line (cf. Fig. 10.3). This branched manifold describes a dynamical system with closed orbits of arbitrary period $(1, 2, 3, \dots)$ rather than just even period. Since there are four branches, the topological entropy is $\log(4)$. There is no contradiction. The number of orbits of period p is $N(p) \sim e^{p\log(4)} \sim 4^p$. The number of orbits of period $2p$ in the original branched manifold is $N(2p) \sim e^{2p\log(2)} \sim 2^{2p} = 4^p$. There is a 1:1 correspondence between orbits of period p in the branched manifold shown in the center of Fig. 10.3 and orbits of period $2p$ in the original branched manifold shown in Fig. 10.2 and on the left in Fig. 10.3. In each case we count a period each time an orbit trajectory crosses a branch line.

The principal difference between the two dynamical systems equivariant under the same symmetry group $R_Z(\pi)$ is the way the flows link the symmetry axis. In Lorenz dynamics, the symmetry axis passes through the saddle at the origin. The flows in neither lobe link this axis—it is only the flows that transit from one lobe to the other, and then back again, which link the axis, once per pair of transitions. On the other hand, all orbits of period $2p$ in Burke and Shaw dynamics link the symmetry axis p times. These topological properties are preserved in the mapping from the cover to the image dynamical system. They are explored in more detail in the following section.

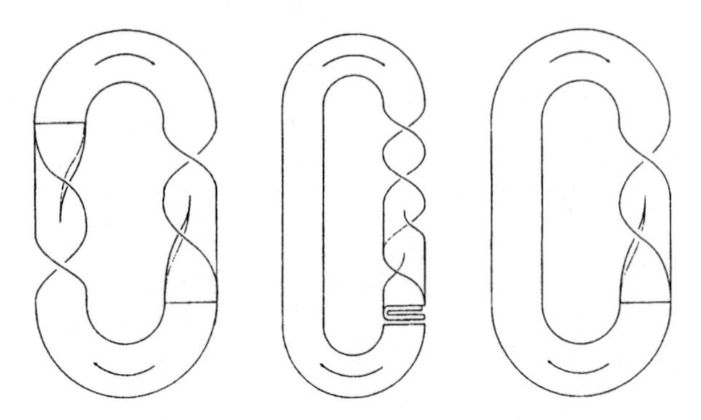

Fig. 10.3 The branched manifold for the Burke and Shaw system (left) can be deformed by sliding one of the two branch lines around until it coincides with the other (center). The resulting branched manifold describes the second iterate of a reverse horseshoe. The $2 \rightarrow 1$ image is shown at the right.

10.4 ROTATION SYMMETRY 2: COVERS

A double cover with symmetry $R_Z(\pi)$ for the Rössler system

$$\begin{aligned} \dot{u} &= -v - w \\ \dot{v} &= u + av \\ \dot{w} &= b + w(u - c) \end{aligned} \qquad (10.10)$$

can be constructed using the equations (10.2) for lifting an image flow to a covering flow. The resulting dynamical system equations are

$$\begin{aligned} \dot{X} &= -\tfrac{1}{2}Y + \tfrac{1}{2\rho}X(2aY^2 - Z) \\ \dot{Y} &= +\tfrac{1}{2}X + \tfrac{1}{2\rho}Y(2aX^2 + Z) \\ \dot{Z} &= b + Z(X^2 - Y^2 - c) \end{aligned} \qquad (10.11)$$

The strange attractors generated by the Rössler equations (10.10) and the twofold covering equations (10.11) are shown in Fig. 10.4. They are projected onto the u–v and the X–Y planes, respectively. We also show the branched manifolds that characterize the two attractors. These branched manifolds have two and four branches.

10.4.1 Topological Index

This is not the only possible double cover of the Rössler dynamical system with $R_Z(\pi)$ rotation symmetry. Many other double covers can be constructed, all invariant under the same symmetry group $R_Z(\pi)$, all topologically inequivalent to the one shown in Fig. 10.4 and with each other. We now explain how this can occur.

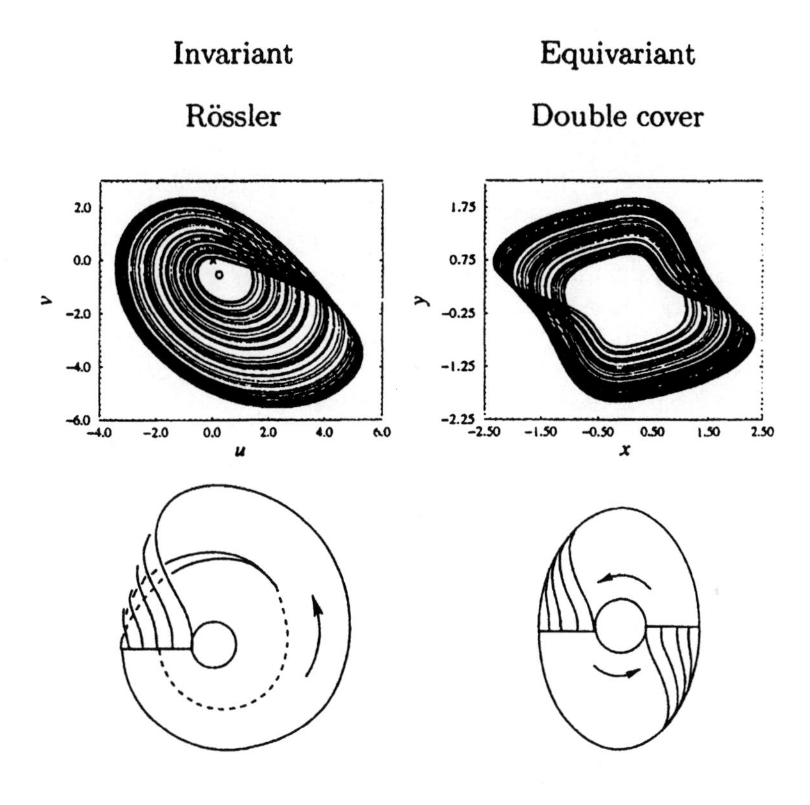

Fig. 10.4 Top: The strange attractor for the Rössler equations is projected onto the u–v plane and that for its double cover is projected onto the X–Y plane. Rotation is counterclockwise in both cases. Parameter values: $(a, b, c) = (0.415, 2.0, 4.0)$. Bottom: Branched manifolds with two and four branches characterize the two strange attractors.

The local diffeomorphism (10.5) mapping $(X, Y, Z) \rightarrow (u, v, w)$ is $2 \rightarrow 1$ everywhere except on the rotation axis $(0, 0, Z) \leftrightarrow (0, 0, w)$, where it is 1:1 and singular. Within the image space there is no singularity, but in the covering space there is. This singularity allows the possibility of constructing several topologically inequivalent covers with the same symmetry group for a single invariant dynamical system.

We illustrate these constructions for an image dynamical system represented by a Smale horseshoe template. The branched manifold is shown in Fig. 10.5. Each branch contains one unstable period-1 orbit. These two orbits are drawn with heavy lines. The period-1 orbits, and the branches that contain them, are indexed by the same labels. In the case shown, this is the global torsion of the orbit and the branch in which it resides: 0 or 1.

Also shown in Fig. 10.5 are four inequivalent axes about which rotational symmetry can be imposed. Each of these axis links the two period-1 orbits in different ways. If we call n_0 and n_1 the linking numbers of the two period-1 orbits 0 and 1 with the symmetry axis (closed by a return at ∞), the four cases shown are (0,0), (0,1), (1,1), and (1,0).

The branched manifolds with rotation symmetry that cover the horseshoe template all have four branches. Now consider a closed orbit in one of the covering branched manifolds. If its linking number with the Z axis is n, the linking number of the image orbit with the w axis is $2n$. If $n = 1$, the orbit in the cover goes around the Z axis once while its image goes around the w axis twice. In particular, if $n_1 = 1$, the period-1 orbit 1 in the Smale horseshoe template cannot be the image of a closed orbit in the covering branched manifold. If the orbit in the covering space does not link the Z axis, its image does not link the w axis, and vice versa.

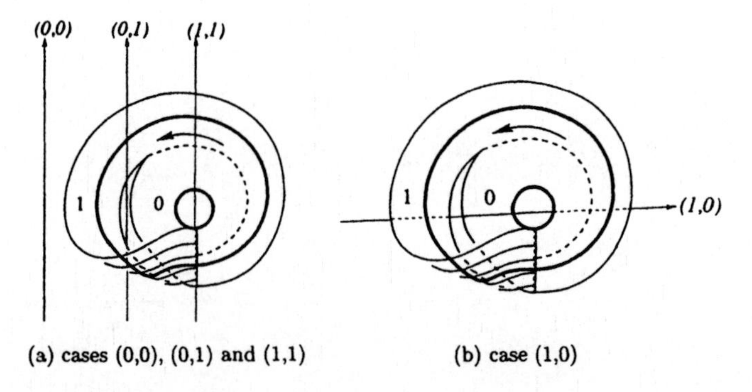

Fig. 10.5 In the image space, the Smale horseshoe template has two branches labeled 0 and 1. Each contains a period one orbit. The symmetry axis $(0, 0, w)$ can be chosen to link the two period one orbits in different ways. (a) Three rotation axes which link the two period one orbits with linking numbers (n_0, n_1) = (0,0), (0,1), and (1,1); (b) rotation axis with linking numbers (1,0).

10.4.2 Covers of Branched Manifolds

In Fig. 10.6 we show four double covers of the Smale horseshoe template $\mathcal{SH}$. Each corresponds to one of the choices of the symmetry axis shown in Fig. 10.5. Each double cover is equivariant under $R_Z(\pi)$. All four double covers are topologically inequivalent. Each of the double covers in this figure has four branches. The branches are labeled by three symbols, each of which takes two values. These are an integer (0 or 1), a letter subscript (l or r), and another symbol ($\hat{}$ or $\bar{}$). The integer identifies whether the branch in the cover is mapped onto the branch 0 or 1 in $\mathcal{SH}$. The letter subscript identifies whether the branch in the cover is on the left- or the right-hand side of the cover. Under $R_Z(\pi)$, $l \to r$ and $r \to l$. The extra symbol has a topological significance. It indicates whether the image links the rotation axis once ($\hat{\ }$) or not at all ($\bar{\ }$).

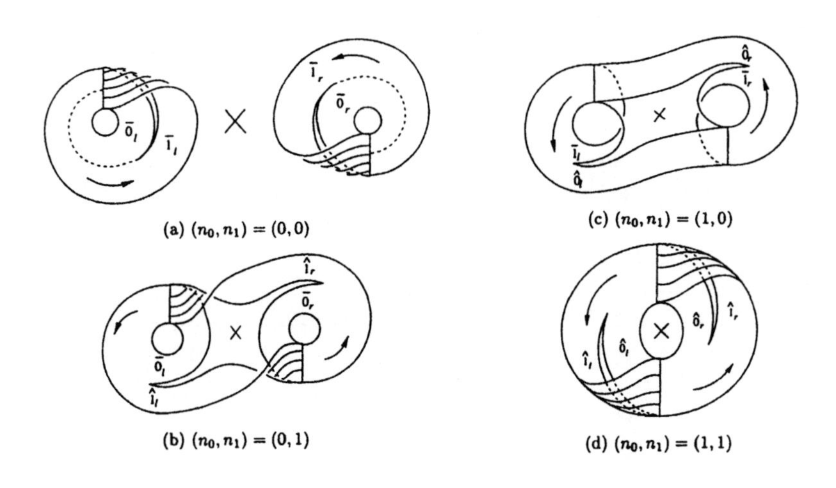

(a) $(n_0, n_1) = (0, 0)$

(b) $(n_0, n_1) = (0, 1)$

(c) $(n_0, n_1) = (1, 0)$

(d) $(n_0, n_1) = (1, 1)$

Fig. 10.6 Four branched manifolds that are double covers of the horseshoe template. All possess rotation symmetry under $R_Z(\pi)$. They differ by the values of the topological index (n_0, n_1). The symmetry axis is perpendicular to the plane of the projection and passes through the x.

We describe in some detail the relation between the covering branched manifold in Fig. 10.6(b) and the image Smale horseshoe template of Fig. 10.5. The covering branched manifold has four branches: one ($\bar{0}_l$) entirely in the left lobe, its image under $R_Z(\pi)$ ($\bar{0}_r$) entirely in the right-hand lobe, one branch ($\hat{1}_l$) extending from the lobe on the left to the one on the right, and its image ($\hat{1}_r$) under $R_Z(\pi)$. Under projection, the image of the two branches $\bar{0}_l$ and $\bar{0}_r$ in the cover template is the branch 0 in the Smale horseshoe template. The branches $\hat{1}_l$ and $\hat{1}_r$ both map to branch 1 in the Smale horseshoe template. The period-1 orbit $\bar{0}_l$ and its symmetry-related counterpart $\bar{0}_r$ do not link the Z axis ($n = 0$). Both map to the period-1 orbit 0 in the horseshoe template, which does not link the w axis ($n_0 = 0$). There is one period-2 orbit $\hat{1}_l\hat{1}_r$ in the covering template. This links the Z axis once. This period-2 orbit maps twice into the period-1 orbit 1 in the Smale horseshoe template ($\hat{1}_l\hat{1}_r \to 11$), which links the w axis twice ($n = 2$). The period-1 orbit 1 in the horseshoe links the w axis once ($n_1 = 1$).

These arguments can be run backwards. Since $n_0 = 0$, the period-1 orbit 0 in the horseshoe lifts to two symmetry-related orbits $\bar{0}_l$ and $\bar{0}_r$, which do not link the Z axis.

Table 10.1 Two period-3 orbits 001 and 011 lifted to different covering systems

Cover	(n_0, n_1)	001	011
$2\mathcal{SH}$	$(0,0)$	$\bar{0}_l\bar{0}_l\bar{1}_l + \bar{0}_r\bar{0}_r\bar{1}_r$	$\bar{0}_l\bar{1}_l\bar{1}_l + \bar{0}_r\bar{1}_r\bar{1}_r$
$\mathcal{L}$	$(0,1)$	$\bar{0}_l\bar{0}_l\hat{1}_l\bar{0}_r\bar{0}_r\hat{1}_r$	$\bar{0}_l\hat{1}_l\hat{1}_r + \bar{0}_r\hat{1}_r\hat{1}_l$
	$(1,0)$	$\hat{0}_l\hat{0}_r\bar{1}_l + \hat{0}_r\hat{0}_l\bar{1}_r$	$\hat{0}_l\bar{1}_r\bar{1}_r\hat{0}_r\bar{1}_l\bar{1}_l$
$\mathcal{D}$	$(1,1)$	$\hat{0}_l\hat{0}_r\hat{1}_l\hat{0}_r\hat{0}_l\hat{1}_r$	$\hat{0}_l\hat{1}_r\hat{1}_l\hat{0}_r\hat{1}_l\hat{1}_r$

Since $n_1 = 1$, the period-1 orbit 1 in the horseshoe lifts to "half a closed orbit" $\hat{1}_l$ or $\hat{1}_r$, which is then closed by adding the complementary symmetry -related segment ($\hat{1}_r$ or $\hat{1}_l$). This closed orbit $\hat{1}_r\hat{1}_l$ in the cover links the Z axis with $n = 1$. That is, the period="2" orbit 11 in the horseshoe, with $n = n_1 + n_1 = 2$, lifts to the period-2 orbit $\hat{1}_r\hat{1}_l$ with $n = 1$.

The relation between cover and image branched manifolds for other choices of the topological index (n_0, n_1) are described similarly. Since the four branched manifolds shown in Fig. 10.6 are not topologically equivalent, the dynamical systems that they characterize are also inequivalent. In particular, the lift with $(n_0, n_1) = (0, 0)$ is disjoint, so of only limited interest.

In Fig. 10.6(d), the double cover with topological index $(n_0, n_1) = (1, 1)$ exhibits dynamics similar to the strange attractor observed for the Duffing and the Burke and Shaw dynamical systems. The four branches in this cover, which we call $\mathcal{D}$, are $(\hat{0}_l, \hat{1}_l, \hat{1}_r, \hat{0}_l)$. The double cover in Fig. 10.6(b) with $(n_0, n_1) = (0, 1)$ is topologically equivalent to the Lorenz dynamical system $\mathcal{L}$. The four branches in $\mathcal{L}$ are $(\bar{0}_l, \hat{1}_l, \hat{1}_r, \bar{0}_l)$. Finally, the double cover in Fig. 10.6(a) with $(n_0, n_1) = (0, 0)$ is topologically equivalent to two disjoint but symmetry-related Smale horseshoes $2\mathcal{SH}$. The four branches in this disconnected double cover are $(\bar{0}_l, \bar{1}_l)$ for the horseshoe on the left and $(\bar{0}_r, \bar{1}_r)$ for the symmetry-related copy on the right.

10.4.3 Covers of Periodic Orbits

Periodic orbits in the image dynamical system can systematically be lifted to periodic orbits in the covering systems. Without stating the rules in general, we provide two simple examples from which the rules can easily be extracted. The lifts of the two period-3 orbits in $\mathcal{SH}$ for the four covering dynamical systems of Fig. 10.6 are given in Table 10.1.

10.5 PEELING: A NEW GLOBAL BIFURCATION

We now describe a new global bifurcation [189]. Choose the rotation axis w so that $(n_0, n_1) = (1, 1)$. Then displace the w axis outward, so that it first intersects the branch 0 of the image horseshoe template $\mathcal{SH}$, then branch 1. The covers are shown in Fig. 10.7. As the rotation axis crosses branch 0 in $\mathcal{SH}$, the outer orbit segments of this branch encircle the w axis once, the inner segments do not. In the cover, the branches $\hat{0}_l, \hat{0}_r$ split into two parts, $\hat{0}_{l,r}$ and $\bar{0}_{l,r}$, depending on where the image occurs under the diffeomorphism (i.e., outside or inside w). The cover branched manifold remains invariant under $R_Z(\pi)$ but now has six branches: $(\bar{0}_l, \hat{0}_l, \hat{1}_l, \hat{1}_r, \hat{0}_r, \bar{0}_r)$ [Fig. 10.7(a)]. This six-branched manifold interpolates between the template $\mathcal{D}$ with four branches and the template $\mathcal{L}$, also with four branches.

A similar perestroika occurs as the rotation axis w crosses branch 1 of $\mathcal{SH}$. The cover branches $\hat{1}_{l,r}$ split into two parts, $\hat{1}_{l,r}$ and $\bar{1}_{l,r}$, depending on where the image occurs under the diffeomorphism. The covering branched manifold once again has six branches [Fig. 10.7(b)]. These are now $(\bar{0}_l, \bar{1}_l, \hat{1}_l, \hat{1}_r, \bar{1}_r, \bar{0}_r)$. This six-branch template interpolates between the Lorenz template $\mathcal{L}$ with four branches and the two disjoint Smale horseshoe templates $2\mathcal{SH}$ with two pairs of branches.

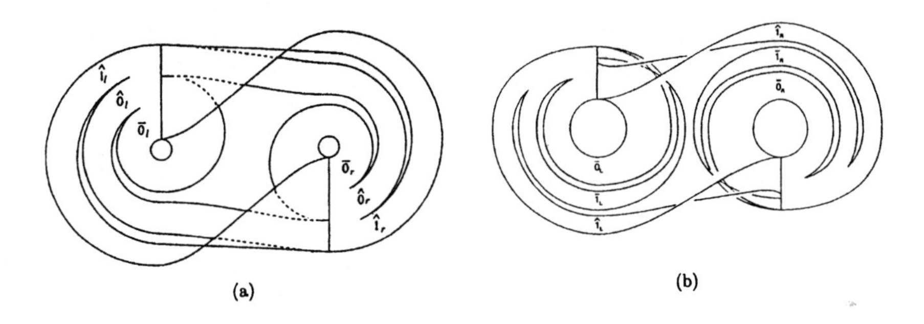

Fig. 10.7 The perestroika of the covers of $\mathcal{SH}$ from the cover with topological index $(n_0, n_1) = (1,1)$ to the cover with $(n_0, n_1) = (0,0)$ produces two six-branch covers. (a) This branched manifold interpolates $\mathcal{D} \to \mathcal{L}$ as the topological index changes from (1,1) to (0,1). (b) This branched manifold interpolates $\mathcal{L} \to 2\mathcal{SH}$ as the topological index changes from (0,1) to (0,0).

10.5.1 Orbit Perestroika

As the w axis is moved across the branches of the image Smale horseshoe template $\mathcal{SH}$, a systematic reorganization takes place in the lifts of the unstable periodic orbits in $\mathcal{SH}$. When the w axis passes through the center of rotation of $\mathcal{SH}$ [structurally stable case, $(n_0, n_1) = (1,1)$], the linking number of this axis with an unstable periodic orbit of period p is its period, p. As w moves outward, it makes a series of p intersections with this unstable periodic orbit. The order in which these intersections occur is determined by kneading theory. At each intersection:

1. The linking number of the orbit with w decreases by one.

2. Two symbols $\hat{}$ in the cover change to $\bar{}$.

We illustrate these ideas with a simple example. This is the perestroika of the lift of the period-4 orbit 0111 in the period-doubling cascade in $\mathcal{SH}$. Its lifts undergo the following perestroika:

Index	Template	Orbit
$(1,1)$	$\mathcal{D}$	$\hat{0}_l\hat{1}_r\hat{1}_l\hat{1}_r + \hat{0}_r\hat{1}_l\hat{1}_r\hat{1}_l$
$(0,1)$	$\mathcal{L}$	$\bar{0}_l\hat{1}_l\hat{1}_r\hat{1}_l\bar{0}_r\hat{1}_r\hat{1}_l\hat{1}_r$
	$\mathcal{L}+2\mathcal{SH}$	$\bar{0}_l\hat{1}_l\bar{1}_r\hat{1}_r + \bar{0}_r\hat{1}_r\bar{1}_l\hat{1}_l$
	$\mathcal{L}+2\mathcal{SH}$	$\bar{0}_l\bar{1}_l\bar{1}_l\hat{1}_l\bar{0}_r\bar{1}_r\bar{1}_r\hat{1}_r$
$(0,0)$	$2\mathcal{SH}$	$\bar{0}_l\bar{1}_l\bar{1}_l\bar{1}_l + \bar{0}_r\bar{1}_r\bar{1}_r\bar{1}_r$

We point out the strong coupling between the left–right symbols l and r and the topological symbols $\hat{}$ and $\bar{}$. That is, $\hat{}$ forces the change $l \rightarrow r$ and $r \rightarrow l$, while $\bar{}$ forces $l \rightarrow l$ and $r \rightarrow r$. We also point out the alternation between lifts of the period-p orbit to two period-p orbits or one period-$2p$ orbit as the axis sweeps through the branches.

10.5.2 Covering Equations

We can exhibit this new type of bifurcation explicitly for covers of the Rössler dynamical system. Instead of displacing the symmetry axis, we equivalently modify the equations by displacing the origin of coordinates $(u,v,w) \rightarrow (u+u_0, v+v_0, w+w_0)$. We first move the origin to the inner fixed point. Next, we displace the origin of coordinates along the u axis by a distance μ. The equations of the image system in the translated coordinates are

$$\begin{aligned}
\dot{u} &= -v-w \\
\dot{v} &= u+av+\mu \\
\dot{w} &= \tilde{b}(u+\mu)+w(u-\tilde{c}+\mu)
\end{aligned} \qquad (10.12)$$

Here $\tilde{b} = w_0$, $\tilde{c} = c - u_0$, and (u_0, v_0, w_0) are determined by $u_0 = -v_0 = aw_0 = (c - \sqrt{c^2 - 4ab})/2$. The covering equations for this dynamical system are

$$\begin{aligned}
\dot{X} &= -\frac{1}{2}Y + \frac{1}{2\rho}\left[X(2aY^2 - Z) + \mu Y\right] \\
\dot{Y} &= +\frac{1}{2}X + \frac{1}{2\rho}\left[Y(2aX^2 + Z) + \mu X\right] \\
\dot{Z} &= \tilde{b}(X^2 - Y^2 + \mu) + Z(X^2 - Y^2 - \tilde{c} + \mu)
\end{aligned} \qquad (10.13)$$

As usual, $\rho = X^2 + Y^2$.

The covering equations (10.13) were integrated for standard values of the control parameters $(a,b,c) = (0.432, 2.0, 4.0)$ and five values of the offset parameter μ. The results are shown in Fig. 10.8. The strange attractors shown in Fig. 10.8(a), (c), and (e) all have four branches and are double covers of the Rössler attractor with topological indices $(n_0, n_1) = (1,1)$, $(0,1)$, and $(0,0)$. The strange attractor in Fig. 10.8(b) has six branches and interpolates between the attractors $\mathcal{D}$ with $(n_0, n_1) = (1,1)$ and $\mathcal{L}$ with $(n_0, n_1) = (0,1)$. The strange attractor in Fig. 10.8(d) also has six branches and interpolates between the attractors $\mathcal{L}$ with $(n_0, n_1) = (0,1)$ and $2\mathcal{SH}$ with $(n_0, n_1) = (0,0)$.

We emphasize again that the placement of the w axis has no effect on the nature of the flow in the image system. The bifurcation takes place only in the covering dynamical system; the image remains undisturbed during displacement of the symmetry axis. The bifurcation is caused by sweeping a splitting axis (the symmetry axis) through the covering flow from inside to outside the flow.

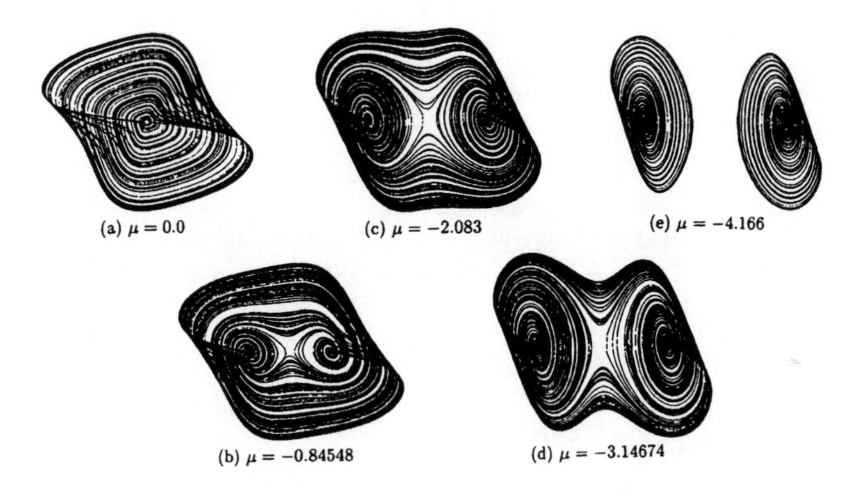

Fig. 10.8 The covering equations of the Rössler equations with $R_Z(\pi)$ symmetry are integrated for control parameters $(a, b, c) = (0.432, 2.0, 4.0)$. The parameter μ which determines the position of the symmetry axis is varied so that the symmetry axis scans through the cover from inside to outside from the cover in (a) to the cover in (e). The covers in (a), (c), and (e) are of types $\mathcal{D}$, $\mathcal{L}$, and $2\mathcal{SH}$. These have good topological indices. The covers in (b) and (d) interpolate between their neighbors.

10.6 INVERSION SYMMETRY: DRIVEN OSCILLATORS

The periodically driven Duffing and van der Pol oscillators are members of a large class of periodically driven nonlinear oscillators with symmetry. We discuss this symmetry now.

10.6.1 Periodically Driven Nonlinear Oscillator

To begin, we consider a two-dimensional nonlinear oscillator that obeys the pair of equations

$$\begin{aligned} \dot{X} &= F(X,Y) \qquad & F(-X,-Y) &= -F(X,Y) \\ \dot{Y} &= G(X,Y) \qquad & G(-X,-Y) &= -G(X,Y) \end{aligned}$$

We specifically assume that both driving functions have odd parity, as indicated.

Next, we assume that the oscillator is driven by an external, periodic source. The source can be added to either equation and has the form $A\cos(\omega t)$. To be specific, we add this source term to the second equation to obtain the equation of a periodically driven nonlinear oscillator:

$$\begin{aligned} \dot{X} &= F(X,Y) \\ \dot{Y} &= G(X,Y) + A\sin(\omega t) \end{aligned} \qquad (10.14)$$

This equation is equivariant under the twofold symmetry group generated by the transformation

$$(X,Y,t) \longrightarrow \left(-X,-Y,t+\frac{1}{2}T\right)$$

Here T is the period of the driving term: $\omega T = 2\pi$. The phase space for this dynamical system is $R^2 \times S^1$. The results described below remain valid provided that the driving term respects the symmetry. That is, if $A\cos(\omega t)$ is replaced by $\sum_n A_n(X,Y)\cos[(2n+1)\omega t] + \sum_n B_n(X,Y)\cos(2n\omega t)$, where the $A_n(X,Y)$ are even and the $B_n(X,Y)$ are odd under inversion.

10.6.2 Embedding in $M^3 \subset R^4$

The symmetry is simpler to discuss if we first convert these driven equations to an autonomous dynamical system with polynomial driving terms. This is simply done by introducing a pair of equations that generate trigonometric functions:

$$\begin{aligned} \dot{X} &= F(X,Y) \\ \dot{Y} &= G(X,Y) + AR \\ \dot{R} &= +\omega S \\ \dot{S} &= -\omega R \end{aligned} \qquad (10.15)$$

The coordinates R and S are not independent. They obey a constraint that we choose to take in the form $R^2 + S^2 = 1$. As a result, the flow (10.15) is constrained to the three-dimensional manifold $M^3 = R^2 \times S^1 \subset R^4$. This flow is invariant under the inversion operation:

$$(X,Y,R,S) \longrightarrow (-X,-Y,-R,-S)$$

The flow is bounded away from the singularity at the origin.

10.6.3 Symmetry Reduction

Image equations can be obtained from (10.15) by modding out the parity operator $\mathcal{P}$. This is done as follows. An integrity basis is constructed from the four variables X, Y, R, S. Since the order of the group is 2, it is necessary to consider monomials of degree only up to 2. The integrity basis consists of all 10 bilinear terms which can be constructed from these four coordinates: $X^2, XY, \ldots, S^2$. It is useful to divide these 10 $\mathcal{P}$-invariant coodinates into three sets as follows:

Physical Coordinates	Cross Terms		Trigonometric Coordinates
$X^2 - Y^2$	RX	RY	$R^2 - S^2$
$2XY$	SX	SY	$2RS$
$X^2 + Y^2$			$R^2 + S^2$

Two types of constraints exist among these 10 coordinates.

First, there are six syzygies among the 10 coordinates. These relations are of the form

$$(X^2)(Y^2) - (XY)^2 = 0$$

Second, there is a constraint on the coordinates r and s defined by $r = R^2 - S^2$ and $s = 2RS$. These coordinates inherit a dynamical constraint from the coordinates R and S. Specifically, $R^2 + S^2 = 1$ forces $r^2 + s^2 = (R^2 - S^2)^2 + (2RS)^2 = (R^2 + S^2)^2 = 1$. In fact, if we take $R = \sin(\omega t)$ and $S = -\cos(\omega t)$, then $r = -\cos(2\omega t)$ and $s = \sin(2\omega t)$. In short, the new trigonometric coordinates r and s have half the period of that of the original trigonometric coordinates R and S.

The 6 + 1 constraints on the 10 bilinear coordinates guarantee that the the symmetry-reduced flow occurs in a three-dimensional phase space $M^3 \sim R^2 \times S^1 \subset R^{10}$. The period of this torus (S^1) is half the period of the torus for the original equations with symmetry.

10.6.4 Image Dynamics

The equations of motion for the three variables $R^2 - S^2, 2RS, R^2 + S^2$ are easily constructed. The third is a constant of motion (= 1). For the first pair, we find (by design!)

$$\frac{d}{dt}\begin{bmatrix} r \\ s \end{bmatrix} = \begin{bmatrix} 2R & -2S \\ 2S & 2R \end{bmatrix}\begin{bmatrix} +\omega S \\ -\omega R \end{bmatrix} = 2\omega\begin{bmatrix} 2RS \\ -(R^2 - S^2) \end{bmatrix} = 2\begin{bmatrix} +\omega s \\ -\omega r \end{bmatrix}$$

There remain two independent variables for which equations can be constructed. It is useful to choose these in the following way:

$$\begin{bmatrix} u \\ v \end{bmatrix} = \begin{bmatrix} R & S \\ -S & R \end{bmatrix}\begin{bmatrix} X \\ Y \end{bmatrix} \tag{10.16}$$

This is a useful choice since the 2×2 matrix is easily invertible (it is orthogonal). Morever, the dynamical equations of motion for these two variables are easily constructed:

$$\frac{d}{dt}\begin{bmatrix} u \\ v \end{bmatrix} = \omega \begin{bmatrix} S & -R \\ R & S \end{bmatrix}\begin{bmatrix} X \\ Y \end{bmatrix} + \begin{bmatrix} R & S \\ -S & R \end{bmatrix}\begin{bmatrix} F(X,Y) \\ G(X,Y) + AR \end{bmatrix} \quad (10.17)$$

$$\begin{aligned} \dot{u} &= \omega(+SX - RY) + (+RF + SG) + ARS \\ \dot{v} &= \omega(+RX + SY) + (-SF + RG) + AR^2 \end{aligned}$$

All terms on the right-hand side of these equations can be expressed in terms of the 10 bilinear coordinates invariant under $\mathcal{P}$.

10.7 DUFFING OSCILLATOR

The Duffing oscillator has already been studied in great detail. Studies of this oscillator are complicated by the twofold symmetry. In particular, the stretching and squeezing mechanisms responsible for creating chaotic behavior involve the twofold iteration of the basic scroll formation. The symmetry reduction carried out for the general nonlinear oscillator in Section 10.6 greatly simplifies study of the Duffing oscillator, since the branched manifold for the image has far fewer branches (by a square root: i.e., 3 rather than 9) than does the original Duffing oscillator. This reduced complexity facilitates studies of the perestroikas of the dynamics as control parameters are changed.

For the Duffing oscillator

$$\begin{aligned} \dot{X} &= F(X,Y) = Y \\ \dot{Y} &= G(X,Y) = -\delta Y - X^3 + X + A\sin(\omega t) \end{aligned}$$

In Fig. 10.9 we show a series of Poincaré sections in the X–Y space for a strange attractor generated by the Duffing equations. The Poincaré sections are equally spaced by $0.1T$ around the torus. The inversion symmetry is evident in sections separated by $\frac{1}{2}T$. Alongside each section for the original attractor we exhibit a section in the u–v space for the attractor that has had the symmetry removed. The periodicity with period $\frac{1}{2}T$ is evident in these sections.

In Fig. 10.10(a) we follow the evolution of the strange attractor in the reduced system through a full period of $\frac{1}{2}T$. We adopt the same procedure used to analyze the development of the *gateau roulé* in Figs. 7.20 to 7.22. We begin with a simple curve, parameterized from 0 to 1 in the first section. As ϕ increases, this is deformed, and at the end of the cycle the curve is mapped back into itself. The return mapping is shown in Fig. 10.10(b). This return map can be used to construct a mapping of a surrounding rectangle into itself. This is done by blowing up the curve in a transverse direction. The resulting mapping is shown in Fig. 10.10(c). The same construction is used to blow the noninvertible logistic map back up into a mapping of the rectangle to itself. This resulting invertible map is the standard Smale horseshoe map.

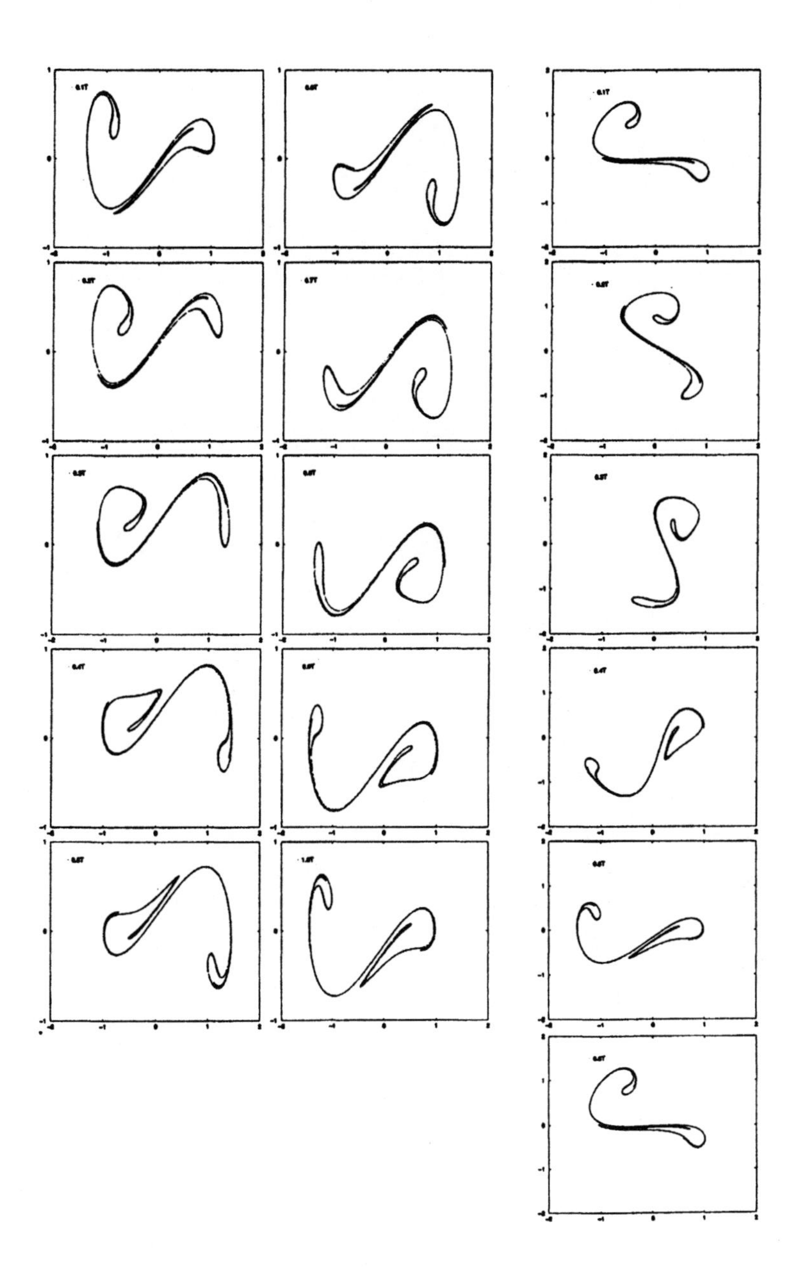

Fig. 10.9 Left: Series of 10 Poincaré sections for the Duffing attractor. The sections are spaced at $0.1T$. The symmetry $(X, Y, t) \rightarrow (-X, -Y, t + \frac{1}{2}T)$ is evident. Right: The $2 \rightarrow 1$ images of these 10 sections are shown after the symmetry has been removed. The reduced dynamical system has period $\frac{1}{2}T$. Parameter values: $\delta = 0.4$, $A = 0.4$, $\omega = 1.0$.

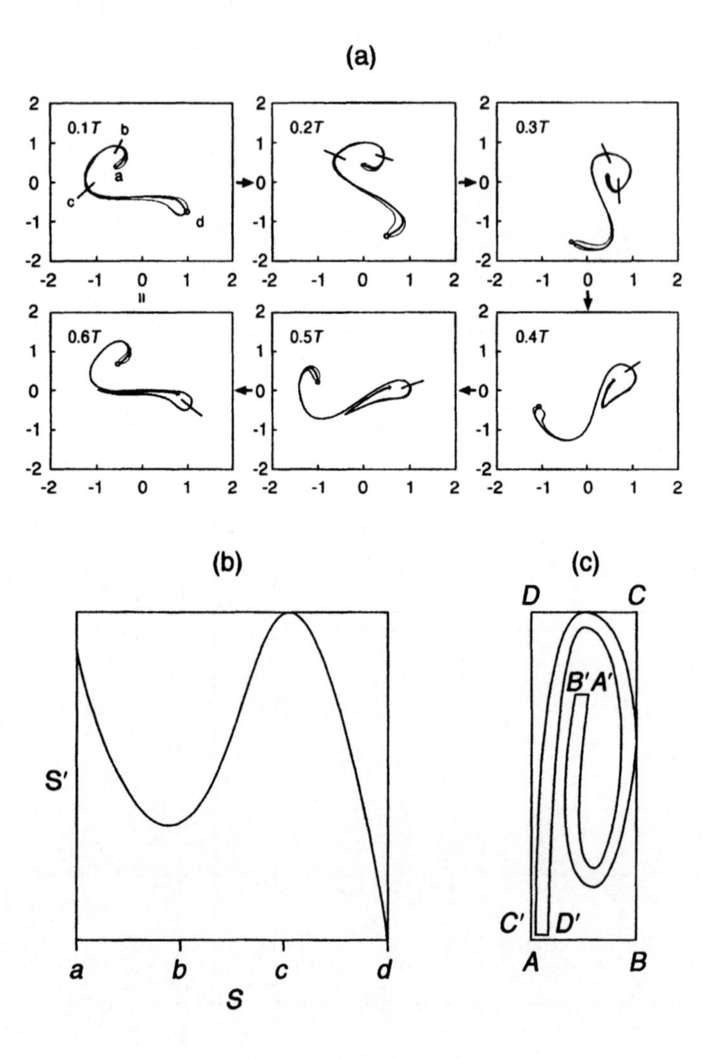

Fig. 10.10 (a) By following the stretching of the reduced attractor through a full period (here $\frac{1}{2}T$), it is possible to determine how an elongated rectangle surrounding the attractor is mapped into itself. (b) This provides a reasonable estimate for the return map. The return map for the system with symmetry is the second iterate of this return map. (c) The return map of the interval can be blown up to provide a mapping of the rectangle back to itself. The stretching and squeezing mechanisms acting on the original Duffing system are the second iterate of the mechanism shown here.

Figures 10.9 and 10.10 have been constructed for particular values of the control parameters in the Duffing equations. As the parameters are varied, the details change. However, they change in a systematic and predictable way. Two control parameters are important for describing the perestroika of this oscillator. These describe the number of branches present at any time and the systematic increase in the global torsion of the branches that are present (cf. Fig. 7.37).

The number of branches present is governed by the amount of stretching which occurs in the phase space during one period. This, in turn, is governed by the strength of the nonlinearity in the original equations. The greater the strength of the nonlinearity, the larger the number of branches present in any strange attractor during the perestroika of the dynamics [192].

The global torsion of the branches present is governed by the period of the driving term. More specifically, the larger the ratio of the external period to the intrinsic period of the undriven oscillator, the greater the torsion of the branches that are present. As this ratio increases through successive integers, the behavior exhibited by the attractor cycles repetitively. An explanation is provided in Fig. 10.11. This figure shows a period-1 "snake." This is the set of orbits of period 1 present as a function of the ratio T/T_r of the driving period to the natural period. The period-1 orbits in this figure are dressed by their local torsion. This integer also acts as a branch label: It identifies the torsion of the branch of the scroll template that describes the flow. As T/T_r increases, old orbits (e.g., 1 and 2) are annihilated in inverse saddle-node bifurcations. The corresponding branches are no longer visited. New period-1 orbits (e.g., 4 and 5) are created in direct saddle-node bifurcations, and the branches that contain them are visited by the flow. In this way, branches $n,\ n+1,\ \ldots$ are replaced by branches $n+2,\ n+3,\ \ldots$ as T/T_r increases by 1.

The snake diagram shows that for any value of T/T_r there is an odd number of period-1 orbits. The strength of the nonlinearity governs the range over which the stable and unstable period-1 orbits exist: The greater the nonlinearity, the longer the orbits exist as a function of T/T_r.

10.8 VAN DER POL OSCILLATOR

The van der Pol oscillator has also been studied in great detail [193, 194]. The twofold symmetry complicates studies of this oscillator as well. The stretching and squeezing mechanisms responsible for creating chaotic behavior involve the twofold iteration of a basic process but a process which is different from that which operates for the Duffing oscillator. Symmetry reduction simplifies the study of this nonlinear oscillator as well. The version of the van der Pol oscillator that we study here is defined by

$$\begin{aligned}\dot{X} &= F(X,Y) &= bY + (c - dY^2)X \\ \dot{Y} &= G(X,Y) &= -X + A\sin(\omega t)\end{aligned}$$

In Fig. 10.12 we show a series of Poincaré sections in the X–Y space for a strange attractor generated by the van der Pol equations. Alongside each section for the original attractor, we exhibit a section in the u–v space for the attractor which has had

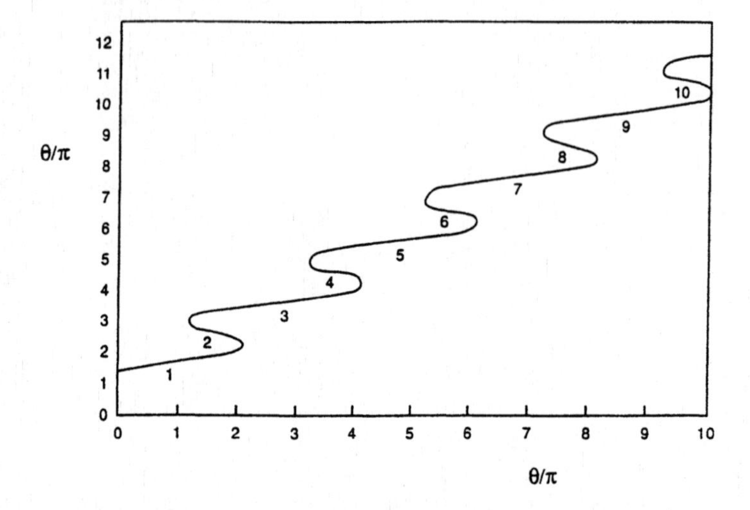

Fig. 10.11 A snake of period-1 orbits runs through the bifurcation diagram for any dynamical system that exhibits the scrolling mechanism. As the ratio T/T_r increases, newer branches with greater local torsion are visited and branches with smaller local torsion are abandoned. The period-1 orbits on these branches are dressed by the local torsion of their branch. The new period-1 orbits (e.g., 4 and 5) are created by saddle-node bifurcations and older period-1 orbits (e.g., 1 and 2) are destroyed by inverse saddle-node bifurcations. The larger the nonlinear coupling, the longer each branch exists as T/T_r increases.

the symmetry removed. The reduction of the period from T in the original equations with symmetry to $\frac{1}{2}T$ in the image equations is evident.

Van der Pol dynamics differs in a fundamental way from Duffing dynamics. In the latter case the stretch-and-squeeze mechanism is similar to a mapping of the interval onto itself—in fact, it is equivalent to a mapping of a (long, thin) rectangle into itself. In the case of van der Pol dynamics, the stretch-and-squeeze mechanism is similar to a mapping of the circle onto itself—in fact, it is equivalent to a mapping of an annulus into itself [195]. By following the stretching of the attractor around one period [$\frac{1}{2}T$; Fig. 10.13(a)], it is possible to construct a rough approximation to a return map of the circle to itself. The return map is shown in Fig. 10.13(b). Finally, the original circle can be blown up in the transverse direction and the return map used to construct a map of the annulus to itself. This map is shown in Fig. 10.13(c).

The branched manifold for the original van der Pol oscillator (at appropriate parameter values) is shown at the top of Fig. 10.14. In this representation, the left- and right-hand edges of the branched manifold must be identified, using periodic boundary conditions. The processes that take place in the first half period, from $t = 0$ to $t = \frac{1}{2}T$, are repeated in the second half period, phase-shifted by π radians. The phase

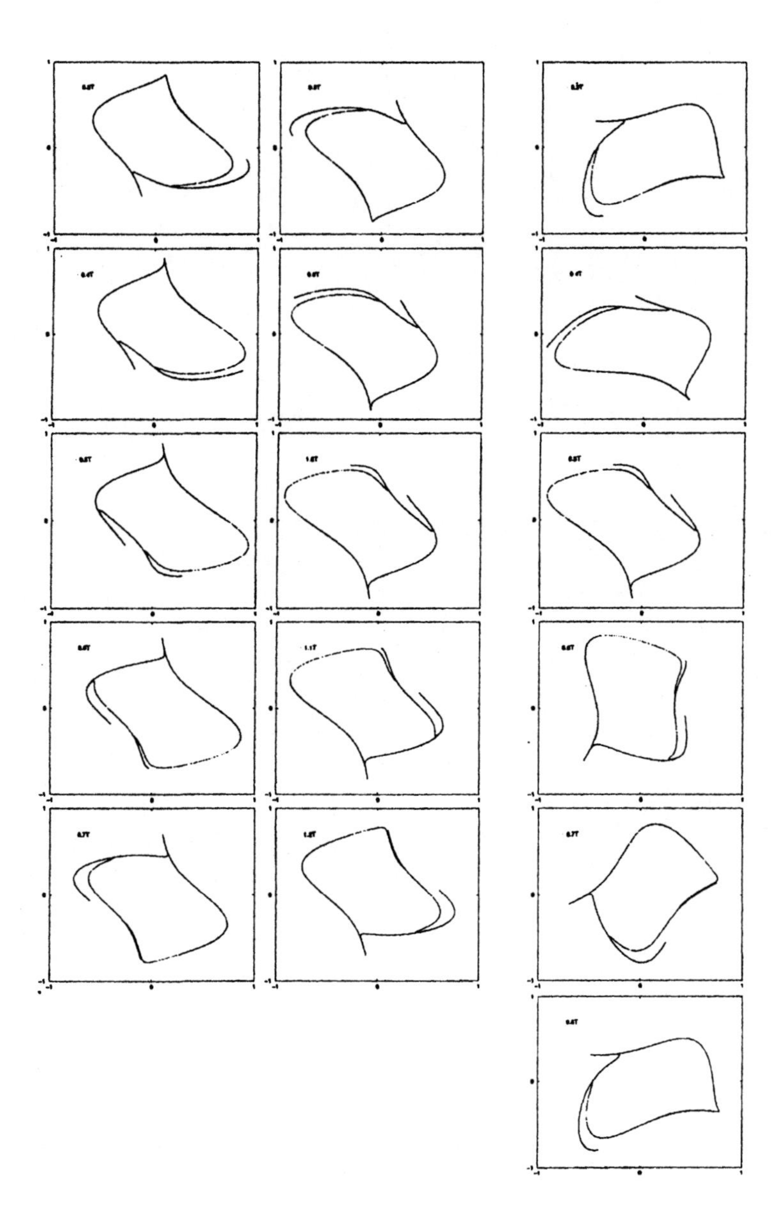

Fig. 10.12 Left: Series of 10 Poincaré sections for the van der Pol attractor. The sections are spaced at $0.1T$. The symmetry $(X, Y, t) \to (-X, -Y, t + \frac{1}{2}T)$ is evident. Right: The $2 \to 1$ images of these 10 sections are shown after the symmetry has been modded out. The reduced dynamical system has period $\frac{1}{2}T$. Parameter values: $b = 0.7$, $c = 1$, $d = 10$, $A = 0.25$, $\omega = \pi/2$.

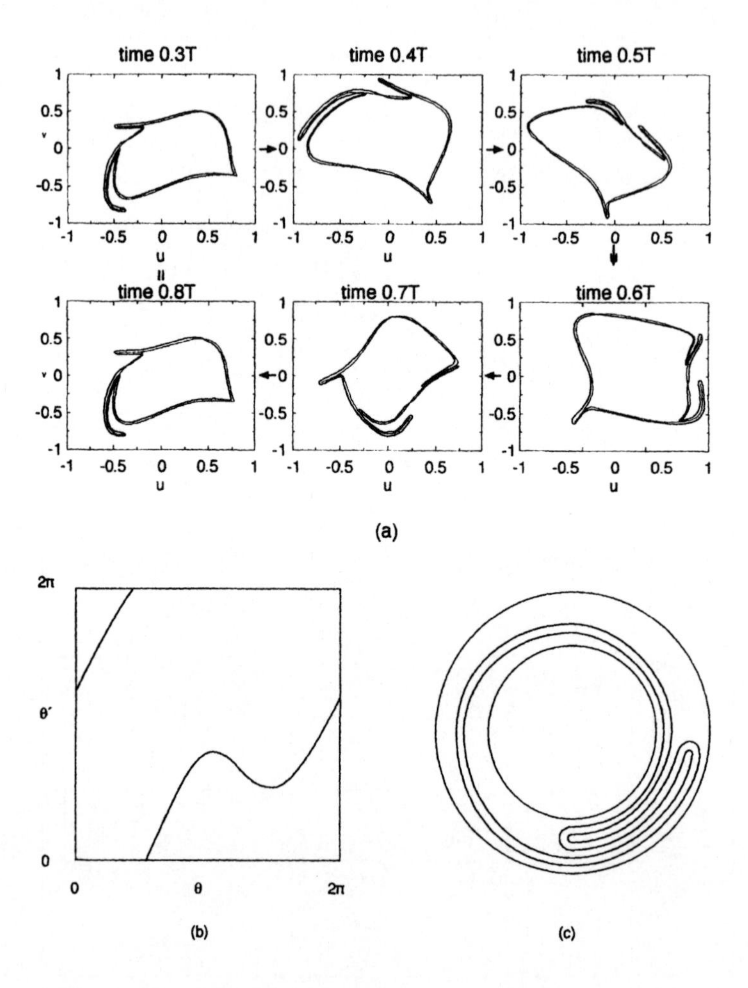

Fig. 10.13 (a) By following the stretching of the attractor through a full period (here $\frac{1}{2}T$), it is possible to determine how a topological circle is mapped to itself under the symmetry reduced flow. (b) This provides a reasonable estimate for the return map. The return map for the system with full symmetry is the second iterate of this return map. (c) The return map of the circle can be blown up to provide a mapping of the annulus back to itself.

shift is equivalent to the map $(X, Y) \rightarrow (-X, -Y)$. When the symmetry is removed (Fig. 10.14, bottom), the stretching and squeezing that takes place during the first half period is extended a further π radians. As a result, the stretching and squeezing that takes place during the second half period repeats that of the first half period. The period of the flow has been reduced to $\frac{1}{2}T$. The simplified branched manifold (bottom) shows three branches, compared to the six of the original branched manifold.

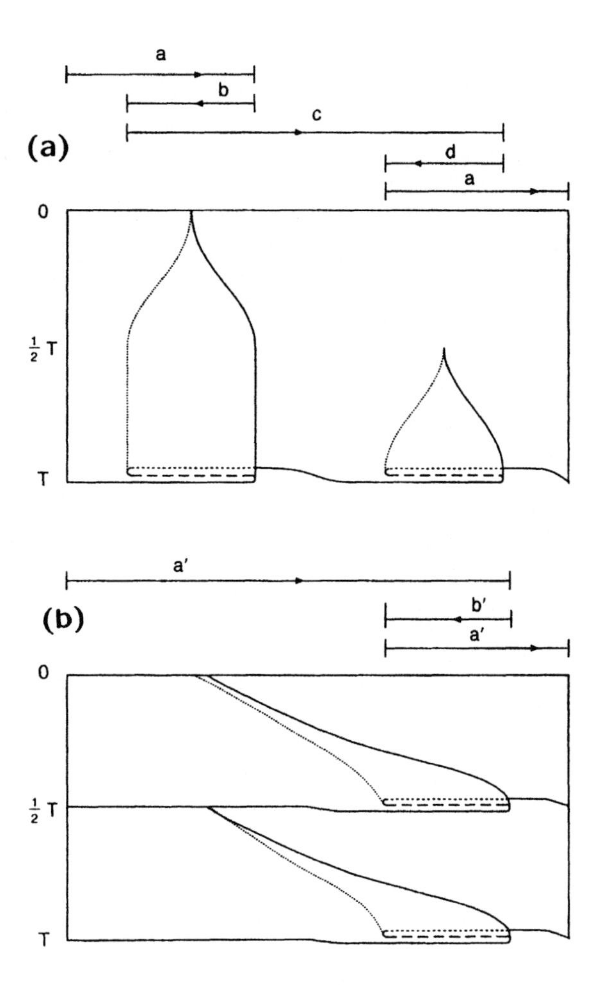

Fig. 10.14 (a) The stretching and squeezing that take place for the van der Pol oscillator are represented by this template with six branches. The left- and right-hand edges are to be identified. (b) After the inversion symmetry has been removed, the stretching and folding is represented by this branched manifold. The basic periodicity has been reduced from T to $\frac{1}{2}T$.

The perestroika of the van der Pol equations is governed by the same constraints as that of the Duffing equations. It is only the topology that is different. Specifically:

- As the ratio T/T_r increases, the global torsion of the branches present in the branched manifold describing the strange attractor increases systematically.

- As the strength of the nonlinear coupling increases, the average number of branches required to describe the strange attractor increases.

The details are summarized in Fig. 10.15. In Fig. 10.15(a) we present a return map of the circle to itself: $\theta \rightarrow \theta'$. In this map we take $0 \leq \theta \leq 2\pi$, with 0 and 2π identified. We should do the same for the image θ', but do not, to make the argument clearer. In this figure, the circumference of the circle is stretched by a factor of approximately 3, with $0 \rightarrow 0$. The period-1 orbits are identified by the intersections of the return map with the diagonals. The branches of the return map are dressed by their local torsion. There are four fixed points when we identify the points at 0 and 2π. The nonlinearity governs the amplitude of the return map (approximately 4π) and the ratio T/T_r governs the "rotation" (here 0). As T/T_r increases, the return map is rigidly shifted upward.

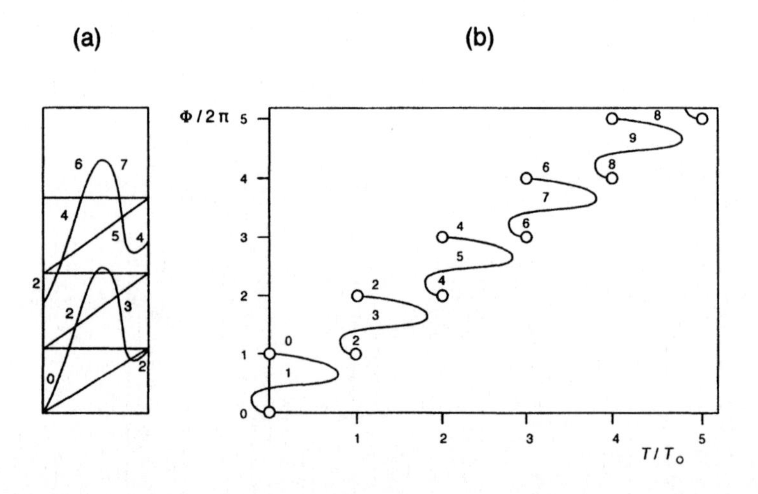

Fig. 10.15 (a) A return map $\theta \rightarrow \theta' = f(\theta)$ of the circle to itself has amplitude governed by the strength of the nonlinearity and rotation (vertical displacement) governed by the ratio of time scales: T/T_r. The branches of the map are dressed by the local torsion of the corresponding period-1 orbit in the flow. The image θ' has not been taken mod 2π for the sake of clarity. (b) The period-1 orbits that exist for any value of T/T_r are indicated in this circle map analog of a snake diagram. Each sinuous curve is a "bubble," difficult to recognize because of the 2π vertical boundary conditions. Unlike the snake, the period-1 orbits mutually create and annihilate.

The evolution of the period-1 orbits as a function of increasing T/T_r is summarized in Fig. 10.15(b). As T/T_r increases above 0, the period-1 orbit at $0 = 2\pi$ and the one near 2π with torsion 2 collide and annihilate in an inverse saddle-node bifurcation. The two fixed points on branches 2 and 3 separate, and by the time $f(0) = 2\pi$, the entire curve has been shifted up by 2π. The return map has been shown for the case $f(0) \simeq 3\pi$ in Fig. 10.15(a). In this case the orientation-preserving branches have torsions 2, 4, and 6, while the orientation-reversing branches have torsions 7 and 5. Figure 10.15(b) for the circle map is the analog of the snake diagram (Fig. 10.11) for the scroll return map of the interval to itself. It consists of a sequence of "bubbles" which do not look like bubbles. For example, the bubble at $T/T_r = 2$ consists of two period-1 orbits with torsions 4 and 5. These orbits are created in a saddle-node bifurcation for $T/T_r \sim \frac{7}{4}\pi$ and annihilate in an inverse saddle-node bifurcation for $T/T_r \sim \frac{11}{4}\pi$. For any value of T/T_r there is only an even number of period-1 orbits. This is in opposition to the Duffing case, in which the snake diagram indicates that only an odd number of period-1 orbits is present for any value of T/T_r. The difference, even (for maps of S^1) or odd (for maps of R^1), is tied intimately to the global topology of a one-dimensional surface. This surface is the intersection of the branched manifold for a strange attractor with a Poincaré section.

10.9 SUMMARY

The relation between dynamical systems that are locally indistinguishable but globally distinct has barely been scratched. We have described only twofold symmetries here—rotation and inversion symmetries—and have found a wealth of riches.

The ability to relate cover and image dynamical systems is useful in both directions. If chaotic intensity data are recorded, it is necessary to determine the spectrum of chaotic amplitude (covering) dynamics that is compatible with the observed intensity (image) dynamics. In the other direction, symmetry adds complications to eliciting the stretching and squeezing mechanisms which act in phase space to generate strange attractors. Modding out the symmetry simplifies the identification of an appropriate branched manifold for both the image and the original dynamical system.

A straightforward algorithm exists to relate cover and image dynamical systems. This algorithm can be implemented at the level of dynamical system equations. The cover–image relations can also be implemented at the level of branched manifolds [84] and at the level of periodic orbits in the attractors.

It is surprising that a single dynamical system can have many inequivalent covers, each with the same symmetry. The different covers are distinguished by a topological index. This index is the set of linking numbers of branches in the image template with the singular set of the local diffeomorphism. When this singular set [for $R_z(\pi)$ it is an axis] does not intersect the image flow, cover branched manifolds have twice as many branches as the image template. If the singular set does intersect the image flow, a new type of global bifurcation takes place.

By modding out the inversion symmetry present in many two-dimensional periodically driven nonlinear oscillators, we essentially take the square root of the dynamics. This greatly facilitated construction of the branched manifolds for the Duffing and van der Pol oscillators.

11

Flows in Higher Dimensions

A doubly discrete classification theory for strange attractors in R^3 has been developed in the previous chapters. We have identified the appropriate measures required for this classification, provided algorithms for extracting these measures from experimental data, and reviewed a number of experiments where the identification and classification have been carried out successfully.

All algorithms depend on the linking numbers of periodic orbits in the flow: in particular, in the strange attractors to be analyzed. As a result, these algorithms do not immediately extend to flows in higher dimensions. In this chapter we indicate some ways in which the classification theory can be extended to strange attractors in higher-dimensional spaces.

11.1 REVIEW OF CLASSIFICATION THEORY IN R^3

The doubly discrete classification theory for strange attractors in three dimensions is based on two constructions:

Branched Manifolds: These are discretely classified by their topology matrices, insertion arrays, and transition matrices.

Basis Sets of Orbits: These are minimal sets of periodic orbits that force the existence of all other orbits in the flow. At the present time this information is available only for the Smale horseshoe branched manifolds, independent of their global torsion, and only up to a finite period.

Two strange attractors are equivalent in the sense that a smooth transformation can be constructed which maps one to the other, only when they possess the same discrete information: They have identical branched manifolds and the same basis sets of orbits. If they have identical branched manifolds but different basis sets of orbits, variation of some control parameters for one of the systems can be found which changes the basis set, by creating and annihilating orbits, until the two share the same basis set. Then they are equivalent.

Both these analyses, to determine a branched manifold and a basis set of orbits, depend essentially on the topological organization of unstable periodic orbits in the flow. In dimensions higher than 3, knots fall apart, all their linking numbers are zero, and their rigid topological organization is simply a wishful thought. The result is that the methods of previous chapters cannot be extended to develop a classification theory for strange attractors in dimensions higher than 3.

In reviewing the results presented in previous chapters, we can see the following historical progression:

Branched Manifold: The initial data analyses were devoted to determining the underlying branched manifold onto which the strange attractor projected under the Birman–Williams identification. At first, only the simple Smale horseshoe was observed. Later, reverse horseshoes, horseshoes with global torsion, and scroll templates with more than two branches were also observed. Some data sets also exhibited a Lorenz tearing mechanism.

Basis Set of Orbits: More refined analyses focused on the spectrum of unstable periodic orbits in chaotic data sets and how these changed as control parameters varied.

Stretch and Squeeze Mechanism: More recent analyses have been devoted to determining the basic stretch and squeeze mechanism which deforms the flow in phase space. This can be illustrated by reviewing the topological underpinnings for the four standard flows introduced in Chapter 3:

Equations	Mechanism
Rössler	Stretch and fold
Lorenz	Tear and squeeze
Duffing	Second iterate of a scroll
van der Pol	Stretch and double fold on a circle, iterated twice

The point to be emphasized is that the properties of periodic orbits were used to tease out information about the underlying mechanism—but at the end of the day it is the mechanism that is of predominant interest. Organization of periodic orbits is the tool: Underlying mechanism is the end result of interest. As a result, it would be profitable to look for ways to determine the underlying mechanism that do not involve the fundamental tool so important in R^3, the topological organization of unstable periodic orbits in strange attractors.

Many of the classification results in R^3 can be reproduced without using periodic orbits as follows.

Poincaré Section: The first step involves construction of a Poincaré section $\mathcal{PS}$ for the flow. In general, the Poincaré section consists of a union of disks, where each disk is the blowup of a branch line in the branched manifold. In the simple case where the flow is embedded in a torus $R^2 \times S^1$, any plane R^2 of constant $\phi \in S^1$ is suitable as a Poincaré section.

Return Maps: The flow provides a return map of the Poincaré section to itself. This map is invertible. Invertible maps are difficult to classify. However, under the Birman–Williams projection, the strange attractor projects to a two-dimensional branched manifold $\mathcal{BM}$, whose intersection with the Poincaré section is one-dimensional: $\mathcal{M}^1 = \mathcal{BM} \cap \mathcal{PS}$. The flow provides a return map $\mathcal{M}^1 \rightarrow \mathcal{M}^1$ of this one-dimensional manifold to itself. This map is noninvertible. Noninvertible maps can be (discretely) classified.

Singularities of Maps: The singularities of the map $\mathcal{M}^1 \rightarrow \mathcal{M}^1$ can easily be classified. If $\mathcal{M}^1$ is an interval, the only irreducible singularity is the fold A_2 (Smale horseshoe, reverse horseshoe, horseshoe with global torsion). Additional nonlocal folds can occur, giving rise to scrolls and more complicated situations. If $\mathcal{M}^1$ is a circle (S^1 or T^1), nondegenerate folds can only occur in pairs. The simplest occurrence of this mechanism takes place in the van der Pol oscillator, from which the twofold symmetry has been eliminated using a local diffeomorphism. Second iterates of the scroll and circle mechanisms (Duffing and van der Pol oscillators, respectively) create additional nonlocal folds in the singular return map, and these folds are organized in specific ways with respect to each other. Once the nonlocal singularity has been identified, the underlying mechanism has been discretely classified. The singular map can be blown back up to an invertible return map of the Poincaré section to itself.

It seems that there is no major impediment to implementing these three processes to classify and identify flows in higher dimensions. In the following sections we sketch how this can be done, emphasizing flows in R^4.

11.2 GENERAL SETUP

For flows in R^n there are two tools to work with. These are the spectrum of Lyapunov exponents and projections into subspaces.

11.2.1 Spectrum of Lyapunov Exponents

At any point in the phase space R^n a flow has n Lyapunov exponents. At a singular point the exponents are the eigenvalues of the linearized flow in the neighborhood of the fixed point. Typically, all n eigenvalues are nonzero. At a generic point in the phase space the flow is nonzero and the spectrum of Lyapunov exponents λ_i satisfies

$$\begin{array}{ccccccccccccc} \lambda_1 & \geq & \lambda_2 & \geq & \cdots & \geq & \lambda_{k_1} & > & \lambda_{k_1+1}=0 & > & \lambda_{k_1+2} & \geq & \cdots & \geq & \lambda_n \\ \epsilon_1 & \geq & \epsilon_2 & \geq & \cdots & \geq & \epsilon_{k_1} & \geq & \epsilon_{k_1+1}=1 & \geq & \epsilon_{k_1+2} & \geq & \cdots & \geq & \epsilon_n \end{array} \tag{11.1}$$

At any typical point in R^n an eigenvector is associated with each eigenvalue. Along the local eigendirections, the flow has the canonical form $\dot{y}_i = \lambda_i y_i$ [cf. Eq. (3.61)]. In principle, we should continue to distinguish between local Lyapunov exponents and their globally averaged values, but we do not.

Associated with each eigendirection is a *partial dimension* [196] ϵ_i, $0 \leq \epsilon_i \leq 1$. The partial dimension describes the smoothness of the attractor along the direction of the ith eigenvector. The attractor is smooth in the stretching directions ($\epsilon_i = 1$ for $1 \leq i \leq k_1$) and in the flow direction ($\epsilon_{k_1+1} = 1$), but fractal in the squeezing directions ($k_1 + 2, \ldots, k_n$).

It is useful to construct the cumulative sum of eigenvalues and weighted eigenvalues, defined as follows:

$$\Lambda_r = \sum_{i=1}^{r} \lambda_i \qquad D_r = \sum_{i=1}^{r} \epsilon_i \lambda_i \tag{11.2}$$

These cumulative sums are illustrated in Fig. 11.1. The important points about these sums are: (1) $\Lambda_n < 0$, and (2) $D_n = 0$. More specifically, Λ_r rises to a maximum value at $r = k_1$, maintains this value for $r = k_1 + 1$, and then decreases, ultimately falling below 0. At the same time, $D_r = \Lambda_r$ for $r = 1, 2, \ldots, k_1, k_1 + 1$, and then decreases more slowly than Λ_r for $r > k_1 + 1$. For some value of $k_2 > 0$, $D_{k_1+k_2} > 0$ and $D_{k_1+k_2+1} = D_{k_1+k_2+2} = \cdots = D_n = 0$.

11.2.2 Double Projection

The information is Fig. 11.1 can be used to carry out two important projections.

11.2.2.1 Inertial Manifold The first projection is from the original phase space R^n (or n-manifold) into a smaller manifold. This manifold is called an *inertial manifold* and denoted $\mathcal{IM}^{k_1+k_2+1}$. Its dimension is determined by the first value of the cumulative sum D_r which falls to zero. Beyond this value of $r = k_1 + k_2 + 1$, all partial dimensions ϵ_i are zero. All the dynamics can be accommodated in a space of dimension $k_1 + k_2 + 1 \leq n$. In this space, k_1 is the number of stretching (unstable) directions, k_2 is the number of squeezing (stable) directions. There is one time flow direction.

Projections into inertial manifolds have already been encountered. Many of the data sets analyzed in Chapter 7 satisfy $\lambda_1 > 0 = \lambda_2 > \lambda_3 \cdots$ and $\lambda_1 + \lambda_2 + \lambda_3 < 0$.

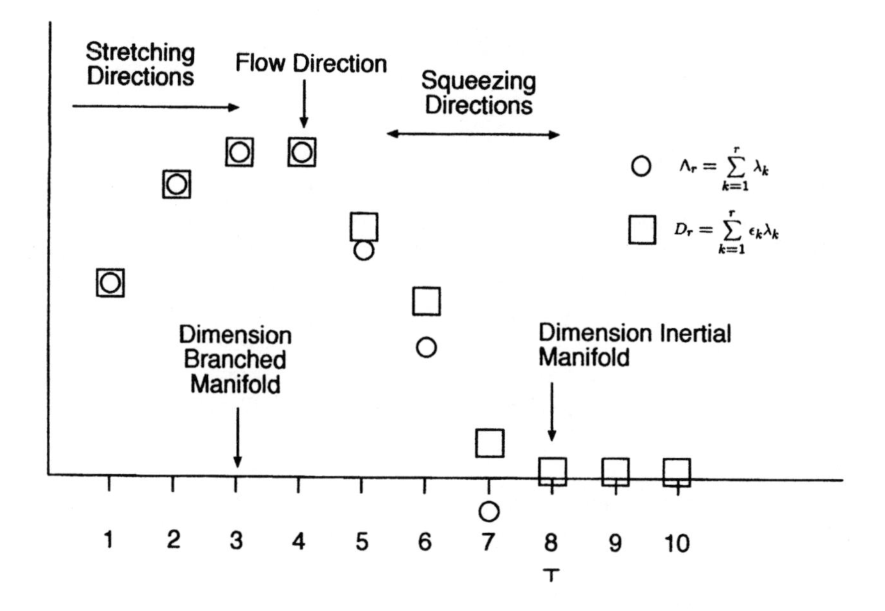

Fig. 11.1 Cumulative sum of Lyapunov exponents $\Lambda_r = \sum_{i=1}^{r} \lambda_i$ (circles) and weighted exponents $D_r = \sum_{i=1}^{r} \epsilon_i \lambda_i$ (squares) is shown for a dynamical system in R^9 with three positive Lyapunov exponents.

For such data sets, the dynamics can be projected into R^3 or some other three-dimensional manifold, so that linking numbers can be computed and the standard topological analysis can be carried out.

The Lyapunov dimension of the strange attractor is the sum of the partial dimensions:

$$d_L = \sum_{i=1}^{n} \epsilon_i = \sum_{i=1}^{k_1+1+k_2} \epsilon_i = k_1 + 1 + (\epsilon_{k_1+2} + \cdots + \epsilon_{k_1+1+k_2}) \tag{11.3}$$

In cases where the inertial manifold is three-dimensional, $D_3 = 0 = \lambda_1 + \epsilon_3 \lambda_3$, so that $\epsilon_3 = -\lambda_1/\lambda_3 = \lambda_1/|\lambda_3|$ and $d_L = 2 + \epsilon_3 = 2 + \lambda_1/|\lambda_3|$. More generally, with k_1 expanding directions and only one squeezing direction, similar arguments give $\epsilon_{k_1+2} = \sum_{i=1}^{k_1} \lambda_i / |\lambda_{k_1+2}|$, which results in the Kaplan–Yorke estimate [138] for the Lyapunov dimension of a strange attractor with one stable direction:

$$d_L = k_1 + 1 + \frac{\sum_{i=1}^{k_1} \lambda_i}{|\lambda_{k_1+2}|}$$

In the general case, the Lyapunov dimension is the sum of the partial dimensions.

11.2.2.2 Manifold with Singularities Perhaps the most important aspect of the Birman–Williams theorem [83, 84] is the identification used to project a chaotic flow down to a branched manifold. This projection can be used in higher dimensions as follows. We identify all points x and y in $\mathcal{IM}^{k_1+k_2+1}$ with the same future:

$$x \sim y \quad \text{if} \quad \lim_{t\to\infty} |x(t) - y(t)| \to 0 \tag{11.4}$$

This projects the flow in $\mathcal{IM}^{k_1+k_2+1}$ down along the remaining k_2 stable directions to a $(k_1 + 1)$-dimensional manifold with singularities. This is a higher-dimensional analog of the branched manifold in R^3. It is not exactly a branched manifold, since its singular set does not describe only stretching and squeezing. The singular sets for branched manifolds in R^3 are splitting points (they describe stretching) and branch lines (squeezing). For the case where $k_1 = 2$ and $k_2 = 1$, the singular manifold is three-dimensional in R^4. The spectrum of dimensions of its singularities is 0, 1, and 2. When two parts of this three-dimensional space meet in R^4, the intersection is generically of dimension $3 + 3 - 4 = 2$. This two-dimensional analog of the branch line is split by a one-dimensional curve. There are components of the singular set that describe neither stretching nor squeezing. Nevertheless, the singular manifold is a useful construct for classifying strange attractors in higher dimensions.

11.3 FLOWS IN R^4

We expend our first tentative steps in classifying strange attractors in higher dimensions in the next-higher dimension: four. If our periodic orbit-free methods are useful here, there is no impediment in extending these methods to yet higher dimensions.

11.3.1 Cyclic Phase Spaces

As a first step, we simplify by assuming that the dynamics takes place in a cyclic phase space with the topology of a torus: $R^3 \times S^1$. We feel that this is not a major restriction on the basis of our experience with three-dimensional dynamical systems. The Rössler dynamical system possesses this characteristic, and by the time we eliminated the symmetries from the remaining standard models, they also possessed this characteristic. The principal point of this assumption is that a simple Poincare section exists.

11.3.2 Floppiness and Rigidity

We study the return mapping on a Poincaré section under the two cases $k_1 = 1, k_2 = 2$ and $k_1 = 2, k_2 = 1$. Typical results are illustrated in Figs. 11.2 and 11.3. In the first case, a cube of initial conditions in the phase space R^3 evolves under a flow with $\lambda_1 = \log(4)$ and $\lambda_3 = \lambda_4 = -\log(3)$. During the evolution, the cube stretches by a factor of 4 in one direction and shrinks by a factor of 3 in each of the two transverse directions (Fig. 11.2). To be mapped back into itself, the long, thin rectangle must be bent. There are many ways in which it can be bent. It is like squeezing toothpaste back into the tube. Specifically, the number of ways of squeezing the long, thin rectangle back into the initial cube is not discretely classifiable.

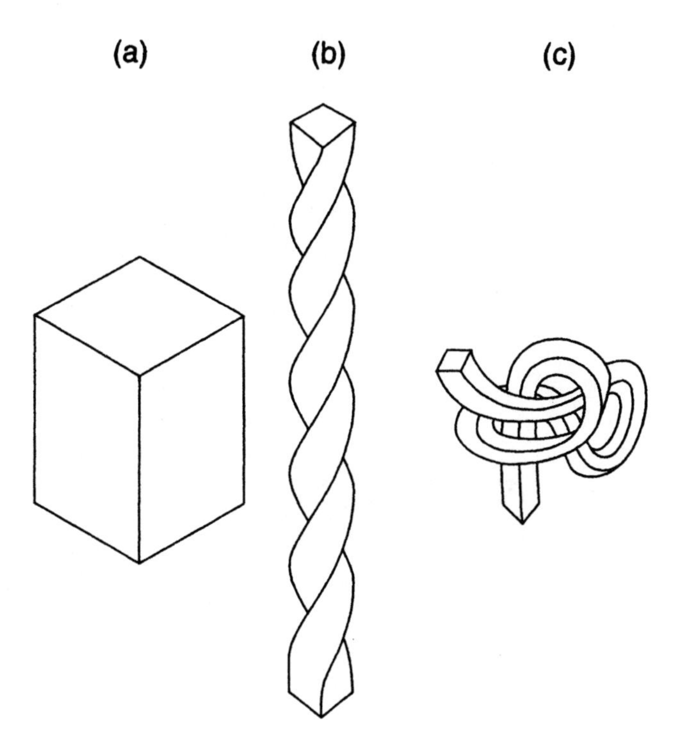

Fig. 11.2 In a flow with one unstable direction [$\lambda_1 = \log(4)$] and two stable directions [$\lambda_3 = \lambda_4 = -\log(3)$], a cube of initial conditions is stretched by a factor of 4 in the unstable direction and shrunk by a factor of 3 in the two stable directions. This elongated tube can be squeezed back into the original cube in (continuously) many different ways. The dynamics is too floppy to be discretely classified.

In the second case we take the Lyapunov exponents to be $\lambda_1 = \lambda_2 = \log(2)$ and $\lambda_4 = -\log(9)$. During the evolution, the initial cube stretches by a factor of 2 in the two unstable directions and shrinks by a factor of 9 in the stable direction. This blanket also needs to be folded to be mapped back into the initial cube. However, in this case there are a discrete number of ways in which the folding can take place. One of these is shown in Fig. 11.3.

Dynamical systems which are too "floppy" in the sense of Fig. 11.2 cannot be discretely classified. However, those that are "rigid" enough can be discretely classified. These results can be made quantitative in terms of the spectrum of Lyapunov exponents. In general, flows with k_1 large (many unstable directions) are discretely classifiable provided that k_2 is sufficiently small.

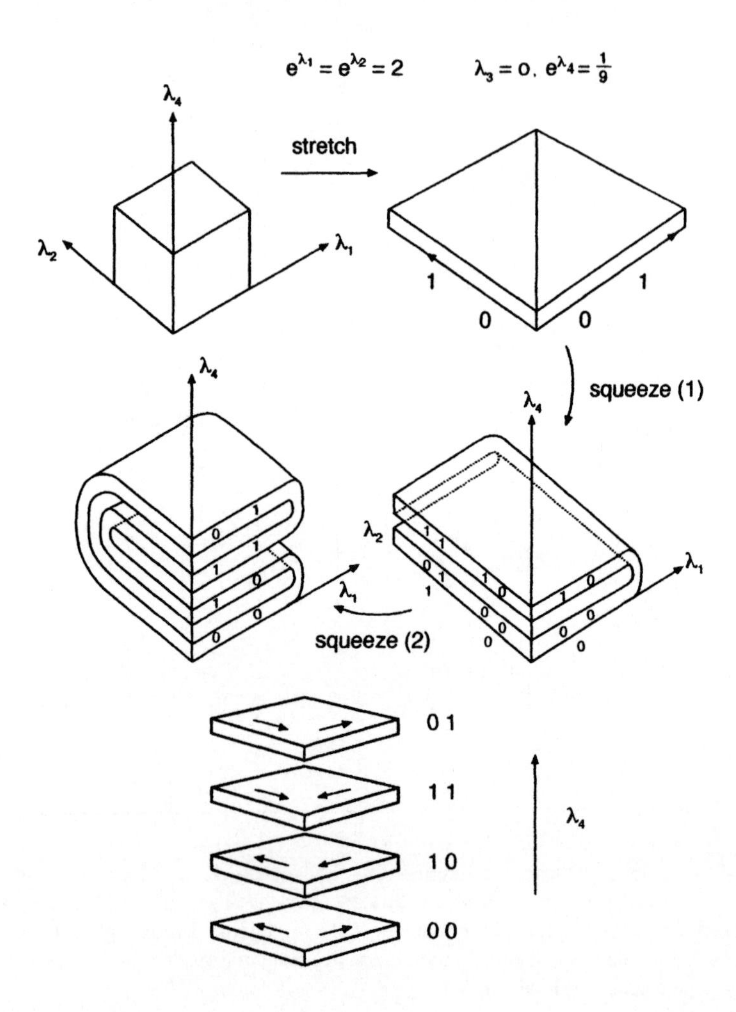

Fig. 11.3 In a flow with two unstable directions $[\lambda_1 = \lambda_2 = \log(2)]$ and one stable direction $[\lambda_4 = -\log(9)]$, a cube of initial conditions is stretched by a factor of 2 in each of the unstable directions and shrunk by a factor of 9 in the stable direction. This blanket shape can be squeezed back into the original cube in (finitely) many different ways. The dynamics is rigid enough to be discretely classified.

11.3.3 Singularities in Return Maps

Our objectives are to provide a discrete classification for dynamical systems in cases where discrete classifications exist. In R^4 they do not exist when there is one unstable direction. We therefore consider here only the case with two unstable directions, shown in Fig. 11.3.

Even in this case there is no discrete classification, since the return map is invertible. To construct a discrete classification, we must go to the limit of noninvertible maps. This means that we must use the second of the two projections described in Section 11.2.2. This projects the strange attractor with dimension $3+\epsilon_4$ to a three-dimensional manifold with singularities. The intersection of this manifold with a Poincaré section is two-dimensional: R^2. The return map induced by the flow generates a mapping of $R^2 \to R^2$ with singularities. It is the existence of singularities that provides the ability to create a discrete classification for the dynamics.

Whitney has told us that there are only two possible generic local singularities in mappings of the plane to itself [181, 197, 198]. These are the fold and the cusp, which have canonical forms:

$$\begin{array}{llrcl} \text{Fold}: & A_2 & x & \rightarrow & x^2 \\ & & y & \rightarrow & y \\ \\ \text{Cusp}: & A_3 & x & \rightarrow & x^3 + xy \\ & & y & \rightarrow & y \end{array} \tag{11.5}$$

We have already encountered the fold many times. Fold lines are defined by the locus of points in phase space where the rank of the Jacobian of the singular mapping drops by 1. The stretching and squeezing mechanisms new to four-dimensional dynamical systems, not possible to encounter in three-dimensional dynamical systems, are due to the cusp singularity.

Figure 11.4 illustrates how the local fold singularity forms the basis for the standard stretch and fold mechanism responsible for creating chaos in many three-dimensional dynamical systems [199]. When the phase space is $R^2 \times S^1$, the Poincaré section is R^2 and the intersection of the strange attractor with a Poincaré section has dimension $1+\epsilon_3$. A Birman–Williams projection maps the strange attractor to a two-dimensional branched manifold. This intersects the Poincaré section in a one-dimensional manifold, which we take as an interval (a piece of R^1) or, in some cases, a circle. The flow provides a mapping of this R^1 to itself which must possess a singularity. The only generic singularities in one-dimensional mappings are folds. Figure 11.4 shows a single fold. Nonlocal folds are also possible. The number of inequivalent stretch and squeeze mechanisms increases very rapidly with the number of nonlocal folds. The global topology (R^1 vs. S^1) constrains the allowed stretch and squeeze mechanisms. We can blow up the fold singularity by expanding against the contracting direction to recreate the horseshoe map.

Figure 11.5 illustrates how the cusp singularity is the next-higher-dimensional analog of the fold. Here the phase space is $R^3 \times S^1$. The strange attractor has two unstable directions and dimension $3 + \epsilon_4$. A Poincaré section is R^3. The intersection of the two has dimension $2 + \epsilon_4$. After the Birman–Williams projection of the flow down to a semiflow, the intersection has dimension 2 (R^2 or a piece of R^2). The flow generates a return map $R^2 \to R^2$. This return map loses information and generates entropy, since the semiflow has no well-defined past. The only local singularities of maps of two-dimensional manifolds to themselves are the fold and the cusp. The cusp deformation is illustrated in Fig. 11.5. The original nonclassifiable invertible

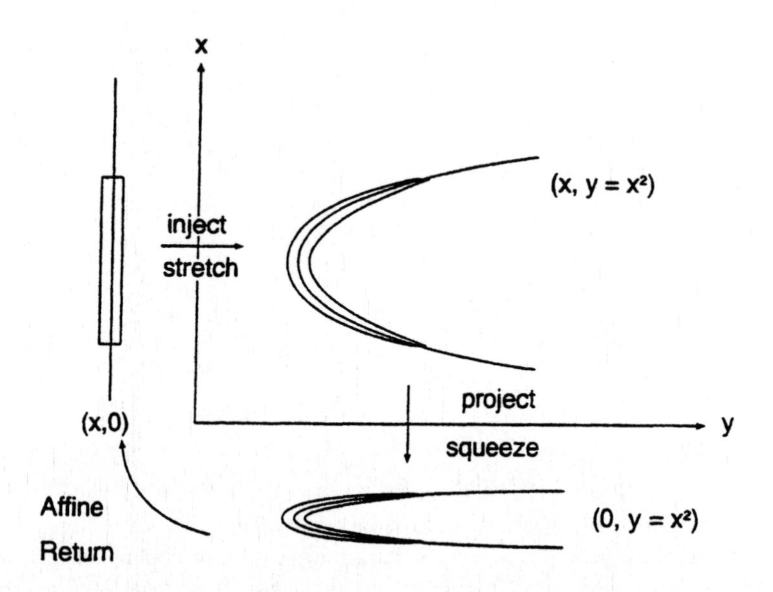

Fig. 11.4 The thin rectangle contains the intersection of a strange attractor in $R^2 \times S^1$ with a Poincaré section. The interval is the intersection with the strange attractor's branched manifold. Under the flow, both are stretched, bent, projected back to themselves, and then returned to their original position by an affine transformation.

mapping with a cusp backbone can be obtained by "blowing up" the cusp by expanding against the contracting direction.

11.4 CUSP BIFURCATION DIAGRAMS

If flows do exist that can be classified by a cusp singularity, they will eventually be found. Identifying such flows may prove difficult. One procedure that can be used to simplify this task is to follow the path used to infer the presence of a fold in a flow that generates chaotic behavior. The bifurcation behavior in the limiting case of a singular map reduces to that of the logistic map. By understanding the bifurcations of the logistic map, we are in a position to identify fold mechanisms in a flow because of the similarity of the bifurcation behavior. For example, in the logistic map we see a period-doubling cascade, followed by chaotic behavior containing periodic windows with width generally increasing as the period decreases to a minimum of $p = 3$, following which the windows become narrower again. Within each window there is a period-doubling cascade. When similar behavior is observed in a flow, we expect

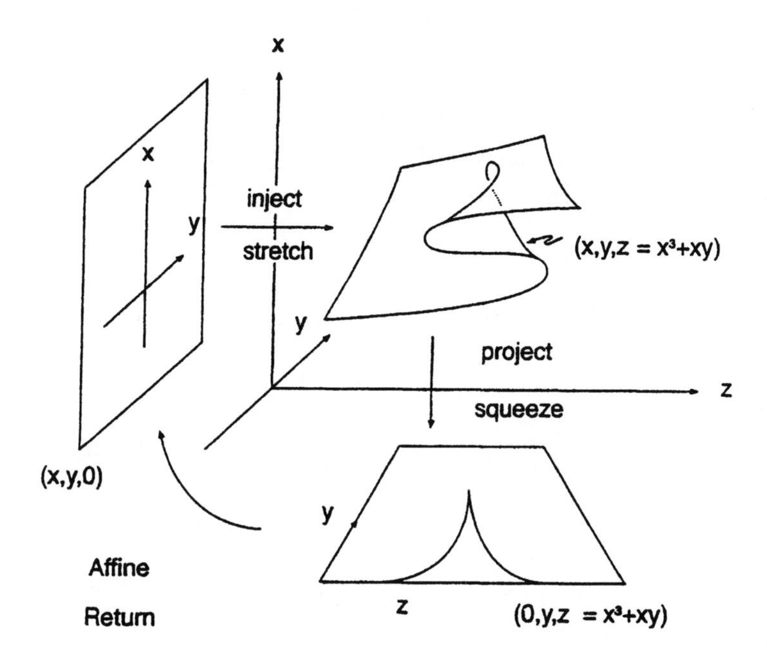

Fig. 11.5 The section of plane R^2 represents the intersection of the branched manifold of a strange attractor in $R^3 \times S^1$ ($\lambda_1 > \lambda_2 > 0$) with a Poincaré section R^3. Under the flow, it is stretched, deformed, and projected to a plane. The plane is mapped back to the initial position by an affine transformation. The nonsingular map with a cusp skeleton can be obtained by fattening up the initial plane section.

that a return map of a Poincaré section back to itself will exhibit an invertible folding structure, whose limit is the noninvertible fold A_2.

On the other hand, alternative mechanisms exhibit alternative bifurcation phenomena. For example, the double folding of a circle generates a different route to chaos. On this route to chaos mode locking occurs before the circle map even becomes noninvertible. The last invariant torus breaks when the circle map loses its invertibility. Thereafter there is chaos, with topological entropy increasing as the coupling strength increases. Within each mode-locked tongue a period-doubling cascade accumulates, and beyond there is chaos within the tongue. This spectrum of bifurcations distinguishes flows whose underlying branched manifold is of Smale horseshoe type from flows whose underlying branched manifold exhibits double folding of a circle S^1. For these reasons, a useful preliminary to studying chaotic flows in higher dimensions involves determining the bifurcations that a singular cusp map can exhibit.

11.4.1 Cusp Return Maps

The cusp return map is illustrated in Fig. 11.5 [199]. A portion of the (x, y) plane is stretched in R^3 as follows:

$$(x, y) \overset{\text{stretch}}{\longrightarrow} [x, y, z = r(x^3 + xy)]$$

The stretching increases as r grows from 0 to 1. The squeezing occurs as this pleated surface is projected down into the (y, z) plane:

$$(x, y, z = x^3 + xy) \overset{\text{squeeze}}{\longrightarrow} (sx, y, x^3 + xy)$$

The squeezing increases as s decreases from 1 to 0. The (y, z) plane $(0, y, x^3 + xy)$ is mapped back to the starting point $(x, y, 0)$ by an affine transformation. This neither stretches nor squeezes: It preserves Euclidean distances. The general affine transformation consists of a rotation and a translation. The result can be expressed

$$\begin{bmatrix} x \\ y \end{bmatrix}' = \begin{bmatrix} a_1 \\ a_2 \end{bmatrix} + \begin{bmatrix} \cos\theta & -\sin\theta \\ \sin\theta & \cos\theta \end{bmatrix} \begin{bmatrix} y \\ x^3 + xy \end{bmatrix} \tag{11.6}$$

Each choice of the angle θ gives a cusp return map, depending on two control parameters which are translations. For example, by choosing $\theta = \pi$ we obtain the map

$$\begin{aligned} x' &= a_1 - y \\ y' &= a_2 - (x^3 + xy) \end{aligned} \tag{11.7}$$

This can be compared to the return map based on the previous singularity A_2: $x' = a - x^2$, which depends on only one coordinate and one control parameter.

11.4.2 Structure in the Control Plane

The logistic map $x' = a - x^2$ is studied most expediently by constructing its bifurcation diagram. This is a plot of the state variable x against the control parameter a. Bifurcation diagrams for the cusp family are more complicated. There are two state variables and two (if we fix θ) control parameters. Bifurcation diagrams in this case involve plots of x or y against either a_1 or a_2 or some linear combinations of the two, or more generally, some path in the control parameter space. There are an infinite number of such paths, making this procedure problematic.

Instead, it is more useful to partition the control parameter plane into regions describing different types of behavior. Such a partition is presented in Fig. 11.6. This diagram at least has the virtue of describing what to expect along any path in phase space.

An attractor exists in a bounded domain in the neighborhood of the origin only for the control parameters within the cusp-shaped region shown in Fig. 11.6. The bounding fold curves describe saddle-node and/or period-doubling bifurcation curves. The region in which a stable period-1 orbit exists is bounded on the right by a Hopf bifurcation curve. On crossing this curve it is possible to encounter invariant tori and

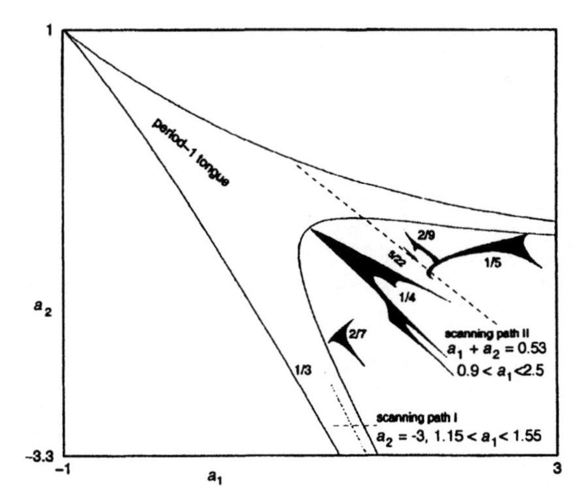

Fig. 11.6 A bounded attractor exists only within the cusp shaped region in the control parameter plane. Attached to the primary Hopf bifurcation curve are secondary tongues. Secondary Hopf bifurcation curves anchor tertiary tongues. The map studied here is (11.6) with $\theta = 0$.

mode-locked tongues. Several low-period tongues are shown in this figure, decreasing from $\frac{2}{7}$ down to $\frac{1}{5}$ along the primary Hopf bifurcation curve. These secondary tongues (counting the stable period-1 region as the primary tongue) are bounded on two sides by fold curves for saddle-node or period-doubling bifurcations. Their third sides are bounded by either secondary Hopf bifurcation curves, period-doubling curves, or a combination of both (cf. the $\frac{1}{4}$ tongue). Higher-order tongues occur. Eventually, a crisis boundary is crossed on the right, and bounded attractors disappear.

With such a partition of the control parameter plane, events observed in a bifurcation diagram are less obscure. Along the path $a_1 + a_2 = 0.53$ shown in Fig. 11.6 a number of different attractors are observed. Eight of these are shown in Fig. 11.7. The path passes through the primary Hopf bifurcation curve, near the $\frac{2}{9}$ tongue, and through the $\frac{5}{22}$ and $\frac{1}{5}$ tongues. The attractors should exhibit properties associated with these tongues. The control parameter values, and a brief description of eight of the attractors encountered along this path, are presented in Table 11.1.

11.4.3 Comparison with the Fold

The fold and the cusp map exhibit both similarities and differences. For both there is always one attractor, the point at infinity. For both, there is only a small, bounded region in control parameter space that supports a bounded attractor. For the fold $x' = a - x^2$ this region is bounded by the two points $a = -\frac{1}{4}$ on the left and $a = 2$ on

Table 11.1 Attractors encountered along the path $a_1 + a_2 = 0.53$ from $(1.50, -0.97)$ to $(2.05, -1.52)$ shown in Fig. 11.7

	a_1	a_2	Description
a	1.500	−0.970	Invariant torus
b	1.750	−1.220	Distorted torus
c	1.818	−1.288	Period 22 orbit
d	1.839	−1.309	22 invariant tori
e	1.850	−1.320	Annular torus
f	1.900	−1.370	Chaos
g	1.980	−1.450	Five chaotic islands
h	2.050	−1.520	Chaos

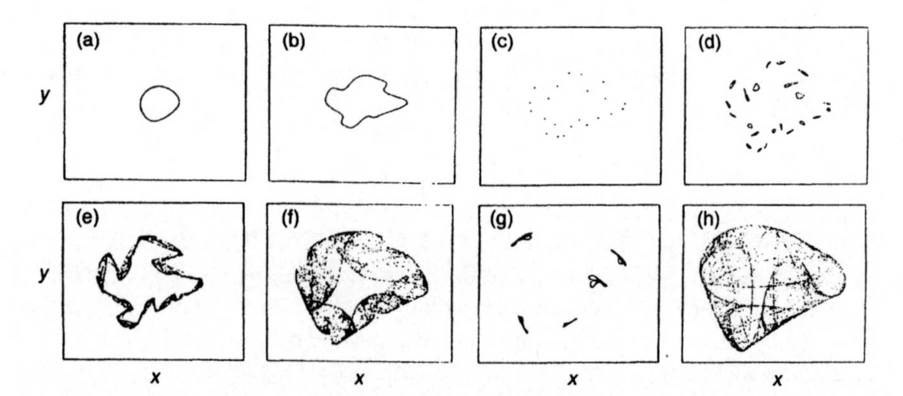

Fig. 11.7 Eight attractors encountered along the path $a_1 + a_2 = 0.53$ in Fig. 11.6. The control parameter values at which they are encountered are provided in Table 11.1. This table also contains a brief description of each attractor.

the right. For the cusp this region is bounded on the left by a saddle-node curve and a period-doubling curve. The right-hand boundary has not been computed analytically.

For the fold map $x \to x' = a - x^2$ control parameter values in the range $-\frac{1}{4} < a < \frac{3}{4}$ support a stable period-1 orbit. This is created in a saddle-node bifurcation on crossing through $a = -\frac{1}{4}$. On transiting control parameter space from $\frac{3}{4}$ to 2 there is a primary period-doubling cascade into chaos, followed by higher-period windows, each period doubling away to chaos ("boxes within boxes"). Finally, there is a crisis that destroys the bounded attractor. Coexisting stable attractors do not exist.

For the cusp, a large triangular open region in the control parameter space supports a period-1 attractor. This region is entered from the left by crossing one of the fold curves emanating from the cusp. One fold curve describes saddle-node bifurcations. This curve separates the regions in control space with one and three real fixed points. The other fold curve describes the transition from an unstable period-2 to a stable period-1 orbit. The right-hand boundary for the stable period-1 attractor is the primary Hopf bifurcation curve. Beyond this boundary there is no longer a stable period-1 orbit, but a variety of other behavior is encountered.

Hopf–Arnol'd resonance tongues p/q $(0 < p/q < \frac{1}{2})$ are anchored to the primary Hopf bifurcation curve. The $\frac{1}{3}$ strong resonance tongue is not. It is encountered along scanning path 1 shown in Fig. 11.6. The domain of the stable period-1 orbit is considered as the primary 0/1 Hopf–Arnol'd resonance tongue. The resonance tongues are typically triangular shaped regions in the control parameter space. Secondary tongues are attached to the primary Hopf bifurcation curve at appropriate points by fold lines that enclose an angle that approaches zero with an asymptotic $(q-2)/2$ power law dependence. Secondary tongues are terminated by secondary Hopf bifurcation curves (and also sometimes by period-doubling cascades). Tertiary tongues are attached to secondary Hopf bifurcation curves, and so on. This sequence of higher-order bifurcations (tongues on tongues) is analogous to, but more complicated than, the complex behavior encountered in the fold singularity (boxes within boxes) as the control parameter is scanned. In addition, the cusp map supports simultaneously coexisting stable attractors, whereas the fold does not.

11.5 NONLOCAL SINGULARITIES

The irreducible singularities that can typically occur in mappings of the plane to itself are the fold and the cusp. Two fold curves emanate from each cusp. It is to be expected that multiple cusps and/or folds can occur in mappings of $R^2 \rightarrow R^2$. In the following two subsections we explore some of the possibilities that can occur when multiple nonlocal cusps are present and when both cusps and folds are present as singularities in the return map.

11.5.1 Multiple Cusps

Multiple cusps can occur in several different ways. We begin with two cusps, then treat three and four.

11.5.1.1 Two Cusps In Fig. 11.8 we show ways that a pair of cusps can occur in mappings $R^2 \rightarrow R^2$. In both cases the cusps are independent. In Fig. 11.8(a) the cusp axes are more or less parallel. The fold lines typically cross transversally. They separate the image plane into open regions with one, three, and five preimages. In Fig. 11.8(b) the two cusp axes are more or less orthogonal, and the fold lines divide the image plane into regions with one, three, and nine preimages. It is also possible to find one cusp inside a second. In this case the inner cusp may deform either of the three sheets of the outer cusp, as shown in Fig. 11.8(c).

When two cusps occur in a return map, they may be uncorrelated, as shown in Fig. 11.8, or they may be correlated. In the latter case they are face to face and share their

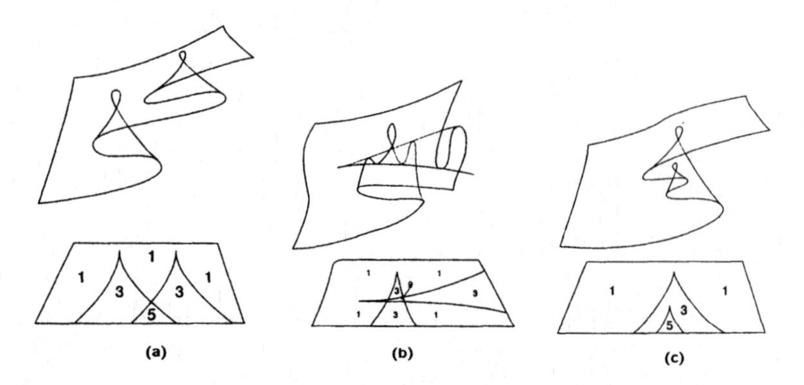

Fig. 11.8 Two nonlocal cusps may occur in different ways in mappings $R^2 \rightarrow R^2$: (a) their axes are parallel; (b) their axes are perpendicular; (c) one is inside the other. In each case the fold lines divide the image plane into open regions with different numbers of preimages.

fold curves. This is illustrated in Fig. 11.9. Here the center of the plane is stretched. Part of the stretched region is pulled out above the plane in one direction, part is pushed down below the plane and in the opposite direction. This deformed plane is projected down to R^2, creating the "smile" singularity. The fold curves divide the image plane into the exterior region on which the map is 1:1 and the interior region where each point has three preimages.

11.5.1.2 Three Cusps Three nonlocal cusps can occur in a return map $R^2 \rightarrow R^2$ as variations on themes shown in Figs. 11.8 and 11.9. They can all be uncorrelated and appear as more complicated forms of those shown in Fig. 11.8. Two may be correlated as in Fig. 11.9, and the third uncorrelated. It is also possible that all three are correlated, each cusp sharing one fold line with each of the others. A multisheeted two-dimensional manifold whose projection into R^2 contains three cusps, each sharing two fold curves, can exist only in R^4. In R^3 such a manifold must exhibit self-intersections. As a result, if the dynamics exists in $R^3 \times S^1$, three correlated cusps will never be seen. It can occur only in higher-dimensional dynamical systems.

Remark: This is the first instance in which the observation of a particular nonlocal singularity has significant implications on the dimensionality of the underlying dynamical system, in particular, on the necessary number of stable directions.

11.5.1.3 Four Cusps Four nonlocal cusps can occur in many ways. They can all be uncorrelated. One or even two pairs can be correlated. In all these cases the fold lines divide the image plane into open regions with more or fewer preimages (odd numbers all). However, it is also possible for all four to be correlated, as shown in Fig. 11.10. The equilibrium surface of Zeeman's "catastrophe machine" exhibits this geometry [183].

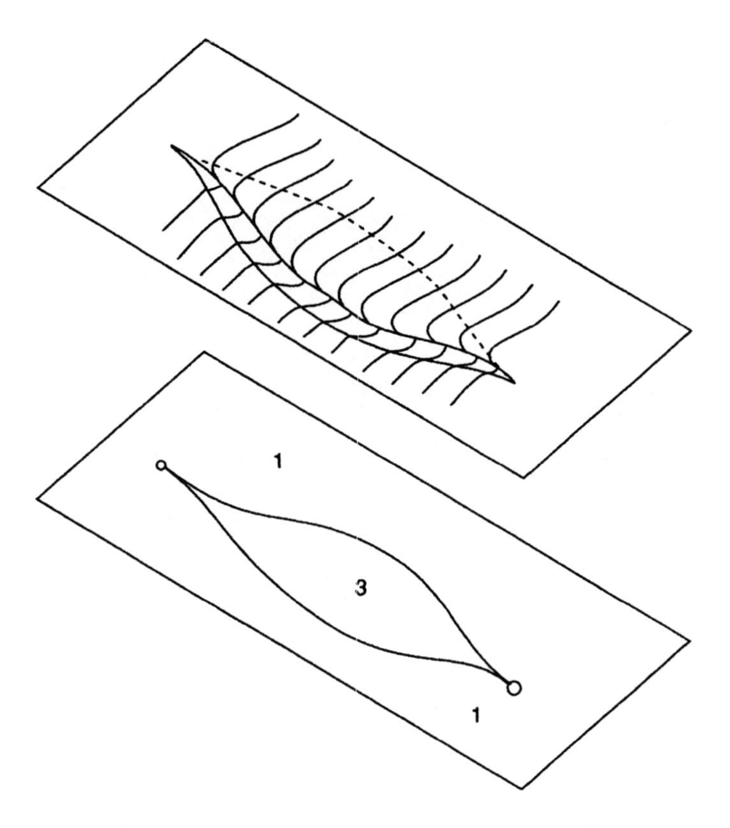

Fig. 11.9 These two correlated cusps share their fold lines. The exterior and interior regions have one and three preimages.

11.5.2 Cusps and Folds

Strange attractors in $R^3 \times S^1$ can also be generated by two successive mappings of the plane to itself, which are simple transverse folds. The mechanism has been illustrated in Fig. 11.3. In this case it even appears that the double fold naturally generates a useful symbolic dynamics.

This, regrettably, seems not to be the case. The double fold is nongeneric [180]. It generates a square corner, which perturbs away. Under arbitrary perturbation, the square corner perturbs to a cusp and a fold line. One possibility is shown in Fig. 11.11. A simple canonical form for the generic double fold is

$$\begin{array}{ccccc} x & \rightarrow & x' & = & x^2 + ay \\ y & \rightarrow & y' & = & y^2 + bx \end{array} \qquad (11.8)$$

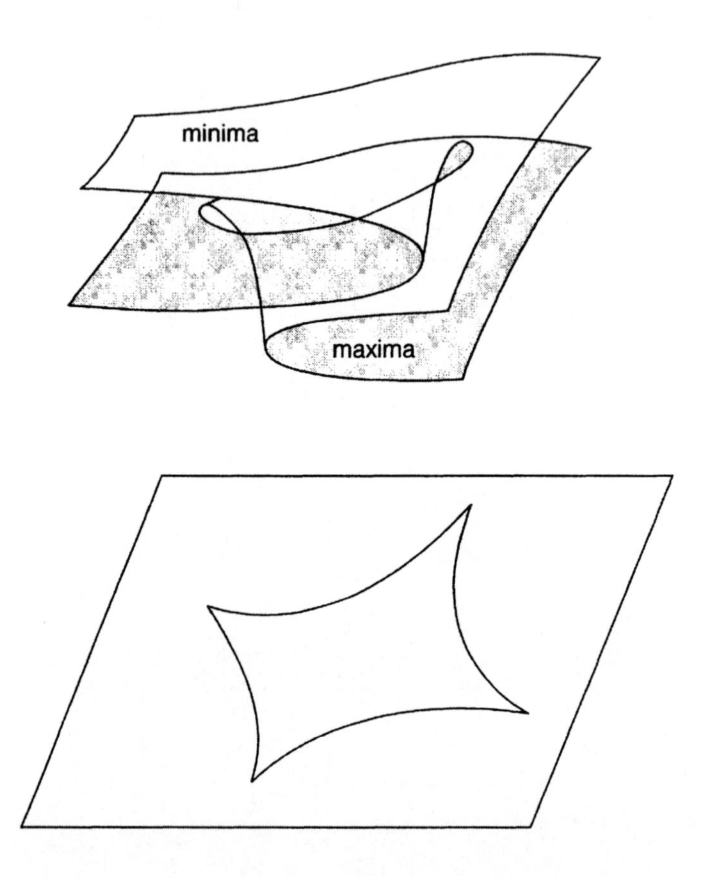

Fig. 11.10 Each cusp shares fold lines with its two nearest neighbors. The multisheeted two-dimensional preimage can exist in $R^2 \times S^1$ without self-intersections.

Outside the smooth fold line there are no preimages. Between the smooth fold and the pair of folds emanating from the cusp there are two preimages. Inside the folds originating from the cusp, there are four preimages.

11.6 GLOBAL BOUNDARY CONDITIONS

Whitney's theorem provides information about the local nature of singularities in mappings of n-dimensional manifolds to themselves, $M^n \to M^n$, for n =1 and 2. We have exploited this to describe the nature of flows in R^3 and R^4 (more accurately, in $R^2 \times S^1$ and $R^3 \times S^1$). Placing local singularities together in a nonlocal way enriches the spectrum of possibilities that can be encountered. The global nature

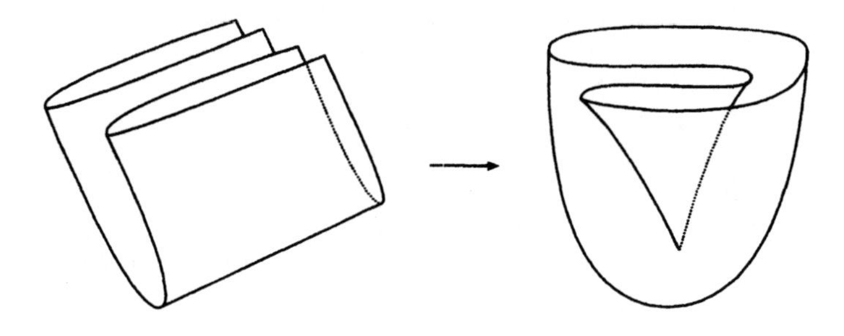

Fig. 11.11 Under perturbation, double folds with degenerate corners perturb to a cusp and its fold curves, and another fold.

of the manifold M^n also enriches the spectrum of possibilities while placing some constraints on the way they can be distributed.

11.6.1 R^1 and S^1 in Three-Dimensional Flows

In three-dimensional flows embedded in a torus $R^2 \times S^1$ only local folds can be encountered. Their nonlocal distribution accounts for the rich variety of behavior we have already seen. Part of this rich variety occurs because the one-dimensional space mapped into itself under the flow can have two distinct topologies. It may be an interval, more properly a connected bounded subset of R^1, or it may be a circle $S^1 = T^1$. Only folds can occur in mappings of these one-dimensional spaces to themselves. In the former case any number of folds can occur. In the latter case, folds must be paired. There are also significant constraints, and differences in the constraints, on contiguous branches of the branched manifold compatible with the return map.

11.6.2 Compact Connected Two-Dimensional Domains

The simplest compact two-dimensional space is the disk, $D^2 \subset R^2$. It is simply connected and topologically equivalent to the square shown in Fig. 11.12(a). Two more interesting compact connected two dimensional spaces are the cylinder $I^1 \times S^1$ [Fig. 11.12(b)] (I^1 an interval) and the Möbius strip [Fig. 11.12(c)]. These are obtained from the square by identifying opposite edges, as indicated. The edges are labeled with arrows. When the edges are identified, the arrows must point in the same direction.

There are many compact connected two-dimensional manifolds [200]. They can be obtained from a bounded domain $D^2 \subset R^2$ in a similar way. The torus T^2 is obtained by identifying opposite edges with parallel arrows. The Klein bottle K^2 is obtained by identifying one pair of opposite edges with parallel arrows and the other pair with antiparallel arrows. The sphere S^2 is obtained from the cylinder by identifying the upper edge with a single point, the north pole, and similarly for the lower edge. This identification is indicated by the * in Fig. 11.12. The real projective plane P^2 is obtained from the sphere by identifying opposite points on the surface. Equivalently, it is obtained from a bounded domain in R^2 by identifying opposite edges with reverse senses.

The Möbius strip, the Klein bottle, and the real projective plane are nonorientable. The Möbius strip can be embedded in R^3 while the other two nonorientable surfaces can be embedded in R^4. The sphere and real projective plane have positive curvature, the other spaces are flat.

11.6.3 Singularities in These Domains

The global geometry of the two-dimensional domain and nonlocal singularities interact in a significant way. This interaction is much richer than in the one-dimensional case, with only R^1 and S^1 to impose conditions on nonlocal folds.

Without going into details, we show two possibilities for the distribution of cusp singularities on the cylinder in Fig. 11.13.

11.6.4 Compact Connected Two-Dimensional Domains

The compact connected two dimensional manifolds presented in Fig. 11.12 by no means exhaust the list of existing compact connected two-dimensional manifolds. Every such manifold without boundary is one of the following:

$$\begin{array}{ccc} & S^2 & \\ T^2 & & P^2 \\ T^2\#T^2 & & P^2\#P^2 \\ T^2\#T^2\#T^2 & & P^2\#P^2\#P^2 \\ \vdots & & \vdots \end{array}$$

The symbol # is to be read "connected sum." The connected sum of two and three tori are illustrated in Fig. 11.14. Connected sums are constructed by removing a disk from each of the two surfaces and gluing the surfaces together where the disks were removed. The connected sum of n tori has n holes. The Klein bottle is hiding in this list as $K^2 = (P^2)^{\#2}$. The Euler characteristics for these surfaces are

$$\begin{array}{lcl} \chi(S^2) & = & +2 \\ \chi(P^2) & = & +1 \\ \chi(T^2) & = & 0 \\ \chi(P^2\#P^2 = K^2) & = & 0 \\ \chi((T^2)^{\#n}) & = & 2-2n \\ \chi((P^2)^{\#n}) & = & 2-n \end{array}$$

As a result, S^2 and P^2 are elliptic ($\chi > 0$), T^2 and K^2 are flat (parabolic, Euclidean, $\chi = 0$), and the rest of these surfaces are hyperbolic ($\chi < 0$).

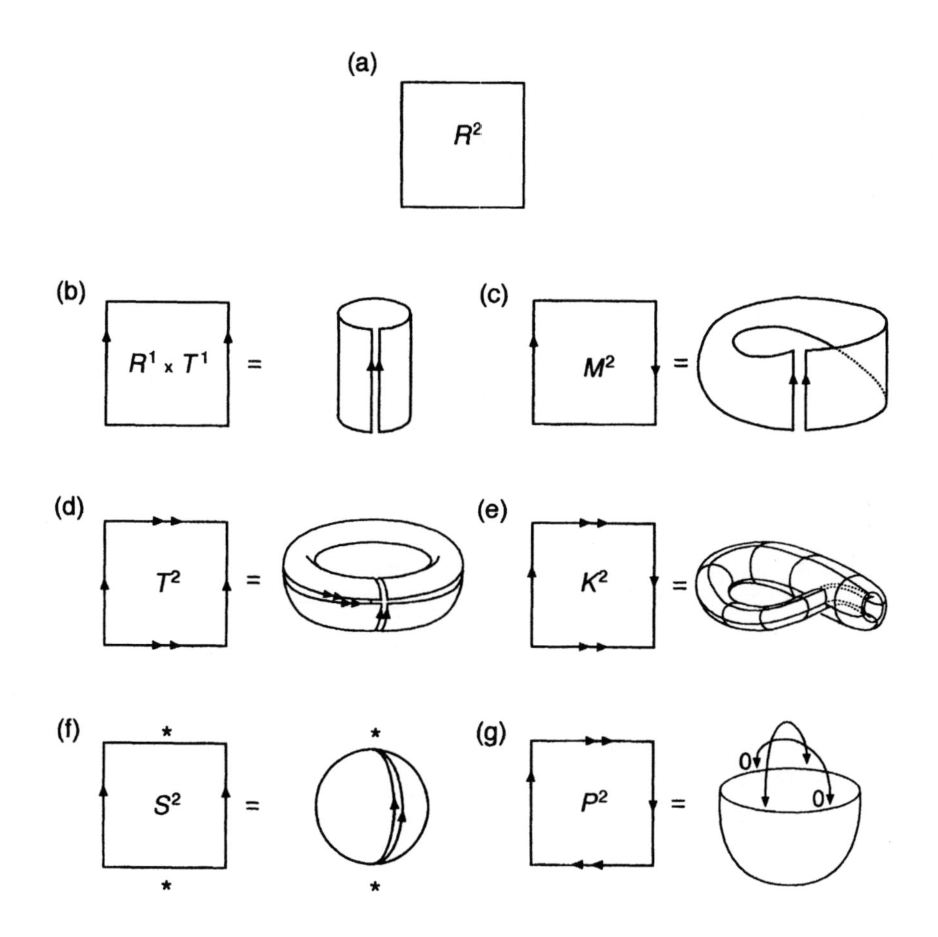

Fig. 11.12 Seven compact connected two-dimensional manifolds. (a) A bounded subset $D^2 \subset R^2$ of the plane is equivalent to a square. (b) The cylinder $R^1 \times S^1$ and (c) the Möbius strip M^2 are obtained by identifying opposite edges of the square with alignment indicated by the arrows. (d) The torus T^2 and (e) the Klein bottle K^2 are obtained by identifying the top and bottom edges of the cylinder in two different orders. (f) The sphere S^2 is obtained from the cylinder by identifying all points (*) in the top edge with one point (north pole), and similarly for the bottom edge. (g) The real projective plane P^2 is obtained by identifying opposite points on the circumference of a hemisphere, or equivalently, opposite points on the sphere S^2. The spaces M^2, K^2, and P^2 are nonorientable. The spaces K^2 and P^2 cannot be embedded in R^3.

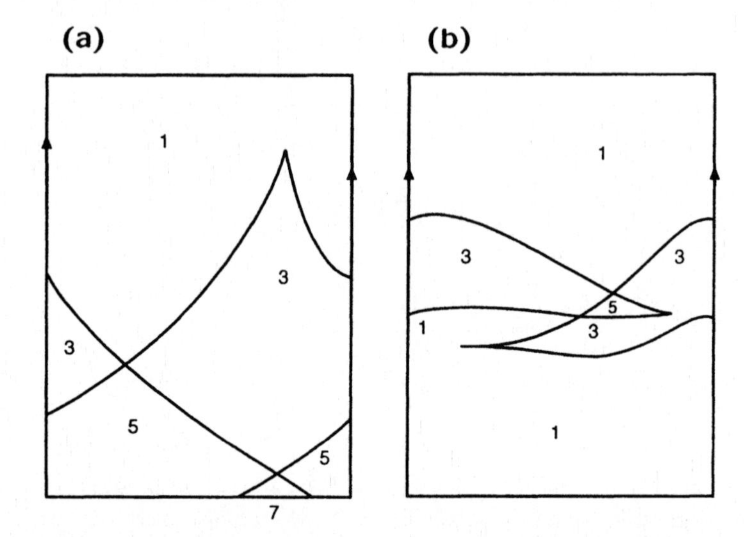

Fig. 11.13 (a) The fold lines from a single cusp with axis parallel to the cylinder axis can wrap around the cylinder to intersect and cause additional folding. (b) Two correlated nonlocal cusps with axes transverse to the cylinder axis can wrap around as shown.

Compact connected two-dimensional domains with boundary are obtained from the surfaces without boundary by removing one or more disks. For example, the cylinder is obtained by removing two disks from the sphere and the Möbius strip is obtained by removing one disk from the real projective plane.

Constructing the interaction between global boundary conditions and nonlocal cusp bifurcations should keep us occupied for some time to come.

11.7 SUMMARY

The classification of higher-dimensional strange attractors is discrete under some conditions and not discrete under others.

The classification is aided by two projections. The first is a projection from the original phase space into the important submanifold in which all the dynamics takes place. In effect, the flow relaxes to an inertial manifold. Once in the manifold, it remains within this manifold forever. The dimension, but not the global topology, of this manifold is determined by the vanishing of a weighted sum involving the Lyapunov exponents and partial dimensions. It is the first value of r for which $D_r = \sum_{i=1}^{r} \epsilon_i \lambda_i = 0$. Here $r = k_1 + k_2 + 1$, where k_1 is the number of unstable directions and k_2 is the number of stable directions in the projected flow.

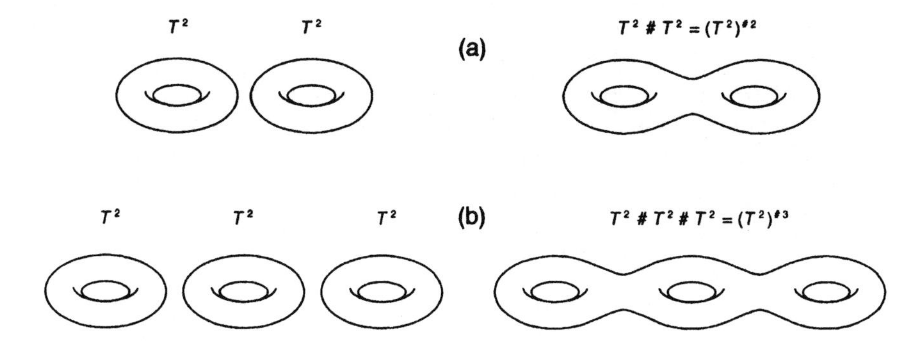

Fig. 11.14 (a) The connected sum of two tori has two holes. (b) The connected sum of three tori has three holes.

The second projection is along the remaining stable directions. It is induced by the Birman–Williams identification. The flow in the inertial manifold is projected down to a semiflow on a $(k_1 + 1)$-dimensional manifold with singularities.

In the case that the inertial manifold has the topology of a torus $\mathcal{IM}^{k_1+k_2+1} \sim R^{k_1+k_2} \times S^1$, the intersection of a Poincaré section with this manifold with singularities is a manifold of dimension k_1. The semiflow induces a map of this space to itself. This map must possess singularities. It can therefore be classified. The only singularities that can occur are those that can be constructed by continuous deformation of a k_1-dimensional manifold within a manifold of dimension $k_1 + k_2$. In other words, there are k_2 additional directions in which the deformation (folding) can occur. If k_2 is small, the classification can be discrete.

For $k_2 = 1$, singularities of cuspoid type can occur for $k_1 \geq 1$. The local fold singularity A_2 occurs for $k_1 = 1$. The fold and cusp A_3 occur for $k_1 = 2$. Higher singularities A_m occur for $k_1 \geq m - 1$. These all lead to discretely classifiable strange attractors.

For $k_2 = 2$, singularities of umbilic type can occur for $k_1 \geq 4$. Singularities of type D_m occur for $k_1 \geq m, m \geq 4$. In particular, the lowest-dimensional strange attractor, which has two folding dimensions and which is discretely classifiable, occurs with a D_4 singularity. This is allowed for $k_2 = 2$, $k_1 \geq 4$, so that the dynamical system must be of dimension at least $4 + 2 + 1 = 7$.

The classification of singularities of mappings is local. This means that they can be given canonical forms in (locally) Euclidean coordinate systems. Multiple singularities can occur in nonlocal settings. In this case the several singularities can be arranged in many different ways. In addition, the space on which the mapping takes place may be locally Euclidean, but it is often not globally Euclidean. We have seen this for the first time in van der Pol dynamics. For flows in $R^4 \simeq R^3 \times S^1$ with $k_1 = 2$ and $k_2 = 1$ there is an exciting spectrum of possibilities. Multiple cusps and folds can occur arranged in a variety of ways, some allowed, some forbidden (e.g., three cusps face to face in certain geometries). The two-dimensional space, which is the intersection of the manifold with singularities and a Poincaré section, may be a domain of R^2 or it may have a more exciting global topology. The interaction of this global topology with the nonlocal arrangement of the singularities is much richer than in the simpler case of three-dimensional flows with $k_1 = k_2 = 1$, only one type of local singularity, multiple A_2 nonlocal singularities, and only two types of global boundary conditions leading to topologies of type R^1 and S^1 in the Poincaré section R^2.

12

Program for Dynamical Systems Theory

Suppose that you, the reader, are asked to teach a course in dynamical systems theory in 25 years. What would you teach? How would you teach it?

The standard response to a request to teach is to reach into the bookshelf above your desk and pull down the "Goldstein" or "Landau and Lifshitz" of the field. Then open the book and follow the table of contents. But what would be in the table of contents of the "standard book" on dynamical systems theory?

This is a particularly delicate question, as (1) 25 years is not that long a time for the development of a new field, and (2) the questions currently being addressed are typically in the nature of small problems, not big picture questions. One way to look at this question is as follows. Two other branches of mathematics with close ties to physics already exist and are reasonably mature: these are Lie group theory and catastrophe theory, or its more all-encompassing big brother, singularity theory.

Lie group theory grew from its infancy, in 1880, to maturity as a still-growing branch of mathematics by about 1960, when it was incorporated into physics in a fundamental way. Singularity theory grew from infancy in 1955 to a mature field by about the early 1980s. The maturation of the second field occurred more quickly in part for the following reason. The first time through (Lie groups), the proper questions to ask weren't known, let alone the answers. The second time through (catastrophes, singularities) there was a better feeling for the proper questions to ask, if only by analogy to the questions asked of Lie groups. And it must certainly be admitted that the most important part of constructing an important new result lies in framing the proper questions.

By this (admittedly hopeful) line of thinking, it should be possible for dynamical systems theory to mature relatively quickly if we use whatever guidance is provided by the development of the two older fields—and guidance they do provide.

This chapter is organized as follows. Thirteen general questions are raised in the following 13 sections. These questions were useful in exploring the structure and properties of Lie group theory and/or singularity theory. The first paragraph of each section points in the briefest of ways to the value of the general question in Lie group theory and singularity theory. In the remainder of the section we take it for granted that the general question is important to address in the development of dynamical systems theory.

This chapter is a pared-down version of a longer work of the same title that can be found on the authors' Web pages. The 13 basic questions are summarized in Table 12.1. For each of the three fields of interest, a brief phrase describes how the question has been useful in influencing the development of the field.

12.1 REDUCTION OF DIMENSION

The complexity of a problem is reduced dramatically if its dimension (number of variables) can be reduced. Both Lie group theory and catastrophe theory provide effective tools for this process [37,68].

A first step in extracting qualitative information from a dynamical system involves reducing its complexity. One way to do this involves reducing its dimension. We come at the dimension reduction problem from two directions: from data to theory; and from theory to predictions. In the first case, suppose that we have data taken from a physical experiment, such as a fluid, laser, or chemical experiment. How do we process the data so that they can be modeled by a set of ordinary differential equations? The most serious constraint is that the most important theorem for ODEs is satisfied everywhere: This is the existence and uniqueness theorem. This problem is the embedding problem. We usually seek the simplest possible embedding; this is the one of minimal dimension. Although there are many embedding theorems,

Table 12.1 These procedures aid the development of three related fields

	Question	Lie Group Theory	Singularity Theory	Dynamical Systems Theory
1.	Reduction of dimension	Reduction to canonical forms	Isolation of important directions	Absorbing manifold $\mathcal{A}$ Inertial manifold $\mathcal{IM}$ Branched manifold $\mathcal{BM}$
2.	Equivalence	Isomorphism, homomorphism	Reduction to canonical forms: • Global • Local	Diffeomorphisms: • Global • Local
3.	Structure theory	Reducible Fully reducible Irreducible Young partitions for representations	State variables Unfolding variables Young partitions for singularities	Reducible Fully reducible Irreducible Young partitions for singularities
4.	Germs	Lie algebras	Degenerate monomials Singular return maps	Branched manifolds Singular return maps
5.	Unfolding	Exponentiation to Lie groups	Universal perturbations	Unfolding of branched manifolds Unfolding of return maps Forcing diagrams
6.	Paths	Homotopy	Routes through control parameter space	Routes to chaos
7.	Rank	Independent invariants Commuting operators	Number of essential variables	Rank of singular return map Number of stable directions
8.	Complex extensions	Classification of complex Lie algebras	Classification of complex germs	Complex fixed-point distributions Singular return maps
9.	Coxeter–Dynkin diagrams	Classification of simple Lie algebras over $\mathbb{C}$	Classification of simple singularities and catastrophes	Simple fixed-point distributions Classification of singular return maps
10.	Real forms	Projection from complex to real Lie algebras Analytic continuation of real Lie algebras	Projection from complex to real germs Analytic continuation of germs	Projection from complex to real fixed-point distributions Analytic continuation of flows Real singular return maps
11.	Local vs. global	Lie algebras $\rightarrow$ universal covering groups	Germs and universal unfoldings Multiple nonlocal singularities	Multiple nonlocal singularities
12.	Cover–image relations	Universal covering groups Homomorphic images (Cartan's theorem)	Riemann sheets Covering singularities	Cover and image dynamical systems Universal image dynamical systems
13.	Symmetry-breaking & restoration	Group-subgroup relations Grand unified theories	Symmetry-restricted singularities	Entrainment and synchronization

these are usually useless, since in applications with real data we almost always find embeddings of much smaller dimension than guaranteed by mathematical theorems.

Going in the other direction, we ask if it is possible to reduce the dimension of dynamical systems equations. We indicate three steps that would be useful. The first does not actually result in reducing the dimension.

12.1.1 Absorbing Manifold

Suppose that we have defined a dynamical system $\dot{x} = f(x)$, $x \in R^N$. Is it possible to find an open set, $\mathcal{A} \subset R^N$, so that whenever the flow enters $\mathcal{A}$, it remains in $\mathcal{A}$? Such a set is called an *absorbing set* or *absorbing manifold*. Lorenz constructed an absorbing set for the truncated version of the Navier–Stokes equations that he studied. This set is an ellipsoid. Lorenz showed that the flow field intersects the boundary of $\mathcal{A}$, $\partial\mathcal{A}$, in the inward direction everywhere. It would be very useful to have an algorithm for constructing an absorbing set $\mathcal{A}$, or its trapping boundary, $\partial\mathcal{A}$, for an arbitrary dynamical system D: $\dot{x} = f(x)$, $x \in R^N$.

12.1.2 Inertial Manifold

An inertial manifold $\mathcal{IM}$ is an n-dimensional manifold embedded in R^N with the following properties:

1. $\mathcal{IM}$ is invariant under the flow. An initial condition in $\mathcal{IM}$ evolves in $\mathcal{IM}$.
2. The flow approaches the inertial manifold $\mathcal{IM}$ exponentially fast.
3. The dynamics in the full space R^N is the same as the dynamics in the inertial manifold $\mathcal{IM}$.

The hope is that $n < N$, in fact, that $n \ll N$.

It would be useful to have an algorithm for constructing an inertial manifold for a dynamical system, or even determining its dimension. One such method was proposed in Section 11.2. It depends on the spectrum of Lyapunov exponents and partial dimensions. Briefly, there exist k_1 unstable directions with positive Lyapunov exponents λ_i $(1 \le i \le k_1)$, one flow direction with $i = k_1 + 1$, and k_2 negative Lyapunov exponents λ_j $(k_1 + 1 < j \le k_1 + k_2 + 1)$, with the property that the sum $D_r = \sum_{i=1}^{r} \lambda_i \epsilon_i = 0$ for $r \ge k_1 + k_2 + 1$.

According to this conjecture, the fastest contraction rate $\lambda_{k_1+k_2+1}$ within the strange invariant set $\mathcal{SA}$ (in the inertial manifold $\mathcal{I}$) is slower than the slowest relaxation rate $\lambda_{k_1+k_2+2}$ toward the inertial manifold from outside that manifold. A more careful treatment involving local Lyapunov exponents is presented by Gilmore and Pei [154].

12.1.3 Branched Manifolds

A second projection within the inertial manifold is possible. By identifying all points $x, y \in \mathcal{IM}$ with the same future using the Birman–Williams identification criterion,

$$x \sim y \qquad \text{if} \qquad \lim |x(t) - y(t)| \overset{t\to\infty}{\longrightarrow} 0$$

we project the flow down along the remaining (weakly) stable directions onto a $(k_1 + 1)$-dimensional surface. This is a manifold almost everywhere. The existence of singularities from the projection prevent the $(k_1 + 1)$-dimensional set from being a manifold. For the original Birman–Williams case in R^3, the two types of singularities are zero- and one-dimensional. The zero-dimensional singularities are splitting points that describe stretching processes. The one-dimensional singularities are branch lines that describe squeezing processes. In higher dimensions there is no such clear dichotomy between the dimensions of singularities and the stretching and squeezing mechanisms.

12.2 EQUIVALENCE

It is extremely useful to be able to tell whether two apparently different systems really are different or are the same but appear different simply because they are viewed from two different coordinate systems. There are standard algorithms both within Lie group theory and singularity theory for addressing this question. The simplest method for showing equivalence involves transforming each structure (group, singularity) to canonical form.

12.2.1 Diffeomorphisms

A diffeomorphism is a mapping of one dynamical system to another which

1. Is 1:1

2. Preserves all differential properties

Two n-dimensional dynamical systems, $\dot{x} = f(x)$ and $\dot{y} = g(y)$, are diffeomorphic if there is a smooth, invertible, differentiable change of variables $y = y(x)$, with the property

$$\frac{dy}{dt} = \frac{\partial y}{\partial x}\frac{dx}{dt} = \frac{\partial y}{\partial x} f(x) = \frac{\partial y}{\partial x} f(x(y)) = g(y)$$

No algorithms exist for finding and constructing such transformations $y = y(x)$. If the transformation is $n \to 1$ $(n > 1)$, it may be locally 1:1 but cannot be globally 1:1. Such a transformation is a local diffeomorphism. These occur in a natural way when the dynamical system has a discrete symmetry group (cf. Chapter 10).

It would be very nice if there were a short, or even discrete, list of canonical forms for dynamical systems. These is no such list. Nor is there even a thought of constructing an algorithm for this, except for the nongeneric case of symplectic flows and transformations to action-angle variables.

12.3 STRUCTURE THEORY

The structure of a Lie group or algebra provides an important piece of information for its eventual classification. For many Lie groups, the structure and properties of its representations are determined by Young partitions $(i_1, i_2, \ldots, i_n)$, with $i_1 \geq i_2 \geq \cdots \geq 0$. The structure and properties of singularities are also determined in part by Young partitions [180].

12.3.1 Reducibility of Dynamical Systems

A dynamical system can be defined as reducible, fully reducible, or irreducible in exact analogy with a Lie group or algebra or their representations. The essential idea was presented in Section 3.4.3. It involves searching for a diffeomorphism

$$(x_1, \ldots x_n) \rightarrow (y_1, \ldots, y_k; z_1, \ldots, z_l) \qquad k + l = n$$

which transforms the dynamical system $\dot{x} = f(x)$ to a simpler form, as shown:

$\dot{y} = g(y; z)$	$\dot{y} = g(y; -)$	
		$\dot{x} = f(x)$
$\dot{z} = h(-; z)$	$\dot{z} = h(-; z)$	
Reducible	Fully reducible	Irreducible

In the form described as reducible, the equation $\dot{z} = h(-; z)$ means that the equations for the motion of the z coordinate are independent of the y coordinates of the point in phase space. However, evolution of the y coordinates depends on *both* the y and z coordinates of a point. A useful approach for studying this dynamical system involves two steps:

1. Solve for the motion of the z coordinates by integrating $\dot{z} = h(-; z)$.
2. Substitute the solution $z = z(t)$ into the other equation, $\dot{y} = g(y, z(t)) = g(y, t)$, and integrate the nonautonomous equation of k variables.

If reducibility simplifies the study of dynamical systems, full reducibility produces a further simplification. That is, the two subsystems $\dot{y} = g(y; -)$ and $\dot{z} = h(-; z)$ can be studied independently. Each subsystem can be investigated for further reducibility. This decomposition eventually terminates in an irreducible subsystem.

In the theory of Lie groups there is an algorithm for determining reducibility, complete reducibility, and irreducibility [68]. Further, irreducible Lie algebras and their groups are (discretely) classifiable—that is, there is a discrete set of simple Lie groups.

It would be very nice to have an algorithm for determining the reducibility, complete reducibility, or irreducibility of a dynamical system. We might remark that all periodically driven dynamical systems are reducible. The details of describing a two-dimensional periodically driven nonlinear oscillator as an autonomous four-dimensional dynamical system that is polynomial and autonomous have been presented in Section 10.6.2. The concept of reducibility crystallizes the idea of replacing periodic by chaotic driving.

The phase space for a periodically driven two-dimensional nonlinear oscillator is $D^2 \times S^1$ (D^2 is the two-dimensional disk). The Rössler (autonomous dynamical) system has an isomorphic phase space. It is likely that every system with a phase space $D^n \times S^1$ ($n \geq 2$) is reducible. If this is true, the topology of the phase space (inertial manifold $\mathcal{IM}$) may provide important information about the existence of diffeomorphisms that effect the reduction of the dynamical system.

12.4 GERMS

The concept of *germ* is simple. A germ is something small and simple but which contains within it all the complexity of some larger system. It is, as it were, a seed that can grow into something big and beautiful if handled properly. The germ for a Lie group is its Lie algebra. The germ for the fold and the cusp catastrophes are the monomials x^3 and x^4, respectively.

Germs can occur in at least two ways in dynamical systems theory. First, they describe the overall structure of a strange attractor $\mathcal{SA}$. Second, they describe the stretching and squeezing mechanisms which generate chaotic behavior.

12.4.1 Branched Manifolds

If we wish to define a germ for a dynamical system that generates chaotic behavior, this germ should encapsulate the properties of strange attractors in a nutshell. Since the only strange attractors we understand well exist in R^3, we should look for inspiration in this low-dimensional case.

We have learned that branched manifolds are accurate caricatures for low-dimensional ($d_L < 3$) strange attractors. In particular, branched manifolds summarize nicely all the information available about the type and organization of the unstable periodic orbits in the strange attractor and more generally about the stretching and squeezing mechanisms that generate chaos.

We therefore propose that branched manifolds be considered germs for dynamical systems that generate chaotic behavior. This implies that we must be able to recover all the important properties of the strange attractor by some appropriate form of perturbation theory on this germ.

12.4.2 Singular Return Maps

This is an appropriate point to reveal that singularity theory can contribute more to our understanding of dynamical systems theory than simply providing an analogy. We assume that it is possible to project a strange attractor to a branched manifold, as described in Section 12.1.3. For simplicity, we assume that the phase space (inertial manifold) is $R^n \times S^1$. We also assume that there are k_1 unstable directions and k_2 stable directions, with $n = k_1 + k_2$. The branched manifold that is the projected image of the strange attractor has dimension $k_1 + 1$.

We now choose a global Poincaré section $\mathcal{P}$ by choosing ϕ = constant ($\phi \in S^1$, $0 \leq \phi < 2\pi$). The intersection of the branched manifold with this Poincaré section has dimension k_1. We assume that this intersection is a simple subset of R^{k_1}, just to avoid complications. The first return map $\phi \to \phi + 2\pi$ is then a mapping $R^{k_1} \to R^{k_1}$. This return map must possess singularities. If it did not, the map would be invertible and would not generate entropy.

To understand the stretching and squeezing mechanisms that generate chaotic behavior it is useful to understand the singularities of mappings $R^{k_1} \to R^{k_1}$. This is more or less the case, but with some subtleties. We elaborate on these subtleties below.

The first subtlety is this. The (local) singularities of mappings $R^k \to R^k$ have been studied, and those of low dimension and low modulus are known [180]. However, all these mapping singularities are discontinuous, whereas the singularities that occur in the branched manifold are created continuously, as the coordinate in S^1 increases smoothly from ϕ to $\phi + 2\pi$ (cf. Figs. 11.4 and 11.5). As a result, we must determine the structure of singularities of mappings $R^{k_1} \to R^{k_1}$ that can be created by continuous deformation of R^{k_1} (or some k_1-dimensional manifold M^{k_1}) in the higher-dimensional space $R^{k_1+k_2}$. That is, the deformation can take place in k_2 additional "deformation directions."

12.5 UNFOLDING

Unfolding a germ is like watering a seed. If the germ contains in embryo all the information for the larger system, the appropriate care and feeding of the germ should result in its budding and flowering. Unfolding a Lie algebra is done by exponentiating it onto (all or part of) the corresponding Lie group. Unfolding the germ of a singularity is accomplished by adding a universal perturbation to it and then transforming the sum back to simple canonical form (cf. Fig. 9.1).

There are two different ways that the idea of unfolding can be applied to strange attractors. They have been described in Chapter 9 for dynamical systems in R^3. First, the branched manifold can be unfolded. This involves determining how previously visited branches can be removed from the branched manifold, or new branches can be added, subject to continuity conditions, as control parameters are varied. Second, as a flow confined to a specific branched manifold (e.g., the Smale horseshoe template) is unfolded, the spectrum of periodic orbits associated with the strange attractor will vary. Forcing diagrams (when they exist) are useful for describing this type of unfolding.

In higher dimensions it will be useful to describe unfoldings of the analogs of branched manifolds and analogs of return maps. We describe briefly how the latter might work. Consider a flow in a phase space $\mathcal{IM} = R^{k_1+k_2} \times S^1$. The flow has k_1 unstable and k_2 stable directions. Under the Birman–Williams projection of the inertial manifold $\mathcal{IM}$ down to the analog of the branched manifold $\mathcal{BM}$, the return map $R^{k_1} \rightarrow R^{k_1}$ generates a singularity that is created by continuous deformation of R^{k_1} in the larger space $R^{k_1+k_2}$. If we define x_i ($i = 1, 2, \ldots, l$) as the germ variables for the singularity, z_j ($j = l + 1, \ldots, k_1$) as the unfolding variables, and y_α ($\alpha = 1, 2, \ldots, k_2$) as the variables that identify the extra deformation directions, then the stretching and squeezing mechanisms can be summarized as follows:

$$\begin{bmatrix} x \\ z \\ y \end{bmatrix} \xrightarrow{\text{stretch}} \begin{bmatrix} x \\ z \\ y + rf(x;z) \end{bmatrix} \xrightarrow{\text{squeeze}} \begin{bmatrix} sx \\ z \\ y + f(x;z) \end{bmatrix} \xrightarrow{\text{affine}} \begin{bmatrix} x \\ z \\ y \end{bmatrix}'$$

where

$$\begin{array}{lll} x_i & i = 1, 2, \ldots, l & \text{Germ variables} \\ z_j & j = l + 1, \ldots, k_1 & \text{Unfolding variables} \\ y_\alpha & \alpha = 1, 2, \ldots, k_2 & \text{Deforming variables} \end{array}$$

In this process:

- The number of germ variables, l (rank of singularity), is equal to the length of the longest row of the Young partition $(i_1, i_2, \ldots, i_k)$, which describes the singularity:
$$l = \text{rank } (i_1, i_2, \ldots, i_k) = i_1$$
- The number of x and z variables is greater than or equal to the dimension of the singularity:
$$k_1 \geq \dim(i_1, i_2, \ldots, i_l)$$
- The number of deforming variables, k_2, is greater than or equal to the rank of the germ:
$$k_2 \geq l = i_1$$
- The affine transformation returns the last column vector to the first, introducing neither stretching nor squeezing. In short, it involves a rotation from $SO(k_1 + k_2)$, followed by a translation.

This procedure was illustrated in Section 11.3. The basic singularity is the fold for flows in $R^2 \times S^1$. This is illustrated in Fig. 11.4. For flows in $R^3 \times S^1$, one possible singularity is the cusp. This deformation was illustrated in Fig. 11.5.

We illustrate this general form for describing the stretching and squeezing deformation for the first nontrivial (umbilic, as opposed to cuspoid) case: D_4:

$$\begin{bmatrix} x_1 \\ x_2 \\ z_3 \\ z_4 \\ y_1 \\ y_2 \end{bmatrix} \longrightarrow \begin{bmatrix} x_1 \\ x_2 \\ z_3 \\ z_4 \\ y_1 + rf_1 \\ y_2 + rf_2 \end{bmatrix} \longrightarrow \begin{bmatrix} sx_1 \\ sx_2 \\ z_3 \\ z_4 \\ y_1 + f_1 \\ y_2 + f_2 \end{bmatrix} \longrightarrow \begin{bmatrix} y_1 + f_1 + a_1 \\ y_2 + f_2 + a_2 \\ z_3 + a_3 \\ z_4 + a_4 \\ sx_1 + a_5 \\ sx_2 + a_6 \end{bmatrix}$$

Here f_1, f_2 describe the D_4 singularity:

$$\begin{aligned} f_1(x;z) &= x_1^2 + x_2^2 + z_3x_1 + z_4x_2 \\ f_2(x;z) &= x_1x_2 \end{aligned}$$

Stretching occurs as r increases from 0 to 1, and squeezing occurs as s decreases from 1 to 0. The branched manifold limit is obtained by setting $s = y_\alpha = (a_5, a_6) = 0$.

This shows that the simplest dynamical system in which chaos is generated by a mechanism more complicated than a cuspoid deformation has dimension 7 [2 germ variables (x) + 2 unfolding variables (z) + 2 deforming dimensions (y) + 1 time flow direction = 7].

12.6 PATHS

Paths through the parameter space of a Lie group are used to probe the structure of the group—in fact, to determine its homotopy group. For example, the one-dimensional Lie algebra R^1 exponentiates to the locally isomorphic but globally inequivalent translation group ($\sim R^1$) and the rotation group $SO(2)$ ($\sim S^1$). Paths through the control parameter space of a singular map are used to reveal very different bifurcation phenomena.

12.6.1 Routes to Chaos

The unfolding of the horseshoe map is summarized by the forcing diagram in Fig. 9.8. The hyperbolic limit occurs at the upper right side of this figure. The *laminar limit,* in which only a stable period-1 orbit is present, occurs at the lower left. When the return map is bent into a horseshoe shape, a series of orbits is created as the system progresses from the laminar to the hyperbolic limit. The progression from laminar to fully chaotic can be represented by a path through this unfolding space.

Every path starting at the laminar limit and progressing to the hyperbolic limit represents a possible route to chaos. Properly speaking, a route to chaos is specified by enumerating a countable sequence of orbits in the order in which they are created. In practice, a route to chaos is specified by a countable sequence of basis sets of periodic orbits.

Several other phenomena described as *routes to chaos* should be interpreted more narrowly. Intermittency and period doubling are phenomena that generally accompany the creation and evolution of a stable orbit on its way to the hyperbolic limit. Mode locking is a precursor to chaotic behavior. It is due to a change in the topology of the return map and it occurs *before* the map becomes nonsingular.

Forcing diagrams exist for three-dimensional dynamical systems with other branched manifolds as caricatures, but they have not yet been constructed. It is not yet clear what the higher-dimensional analog is.

12.7 RANK

Rank is a concept that comes from the study of *linear* systems (i.e., matrices). It is useful for the study of locally linearizable systems. In many instances, rank leads to a further reduction in the dimension of a system.

For a Lie algebra, the rank l is related to the number of independent invariant functions that can be defined on the algebra. In general,

$$\text{rank} \quad l \leq k_2 \quad \# \text{ simultaneously commuting operators}$$

For semisimple Lie algebras this bound is saturated ($l = k_2$), and rank is much smaller than dimension:

$$\text{rank} \sim (\text{dimension})^{1/2}$$

For example, $SU(n)$ has rank $n - 1$ and dimension $n^2 - 1$. For singularities, the rank is the number of essential variables in the germ of the singularity. The rank of the cuspoid family of singularities A_k ($k \geq 2$) is 1, while for the umbilic family D_k ($k \geq 4$) the rank is 2.

12.7.1 Stretching and Squeezing

Rank can also be defined for dynamical systems, at least for the strange attractors which they generate. Specifically, the rank of a strange attractor is the rank of the singularity that occurs in its branched manifold. In particular, all strange attractors with $d_L < 3$ have rank 1.

For an inertial manifold $R^n \times S^1$ containing a strange attractor with k_1 unstable and k_2 stable directions, the branched manifold has dimension $k_1 + 1$ and its intersection with a constant phase plane has dimension k_1. The return map must contain a singularity. This is a singularity of maps $R^{k_1} \rightarrow R^{k_1}$ that can only be obtained by continuous deformation of the k_1-dimensional manifold R^{k_1} in the larger space $R^{k_1+k_2}$. The rank, l, of the strange attractor is the rank of this singularity. It is bounded above by k_2:

$$\text{rank} \quad l \leq k_2 \quad \# \text{ stable directions}$$

This inequality is very reminiscent of the inequality between the rank of a Lie algebra and its number of simultaneously commuting generators. In this case, the bound is saturated for (semi)simple algebras. These are the algebras that can be discretely classified.

It is not unreasonable to anticipate an analogy between the contracting directions in a strange attractor and the commuting operators of a Lie algebra. This leads to the following conjecture.

Conjecture: Strange attractors which obey

$$\text{rank} = \# \text{ contracting directions}$$

are discretely classifiable.

Cuspoid (A_k) and umbilic (D_k) singularities have rank 1 and 2. If this conjecture is true, strange attractors generated by cuspoid and umbilic singularites are rigidly classifiable only if they possess one and two contracting directions in their inertial manifolds, respectively.

12.8 COMPLEX EXTENSIONS

When we attempt to classify Lie algebras and singularities, we encounter algebraic equations. These cannot always be solved completely over the field of real numbers. It then becomes necessary to solve these equations over the field of complex numbers and worry at a later point about possible projections back to the real field. Complex extensions enter dynamical systems theory in (at least) two ways: distributions of fixed points, and stretching and folding structures.

12.8.1 Fixed-Point Distributions

The properties of a dynamical system are determined to a great extent by the number and types of their fixed points. A useful first step in the analysis of a dynamical system therefore involves a search for solutions of the equations $\dot{x}_i = f_i(x) = 0$. In many instances the forcing functions $f_i(x)$ are polynomial in the variables x_i. In such cases the equations $f_i(x) = 0$ have only a finite number of solutions (Bezout's theorem [69]). However, the number of (real) solutions may depend on control parameter values. For example, the Rössler equations have 0 or 2 nondegenerate fixed points, while the numbers are 1 and 3 for the Lorenz equations, depending on control parameter values. Fixed points scatter from the real axis into the complex plane as a function of control parameter values.

Algorithms exist for locating all the fixed points of a set of polynomial equations $f_i(x) = 0$. The number of fixed points is invariant if the solution set is over the field of complex numbers. Thus, the Rössler and Lorenz equations always have 2 and 3 nondegenerate fixed points, respectively. These dynamical systems are therefore of type A_2 and A_3.

12.8.2 Singular Return Maps

The stretching and squeezing mechanisms responsible for generating chaos in a strange attractor $\mathcal{SA} \subset \mathcal{IM} = R^{k_1+k_2} \times S^1$ are summarized by a singularity in mappings $R^{k_1} \to R^{k_1}$ which can be created by continuous deformation of R^{k_1} in $R^{k_1+k_2}$. The first step in the classification of these singularities involves classification of their complex extensions. For example, for a singularity of rank 2 with $k_1 = 4$ and $k_2 = 2$, the germ of the complex extension D_4 is given by $x_1 \to f_1(x_1, x_2) = x_1^2 + x_2^2$ and $x_2 \to f_2(x_1, x_2) = x_1 x_2$. This complex germ has two inequivalent real forms, given by

$$D_{\pm 4}: \quad \begin{array}{ccccc} x_1 & \to & y_1 & = & x_1^2 \pm x_2^2 \\ x_2 & \to & y_2 & = & x_1 x_2 \end{array}$$

12.9 COXETER–DYNKIN DIAGRAMS

For unknown reasons, these diagrams turn up again and again in branches of mathematics that are closely related to physics. They occur in the classification of regular polyhedra, discrete groups generated by reflections, caustics, Lagrangian singularities, Lie algebras, catastrophe functions, and singular maps [13].

In particular, there is a 1:1 correspondence between Coxeter–Dynkin diagrams and complex extensions of simple Lie algebras. There is also a $1-1$ correspondence between Coxeter-Dynkin diagrams and complex extensions of free and boundary catastrophes. These comments suggest that we have a right to expect that they also play an important role in dynamical systems theory. In this expectation we will not be disappointed.

Coxeter–Dynkin diagrams occur in dynamical systems theory in (at least) two ways: determination of fixed points and classification of singular return maps.

12.9.1 Fixed-Point Distributions

The fixed-point distributions for a dynamical system $\dot{x}_i = f_i(x) = 0$ are determined algorithmically. In the generic nondegenerate case they can lie along a one-dimensional manifold; a two-dimensional manifold; ... ; an l-dimensional manifold, where l is the rank of some singularity. They can be given canonical locations in these manifolds (zero-modal case) or their locations can depend on $1, 2, \ldots, m$ (modality) parameters.

The rank 1 fixed-point distributions are described by the Coxeter–Dynkin diagrams for cuspoid singularities, A_l. The diagrams that describe the Rössler and Lorenz systems are A_2 and A_3. Flows for dynamical systems with fixed-point distributions $A_l, l = 2, 3, 4, 5$ are shown in Fig. 12.1.

The rank 2 fixed-point distributions which are zero modal are described by Coxeter–Dynkin diagrams D_l and E_6, E_7, E_8. Fixed-point distributions of rank 2 and low modality have been classified.

Flows associated with fixed-point distributions of type $D_l, l = 4$ and 5, are shown in Fig. 12.2. Flow of type D_4 is equivalent to one of the triple covers of the Rössler system (A_2).

12.9.2 Singular Return Maps

Stretching and squeezing is classified by singular return maps. As a result, the classification of singularities by Coxeter–Dynkin diagrams can be applied immediately to the classification of singular first return maps $R^{k_1} \rightarrow R^{k_1}$, which are obtained by continuous deformation of R^{k_1} in $R^{k_1+k_2}$. The results are:

- **Rank 1 singularities**: A_l

- **Rank 2, zero modal singularities**: D_l and E_6, E_7, E_8

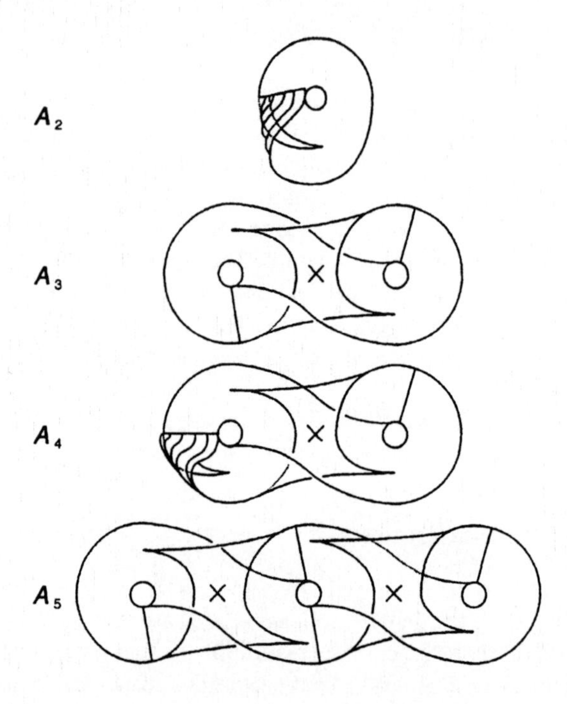

Fig. 12.1 Some of the flows associated with fixed-point distributions of type A_l, $l = 2, 3, 4, 5$.

12.10 REAL FORMS

The classification theory for simple Lie algebras and simple singularity germs can proceed when we extend the field from real to complex numbers, so that certain algebraic equations can be solved. Several different real systems (Lie algebras, germs) can have the same complex extension. These different real forms are essentially analytic continuations of each other. Once again, in dynamical systems theory, projections from complex to real forms can occur at two levels.

12.10.1 Stability of Fixed Points

Coxeter–Dynkin diagrams define the distribution of fixed points of a dynamical system. They provide no information about their type and stability properties. In Fig. 12.3 we show two fixed-point distributions of type A_3. In one, two unstable foci are separated by a regular saddle. The strange attractor that this real form of A_3 organizes is topologically equivalent to the Lorenz attractor (with rotation or inversion symmetry). The other real form has an unstable focus sandwiched between two regular saddles. This real form of A_3 organizes a strange attractor that is very similar to the

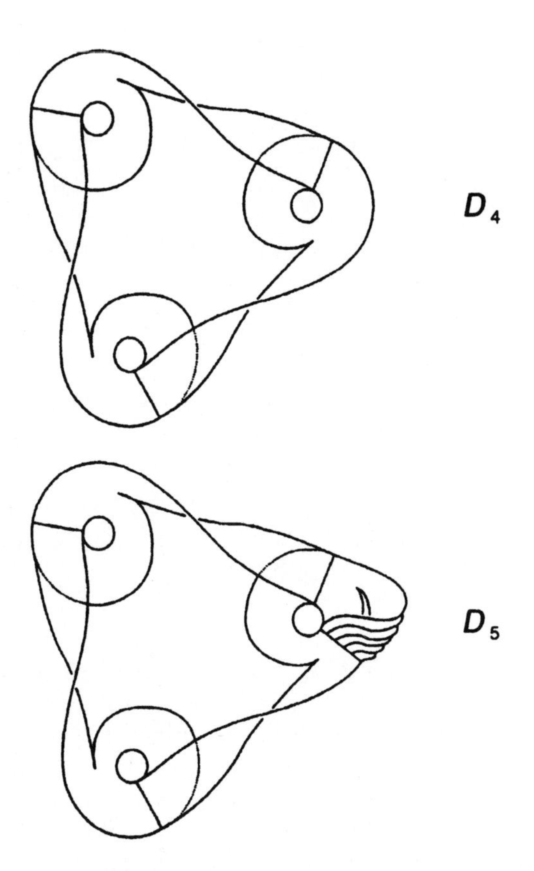

Fig. 12.2 Some of the flows associated with fixed-point distributions of type D_4 and D_5.

attractors generated by the periodically driven Duffing equations (nonautonomous) and the Burke and Shaw attractor (autonomous) for certain parameter values.

Both real forms (of the strange attractors) are topologically inequivalent covers of the Rössler attractor. They have the same symmetry group $[R_z(\pi)]$ but different topological indices.

12.10.2 Singular Return Maps

Germs of return maps are classifiable over the complex field. For rank-1 return maps the different real forms produce equivalent bifurcation diagrams. However, for rank-2 germs this is no longer the case.

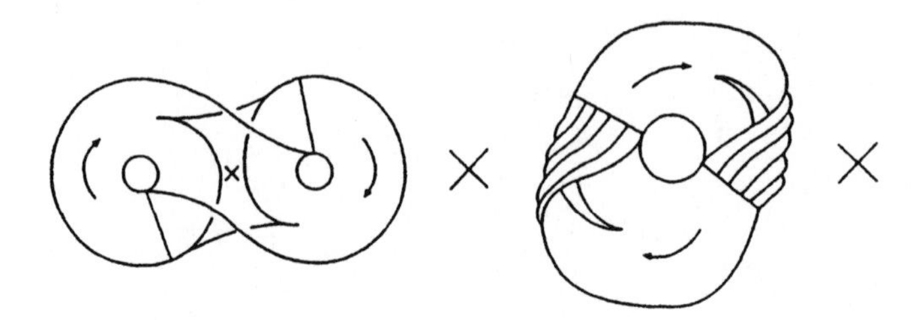

Fig. 12.3 Two real forms of the fixed-point distribution of type A_3 organize flows whose structure is shown by the branched manifolds: (a) Lorenz attractor; (b) Duffing attractor, Burke and Shaw attractor. The two branched manifolds are inequivalent.

12.11 LOCAL VS. GLOBAL CLASSIFICATION

A great deal of classification theory is carried out locally. This is simply a reflection of the power of the Taylor series expansion. Extending local to global presents a multiplicity of problems. For example, the well-known Lie groups $SO(3)$ and $SU(2)$ are locally equivalent, since they have isomorphic Lie algebras. They are not globally equivalent: $SO(3)$ is a $2 \to 1$ locally faithful image of $SU(2)$. By the same token, Whitney's theorem tells us that the only local singularities that can occur in mappings of the plane to itself are the fold and the cusp. If several nonlocal cusps occur, the theorem does not determine how they can be arranged among themselves.

The idea of extending from local to global can occur in several different ways in dynamical systems theory. In this section we concentrate on extensions involving singularities of return mappings of the local type $R^{k_1} \to R^{k_1}$ in $R^{k_1+k_2}$.

12.11.1 Nonlocal Folds

A single fold in the return map $R^1 \to R^1$ in the phase space $R^2 \times S^1$ leads to a Smale horseshoe mechanism. Multiple nonlocal folds can also occur. These can describe a variety of different stretching and squeezing mechanisms. Several of them were analyzed in Chapter 7.

12.11.2 Nonlocal Cusps

A single cusp singularity in the return map $R^2 \to R^2$ in the phase space $R^3 \times S^1$ leads to a rich bifurcation diagram, as illustrated in Fig. 11.6 and 11.7. More than one nonlocal cusp can occur in a variety of ways, as illustrated in Figs. 11.8 to 11.10, and in a variety of two-dimensional manifolds, as illustrated in Figs. 11.12 to 11.14.

12.12 COVER–IMAGE RELATIONS

The prototype cover–image relation are the two Riemann sheets of the complex square-root function $f(z) = \sqrt{z}$ that cover the complex z plane. The cover–image relation for Lie groups is illustrated in Fig. 12.4(a). All Lie groups $G_1,\ G_2,\ \ldots$ with the same Lie algebra A are listed. To this Lie algebra there corresponds a unique universal covering group $\overline{G}$ that is simply connected. This covering group has discrete invariant subgroups $D_1,\ D_2,\ \ldots$. Each Lie group G_i with Lie algebra A is uniquely the quotient of the universal covering group by one of its discrete invariant subgroups of the form $G_i = \overline{G}/D_i$.

Two dynamical systems $D_1,\ D_2$ are equivalent if there is a global diffeomorphism $y = y(x)$ that maps one into the other. If the diffeomorphism is local but not global, the dynamical systems can be locally identical but not globally identical. If the dynamical systems have discrete symmetry groups $G_1,\ G_2$, it is possible to construct a (universal) image dynamical system $\underline{D}$ by modding out the symmetry: $\underline{D} = D_1/G_1 = D_2/G_2$, as shown also in Fig. 12.4(b). In this case the classification program for dynamical systems reduces to a number of smaller programs: Classify inequivalent universal image dynamical systems $\underline{D}$; identify discrete symmetry groups G_i; identify appropriate topological indices, the local diffeomorphism, and the lift or covering dynamical system(s) $D_i =$ "$\underline{D} \otimes G_i$."

12.13 SYMMETRY BREAKING AND RESTORATION

Symmetry and its breaking and restoration play fundamental roles in physics. Group theory is the natural vehicle for discussions of symmetry breaking and restoration. However, these concepts also play an important role in singularity theory. For example, a typical one-parameter path $[a(t), b(t)]$ in the control space of the cusp catastrophe $V(x; a, b) = \frac{1}{4}x^4 + \frac{1}{2}a(t)x^2 + b(t)x$ will unhinge the cusp singularity, but imposition of symmetry under $x \to -x$ will restore the cusp singularity. Restoration of symmetry will play an important role in dynamical systems theory through entrainment and synchronization. We now propose how this might occur.

12.13.1 Entrainment and Synchronization

A standard progression in the sciences is: Propose a theory; design an experiment; take data; test the theory. In practice, this often reduces to proposing a general linear model, fitting the model, and then testing whether the model is any good. This last phase, the *goodness-of-fit* test, is usually implemented by a χ^2 or similar test.

A similar sequence of events can be carried out for nonlinear systems. This parallels the procedure for linear models, except for the last step. There is at present no nonlinear analog of the χ^2 goodness-of-fit test for linear models.

Qualitative tests for nonlinear models have been proposed. The most interesting such test is based on an observation made by Huyghens [105] 300 years ago. He observed that two similar clocks that keep slightly different time will synchronize

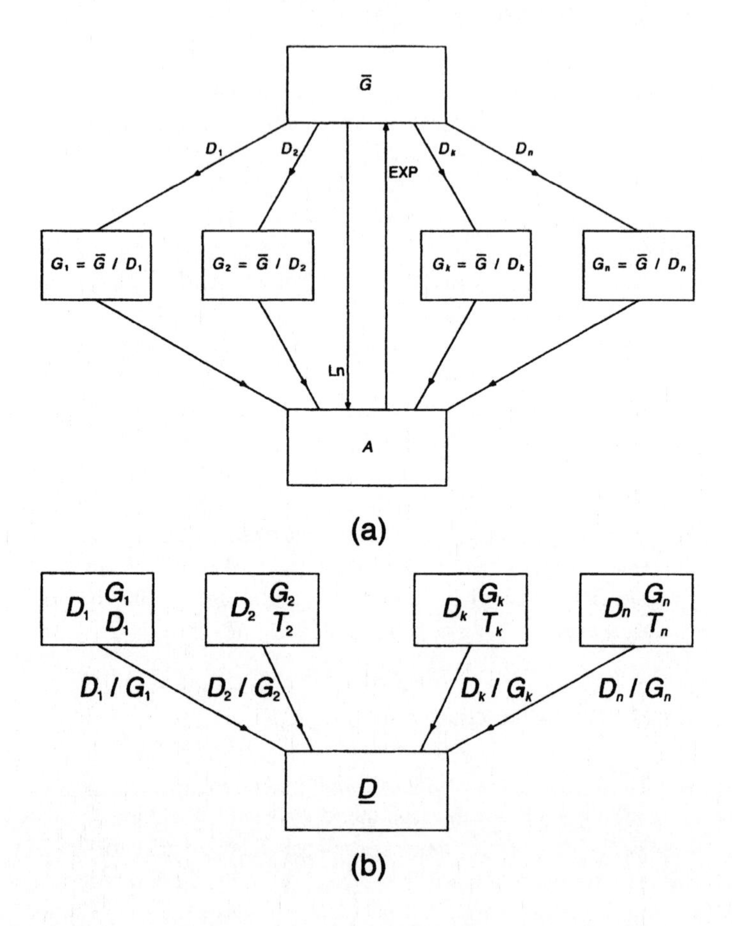

Fig. 12.4 (a) There Is a unique correspondence between Lie algebras and simply connected Lie groups, $\overline{G}$. Every Lie group with the same Lie algebra is a homomorphic image of $\overline{G}$, obtained by forming the quotient $\overline{G}/D_i$, where D_i is one of the subgroups of the maximal discrete invariant subgroup D of $\overline{G}$. (b) Is There a correspondence between dynamical systems D_i with symmetry group G_i and universal image dynamical systems $\underline{D}$ without symmetry?

their timekeeping if placed sufficiently close together on a wall that couples their internal mechanisms.

This has been developed into a qualitative entrainment test as follows. Suppose that data are generated by a physical system under the dynamics $\dot{x} = f(x; c')$, where $x \in R^n$ is the physical state variable and $c \in R^k$ are control parameters. Assume that a model for this physical system has the form $\dot{y} = f(y; c)$. The vector $y \in R^n$ is assumed to be the model surrogate for the physical state x. The functions f in the two dynamics are the same, but evaluated at different control parameter values c' and c.

The two dynamical systems

$$\begin{array}{rcll} \dot{x} & = & f(x;c') & \text{Fully reducible} \\ \dot{y} & = & f(y;c) & \text{Independent} \end{array}$$

are fully reducible.

If the two dynamical systems are allowed to evolve, even starting from the same initial condition, $x(t)$ and $y(t)$ will eventually diverge to measurably different values. This is true no matter how close the model $f(y;c)$ is to the real dynamics $f(x;c')$. That is, this will occur no matter how small the difference $c - c'$ is. To test for goodness of fit, it is useful to add a small term to the model that decreases y when it is too large, and increases y when it is too small. A simple perturbation with this property depends on the difference $y - x$. The dynamics is then represented by the equations with a simple linear one-way coupling term:

$$\begin{array}{rcllll} \dot{x} & = & f(x;c') & & & \text{Reducible} \\ \dot{y} & = & f(y;c) & - & \lambda(y-x) & \text{Entrain} \end{array}$$

This dynamical system is reducible. If c is close to the real value c', the data $x(t)$ will *entrain* [i.e., $y(t) - x(t) \to 0$] the model output $y(t)$ for small values of the coupling constant λ. The general form of the entrainment region is shown, both with and without noise, in Fig. 6.14.

It would be very useful if the entrainment test could be made quantitative in the control parameter difference $c' - c$, the strength of the coupling constant λ, and the noise level. This quantitative test would then be the analog for nonlinear systems of the χ^2 test for linear systems.

If feedback between the two dynamical systems is possible, the coupled equations

$$\begin{array}{rcllll} \dot{x} & = & f(x;c') & - & \lambda(x-y) & \text{Irreducible} \\ \dot{y} & = & f(y;c) & - & \lambda(y-x) & \text{Synchronize} \end{array}$$

are irreducible. It is useful to introduce sum and difference coordinates $u = x+y$ and $v = x-y$. For $c-c'$ small and λ large, the flow projects into the n-dimensional inertial manifold $\mathcal{IM}$ defined by $v = 0$. The two systems are then said to be *synchronized*. As λ is decreased, a symmetry breaking bifurcation $v \neq 0$ will occur transverse to $\mathcal{IM}$. Entrainment and synchronization can be considered as symmetry-restoring bifurcations as a function of increasing interaction strength λ.

12.14 SUMMARY

It is a truism that 90 percent of solving a problem is formulating the right question. Our long-range objective is to formulate a theory for dynamical systems. This objective can be reached the more quickly, the sooner we formulate the right questions. We have taken the opportunity in this chapter to point out that there are two fields of mathematics, both useful for the sciences, which are related to a greater or lesser extent with dynamical systems. These fields have reached some level of maturity by formulating and responding to a useful set of questions. Further, the questions asked in these two fields are surprisingly similar. We would not have a soul if we did not believe that a similar set of questions would be useful to frame for dynamical systems theory. For this reason we have presented a set of 13 questions that have come up somehow or other, and been found useful, in the development of Lie group theory and of singularity theory. It is our hope that framing these questions in the language of dynamical systems theory will facilitate maturation of this field.

Appendix A *Determining Templates from Topological Invariants*

A.1 THE FUNDAMENTAL PROBLEM

Given a chaotic attractor that has been reconstructed in an embedding phase space from experimental time series or from numerical simulations, the central problem of topological analysis is to determine a template describing the structure of this attractor. Mathematically, this problem is well defined, at least for hyperbolic systems: The template is specified by the branched manifold and the semiflow that result from the Birman–Williams projection discussed in Section 5.4 (i.e., by identifying points having the same asymptotic future). However, there are difficulties in applying this definition directly to experimental systems. That the Birman–Williams theorem holds only for hyperbolic systems is not an issue: We can always find hyperbolic sets (e.g., sets of unstable periodic orbits) approximating well the chaotic attractor. However, determining stable and unstable directions from experimental data contaminated by noise is a difficult problem. And even if we could compute these directions reliably, no simple procedure for extracting the template from this distribution of directions in phase space is currently known.

Thus, topological analysis of experimental attractors should be based on an alternate definition of templates. Chaotic attractors live in phase space. Templates live in an abstract space of equivalence classes, obtained by squeezing the state space along the stable direction (Birman–Williams reduction). What provides a bridge between these two worlds is that the Birman–Williams reduction can be viewed as resulting from a smooth deformation which preserves periodic orbits and does not induce crossings between them. Hence, knot and braid invariants of any set of periodic orbits in the real phase space and of its counterpart in the abstract space are identical.

Given a set of unstable periodic orbits extracted from a chaotic attractor, we accordingly say that a branched manifold is a template for the attractor if it can hold the braid made of the observed periodic orbits. A key step of topological analysis is thus to determine *the simplest template that has a set of periodic orbits whose topological invariants are identical to the invariants measured experimentally*. Although there are simple cases where the structure of the template can be guessed (e.g., when the system is highly dissipative and the template is a standard horseshoe template), this is generally not a simple problem. In this Appendix, we provide technical details about how it can be solved systematically (i.e., by implementing a computer algorithm). For simplicity, we restrict ourselves to the case of templates with a single branch line, but the tools detailed here can be generalized to more complicated configurations.

The input data of the algorithm consist of a set of topological invariants characterizing the intertwining of periodic orbits, the output is a few integers describing the structure of the simplest template compatible with these invariants. The invariants generally used are the topological period (i.e., the number of intersections with a reference Poincaré section), the self-linking number, and linking numbers. When periodic orbits are naturally presented as braids (for example, the phase space is $\mathbb{R} \times S^1$ and has a hole in the middle), more powerful invariants can be used, such as (self-) relative rotation rates. When available, the information contained in the local torsion of an orbit is also useful. It should be noted that these invariants are not complete invariants: In principle, they may take the same value on two links that are not isotopic. In practice, they are very efficient for classifying periodic orbits. For example, two horseshoe orbits of period lower than or equal to 10 have the same spectrum of self-relative rotation rates only if they are isotopic. This is because only a limited set of different braid types can be realized among the periodic orbits of a given template.[1]

In Chapter 5, we showed that the topological invariants of any periodic orbit living on a branched manifold are easily computed if a description of this manifold is given.

[1] It has been proven that certain templates have all possible knot types among their periodic orbits [201]. However, this does not preclude that the number of different braid types for a period p is limited. Indeed, a given knot type is not realized for every possible period.

As discussed in Section A.2, simple invariants such as relative rotation rates can even be expressed analytically as a function of a small set of integers describing the structure of the template, which have been presented in Section 5.6. This allows one to design the template-finding problem as an inverse problem, where the topological invariants are given and the characteristic numbers of the template are computed from them. How to handle this inverse problem is presented in Section A.3. One of the advantages of topological analysis is that the template-finding problem is overdetermined: Only a few topological invariants are needed to determine a candidate template, and the remaining ones can be used to validate the solution by verifying that they are consistent with it. Finally, we explain in Section A.4 how topological analysis can be used to extract information about the symbolic dynamics of a chaotic attractor and to construct a generating partition for it. This method provides an interesting alternative to the techniques based on homoclinic tangencies that have been outlined in Section 2.11.2.

A.2 FROM TEMPLATE MATRICES TO TOPOLOGICAL INVARIANTS

A.2.1 Classification of Periodic Orbits by Symbolic Names

Periodic orbits on a template are uniquely identified by a symbolic name that specifies their itinerary on the branched manifold. Once distinct symbols have been assigned to the branches of a template, the symbolic name of an orbit is obtained by concatenating the symbols associated with successively visited branches. Because the return map from the branch line to itself is expansive (trajectories originating from one branch cover one or more branches on the next round), two different periodic orbits have different symbolic names. If an n-branch template is *fully expansive* (i.e., the return map is conjugate to a full shift), there is a 1:1 correspondence between periodic orbits of the template and nonrepeating words on n symbols (up to a cyclic permutation).

The symbolic name of a periodic orbit contains all the information required to compute its topological invariants. This is illustrated by Fig. A.1, which shows the period-4 orbit 0111 of the Smale's horseshoe template. Each intersection of the orbit with the branch line is a periodic point and is associated with one of the four cyclic permutations of the symbolic name. Taking into account whether branches have odd or even torsion (i.e., whether the corresponding branch of the return map is orientation-reversing or orientation-preserving), the order in which these periodic points are found along the branch line can be determined easily using kneading theory, as has been discussed in Sections 2.7 and 5.9 (compare Figs. A.1 and 2.15).

As can be seen in Fig. A.1, the braid associated with the orbit is simply obtained by connecting periodic points Σ_i to their images $\sigma\Sigma_i$ with segments following the semi-flow on the branched manifold. From this braid, all the topological invariants can be obtained. In the example shown, it is straightforward to verify that the self-linking number of the orbit is 5 by counting the number of crossings of this braid. We now briefly describe how this can be computed systematically.

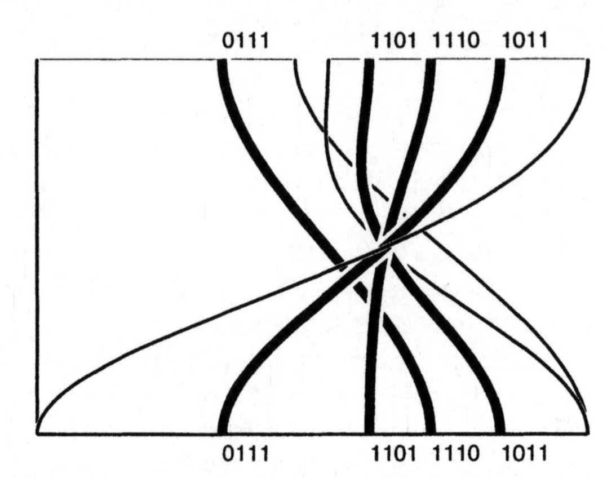

Fig. A.1 Topological structure of the period-4 orbit 0111 of the Smale's horseshoe template. Only the relevant part (the branched one) of the template is shown. The top and bottom lines should be identified and correspond to the branch line.

A.2.2 Algebraic Description of a Template

We want to compute linking numbers, relative rotation rates and local torsions. Since linking numbers can be expressed as a sum of relative rotation rates (Section 4.4) we only need to consider the computation of a relative rotation rate, which measures the average number of rotations per period of two orbits originating from two periodic points around each other (Section 4.3). First, we recall how the structure of the template can be described by a few integers, as is illustrated in Fig. A.2. These integer numbers are

- **Torsions:** t_i indicates the twist of branch i in half-turns. This is the local torsion of the period-1 orbit $\bar{i}$ attached to branch i.

- **Linking numbers:** $L(i, j)$ is the number of rotations of branch i around branch j. This is the linking number of the period-1 orbits with itineraries $\bar{i}$ and $\bar{j}$.

- **Layering coefficients:** assuming that $i < j$, $l_{ij} = 1$ if branch j is stacked on branch i (i.e., is closer to the reader than branch i), and $l_{ij} = -1$ otherwise. The matrix (l_{ij}) is symmetric.

It is convenient to group torsions and linking numbers in a *template matrix* (t_{ij}), with the diagonal and off-diagonal elements being given by $t_{ii} = t_i$ and $t_{ij} = 2L(i, j)$, respectively. What makes the template matrix (t_{ij}) and the layering matrix (l_{ij}) so useful is that the crossing number of a braid such as shown in Fig. A.1 can easily be computed from the values of their elements, as well as the local torsion of any periodic orbit.

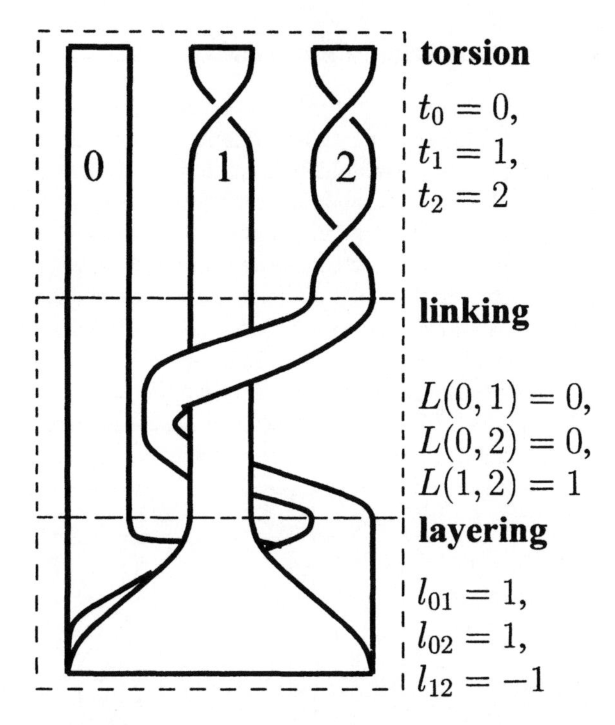

Fig. A.2 The structure of a template can be described by a few integer numbers which describe (1) how branches are twisted (torsions), (2) how they link each other (linking numbers), and (3) how they are stacked on the branch line (layering coefficients). This is the only information required to compute local torsions and relative rotation rates of any knot or link formed by periodic orbits on a template.

A.2.3 Local Torsion

The local torsion is by far the simplest invariant that can be computed in this way. On a branched manifold, the direction transverse to the semi-flow represents the direction of the unstable manifold. Each time a periodic orbit visits a branch, its unstable manifold rotates by the same number of half-turns as the branch. This number of half-turns is nothing but the diagonal element t_{ii}.

If we denote by $n_i(\Sigma)$ the number of occurrences of symbol i in a symbolic name Σ, the local torsion of the orbit associated with Σ is thus the sum of the torsions of the branches visited:

$$t(\Sigma) = \sum_{i=0}^{n-1} n_i(\Sigma)\, t_{ii} \tag{A.1}$$

For example, $t(0111) = t_{00} + 3t_{11}$. Note that expression (A.1) is completely linear in the diagonal elements t_{ii}.

A.2.4 Relative Rotation Rates: Examples

The computation of a relative rotation rate $R(\Sigma, \Sigma')$, where Σ and Σ' are the symbolic names of the periodic points used as initial conditions, is more involved than that of local torsions. In order to understand the building blocks of the general procedure presented in Section A.2.5, we now study the structure of crossings encountered in the computation of two self-relative rotation rates of the period-4 orbit 0111. This computation is illustrated in Fig. A.3, which has been designed so that the link between the template characteristic numbers and the relative rotation rate appears explicitly. It shows the two strands whose relative rotation rate is to be measured traversing four copies of the underlying template before returning to their initial location (cf. see also Fig. 4.16)).[2] For simplicity, we have chosen to begin the diagram at a time where the two orbits are on different branches (this is always possible unless the two points have the same symbolic name and hence are identical). As we discuss below, different types of crossings occur depending on which branches are visited.

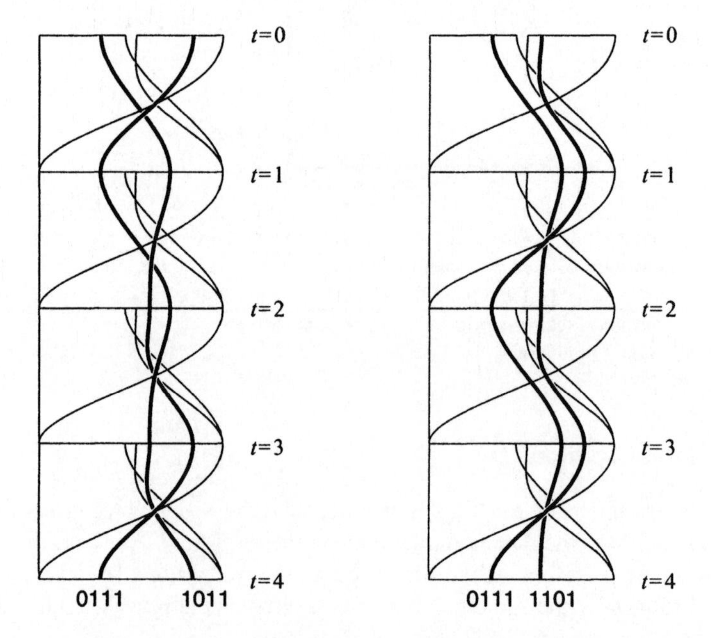

Fig. A.3 Two examples of relative rotation rate computations. (a) $R(0111, 1011)$; (b) $R(0111, 1101)$. The times indicated are measured in units of the natural period.

Let us follow the two orbits shown in Fig. A.3(a) from top to bottom. In the first period (from $t = 0$ to $t = 1$), one orbit goes from branch 0 to branch 1 as it follows the path $0111 \to 1110$ while the other one goes from branch 1 at 1011 to branch 0

[2]More generally, two orbits of periods p and q should be followed for a number of periods equal to the least common multiple of p and q.

at 0111. Thus, they must cross at least once. The sign of this crossing is determined by the stacking order of the two branches. Since branch 1 is here in front of branch 0, it is positive. It is easy to see that this contribution to the total crossing number is given exactly by the layering coefficient l_{01}. As for linking numbers $L(i, j)$, we see in Fig. A.2 that each turn of one branch around the other adds two crossings. This shows the existence of a term $c_L = 2L(0,1) = t_{01}$ in the expression of the total crossing number. In the present example chosen, the linking number $L(0,1)$ is zero and there is no contribution from the linking part of the template.

The analysis of the second period (from $t = 1$ to $t = 2$) is more complicated. Starting from different branches, the two trajectories end in branch 1. Whether or not a crossing occurs in this part of the diagram depends on the relative order of the two periodic points at $t = 2$. This could be solved using kneading theory but this information is in fact not needed, as we shall see at the end of the computation. We delay for now the computation of the contribution of the second period. Independently, we note again that a nonzero linking number $L(0,1)$ would add t_{01} crossings.

The contribution of the third period is straightforward, as the two orbits stay on branch 1. Since trajectories do not intersect on the manifold, crossings in the plane projection occur only if the branch is twisted: The number of crossings is given by the local torsion t_1. Interestingly, the contribution of the third period is thus t_{11}, which is similar to the two t_{01} terms that were encountered in the first two periods.

In the fourth and last period, the two orbits start from branch 1 and end in separate branches. As in the third period, the only contribution is a trivial term t_{11} stemming from the common path on branch 1. We are now in a position to determine the number of crossings occuring in the second period due to periodic point ordering. Between $t = 1$ and $t = 4$, the order of the two orbits on the branch line changes. Since the orbits pass twice through branch 1, no exchange of the strands due to torsion of branch 1 can occur. Linking never modifies the order of the two orbits since it contributes crossings by pairs. This shows that there must be a crossing occuring in the layering part of the second period, corresponding to a term l_{01}.

If we sum the different contributions, we get a crossing number

$$c(0111, 1011) = (t_{01} + l_{01}) + (t_{01} + l_{01}) + t_{11} + t_{11} \tag{A.2}$$

For the horseshoe template described by $t_{01} = 0$, $l_{01} = 1$, and $t_{11} = 1$, expression (A.2) yields a crossing number of four, which is consistent with Fig. A.3(a). This corresponds to a rotation of two turns over four periods, hence to a relative rotation rate of $\frac{1}{2}$.

Consider now the example of Fig. A.2(b), which illustrates the computation of $R(0111, 1101)$. Starting from the top, the first period is similar to the second period of Fig. A.3(a): The two orbits start from different branches and end in the same branch. They are again in different branches at $t = 2$. Between $t = 0$ and $t = 2$, the two orbits are exchanged, which is explained by the single twist occuring in the second period. Hence, there is no crossing in the layering part in the first period, and the crossing number for the first two periods is $t_{01} + t_{11}$. Noting that the computation for the last two periods is exactly the same as for the first two ones, but with the two strands swapped, we obtain

$$c(0111, 1101) = t_{01} + t_{11} + t_{01} + t_{11} \tag{A.3}$$

Again, expression (A.3) correctly predicts the two crossings of Fig. A.3(b), which correspond to a relative rotation rate of $\frac{1}{4}$. It is important to note that unlike in the computation of Fig. A.3(a), the parity of t_{11} must here be explicitely taken into account. If it is odd, there is no crossing in the layering part of the first period. If t_{11} is even, there must be one crossing in this area to account for the exchange of the two orbits between $t = 0$ and $t = 2$.

A.2.5 Relative Rotation Rates: General Case

With these two examples in mind, we now present a general procedure for computing the relative rotation rate $R(\Sigma, \Sigma')$ of orbits starting from two initial conditions specified by symbolic names $\Sigma = s_1 \ldots s_p$ and $\Sigma' = s'_1 \ldots s'_q$, of lengths p and q.

A relative rotation rate is computed by following the two orbits until they return simultaneously to their initial conditions. This first occurs after l periods, where l is the least common multiple of the two periods p and q. Accordingly, we first increase the length of the two symbolic names to l by repeating them as appropriate. To make the description easier, we also shift simultaneously the two names until their first symbols differ (modifying the starting point of a closed loop obviously does not change the final result). For example, $\Sigma = 0111$ and $\Sigma' = 011$ are changed to $1011 = \sigma^3\, 0111$ and $011 = \sigma^3\, 011$. The symbolic words obtained are shown in the second and third rows of Fig. A.4.

Blocks	A	B		C		D	E					
0111	1	0	1	1	1	0	1	1	1	0	1	1
011	0	1	1	0	1	1	0	1	1	0	1	1
Linking	t_{01}	t_{01}		t_{01}		t_{01}	t_{01}					
Torsion			t_{11}		t_{11}			t_{11}	t_{11}	t_{00}	t_{11}	t_{11}
Layering	1		$1-\pi_1$		$1-\pi_1$	1						π_0

Fig. A.4 Computation of $R(0111, 011)$. The length of the two symbolic names is first increased to 12, which is the least common multiple of 3 and 4. Then, the two names are shifted simultaneously until the leading symbols differ. Then the two sequences are divided into blocks such that symbols differ only at the beginning of a block. We find here five blocks, labeled A through E.

Then, the two length-l sequences are divided into blocks, as illustrated in Fig. A.4. The division follows a simple rule: A new block is started at each index i for which the symbols s_i and s'_i differ. This step is related to our previous observation that when two orbits come from different branches, whether or not they cross in the layering part can only be determined the next time they intersect the branch line in different branches. Thus, each block contains all the information needed to determine whether one crossing has occured in the layering part.

The contribution of the linking and torsion parts of the template to the total crossing number is easy to compute. In both cases the contribution of symbols in the ith position is $t_{s_i s'_i}$, as shown in Fig. A.4 where for clarity we have separated the linking terms (fourth row) from the torsion terms (fifth row). The number of crossings occuring in the layering part is slightly more involved, because blocks must be analyzed as a whole. There is at most one such crossing per block and its contribution is then $l_{s_i s'_i}$, by definition. Whether a crossing actually occurs is determined as follows.

When the two orbits intersect the branch line in different branches, their relative position is easily read from the branch symbols: The smallest symbol corresponds to the leftmost orbit. The relative position of the two orbits is modified at each crossing, be it positive or negative. Thus, the crossing number of a block is odd or even depending on whether the relative positions of the two orbits along the branch line are modified across the block. This can be determined by comparing the ordering of the leading symbols of the block with that of the leading symbols of next block. In Fig. A.4, the first four blocks have necessarily an odd crossing number while block E should have even parity. Only torsion and layering crossings determine the parity of a block, since linking crossings come in pairs. Since the sum of the torsion terms and hence its parity are easily computed, whether a layering crossing occurs inside the block can be decided by comparing the parities of the total crossing number of the block and of the sum of torsion terms. If the two parities are identical, there is no layering crossing. If they have different parities, one layering crossing must be introduced so that the sum of torsion and layering crossings has the correct parity. For example, if the orbits are exchanged across the block (odd block parity) and if there is an even number of torsion crossings, this indicates the occurence of a layering crossing.

This rule can be implemented in equations by using a parity function $\pi(x)$ such that $\pi(x) = 1(0)$ if x is odd (even). The parity of the torsion terms is obtained by applying this function to the sum of torsion terms computed with modulo 2 arithmetics. For example, one has $\pi(3t_{00} + 4t_{11}) = \pi(t_{00})$. In general, the parity π_t of the sum of torsion terms can be expressed as 0, $\pi(t_{00})$, $\pi(t_{11})$, or $\pi(t_{00} + t_{11})$. Denoting the parity of the block $\mathcal{B}$ by $\pi_{\mathcal{B}}$, the layering term in the expression for the total crossing number of the block is

$$c_t = \begin{cases} \pi_t \times l_{s_i s'_i} & \text{if } \pi_{\mathcal{B}} = 0 \\ (1 - \pi_t) \times l_{s_i s'_i} & \text{if } \pi_{\mathcal{B}} = 1 \end{cases} \tag{A.4}$$

With the rules established above, we can write the expression of the crossing number of any block, and hence of the relative rotation rate corresponding to two arbitrary symbolic names. For example, the crossing numbers for the five blocks of Fig. A.4

are given by

$$
\begin{array}{llll}
c_A = & t_{01}+ & & l_{01} \\
c_B = & t_{01}+ & t_{11}+ & [1-\pi(t_{11})] \times l_{01} \\
c_C = & t_{01}+ & t_{11}+ & [1-\pi(t_{11})] \times l_{01} \\
c_D = & t_{01}+ & & l_{01} \\
c_E = & t_{01}+ & t_{00}+4t_{11}+ & \pi(t_{00}) \times l_{01}
\end{array}
\quad (A.5)
$$

where the expressions have been laid out so as to separate linking, torsion, and layering terms (shown from left to right). The total crossing number for the diagram of Fig. A.4 is the sum of the five crossing numbers in (A.5):

$$c(0111,011) = 5t_{01} + t_{00} + 6t_{11} + [4 + \pi(t_{00}) - 2\pi(t_{11})] \times l_{01} \quad (A.6)$$

and the corresponding relative rotation rate is $R(0111,011) = c(0111,011)/24$, since the crossing number is computed on 12 periods and two crossings make one turn. As a last example, we compute the general expression of the four self-relative rotation rates of the 0111 orbit. One of them is trivially zero, two correspond to the diagram of Fig. A.3(a), and the last one is associated with the diagram of Fig. A.3(b).

The reader can verify that with the rule described above, expression (A.2) is recovered for the diagram of Fig. A.3(a) and that the crossing number of the diagram in Fig. A.3(a) is given by $2t_{01}+2t_{11}+2[1-\pi(t_{11})]\times l_{01}$, which generalizes expression (A.3) for arbitrary branch parities. The expression for the self-linking number of the 0111 orbit is obtained by summing the self-relative rotation rates:

$$slk(0111) = 3t_{01} + 3t_{11} + [3 - \pi(t_{11})] \times l_{01} \quad (A.7)$$

If the horseshoe characteristic numbers $t_{01} = 0$, $t_{11} = 1$, and $l_{01} = 1$ are inserted into Eq. (A.7), we recover the value of 5 that can be read from Fig. A.1. It is easy to see that for a two-branch template, the general expression for a linking number or relative rotation rate I of a link $\mathcal{L}$ is

$$
\begin{aligned}
I = {} & \alpha_{01}t_{01} + (\alpha_{00}t_{00} + \alpha_{11}t_{11}) \\
& + [\gamma + \beta_0\pi(t_{00}) + \beta_1\pi(t_{11}) + \beta_{01}\pi(t_{00}+t_{11})] \times l_{01}
\end{aligned}
\quad (A.8)
$$

where the coefficients α_{ij}, β_i, β_{01} and γ depend on the symbolic itineraries of the orbits involved in $\mathcal{L}$ and the t_{ij} and l_{01} describe the template structure. The factor of l_{01} has been written in a form directly related to the computational rules described above. It can also be rewritten as the most general polynomial function of the two logical variables $\pi_i = \pi(t_{ii})$, which satisfy $\pi_i^2 = \pi_i$. Indeed,

$$
\begin{aligned}
& \gamma + \beta_0\pi(t_{00}) + \beta_1\pi(t_{11}) + \beta_{01}\pi(t_{00}+t_{11}) \\
& = \gamma + \beta_0\pi_0 + \beta_1\pi_1 + \beta_{01}(\pi_0 - \pi_1)^2 \\
& = \gamma + (\beta_0 + \beta_{01})\pi_0 + (\beta_1 + \beta_{01})\pi_1 - 2\beta_{01}\pi_0\pi_1
\end{aligned}
\quad (A.9)
$$

Thus, the general expression for a topological invariant of a link on an n-branch template is

$$\begin{aligned} I &= \sum_{i \leq j} \alpha_{ij} t_{ij} + \sum_{i<j} \Pi_{ij}(\pi_0, \pi_1, \ldots, \pi_{n-1}) \times l_{ij} \\ &= L(t_{ij}) + N(t_{kk}, l_{ij}) \end{aligned} \tag{A.10}$$

where $\Pi_{ij}(\pi_0, \pi_1, \ldots, \pi_{n-1})$ is the most general polynomial function in n logical variables $\pi_0, \pi_1, \ldots, \pi_{n-1}$.[3] Note that the expression (A.1) for local torsions has a structure similar to (A.10), but with a zero nonlinear part N and a linear part L depending only on the t_{ii}. The α_{ij} and the coefficients of the polynomial Π_{ij} depend only on symbolic itineraries and not on the structure of the branched manifold. Their values for the lowest-period orbits of a two-branch template are given in Table A.1. In fact, the α_{ij} depend on symbol counts only and not on the precise order in which symbols appear in the symbolic itinerary, as the polynomials Π_{ij} do. If $n_i(\Sigma)$ denotes the number of symbols i in a symbolic name Σ, the expression for the self-linking number of the corresponding orbit has the coefficients

$$\alpha_{ij}(\Sigma) = \begin{cases} n_i(\Sigma) n_j(\Sigma) & \text{if } i \neq j \\ n_i(\Sigma)[n_i(\Sigma) - 1]/2 & \text{if } i = j \end{cases} \tag{A.11}$$

as is easily found from the computational rules, whereas for the linking number of two orbits with itineraries Σ and Σ' we have

$$\alpha_{ij}(\Sigma, \Sigma') = \begin{cases} [n_i(\Sigma) n_j(\Sigma') + n_j(\Sigma) n_i(\Sigma')]/2 & \text{if } i \neq j \\ n_i(\Sigma) n_i(\Sigma')/2 & \text{if } i = j \end{cases} \tag{A.12}$$

The linking number

$$\begin{aligned} lk(0001, 0111) &= 5t_{01} + \frac{3}{2} t_{00} + \frac{3}{2} t_{11} \\ &\quad + (1 + \frac{3}{2}\pi_0 + \frac{3}{2}\pi_1 - 2\pi_0\pi_1) \times l_{01} \end{aligned} \tag{A.13}$$

is an example of an invariant with a polynomial Π_{01} of degree 2.

An extremely important result of the present section is the particular structure of the general expression (A.10). With respect to the template characteristic numbers t_{ij} and l_{ij}, it naturally splits into a linear part L that depends only on the template matrix elements t_{ij} and a nonlinear part N containing terms such as $\pi(t_{kk}) \times l_{ij}$. When the branch parities $\pi(t_{kk})$ can be considered as fixed, for example when they can be determined independently of topological analysis, expression (A.10) thus becomes a completely linear function of the t_{ij} and l_{ij}. In Section A.3, this will prove to be a very useful property.

[3] Π_{ij} is a polynomial of degree at most n, with its monomials of degree at most one in each variable.

Table A.1 Coefficients defining expressions of self-linking numbers for the lowest-period orbits of a two-branch template.

Orbit	α_{01}	α_{00}	α_{11}	Π_{01}
0, 1	0	0	0	0
01	1	0	0	1
001	2	1	0	$2 - \pi_0$
011	2	0	1	$2 - \pi_1$
0001	3	3	0	$3 - \pi_0$
0011	4	1	1	$3 - \pi_0 - \pi_1$
0111	3	0	3	$3 - \pi_1$
00001	4	6	0	$4 - 2\pi_0$
00011	6	3	1	$4 - \pi_0 - \pi_1$
00101	6	3	1	$6 - \pi_0 - \pi_1$
00111	6	1	3	$4 - \pi_0 - \pi_1$
01011	6	1	3	$6 - \pi_0 - \pi_1$
01111	4	0	6	$4 - 2\pi_1$

A.3 IDENTIFYING TEMPLATES FROM INVARIANTS

In Section A.2.5, we have computed expressions for the topological invariants as functions of the template and layering matrices. We now consider the inverse problem: Topological invariants of periodic orbits have been measured and we want to determine the simplest template that is consistent with these invariants. More precisely, we want to find (1) a branched manifold and (2) a set of symbolic itineraries such that the corresponding orbits on the branched manifold have exactly the same invariants as the experimental orbits. This involves solving equations of the form (A.10) for the template and layering matrix elements, which is greatly simplified by the quasilinear structure of these expressions.

However, such equations must be written before they can be solved. Since they depend on symbolic itineraries, they can be written only if symbolic names have been assigned to periodic orbits, at least tentatively: The problems of identifying a symbolic dynamics for the periodic orbits and the branches of a template that reproduces the spectrum of topological invariants must generally be solved simultaneously. Thus, two important cases must be distinguished according to whether or not a symbolic coding is available for the attractor under study.

A.3.1 Using an Independent Symbolic Coding

The ideal situation is when a symbolic coding is available prior to topological analysis. This is often the case when the system is highly dissipative and there is a Poincaré section whose return map can be approximated by a one-dimensional map. Figure A.5 shows two such maps describing the dynamics of the modulated CO_2 laser and of the Nd:YAG laser described in Section 7.5. Although the data in Fig. A.5(b) are considerably noisier than in Fig. A.5(a), a one-dimensional structure can still be discerned. The results from one-dimensional symbolic dynamics can then be applied to construct a generating partition and encode chaotic trajectories as well as periodic orbits. For example, the symbolic itinerary of the period-7 orbit shown in Fig. A.5(a) is easily read to be 0101011.

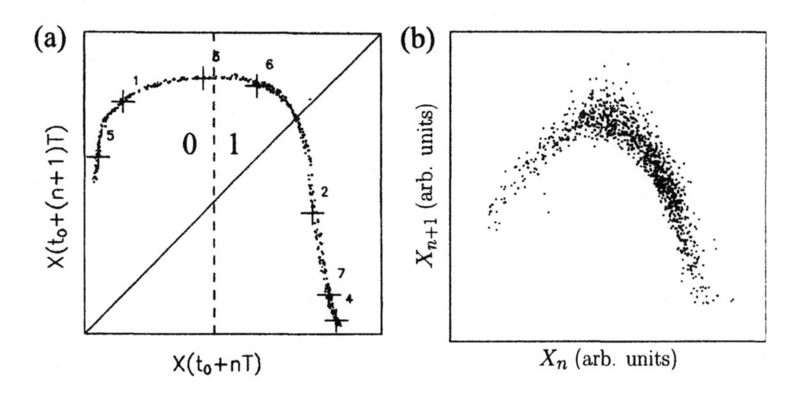

Fig. A.5 (a) Stroboscopic return map $X[t_0 + (n+1)T]$ vs. $X(t_0 + nT)$, where $X(t)$ is a time series from a CO_2 laser with modulated losses, T is the modulation period and t_0 is an arbitrary time. The canonical partition based on the critical point can be used to assign symbolic names to periodic orbits, as is illustrated with points belonging to the orbit 0101011. (b) Similar return map for a Nd:YAG laser.

The problem of constructing a symbolic dynamics then decouples from the problem of determining the template structure. This simplifies the analysis in several ways:

- Since each periodic orbit is naturally associated with a unique symbolic name, the equation corresponding to each measured invariant can be written directly.
- The number of branches of the template is given by the number of monotonic branches of the return map (i.e., the number of symbols needed to describe the symbolic dynamics).
- Branch parities can be determined directly from the structure of the return map. Orientation-preserving (orientation-reversing) branches correspond to a parity $\pi_i = 0$ ($\pi_i = 1$). Since the π_i are obtained independently of the other template indices, they become parameters of the equations (A.10). These equations are then completely linear in the t_{ij} and l_{ij} and are easily solved.

For example, assume that we are studying a system whose dynamics is governed by an underlying horseshoe and is described by a return map similar to that of Fig. A.5. Because they exist at any parameter value beyond the period-doubling cascade, we consider the orbits of periods 1, 2, 4, and 8 created in this cascade. Their symbolic names are 1, 01, 0111, and 01010111 and their self-linking and linking numbers are

Table A.2 Self-linking (diagonal entries) and linking numbers (off diagonal entries) of a small set of orbits of symbolic names 1, 01, 0111, and 01010111. These are the first periodic orbits created in a period-doubling cascade leading to a horseshoe.

Orbit	1	01	0111	01010111
1	0			
01	1	1		
0111	2	3	5	
01010111	4	6	13	23

given in Table A.2. To show that even simple invariants carry useful information, we do not use torsions (which are a little more difficult to measure than linking numbers) nor rotation rates in this example, even though these invariants can prove useful. In particular, torsions lead directly to linear equations and should be measured when possible. The five equations corresponding to the five invariants of the three lowest-period orbits are

$$lk(1,01) = \frac{1}{2}t_{01} + \frac{1}{2}t_{11} + \frac{\pi_1}{2} \times l_{01} = 1 \qquad \text{(A.14a)}$$

$$slk(01) = t_{01} + l_{01} = 1 \qquad \text{(A.14b)}$$

$$lk(1,0111) = \frac{1}{2}t_{01} + \frac{3}{2}t_{11} + \frac{\pi_1}{2} \times l_{01} = 2 \qquad \text{(A.14c)}$$

$$lk(01,0111) = 2t_{01} + \frac{1}{2}t_{00} + \frac{3}{2}t_{11} + \frac{1}{2}(\pi_0 + \pi_1) \times l_{01} = 3 \qquad \text{(A.14d)}$$

$$slk(0111) = 3t_{01} + 3t_{11} + (3 - \pi_1) \times l_{01} = 5 \qquad \text{(A.14e)}$$

For completeness we have written these equations in their general form, but the branch parities π_i should be replaced with their values $\pi_0 = 0$ and $\pi_1 = 1$ determined from the structure of the return map. Equations (A.14) have a unique overdetermined solution

$$t_{01} = 0 \quad t_{00} = 0 \quad t_{11} = 1 \quad l_{01} = 1 \qquad \text{(A.15)}$$

which corresponds to the standard horseshoe template (and is consistent with the π_i assumed). Note that the problem is intrisically overdetermined: The number n of invariants, hence of equations, can in principle be arbitrarily large whereas four numbers suffice to specify the basic layout of the template. That a solution exists when there are more equations than unkowns indicates that invariants are not independent; they all are signatures of the same stretching and folding mechanisms. This is what allows us to predict invariants whose companion equations have not been used to determine the template. For example, the self-linking number of the period-8 orbit is

$$slk(01010111) = 15t_{01} + 3t_{00} + 10t_{11} + (13 - \pi_0) \times l_{01} \qquad \text{(A.16)}$$

With solution (A.15), this expression correctly predicts the measured value of 23, which increases our confidence that the template found correctly describes the dynamics. Similarly, the local torsions of the orbits 1, 01, 0111, and 01010111 should be equal to 1, 1, 3, and 5, respectively.

In most cases, a template can be determined from only a subset of the invariants measured. An important step of topological analysis is to validate this template by checking that all the other invariants are correctly predicted by this template. When this is the case, we can be confident that the solution found describes faithfully all the information we have extracted from the time series. However, it sometimes happens when characterizing experimental systems that a small number of invariants do not match the template found, whereas all other invariants do. This can occur when spurious crossings have been introduced due to noise or imperfect localization of the orbit. Thus, invariants whose value is difficult to ascertain (e.g., different representatives of the orbits give different values of the invariant) should generally not be taken into account in the analysis, at least in a first step.

Since the horseshoe template is commonly associated with return maps such as that of Fig. A.5, it is tempting to believe that the result could have been guessed directly from the return map. This is not so, as the two return maps shown in Fig. A.5 correspond to different templates. In the regimes shown in Fig. A.5, the CO_2 laser is associated with a standard horseshoe template [89] whereas the dynamics of the Nd:YAG laser is described by a reverse horseshoe template [136]. It is instructive to compare the invariants of the orbits of period 1, 2, and 4 in Tables A.2 and 7.8.

A.3.2 Simultaneous Determination of Symbolic Names and Template

Often, it is difficult to construct a generating partition. We then have no information about the number of symbols, symbolic names of orbits, and branch parities. In this case, template determination is more involved than in Section A.3.1 but still can be carried out by systematic methods. One possible strategy is to progressively increase the number of branches until a consistent solution is found and, for each number of branches, to decompose the problem into subproblems with fixed branch parities and symbolic names.

First, we iterate over configurations of the branch parities. Since $\pi_i = \pi(t_{ii}) \in \{0, 1\}$, there are a priori 2^n such configurations for an n-branch template. If the local torsion of a period-1 orbit has been measured, the parity of the corresponding branch is known and the number of configurations is reduced. For each set of parities, we consider a small set of low-period orbits and iterate over all their possible symbolic names.

For each set of symbolic names, parities and names are fixed as in Section A.3.1 and we are left with equations that are linear in t_{ij} and l_{ij}, one for each invariant available in the working set. When solving each equation set for the t_{ij} and l_{ij}, there are four possible results:

- There is no solution; equations are not consistent. This indicates that one of the parities or symbolic names assumed is incorrect and that the search must proceed to the next configuration.

- Linear equations have a solution, but it is invalid because it cannot be interpreted geometrically. For example, $l_{ij} \notin \{-1, 1\}$, an off-diagonal term t_{ij} is odd, or the t_{ii} found are not consistent with the parities π_i assumed. Again, the current configuration must be skipped.

- There is a unique solution to the working equation set and it can be translated into a candidate template, whose validity must be checked by comparing the invariants of the remaining orbits with the values predicted by that template.

- The problem is underdetermined. The equations are consistent but specify a family of solutions rather than a single solution.

In the latter case, we add one orbit to the working orbit set to increase the number of constraints and for each possible symbolic name of the new orbit, we analyze the corresponding equation set as described above. Eventually, the working set contains enough equations so that they have a unique solution or no solution at all. Thus, the search proceeds along a tree whose branches are the possible choices of symbolic names, going down one level (adding one orbit) when the problem is underdetermined. Leaves of the tree correspond to equation sets having one or zero solutions. Leaves with one solution contain candidate templates whose validity must be checked.

Template validation can easily be integrated into this search scheme. When a consistent solution is found, we then keep adding orbits as in the underdetermined case, until either an inconsistency is found or all orbits and their invariants have been taken into account. In the latter case, the template obtained describes correctly the global organization of the orbits as measured by their invariants and is an acceptable solution of the problem. In practice, the initial search and the validation phases can be implemented slightly differently for performance reasons.

As a simple example, let us return to the case study of Table A.2 and assume that no information about the symbolic dynamics is available. Thus, we must consider all possible branch parities and symbolic names. Only half of the configuration space of symbolic names needs to be explored because a template has two different presentations, which correspond to its front and rear views. Consequently, the transformation $i \leftrightarrow n-1-i$ can applied to symbolic names of an n-symbol configuration without modifying the equations and the solution [cf. also Eq. (5.6)]. For every two-symbol configuration where the period-1 orbit has symbolic name 1, there is a mirror configuration where it has symbolic name 0. Thus, we may restrict the search to configurations where the symbolic name of the first orbit is 1. Since there is only one period-2 orbit, there are only three different sets of symbolic names for the first three orbits, which are $\{1, 01, 0001\}$, $\{1, 01, 0011\}$, and $\{1, 01, 0111\}$. There are 30 symbolic names on 2 symbols and thus 90 symbolic name configurations $(\mathcal{N}_i)_{i=1..4}$ for the four orbits of Table A.2. Thanks to the tree structure of the search, only a small fraction of these configurations is actually explored.

The output of the search is displayed in Table A.3. When only the first two orbits are used, the equation set is underdetermined in all configurations. With three orbits, 10 out of the 12 possible configurations lead to no solution or to an invalid solution. Each of the two remaining configurations yields a different candidate template. The first solution (for $\pi_0 = 0, \pi_1 = 1, \mathcal{N}_3 = 0111$) is the standard horseshoe template, the second one (for $\pi_0 = 1, \pi_1 = 1, \mathcal{N}_3 = 0111$) differs from the horseshoe only by a branch torsion t_{00} that is -1 instead of 0: The branch parity π_0, which was imposed in Section A.3.1, remains undetermined at this stage. Indeed, the first three orbits of Table A.2 mostly visit branch 1 and not enough information about branch 0 can be extracted from their invariants. Note that the symbolic name of the period-4 orbit is correctly identified in both cases.

Finally, only the horseshoe solution survives when the invariants of the period-8 orbit are taken into account, but only when this orbit is assigned the period-8 symbolic name 01010111. In this example, we have explored 80 out of the 360 possible configurations with fixed parities and symbolic names. This number would have been significantly lower if additional invariants such as torsions and rotation rates had been used, providing extra information. For example, knowing that the local torsion of the period-2 orbit is 1 would have allowed us to eliminate the template with $t_{00} = -1$ already at the two-orbit stage, since it predicts a value of 0 for this invariant. This is a general property that if some invariants of low-period orbits could not be measured, the same information can be extracted from invariants of higher-period orbits, at the cost of having to explore more configurations. This is consequence of the template-finding problem being extremely overdetermined.

Table A.3 Search for a template compatible with invariants of Table A.2. For each set of branch parities, equations are generated for the different symbolic name configurations and solved, first for three orbits, then for four orbits if successful. Possible outcomes are: IS (invalid solution); NS (no solution); a template represented by the quartet $[t_{01}, t_{00}, t_{11}, l_{01}]$.

π_0, π_1	$\mathcal{N}_3$	outcome	$\mathcal{N}_4$	outcome
0, 0	0001	IS		
	0011	IS		
	0111	IS		
0, 1	0001	NS		
	0011	IS		
	0111	[0, 0, 1, 1]	01010111	[0, 0, 1, 1]
			other names	NS
1, 0	0001	NS		
	0011	IS		
	0111	NS		
1, 1	0001	NS		
	0011	IS		
	0111	[0, −1, 1, 1]	any name	NS

Thus, we conclude that the horseshoe template is the only two-branch template compatible with the topological invariants of Table A.2. Remarkably, this result is obtained without using any preexisting symbolic encoding. Moreover, we have correctly identified the symbolic names of the orbits as being 1, 01, 0111, and 01010111 (in fact, the three last names were found relatively to the period-1 name). This suggests that symbolic names are not arbitrary but reflect a geometric organization that can be captured through topological invariants

This holds more generally than for the orbits of Table A.2, as illustrated by Table A.4. It lists the invariants of the lowest-period orbits of a set of about 1600 periodic orbits extracted from a horseshoe-type attractor of a laser model [49]. Although the invariants listed (topological period, self-linking number and local torsion) are very basic, they prove to be genuine fingerprints: Except for orbits 9b and 9c, there is a single horseshoe orbit matching the measured invariants. Since the corresponding symbolic name identifies the only possible projection of the orbit on the horseshoe template, this name can be assigned to the experimental orbit. For example, there is a single period-7 horseshoe orbit with a self-linking number of 16 and with even torsion: it is the orbit with itinerary 0101011. When we detect an orbit with the same invariants in a horseshoe-type attractor, it is natural to assume that this is the orbit originating from the orbit 0101011 of the hyperbolic horshoe.

Table A.4 Basic topological properties of orbits of period up to 9 extracted from numerical simulations of a laser model. The invariants listed are: period, self-linking number, and torsion. Symbolic names of horseshoe orbits with identical invariants are also displayed. Note that except for orbits 9b and 9c, there is only one possible symbolic name.

Orbit	Invariants	Names	Orbit	Invariants	Names
1a	1, 0, 1	1	8b	8, 21, 5	01011011
2a	2, 1, 1	01	8c	8, 25, 7	01111111
4a	4, 5, 3	0111	8d	8, 25, 6	01011111
5a	5, 8, 3	01011	8e	8, 23, 5	01010111
5b	5, 8, 4	01111	9a	9, 28, 7	011011111
6a	6, 13, 5	011111	9b	9, 28, 6	010110111, 010111011
6b	6, 13, 4	010111	9c	9, 28, 6	010110111, 010111011
7a	7, 16, 5	0110111	9d	9, 28, 5	010101011
7b	7, 16, 4	0101011	9e	9, 30, 7	011101111
7c	7, 18, 6	0111111	9f	9, 32, 8	011111111
7d	7, 18, 5	0101111	9g	9, 30, 6	010101111
8a	8, 21, 6	01101111	9h	9, 32, 7	010111111

There is a greater indeterminacy in the symbolic name of higher-order orbits, which can more rarely than low-period orbits be associated with a unique symbolic name after topological analysis is completed (i.e., their invariants are matched by several different template orbits). This is also the case when templates more complicated than the Smale's horseshoe are considered, in particular in the presence of symmetries. Yet there is in the worst case only a handful of possible symbolic names. Sometimes the indeterminacy is merely due to experimental constraints: an orbit whose linking number with the orbit studied distinguishes two candidate names has not been detected. Sometimes the indeterminacy has more fundamental roots, such as the exchange phenomena mentioned in Section 9.3.2. A systematic study of these questions has still to be carried out.

That topological analysis cannot always distinguish between possible symbolic names of a given orbit does not necessarily mean that all these names, selected by purely algebraic criteria, are equally valid. One important property of symbolic dynamics is that symbolic sequences that are close in symbol space should be associated with points that are close in phase space. This has not been addressed at all in the techniques described above: topological invariants are insensitive to isotopic deformations and thus are poor probes of continuity.

To enforce continuity, we must be able to link the symbolic dynamical information extracted from periodic orbits with spatial information. This is where generating partitions come into play, as they are the natural tools for converting spatial information (position) into symbolic dynamical information (symbolic sequence). However, a generating partition might not be available, especially in experimental systems to which the method based on homoclinic tangencies is difficult to apply. It is then tempting to test the relevance of the symbolic names selected by topological analysis by verifying that they are consistent with a symbolic encoding satisfying continuity properties. That topological invariants of a periodic orbit carry important information about its symbolic dynamics is indeed the starting point of a method for constructing generating partitions proposed by Plumecoq and Lefranc [49, 58], which we briefly present in Section A.4.

A.4 CONSTRUCTING GENERATING PARTITIONS

A.4.1 Symbolic Encoding as an Interpolation Process

Periodic orbits whose symbolic name has been unambiguously identified are natural bridges between phase space and symbol space: they are associated with a unique symbolic sequence and they are precisely localized in phase space. In fact, the problem of finding the symbolic sequence associated with the orbit of a given point is trivial if that point coincides with one of the periodic points whose symbol sequence has been identified with topological analysis. Of course we would like to solve this problem also for an arbitrary point of phase space.

Computing the value of a function at an arbitrary location from values known at a given set of points is a standard mathematical problem, known as *interpolation*. Many interpolation procedures are designed so that the resulting function is continous: The closer the points, the closer the values of the interpolated function. Similarly, generating partitions usually have a simple structure, with a border line whose length is of the order of the attractor diameter. The closer the two points are, the more likely they are associated with the same symbol except if they are located in a small region around the border.

Accordingly, the following interpolation scheme was used by Plumecoq and Lefranc [49]. The partition is parameterized by a set of reference points R_i whose symbolic sequence Σ_i is assumed to be known exactly. The interpolation data are the pairs (R_i, s_i) where s_i is the leading symbol of Σ_i. The symbol associated with an arbitrary point of phase space is defined to be that associated with the closest reference point. This is a piecewise-constant interpolation scheme where the interpolated function (i.e., the symbol) is equal to the value at the reference point in an entire cell around the reference point. The partition is initially based on a small set of periodic points and then refined by progressively inserting higher-period orbits in the set of reference points.

To illustrate this technique, we return to the chaotic attractor whose low-period orbits were listed in Table A.4. As mentioned above, its topological structure is described by a standard horsehoe and a set of about 1600 periodic orbits of period up to 32 is available [49]. These orbits were selected so that they provide us with a good cover of the strange attractor: The maximal distance betwen an arbitrary point of the attractor and the closest periodic point is smaller than 0.1% of the attractor width. Obviously, such a resolution is much higher than can achieved in experiments and, in fact, is useless except to test the validity of topological analysis down to extremely small scales, as here.

The simplest partition that can be constructed from this orbit set is based on the three periodic points belonging to the period-1 orbit 1 and to the period-2 orbit 01. Whereas the symbol associated with the period-1 orbit can only be 1, there are two possible choices for labeling the two periodic points of orbit 01: One point should have symbolic sequence 01 and the other point 10, but the two sequences can be exchanged. These two choices lead to different, but dynamically equivalent partitions [58]. Here, we arbitrarily assign the sequence 01 to the leftmost period-2 point, which thus defines the region of symbol 0 (cf. Fig. A.6).

According to the interpolation scheme, points of the section plane are associated with symbol 0 (resp., 1) if the closest reference point has leading symbol 0 (resp., 1). For the partition of Fig. A.6, the border line is easily constructed with the perpendicular bisectors of the triangle made of the three points. For an arbitrary number of points, it is obtained from a Delaunay triangulation of the set of reference points [49].

The initial partition shown in Fig. A.6, hereafter denoted by Γ_i, is rough and not reliable except in a close neighborhood of each of the three points. To refine it, information must be extracted from higher-period orbits by adding them to the set of reference points. In a first stage, only orbits having a single topological name are considered, to avoid any ambiguity. As with the period-2 orbit, however, a decision is to be made for each orbit inserted into the reference set. There are p cyclic permutations of the topological name of a period-p orbit and there are p ways to assign these p permutations to the p points of the periodic orbit. Which one is chosen can be specified, for example, by the symbolic sequence associated with the leftmost periodic point (as we did for the period-2 orbit).

If an arbitrary permutation is selected for each orbit, the resulting partition can be so convoluted as to be useless, as most of the section plane will be close to the border.

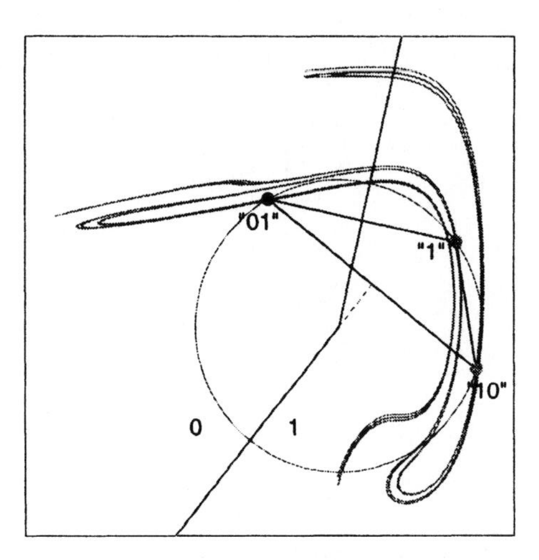

Fig. A.6 Partition of a section plane of a laser model based on three reference points belonging to orbits 1 and 01. The leftmost period-2 point is arbitrarily assigned symbol 0 while the other one is associated with 1. The section plane is divided into two regions depending on whether the closest reference point is associated with symbol 0 or 1. Reprinted from Plumecoq and Lefranc [49] with permission from Elsevier Science.

The description of such a partition would require an enormous amount of information and the encoding of a chaotic trajectory would be extremely sensitive to noise. It is thus important to ensure that the structure of the current partition is modified as little as possible when a periodic orbit is inserted in the reference set, in order to preserve the simplicity of the partition of Fig. A.6.

For example, let us add the period-4 orbit of Table A.4 to the initial partition Γ_i. Figure A.7(a) shows that Γ_i encodes this orbit as 0111, which matches exactly the topological name of the orbit. To modify this partition the least, it is natural to choose the symbolic sequences associated to the four period-4 points so that the symbols attached to them are those already given by Γ_i. Once symbolic names of the period-4 points are fixed, these points can be inserted in the triangulation and the border line is updated accordingly. Comparing Figs. A.7(a) and A.7(b), we see that the partition border has been slightly displaced but that its global structure has not changed much. The topological name of an orbit about to be inserted does not always match the name computed from the current partition but there is usually a permutation choice for which the partition is much less modified than for others, and simple criteria can be designed to select it [49].

After each insertion of an orbit in the partition, the uncertainty about the position of the partition border is reduced. The partition of Fig. A.7(b) does not improve much on that of Fig. A.7(b) but after low-period orbits have been incorporated [Fig. A.8(a)], a much better estimate of the final partition is available. Fig. A.8(b) shows the partition that results from inserting all orbits with a unique topological name up to period 32.

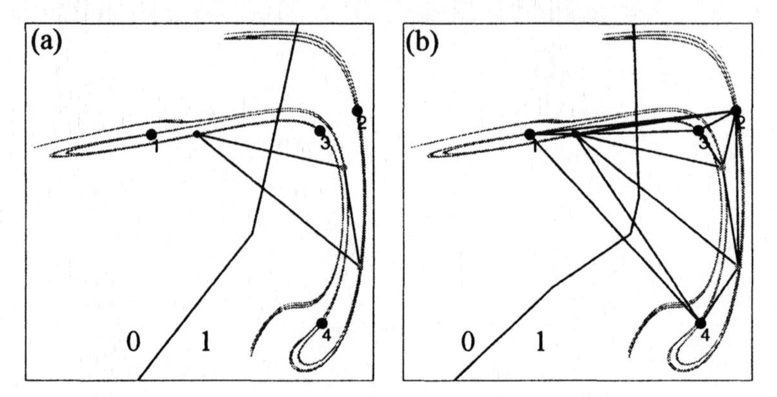

Fig. A.7 Insertion of a period-4 in the partition of Fig. A.6. (a) Initial partition with the four periodic points of orbit 0111; (b) period-4 points are inserted in the triangulation with symbols chosen so as to modify the partition the least. The partition border is updated. Reprinted from Plumecoq and Lefranc [49] with permission from Elsevier Science.

It can be seen that the border is tightly bracketed and that the precision achieved is more than is needed for practical purposes.

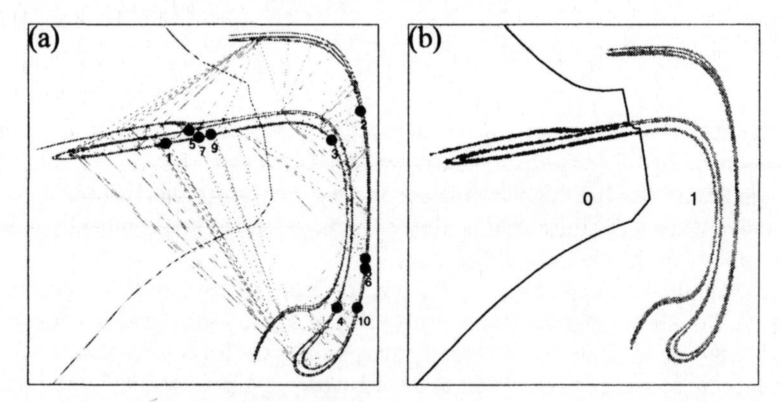

Fig. A.8 (a) The partition under construction: Orbits with unique topological names up to period 9 have been added and a period-10 orbit is being considered for insertion. (b) The partition obtained from all orbits with a unique topological name up to period 32. The partition border is already localized with a very good precision. Reprinted from Plumecoq and Lefranc [49] with permission from Elsevier Science.

When the border is sufficiently well localized, the partition can be used with confidence in regions of phase space far enough from the border. For each orbit not yet identified, it is then possible to determine the name given by the partition and to estimate which parts of this name are reliable, in other words to determine a partial print of the real name. This can be used to discard the symbolic names determined from topological analysis that do contain the trusted substrings of the name indicated by the partition. Often, this also allows us to discard a possible name of another orbit when the linking number computed from this name and from the remaining names of the orbit under study no longer matches the value measured experimentally.

Alternatively discarding symbolic names that are no longer compatible with the current partition and names that are no longer compatible with measured topological invariants, all the orbits can be progressively associated with a unique name and inserted in the partition. The final partition is not shown here because it is indistiguishable to the naked eye from the partition in Fig. A.8(b) but its border is localized with a much better precision, of the order of 0.1% of the attractor width. It provides by construction a symbolic encoding that is both consistent with the topological structure of the set of periodic orbits and continuous: Points that are close in the section plane are associated to symbolic sequences that are close in the symbol space. Given the high number of periodic orbits in our example, it is quite remarkable that the simple rules we have followed naturally select a single name for each orbit: this seems to support the existence of a well-defined symbolic encoding. This numerical experiment is also a stringent test of the validity of topological analysis: The final partition provides explicitly a projection of the unstable periodic orbits embedded in the chaotic attractor onto a branched manifold. A set of about 1600 periodic orbits is characterized by 1600 self-linking number, over one million linking numbers, and many more relative rotation rates. That all these numbers can be reproduced from the four integers describing the template shows clearly that chaos is an extremely organized dynamical state.

Finally, we note that there are two important situations where a symbolic encoding can be obtained rigorously: the hyperbolic case (e.g., horseshoe map) and the one-dimensional case (e.g., logistic map). A very important property of the method described here is that it is continuously connected to both situations. In the hyperbolic case, template theory is rigorous and the decomposition into branches coincides with a canonical Markov partition. Assume that by varying a control parameter, a system is brought from a hyperbolic to a nonhyperbolic regime. Each orbit not destroyed by an inverse sadlle-node or period-doubling bifurcation has the same invariants as in the hyperbolic case. Thus, when the symbolic name of an orbit can be identified solely from its knot invariants, it must be identical in the hyperbolic and nonhyperbolic cases, and remain the same on the entire domain of existence of the orbit. Similarly, dissipation can be increased to infinity without modifying topological invariants (this is equivalent to the Birman–Williams reduction). Again, the natural coding on the template (decomposition into branches) is consistent with the canonical one-dimensional coding for the one-dimensional return map of the branch line into itself, which becomes closer and closer to the return map of the system itself in the limit of infinite dissipation. We thus expect the techniques presented here to be useful to gain some insight into fundamental aspects of symbolic dynamics.

A.4.2 Generating partitions for Experimental Data

The large number of orbits used in these simulations should not lead us to believe that topological methods for constructing partitions can only be used in numerical simulations. Actually, only very modest resolutions are usually needed to characterize the symbolic dynamics of an attractor. It was noted in [58] that because trajectories near the border are very weakly unstable, a precision of only a few percent (to be compared to the 0.01% achieved in numerical comparisons) is sufficient to obtain the correct list of forbidden sequences up to period 19 for the attractor studied in this section.

In fact, the method described in this section was first applied to an experimental system, a CO_2 laser with modulated losses [202], before its validity was more carefully tested in numerical simulations [49, 58]. Figure A.9 shows a generating partition obtained for that system, with orbits of period up to 12. Note that the periodic orbits detected experimentally do not cover well the entire attractor. However, they are found in abundance in the neighborhood of the partition, which is precisely the place where they are needed. This is because orbits are much less unstable in this region than elsewhere and thus are more easily detected. This example is interesting because a technique commonly used for constructing empirically generating partitions is to connect the most apparent folds of the Poincaré section of the attractor. As can be seen in Fig. A.9, this is not completely exact: The partition border as determined from topological analysis passes very near these folds but not exactly through them. This should not be surprising: the border is supposed to pass through the primary folds of the infinitely iterated return map, not those of the time-one map.

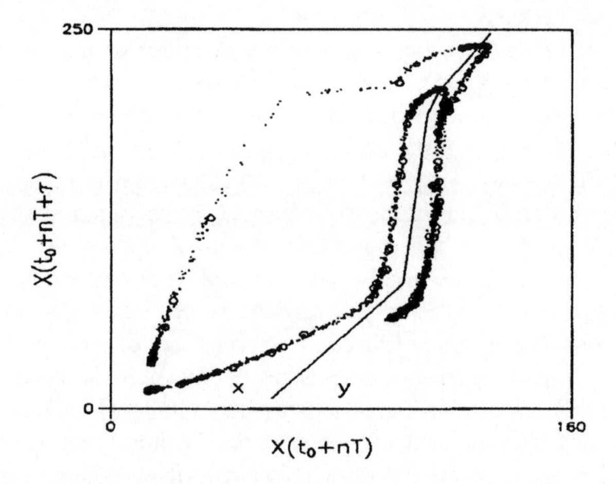

Fig. A.9 Section plane of an attractor observed in a CO_2 laser with intracavity losses modulated at about 380 kHz. Points represent chaotic trajectories, and circles periodic points of period up to 12 extracted from the time series. The generating partition shown has been obtained with topological analysis.

A.4.3 Comparison with Methods Based on Homoclinic Tangencies

The same problem — constructing a generating partition — can be solved by two completely different approaches, the method described in this section and that based on homoclinic tangencies (Section 2.11). It is only natural to check whether the results obtained by the two methods are consistent. To make this test as clear-cut as possible, a set of about 750 periodic orbits of period up to 64 was detected and carefully selected in order to provide a cover of the neighborhood of the border with a resolution of about 10^{-4} of the attractor width [49,58]. The border of the resulting partition, whose enlarged views are shown in Fig. A.10, is thus localized with the same precision. Once again, the regular structure of the partition obtained shows that topological analysis describes accurately the geometric structure of the attractor down to very small scales.

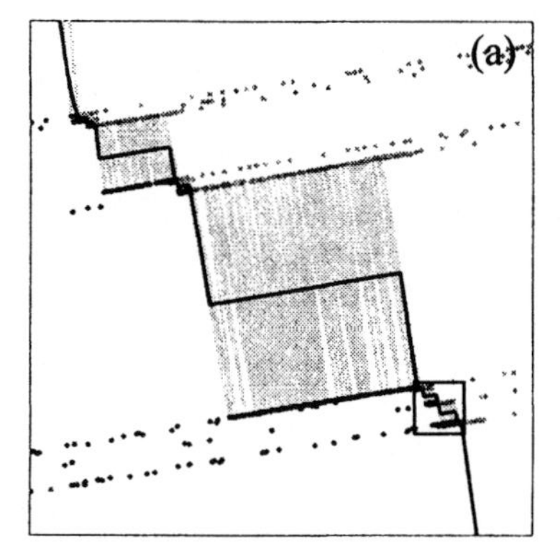

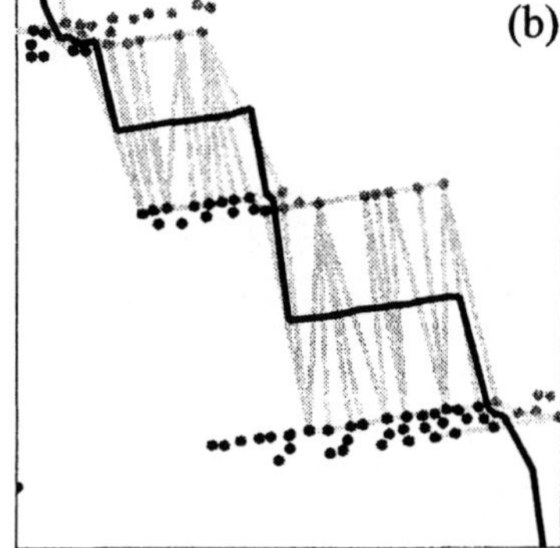

Fig. A.10 (a) Enlarged view of the border of the partition obtained using a high-resolution cover of the border region by periodic points (the enclosing box has a side of length 5×10^{-2} in units of the attractor width). The triangles shown are those defining the partition border. The higher density of periodic points in the region of the border is due to the orbit selection algorithm. (b) Enlarged view of the small square of size 5×10^{-3} displayed in the bottom part of (a). The linewidth used to draw the border line and the periodic points is 5×10^{-5}. Reprinted from Plumecoq and Lefranc [49] with permission from Elsevier Science.

The same attractor was analyzed with the method of homoclinic tangencies, and a generating partition was built. The comparison with the partition obtained from topological analysis is illustrated in Fig. A.11 [58]. The agreement is seen to be excellent: all homoclinic tangencies are located inside the triangles that enclose the border of the partition. This not only shows that the partition line is well approximated by the topological approach, but also that the error bounds it provides strictly hold. These results give strong evidence that algorithms based on topological analysis and on the structure of homoclinic tangencies converge to the same answer. The theoretical tools on which these two methods rely are so utterly different that this agreement is fascinating: Chaotic dynamics has many faces. Because of the (at least superficially) complete independence of the two approaches, this not only supports the validity of the topological approach, but also provides an additional confirmation of the correctness of methods based on homoclinic tangencies.

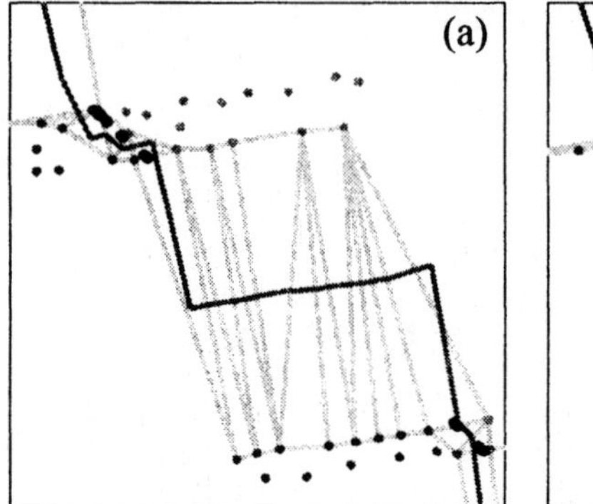

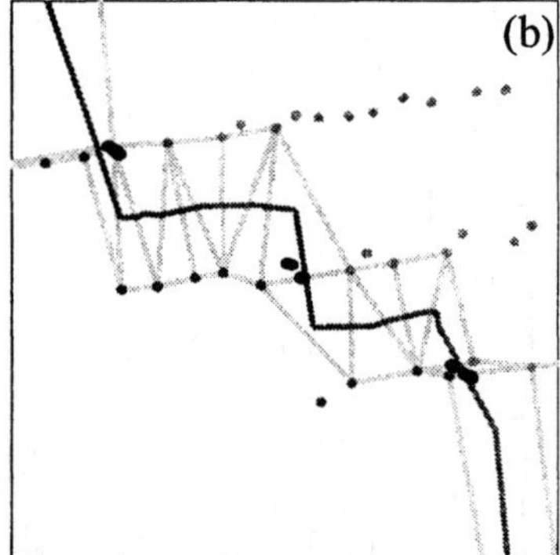

Fig. A.11 Enlarged views of two neighborhoods of the border line for the high-resolution partition of Fig. A.10. The side of the two represented squares is 2.5×10^{-3}, in units of the attractor width. Light dots represent periodic points parameterizing the partition. Dark dots indicate homoclinic tangencies (the angle between the two invariant manifolds at these points is below 2×10^{-4} radians). The border line and the triangles enclosing it are also represented. The linewidth used to draw the border line is 2.5×10^{-5}. Reprinted from Plumecoq and Lefranc [58] with permission from Elsevier Science.

A.4.4 Symbolic Dynamics on Three Symbols

A difficult problem when constructing a symbolic encoding is to determine how many different symbols are needed to describe faithfully the dynamics. A full shift on n symbols has topological entropy $h_T = \ln n$ (in units of the inverse fundamental period), thus the number of symbols should at least be larger than $\exp(h_T)$. If topological entropy cannot be easily computed, metric entropy, which is a lower bound of topological entropy, can be estimated from the greatest Lyapunov exponent. However, there can be situations where the correct number of symbols is larger than the minimum indicated by topological entropy considerations because there are many forbidden symbolic sequences.

Besides their robustness, a clear-cut advantage of methods based on topological analysis is that the correct number of symbols is automatically obtained from the preliminary template analysis: this is simply the number of branches of the simplest template that describes the topological organization of the data. To illustrate this, we consider the example of the Duffing oscillator described in Section 10.7, at the parameter values $\delta = 0.25$, $A = 0.4$, and $\omega = 2\pi$.

To simplify the analysis, we take advantage of the invariance of the Duffing equations under the transformation $x \to -x, y \to -y, \phi \to \phi + \pi$, where $\phi(t) = \omega t$ (mod 2π). Thus, the Poincaré sections at $\phi = 0$ and $\phi = \pi$ are identical modulo an inversion around the origin. We consider the reduced dynamical system where the Duffing equations are integrated over half a period and the $(X, \dot{X})$ plane is then rotated by π radians around the origin.

In [58], a set of 1326 orbits of periods up to 36 was analyzed by the techniques outlined in this Appendix. Two possible three-branch templates were found for this reduced system, hence we know that the dynamics is on three symbols. The first template is described by matrices:

$$t^{D_1} = \begin{pmatrix} 1 & 2 & 2 \\ 2 & 2 & 2 \\ 2 & 2 & 3 \end{pmatrix} \quad l^{D_1} = \begin{pmatrix} 0 & -1 & -1 \\ -1 & 0 & 1 \\ -1 & 1 & 0 \end{pmatrix} \tag{A.17}$$

and corresponds to the spiral template identified as a building block for the Duffing template by Gilmore and McCallum [93], with an additional half-twist stemming from the construction of the reduced system. The other one is an S-shaped template, whose matrices are:

$$t^{D_2} = \begin{pmatrix} 2 & 2 & 2 \\ 2 & 1 & 2 \\ 2 & 2 & 2 \end{pmatrix}, \quad l^{D_2} = \begin{pmatrix} 0 & -1 & -1 \\ -1 & 0 & -1 \\ -1 & -1 & 0 \end{pmatrix} \tag{A.18}$$

These two solutions are algebraically compatible with the input data; each template has a set of periodic orbits having exactly the same invariants as the experimental ones. This indeterminacy is intriguing, as the two solutions have quite different topological structures. However, the partition-building algorithm converges for template (A.17) but not for (A.18) [58]. This could mean that there is no symbolic encoding that simultaneously is continuous in the section plane and reproduces the symbolic dynamics associated with template (A.18). In this case, only solution (A.17) is dynamically relevant. Constructing partitions might thus be helpful to discriminate algebraic solutions of the template-finding problem.

The partition obtained for template (A.17) is shown in Fig. A.12. Its border is localized with a precision of 10^{-3}. It reproduces Fig. 4 of Ref. [54], where the partition had been computed by following lines of homoclinic tangencies. Whereas the partition of Fig. A.12 is the simplest and most natural solution of the template-based algorithm, it was noted in [54] that some homoclinic tangencies involved in this partition are not primary and that heuristic considerations had to be made to select these particular tangencies. This shows that the topological approach provides us with additional information that cannot be obtained from the study of homoclinic tangencies, and is thus a powerful tool for selecting the correct lines of tangencies.

A.5 SUMMARY

The fundamental problem of topological analysis is to determine the simplest template carrying a set of orbits with the same topological invariants as periodic orbits extracted from a chaotic attractor. While this is primarily a topological and geometrical problem, it can be solved by algebraic methods.

Indeed, the structure of templates can be described by a few integer numbers, grouped in the template and layering matrices. Moreover, invariants of orbits lying on the branched manifold can be expressed analytically as a function of the matrix elements describing the template. The structure of these equations is sufficiently

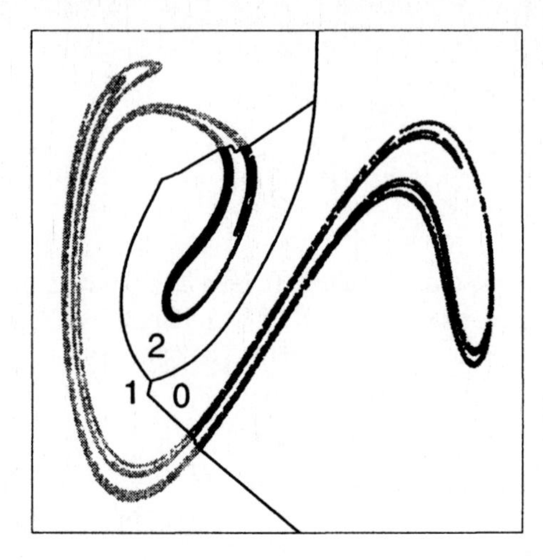

Fig. A.12 Generating partition obtained from topological analysis for the Duffing attractor. Reprinted from Plumecoq and Lefranc [58] with permission from Elsevier Science.

simple that they become linear under some conditions, in particular when a symbolic coding is available prior to topological analysis. In the general case, the problems of determining a branched manifold and a symbolic dynamics for the periodic orbits must be solved simultaneously. A tree search of possible symbolic names then allows us to decompose the problem into linear problems. Each consistent linear equation set with a unique solution yields both a solution of the template-finding problem and a set of possible symbolic names for each orbit.

The information extracted from the topological invariants about the symbolic dynamics can be used to construct generating partitions and encode arbitrary trajectories on the attractor. This is basically an interpolation problem, where periodic orbits and their associated symbolic sequence are the interpolation data. Partitions so obtained can be shown to be identical to those determined using homoclinic tangencies, to a high degree of accuracy. A distinctive advantage of this approach is that the solution is progressively refined using higher-period orbits, which makes it robust to noise, and suitable for analyzing experimental time series.

References

1. G. B. Mindlin, H. G. Solari, M. A. Natiello, R. Gilmore, and X.-J. Hou. Topological analysis of chaotic time series data from Belousov–Zhabotinski reaction. *J. Nonlinear Sci.*, 1:147–173, 1991.

2. R. Gilmore. Topological analysis of chaotic dynamical systems. *Rev. Mod. Phys.*, 70:1455–1530, 1998.

3. F. T. Arecchi, R. Meucci, G. P. Puccioni, and J. R. Tredicce. Experimental evidence of subharmonic bifurcations, multistability, and turbulence in a Q-switched gas laser. *Phys. Rev. Lett.*, 49(17):1217–1220, 1982.

4. R. S. Gioggia and N. B. Abraham. Routes to chaotic output from a single mode, dc-excited laser. *Phys. Rev. Lett.*, 51(8):650–653, 1983.

5. G. P. Puccioni, A. Poggi, W. Gadomski, J. R. Tredicce, and F. T. Arecchi. Measurement of the formation and evolution of a strange attractor in a laser. *Phys. Rev. Lett.*, 55(4):339–342, 1985.

6. J. R. Tredicce, F. T. Arecchi, G. P. Puccioni, A. Poggi, and W. Gadomski. Dynamic behavior and onset of low-dimensional chaos in a modulated homogeneously broadened single-mode laser: Experiments and theory. *Phys. Rev. A*, 34(3):2073–2081, 1986.

7. J. R. Tredicce, N. B. Abraham, G. P. Puccioni, and F. T. Arecchi. On chaos in lasers with modulated parameters: a comparative analysis. *Opt. Commun.*, 55(2):131–134, 1985.

8. J. R. Tredicce, F. T. Arecchi, G. L. Lippi, and G. P. Puccioni. Instabilities in lasers with an injected signal. *J. Opt. Soc. Am. B*, 2(1):173–183, 1985.

9. T. Midavaine, D. Dangoisse, and P. Glorieux. Observation of chaos in a frequency-modulated CO_2 laser. *Phys. Rev. Lett.*, 55(19):1989–1992, 1985.

10. I. I. Matorin, A. S. Pikovskii, and Ya. I. Khanin. Multistability and autostochasticity in a laser with delayed-response active medium subjected to periodic loss modulation. *Sov. J. Quantum Electron.*, 14(10):1401–1405, 1984.

11. H. G. Solari, E. Eschenazi, R. Gilmore, and J. R. Tredicce. Influence of coexisting attractors on the dynamics of a laser system. *Opt. Commun.*, 64(1):49–53, 1987.

12. H. G. Solari and R. Gilmore. Relative rotation rates for driven dynamical systems. *Phys. Rev. A*, 37(8):3096–3109, 1988.

13. V. I. Arnol'd. *Catastrophe Theory, 2nd ed.* Springer-Verlag, New York, 1986.

14. E. Eschenazi, H. G. Solari, and R. Gilmore. Basins of attraction in driven dynamical systems. *Phys. Rev. A*, 39(5):2609–2627, 1989.

15. M. J. Feigenbaum. Qualitative universality for a class of nonlinear transformations. *J. Stat. Phys.*, 19:25–52, 1978.

16. M. J. Feigenbaum. The universal properties of nonlinear transformations. *J. Stat. Phys.*, 21(6):669–706, 1979.

17. M. J. Feigenbaum. Universal behavior in nonlinear systems. *Los Alamos Science*, 1(1):4–27, 1980.

18. M. J. Feigenbaum. Universal behavior in nonlinear systems. *Physics D*, 7:16–39, 1983.

19. E. N. Lorenz. Noisy periodicity and reverse bifurcation. *Ann. N.Y. Acad. Sci.*, 357:282–293, 1980.

20. C. Grebogi and E. Ott. Crisis, sudden changes in chaotic attractors and transient chaos. *Physica D*, 7:181–200, 1983.

21. S. E. Newhouse. Diffeomorphisms with infinitely many sinks. *Topology*, 13(1):9–18, 1974.

22. K. T. Alligood. A canonical partition of the periodic orbits of chaotic maps. *Trans. Am. Math. Soc.*, 292(2):713–719, 1985.

23. K. T. Alligood, T. D. Sauer, and J. A. Yorke. *Chaos: An Introduction to Dynamical Systems.* Springer-Verlag, New York, 1997.

24. S. Smale. Differentiable dynamical systems. *Bull. Am. Math. Soc.*, 73:747–817, 1967.

25. B. Wedding, A. Gasch, and D. Jaeger. Chaos observed in a Fabry Perot interferometer with quadratic nonlinear medium. *Phys. Lett. A*, 105(3):105–109, 1984.

26. F. Waldner, R. Badii, D. R. Barberis, G. Broggi, W. Floeder, P. F. Meier, R. Stoop, M. Warden, and H. Yamazaki. Route to chaos by irregular periods. *J. Magn. Mater.*, 54–57(part 3):1135–1136, 1986.

27. M. F. Bocko, D. H. Douglas, and H. H. Fruchty. Bounded regions of chaotic behavior in the control parameter space of a driven non-linear resonator. *Phys. Lett. A*, 104(8):388–390, 1984.

28. T. T. Klinker, W. Mayer-Ilse, and W. Lauterborn. Period doubling and chaotic behavior in a driven toda oscillator. *Phys. Lett. A*, 101(8):371–375, 1984.

29. I. I. Satija, A. R. Bishop, and K. Fesser. Chaos in a damped and driven toda system. *Phys. Lett. A*, 112(5):183–187, 1985.

30. R. van Buskirk and C. Jeffries. Observation of chaotic dynamics of coupled nonlinear oscillators. *Phys. Rev. A*, 31(5):3332–3357, 1985.

31. I. B. Schwartz and H. L. Smith. Infinite subharmonic bifurcation in an SEIR epidemic model. *J. Math. Biol.*, 18(3):233–253, 1983.

32. N. B. Tufillaro, T. A. Abbott, and J. P. Reilly. *An Experimental Approach to Nonlinear Dynamics and Chaos*. Addison-Wesley, Reading, MA, 1992.

33. H. Poincaré. *Les Methodes nouvelle de la mécanique céleste*. Gauthier-Villars, Paris, 1892.

34. P. Cvitanović. *Universality in Chaos*. Adam Hilger, Boston, 1984.

35. V.I. Arnol'd, S.M.Gusein-Zade, and A.N.Varchenko. *Singularities of differentiable mappings I*. Birkhäuser, 1985.

36. V. I. Arnol'd. Singularities of smooth mappings. *Russ. Math. Surv.*, 23(1):1–43, 1968.

37. R. Gilmore. *Catastrophe Theory for Scientists and Engineers*. Wiley, New York, 1981. Reprinted by Dover, New York, 1993.

38. A. Katok and B. Hasselblatt. *Introduction to the Modern Theory of Dynamical Systems*. Cambridge University Press, Cambridge, 1995.

39. V.I. Oseledec. A multiplicative ergodic theorem: Lyapunov characteristic numbers for dynamical systems. *Trans. Mosc. Math. Soc.*, 19:197–231, 1968.

40. R. L. Devaney. *An Introduction of Chaotic Dynamical Systems*. Addison-Wesley, Reading, MA, 1989.

41. N. Metropolis, M. L. Stein, and P. R. Stein. On finite limit sets for transformations of the unit interval. *J. Combinatorial Theory A*, 15(1):25–44, 1973.

42. P. J. Holmes. Bifurcation sequences in horseshoe maps: infinitely many routes to chaos. *Phys. Lett. A*, 104(6,7):299–302, 1984.

43. D. Lind and B. Marcus. *Symbolic Dynamics and Coding*. Cambridge University Press, Cambridge, 1995.

44. C. Shannon. A mathematical theory of communication. *Bell Syst. Tech. J.*, 27:379–423, 1948.

45. C. Shannon. A mathematical theory of communication. *Bell Syst. Tech. J.*, 27:623–656, 1948.

46. T.-Y. Li and J. A. Yorke. Period three implies chaos. *Am. Math. Mon.*, 82(12):985–992, 1975.

47. A. N. Sarkovskii. Coexistence of cycles of a continuous map of a line into itself. *Ukr. Mat. Z.*, 16(1):61–71, 1964.

48. M. Hénon. A two-dimensional mapping with a strange attractor. *Commun. Math. Phys.*, 50:69–77, 1976.

49. Jérôme Plumecoq and Marc Lefranc. From template analysis to generating partitions I: Periodic orbits, knots and symbolic encodings. *Physica D*, 144:231–258, 2000.

50. P. Grassberger and H. Kantz. Generating partitions for the dissipative Hénon map. *Phys. Lett. A*, 113(5):235–238, 1985.

51. P. Grassberger, H. Kantz, and U. Moenig. On the symbolic dynamics of the Hénon map. *J. Phys. A*, 22:5217–5230, 1989.

52. P. Cvitanović, G. H. Gunaratne, and I. Procaccia. Topological and metric properties of Hénon-type strange attractors. *Phys. Rev. A*, 38:1503–1520, 1988.

53. G. D'Alessandro, P. Grassberger, S. Isola, and A. Politi. On the topology of the Hénon map. *J. Phys. A: Math. Gen.*, 23:5285–5294, 1990.

54. F. Giovannini and A. Politi. Homoclinic tangencies, generating partitions and curvature of invariant manifolds. *J. Phys. A: Math. Gen.*, 24:1837–1848, 1991.

55. F. Giovannini and A. Politi. Generating partitions in hénon-type maps. *Phys. Lett. A*, 161:332–336, 1992.

56. L. Jaeger and H. Kantz. Homoclinic tangencies and non-normal Jacobians - effects of noise in nonhyperbolic chaotic systems. *Physica D*, 105:79–96, 1997.

57. Lars Jaeger and Holger Kantz. Structure of generating partitions for two-dimensional maps. *J. Phys. A: Math. Gen.*, 30:L567–576, 1997.

58. Jérôme Plumecoq and Marc Lefranc. From template analysis to generating partitions II: Characterization of the symbolic encodings. *Physica D*, 144:259–278, 2000.

59. V. I. Arnol'd. Small denominators I: Mappings of the circumference into itself. *A.M.S. Transl. Series 2*, 46:213–284, 1965.

60. J. Guckenheimer and P. H. Holmes. *Nonlinear Oscillators, Dynamical Systems and Bifurcations of Vector Fields*. Springer-Verlag, New York, 1986.

61. V. I. Arnol'd. *Ordinary Differential Equations*. MIT Press, Cambridge, MA, 1989.

62. G. Duffing. *Erzwungene Schwingungen bei Veränderlicher Eigenfrequenz und ihre Technische Bedeutung*. Vieweg-Verlag, Braunschweig, Germany, 1918.

63. B. van der Pol. Forced oscillations in a circuit with nonlinear resistance (reception with reactive triode). *Philos. Mag. (7)*, 3:65–80, 1927.

64. E. N. Lorenz. Deterministic nonperiodic flows. *J. Atmos. Sci.*, 20(3):130–141, 1963.

65. B. Saltzman. Finite amplitude free convection as an initial value problem I. *J. Atmos. Sci.*, 19(7):329–341, 1962.

66. O. E. Rössler. An equation for continuous chaos. *Phys. Lett. A*, 57(5):397–398, 1976.

67. M. C. Mackey and L. Glass. Oscillation and chaos in physiological control systems. *Science*, 197:287–289, 1977.

68. R. Gilmore. *Lie Groups, Lie Algebras, and Some of Their Applications*. Wiley, New York, 1974. Reprinted by Krieger, Melbourne, FL, 1992.

69. D. A. Cox, J. B. Little, and D. O'Shea. *Ideals, Varieties and Algorithms*. Springer-Verlag, New York, 1996.

70. L. H. Kaufmann. *Knots and Physics*. World Scientific, Singapore, 1991.

71. D. Rolfsen. *Knots and Links*. Publish or Perish, Berkeley, CA, 1976.

72. K. Reidemeister. *Knotentheorie*. Julius Springer, Berlin, 1932.

73. J. W. Alexander. Topological invariants of knots and links. *Trans. Am. Math. Soc.*, 30:275–306, 1923.

74. E. Artin. Theory of braids. *Ann. Math.*, 48(2):101–126, 1947.

75. A. Katok. Liapunov exponents, entropy and periodic orbits for diffeomorphisms. *Publ. Math. IHES*, 51:137–174, 1980.

76. P. Boyland. *Braid Types and a Topological Method of Proving Positive Topological Entropy*. Department of Mathematics, Boston University, Boston, 1984.

77. K. Jänlich. *Topology*. Springer-Verlag, Berlin, 1984.

78. N. B. Tufillaro, R. Holzner, L. Flepp, R. Brun, M. Finardi, and R. Badii. Template analysis for a chaotic NMR laser. *Phys. Rev. A*, 44(8):R4786–R4788, 1991.

79. P. J. Holmes and R. F. Williams. Knotted periodic orbits in suspensions of smale's horseshoe: torus knots and bifurcation sequences. *Arch. Rat. Mech. Anal.*, 90:115–194, 1985.

80. P. J. Holmes. Knotted periodic orbits in suspensions of smale's horseshoe: period multiplying and cabled knots. *Physica D*, 21:7–41, 1986.

81. P. J. Holmes. Knots and orbit genealogies in nonlinear oscillators. In T. Bedford and J. Swift, editors, *New directions in dynamical systems*, pages 150–191, Cambridge, 1988. Cambridge University Press.

82. P. J. Holmes. Knotted period orbits in suspensions of smale's horseshoe: extended families and bifurcation sequences. *Physica D*, 40:42–64, 1989.

83. J. S. Birman and R. F. Williams. Knotted periodic orbits in dynamical systems I: Lorenz's equations. *Topology*, 22:47–82, 1983.

84. J. Birman and R. Williams. Knotted periodic orbits in dynamical systems II: knot holders for fibered knots. *Cont. Math.*, 20:1–60, 1983.

85. G. B. Mindlin, X.-J. Hou, H. G. Solari, R. Gilmore, and N. B. Tufillaro. Classification of strange attractors by integers. *Phys. Rev. Lett.*, 64(20):2350–2353, 1990.

86. S. K. Scott. *Chemical Chaos*. Oxford University Press, London, 1991.

87. R. W. Ghrist, P. J. Holmes, and M. C. Sullivan. *Knots and Links in Three-Dimensional Flows*, volume 1654 of *Lecture Notes in Mathematics*. Springer, Berlin, 1997.

88. F. Papoff, A. Fioretti, E. Arimondo, G. B. Mindlin, H. G. Solari, and R. Gilmore. Structure of chaos in the laser with a saturable absorber. *Phys. Rev. Lett.*, 68:1128–1131, 1992.

89. M. Lefranc and P. Glorieux. Topological analysis of chaotic signals from a CO_2 laser with modulated losses. *Int. J. Bifurcation Chaos Appl. Sci. Eng.*, 3:643–649, 1993.

90. C. Letellier, L. Le Sceller, P. Dutertre, G. Gouesbet, Z. Fei, and J. L. Hudson. Topological characterization and global vector field reconstruction of an experimental electrochemical system. *J. Phys. Chem.*, 99:7016–7027, 1995.

91. N. B. Tufillaro, P. Wyckoff, R. Brown, T. Schreiber, and T. Molteno. Topological time series analysis of a string experiment and its synchronized model. *Phys. Rev. E*, 51(3):164–174, 1995.

92. C. Letellier, G. Gouesbet, F. Soufi, J. R. Buchler, and Z. Kolláth. Chaos in variable stars: topological analysis of W Vir model pulsations. *Chaos*, 6:466–476, 1996.

93. R. Gilmore and J. W. L. McCallum. Structure in the bifurcation diagram of the Duffing oscillator. *Phys. Rev. E*, 51:935–956, 1995.

94. R. Shaw. Strange attractors, chaotic behavior and information flow. *Z. Naturforsch. A*, 36a(1):80–112, 1981.

95. J. M. T. Thompson and H. B. Stewart. *Nonlinear Dynamics and Chaos*. Wiley, London, 1986.

96. J. Franks and R. F. Williams. Entropy and knots. *Trans. Am. Math. Soc.*, 291(1):241–253, 1985.

97. L. Kocarev, Z. Tasev, and D. Dimovski. Topological description of a chaotic attractor with spiral structure. *Phys. Lett. A*, 190(5,6):399–402, 1994.

98. T. Shimizu and N. Morioka. On the bifurcation of a symmetric limit cycle to an asymmeteric one in a simple model. *Phys. Lett. A*, 76(3,4):201–204, 1980.

99. W. H. Press, B. P. Flannery, S. A. Teukolsky, and W. T. Vetterling. *Numerical Recipes*. Cambridge University Press, Cambridge, 1986.

100. H. D. I. Abarbanel, R. Brown, J. J. Sidorowich, and L. Sh. Tsimring. The analysis of observed chaotic data in physical systems. *Rev. Mod. Phys.*, 65(4):1331–1392, 1993.

101. H. D. I. Abarbanel. *Analysis of Observed Chaotic Data*. Springer-Verlag, New York, 1996.

102. B. Efron. Computers and the theory of statistics: thinking the unthinkable. *SIAM Rev.*, 21(4):460–480, 1979.

103. H. Fujisaka and T. Yamada. Stability theory of synchronized motion in coupled oscillator systems. *Prog. Theor. Phys.*, 69(1):32–47, 1983.

104. R. Brown, N. F. Rulkov, and E. R. Tracy. Modeling and synchronizing chaotic systems from time-series data. *Phys. Rev. E*, 49(5):3784–3800, 1994.

105. E. A. Jackson. *Perspectives in Nonlinear Dynamics*. Cambridge University Press, Cambridge, 1990.

106. L. M. Pecora and T. M. Carroll. Synchronization in chaotic systems. *Phys. Rev. Lett.*, 64(8):821–824, 1990.

107. A. V. Oppenheim and R. W. Schafer. *Discrete-Time Signal Processing*. Prentice Hall, Englewood Cliffs, NJ, 1989.

108. S. Hammel. A noise reduction method for chaotic systems. *Phys. Lett. A*, 148(8,9):421–428, 1990.

109. T. Sauer. A noise reduction method for signals from nonlinear systems. *Physica D*, 58(1-4):193–201, 1992.

110. D. S. Broomhead and G. P. King. Extracting qualitative dynamics from experimental data. *Physica D*, 20(2,3):217–236, 1986.

111. G. B. Mindlin and H. G. Solari. Topologically inequivalent embeddings. *Phys. Rev. E*, 52:1497–1502, 1995.

112. A. M. Mancho, A. A. Duarte, and G. B. Mindlin. Time delay embeddings and the structure of flows. *Phys. Lett. A*, 221(3-4):181–186, 1996.

113. H. Whitney. Differentiable manifolds. *Ann. Math.*, 37(3):654–680, 1936.

114. N. H. Packard, J. P. Crutchfield, J. D. Farmer, and R. S. Shaw. Geometry from a time series. *Phys. Rev. Lett.*, 45(9):712–715, 1980.

115. F. Takens. Detecting strange attractors in turbulence. In D. A. Rand and L.-S. Young, editors, *Lecture Notes in Mathematics, Vol. 898*, pages 366–381, New York, 1981. Springer-Verlag.

116. T. Sauer, J. A. Yorke, and M. Casdagli. Embedology. *J. Stat. Phys.*, 65(3-4):579–616, 1991.

117. N. B. Tufillaro, H. G. Solari, and R. Gilmore. Relative rotation rates: Fingerprints for strange attractors. *Phys. Rev. A*, 41(10):5717–5720, 1990.

118. J.-P. Eckmann and D. Ruelle. Fundamental limitations for estimating dimensions and Lyapunov exponents in dynamical systems. *Physica D*, 56(2-3):185–187, 1987.

119. C. Gilmore. A new approach to testing for chaos, with applications in finance and economics. *Int. J. Bifurcation Chaos Appl. Sci. Eng.*, 3:583–587, 1993.

120. C. Gilmore. A new test for chaos. *J. Econ. Behav. Organ.*, 22:209–237, 1993.

121. P. Collet and J.-P. Eckmann. *Iterated Maps on the Interval as Dynamical Systems*. Birkhäuser, Boston, 1980.

122. N. R. Draper and H. Smith. *Applied Regression Analysis*. Wiley, New York, 1966.

123. B. P. Belousov. A periodic reaction and its mechanism. *Sb. Ref. Radiat. Med.*, page 145, 1958.

124. B. P. Belousov. A periodic reaction and its mechanism. In R. J. Field and M. Burger, editors, *Oscillations and Traveling Waves in Chemical Systems*, pages 605–613, New York, 1985. Wiley.

125. A. M. Zhabotinskii. *Oscillatory Processes in Biological and Chemical Systems*. Science Publishing House, Moscow, 1967.

126. J. C. Roux, R. H. Simoyi, and H. L. Swinney. Observation of a strange attractor. *Physica D*, 8:257–266, 1983.

127. K. Coffman, W. D. McCormack, Z. Noszticzius, R. H. Simoyi, and H. L. Swinney. Universality, multiplicity, and the effect of iron impurities in the Belousov-Zhabotinskii reaction. *J. Chem. Phys.*, 86(1):119–129, 1987.

128. D. P. Lathrop and E. J. Kostelich. Characterization of an experimental strange attractor by periodic orbits. *Phys. Rev. A*, 40(7):4028–4031, 1989.

129. F. Papoff. Thesis. University of Pisa, Pisa, 1992.

130. A. Fioretti, F. Molesti, B. Zambon, E. Arimondo, and F. Papoff. Topological analysis of laser with saturable absorber in experiments and models. *Int. J. Bifurcation Chaos Appl. Sci. Eng.*, 3:559–564, 1993.

131. F. Papoff, A. Fioretti, E. Arimondo, G. B. Mindlin, H. G. Solari, and R. Gilmore. Structure of chaos in the laser with saturable absorber. *Phys. Rev. Lett.*, 68(8):1128–1131, 1991.

132. N. B. Tufillaro, P. Wyckoff, R. Brown, T. Schreiber, and T. Molteno. Topological time series analysis of a string and its synchronized model. *Phys. Rev. E*, 51(1):164–174, 1995.

133. H. G. Solari, M. A. Natiello, and M. Vázquez. Braids on the Poincaré section: a laser example. *Phys. Rev. E*, 54(4):3185–3195, 1996.

134. G. Boulant, M. Lefranc, S. Bielawski, and D. Derozier. A non-horseshoe template in a chaotic laser model. *Int. J. Birfurcation Chaos Appl. Sci. Eng.*, 8(5):965–975, 1998.

135. R. L. Davidchack, Y.-C. Lai, E. M. Bollt, and M. Dhamala. Estimating generating partitions of chaotic systems by unstable periodic orbits. *Phys. Rev. E*, 61(2):1353–1356, 2000.

136. G. Boulant, S. Bielawski, D. Derozier, and M. Lefranc. Experimental observation of a chaotic attractor with a reverse horsehoe topological structure. *Phys. Rev. E*, 55(4):R3801–3804, 1997.

137. G. Boulant, M. Lefranc, S. Bielawski, and D. Derozier. Horseshoe templates with global torsion in a driven laser. *Phys. Rev. E*, 55(5):5082–5091, 1997.

138. J. L. Kaplan and J. A. Yorke. Chaotic behavior of multidimensisonal difference equations. In H. O. Pietgen and H. O. Walther, editors, *Functional Difference Equations and the Approximation of Fixed Points, Lecture Notes in Math, Vol 730*, pages 204–227, New York, 1979. Springer-Verlag.

139. G. Boulant, J. Plumecoq, S. Bielawski, D. Derozier, and M. Lefranc. Model validation and symbolic dynamics of chaotic lasers using template analysis. In M. Dong, W. Ditto, L. Pecora, M. Spano, and S. Vohra, editors, *Proceedings of the 4th Experimental Chaos Conference*, pages 121–126, Singapore, 1998. World Scientific.

140. C. Letellier, P. Dutertre, and B. Maheu. Unstable periodic orbits and templates of the Rössler system: toward a systematic topological characterization. *Chaos*, 5:271–282, 1995.

141. A. L. Hodgkin, A. F. Huxley, and B. Katz. Ionic currents underlying activity in the giant axon of the squid. *Arch. Sci. Physiol.*, 3:129–150, 1949.

142. A. L. Hodgkin and B. Katz. The effect of temperature on the electrical activity of the giant axon of the squid. *J. Physiol.*, 109:240–249, 1949.

143. A. L. Hodgkin. The ionic basis of electrical activity in nerve and muscle. *Biol. Rev.*, 26:339–409, 1951.

144. A. L. Hodgkin, A. F. Huxley, and B. Katz. Measurement of current-voltage relations in the membrane of the giant axon of *loligo*. *J. Physiol.*, 116:424–448, 1952.

145. A. L. Hodgkin and A. F. Huxley. Currents carried by sodium and potassium ions through the membrane of the giant axon of *loligo*. *J. Physiol.*, 116:449–472, 1952.

146. A. L. Hodgkin and A. F. Huxley. The components of membrane conductance in the giant axon of *loligo*. *J. Physiol.*, 116:473–496, 1952.

147. A. L. Hodgkin and A. F. Huxley. The dual effect of membrane potential on sodium conductance in the giant axon of *loligo*. *J. Physiol.*, 116:497–506, 1952.

148. A. L. Hodgkin and A. F. Huxley. A quantitative description of membrane current and its application to conduction and excitation in nerve. *J. Physiol.*, 117:500–544, 1952.

149. H. A. Braun, H. Wissing, K. Shäfer, and M. C. Hirsch. Oscillation and noise determine signal transduction in shark multimodal sensory cells. *Nature*, 367(6460):270–273, 1994.

150. K. Shäfer, H. A. Braun, R. C. Peters, and F. Bretschnieder. Periodic firing pattern in afferent discharges from electroreceptor organs of catfish. *Pflügers Arch.*, 429:378–385, 1995.

151. H. A. Braun, K. Shäfer, K. Voigt, R. Peters, F. Breitschneider, X. Pei, L. Wilkens, and F. Moss. Low dimensional dynamics in sensory biology. I. Thermally sensitive electroreceptors of the catfish. *J. Comp. Neurosci.*, 4(4):335–347, 1996.

152. H. A. Braun, M. T. Huber, M. Dewald, K. Shäfer, and K. Voigt. Computer simulations of neuronal signal transduction: the role of nonlinear mechanisms and noise. *Int. J. Bifurcation and Chaos*, 8(5):881–889, 1998.

153. R. Gilmore, X. Pei, and F. Moss. Topological analysis of chaos in neural spike train bursts. *Chaos*, 9(3):812–817, 1999.

154. R. Gilmore and X. Pei. The topology and organization of unstable periodic orbits in hodgkin-huxley models of receptors with subthreshold oscillations. In F. Moss and S. Gielen, editors, *Handbook of Biological Physics, Vol. 4. Neuro-Informatics and Neural Modeling*, pages 155–203, Amsterdam, 2001. North-Holland.

155. H. Haken. Analogy between higher instabilities in fluids and lasers. *Phys. Lett. A*, 53(1):77–78, 1975.

156. N. M. Lawandy and G. A. Koepf. Relaxation oscillations in optically pumped molecular lasers. *IEEE J. Quantum Electron.*, 16(7):701–703, 1980.

157. M. Lefebvre, D. Dangoise, and P. Glorieux. Transients in a far-infrared laser. *Phys. Rev. A*, 29(2):758–767, 1984.

158. C. O. Weiss and W. Klisch. On observability of Lorenz instabilities in lasers. *Opt. Commun.*, 51(1):47–48, 1984.

159. N. B. Abraham, D. Dangoise, P. Glorieux, and P. Mandel. Observation of undamped pulsations in a low-pressure, far-infrared laser and comparison with a simple theoretical model. *J. Opt. Soc. Am. B*, 2(1):23–34, 1985.

160. R. G. Harrison, A. Al-Saidi, and D. J. Biswas. Observation of instabilities and chaos in a homogeneously broadened single mode and multimode midinfrared Raman laser. *IEEE J. Quantum Electron.*, 21(9):1491–1497, 1985.

161. R. G. Harrison and D. J. Biswas. Demonstration of self-pulsing instability and transitions to chaos in single-mode and multimode homogeneously broadened laser. *Phys. Rev. Lett.*, 55(1):63–66, 1985.

162. C. O. Weiss. Observation of instabilities and chaos in optically pumped far-infrared lasers. *J. Opt. Soc. Am. B*, 2(1):137–140, 1985.

163. C. O. Weiss and J. Brock. Evidence for lorenz-type chaos in a laser. *Phys. Rev. Lett.*, 57(22):2804–2806, 1986.

164. D. J. Biswas, R. G. Harrison, C. O. Weiss, W. Klische, D. Dangoise, P. Glorieux, and N. M. Lawandy. Experimental observations of single mode laser instabilities in optically pumped molecular lasers. In F. T. Arecchi and R. G. Harrison, editors, *Instabilities and Chaos in Quantum Optics*, volume 34 of *Springer Series in Synergetics*, pages 109–121, New York, 1987. Springer-Verlag.

165. N. B. Abraham, P. Mandel, and L. M. Narducci. Dynamical instabilities and pulsations in lasers. In E. Wolf, editor, *Progress in Optics XXV*, pages 1–190, Amsterdam, 1988. North-Holland.

166. L. M. Narducci and N. B. Abraham. *Laser Physics and Laser Instabilities*. World Scientific, Singapore, 1988.

167. C. O. Weiss and R. Vilaseca. *Dynamics of Lasers*. VCH, New York, 1991.

168. F. T. Arecchi and R. G. Harrison. *Selected Papers on Optical Chaos, SPIE Milestone Series, Vol. MS 75*. SPIE, Bellingham, WA, 1993.

169. Ya. I. Khanin. *Principles of Laser Dynamics*. North-Holland, Amsterdam, 1995.

170. E. Roldán, G. J. de Valcárcel, R. Vilaseca, R. Corbálan, V. J. Martinez, and R. Gilmore. The dynamics of optically pumped molecular lasers. on its relation with the lorenz-haken model. *Quantum Semiclassic. Opt.*, 9:R1–R35, 1997.

171. M. A. Dupertuis, R. R. E. Salomaa, and M. R. Siegrist. The conditions for Lorenz chaos in an optically pumped far-infrared laser. *Opt. Commun.*, 57(6):410–414, 1986.

172. R. Gilmore, R. Vilaseca, R. Corbálan, and E. Roldán. Topological analysis of chaos in the optically pumped laser. *Phys. Rev. E*, 55(3):2479–2487, 1997.

173. H. F. Creveling, J. F. De Paz, J. Y. Baladi, and R. J. Schoenhals. Stability characteristics of a single-phase free convection loop. *J. Fluid Mech.*, 67(1):65–84, 1975.

174. M. Gorman, P. J. Widmann, and K. A. Robins. Chaotic flow regimes in a convection loop. *Phys. Rev. Lett.*, 52(25):2241–2244, 1984.

175. M. Gorman, P. J. Widmann, and K. A. Robins. Nonlinear dynamics of a convection loop: a quantitative comparison of experiment with theory. *Physica D*, 19(2):255–267, 1986.

176. J. Singer, Y.-Z. Wang, and H. Bau. Controlling a chaotic system. *Phys. Rev. Lett.*, 66(9):1123–1125, 1991.

177. E. Ott. *Chaos in Dynamical Systems*. Cambridge University Press, Cambridge, 1993.

178. H. G. Solari, M. A. Natiello, and G. B. Mindlin. *Nonlinear Dynamics: A Two-Way Trip from Physics to Math*. IOP Publishers, London, 1996.

179. V. I. Arnol'd. Critical points of smooth functions and their normal forms. *Russ. Math. Surv.*, 30(5):1–75, 1975.

180. V. I. Arnol'd, A. Varchenko, and S. Gousein-Zadé. *Singularités des applications differentiables*. Editions Mir, Moscow, 1986.

181. M. Golubitsky and V. Guillemin. *Stable Mappings and Their Singularities*. Springer-Verlag, New York, 1973.

182. R. Thom. *Structural Stability and Morphogenesis*. Benjamin/Addison-Wesley, New York, 1975.

183. E. C. Zeeman. *Catastrophe Theory: Selected Papers (1972–1977)*. Addison-Wesley, Reading, MA, 1977.

184. T. Poston and I. N. Stewart. *Catastrophe Theory and Its Applications*. Pitman, London, 1978.

185. G. B. Mindlin, R. Lopez-Ruiz, H. G. Solari, and R. Gilmore. Horseshoe implications. *Phys. Rev. E*, 48(6):4297–4304, 1993.

186. T. Hall. Weak universality in two-dimensional transitions to chaos. *Phys. Rev. Lett.*, 71(1):58–61, 1993.

187. T. Hall. The creation of horseshoes. *Nonlinearity*, 7:861–924, 1994.

188. M. A. Natiello and H. G. Solari. Remarks on braid theory and the characterisation of periodic orbits. *J. Knot Theory Ramifications*, 3:511–529, 1994.

189. C. Letellier and R. Gilmore. Covering dynamical systems: Twofold covers. *Phys. Rev. E*, 63(1):1 0162006–1 – 016206–10, 2001.

190. R. Miranda and E. Stone. The proto lorenz system. *Phys. Lett. A*, 178(1,2):105–113, 1993.

191. C. Letellier, P. Dutertre, and G. Gouesbet. Characterization of the Lorenz system, taking into account the equivariance of the vector field. *Phys. Rev. E*, 49:3492–3495, 1994.

192. C. Letellier, G. Gouesbet, and N. F. Rulkov. Topological analysis of chaos in equivariant electronic circuits. *Int. J. Bifurcation Chaos Appl. Sci. Eng.*, 6:2531–2555, 1996.

193. M. L. Cartwright and J. E. Littlewood. On nonlinear differential equations of the second order, I: The equation $\ddot{y} + k(1 - y^2)\dot{y} + y = b\lambda k\cos(\lambda t + a)$, k large. *J. Lond. Math. Soc.*, 20:180–189, 1945.

194. N. Levinson. A second-order differential equation with singular solutions. *Ann. Math.*, 50:127–153, 1949.

195. P. J. Holmes. Knotted periodic orbits in suspensions of annulus maps. *Proc. Roy. Soc. Lond.*, A411:351–378, 1987.

196. J.-P. Eckmann and D. Ruelle. Ergodic theory of chaos and strange attractors. *Rev. Mod. Phys.*, 57(3):617–656, 1985.

197. H. Whitney. On singularities of mappings of Euclidean spaces, I. Mappings of the plane into the plane. *Ann. Math.*, 62(3):374–410, 1955.

198. H. Whitney. Singularities of mappings of euclidean spaces. In *Symposium Internacional de Topologia Algebraica, Mexico, 1956*, pages 285–301, Mexico, 1958. La Universidad Nacional Autonoma.

199. R. Gilmore, R. Vilaseca, R. Corbálan, and E. Roldán. Topological analysis of chaos in the optically pumped laser. *Phys. Rev. E*, 55(3):2479–2487, 1997.

200. J. R. Weeks. *The Shape of Space.* Marcel Dekker, New York, 1985.

201. R. Ghrist and P. Holmes. An ODE whose solution contains all knots and links. *Int. J. Bifurcation and Chaos*, 6(5):779–800, 1996.

202. M. Lefranc, P. Glorieux, F. Papoff, F. Molesti, and E. Arimondo. Combining topological analysis and symbolic dynamics to describe a strange attractor and its crises. *Phys. Rev. Lett.*, 73:1364–1367, 1994.

Index